SPACECRAFT SYSTEMS ENGINEERING

SPACECRAFT SYSTEMS ENGINEERING

Fourth Edition

Edited by

Peter Fortescue
University of Southampton, UK

Graham Swinerd
University of Southampton, UK

John Stark
Queen Mary, University of London, UK

A John Wiley & Sons, Ltd., Publication

This edition first published 2011
© 2011, John Wiley & Sons, Ltd
First Edition published in 1991, Second Edition published in 1995, Third Edition published in 2003

Registered office
John Wiley & Sons Ltd, The Atrium, Southern Gate, Chichester, West Sussex, PO19 8SQ, United Kingdom

For details of our global editorial offices, for customer services and for information about how to apply for permission to reuse the copyright material in this book please see our website at www.wiley.com.

Library of Congress Cataloging-in-Publication Data

Spacecraft systems engineering / edited by Peter Fortescue, Graham Swinerd, John Stark.—4th ed.
  p. cm.
  Includes bibliographical references and index.
  ISBN 978-0-470-75012-4 (hardback)
 1. Space vehicles—Design and construction. 2. Astronautics—Systems engineering.
I. Fortescue, Peter W. II. Swinerd, Graham. III. Stark, John.
  TL875.S68 2011
  629.47′4—dc22

                                                              2011015486

A catalogue record for this book is available from the British Library.

Print ISBN: 9780470750124
ePDF ISBN: 9781119971016
oBook ISBN: 9781119971009
ePub ISBN: 9781119978367
Mobi ISBN: 9781119978374

Set in 10/12pt Times by Laserwords Private Limited, Chennai, India.
Printed and bound by CPI Group (UK) Ltd, Croydon, CR0 4YY
C9780470750124_260924

Dedicated to the memory of
Nicky Skinner
1978–2011

CONTENTS

15 SPACECRAFT MECHANISMS
Guglielmo S. Aglietti

16 SPACECRAFT ELECTROMAGNETIC COMPATIBILITY ENGINEERING
Ken M. Redford

17 ASSEMBLY, INTEGRATION AND VERIFICATION
Terry Ransome

18 SMALL SATELLITE ENGINEERING AND APPLICATIONS
Martin N. Sweeting and Craig I. Underwood

19 PRODUCT ASSURANCE
Geoffrey Hall

20 SPACECRAFT SYSTEM ENGINEERING
Adrian R. L. Tatnall, John B. Farrow, Massimo Bandecchi and C. Richard Francis

LIST OF CONTRIBUTORS

EDITORS

Peter W. Fortescue
*Aeronautics and Astronautics, Faculty of
Engineering and the Environment, University
of Southampton, UK (retired)*

Graham G. Swinerd
*Aeronautics and Astronautics, Faculty of
Engineering and the Environment, University
of Southampton, UK*

John P. W. Stark
*School of Engineering and Materials Science,
Queen Mary, University of London, UK*

AUTHORS

Guglielmo S. Aglietti
*Aeronautics and Astronautics,
Faculty of Engineering and the Environment,
University of Southampton, UK*

Massimo Bandecchi
*European Space Research and Technology
Centre (ESTEC), European Space Agency,
The Netherlands*

Franck Chatel
*German Space Operations Center (GSOC),
Oberpfaffenhofen, Germany*

Graham E. Dorrington
*School of Engineering and Materials Science,
Queen Mary, University of London, UK*

John B. Farrow
*International Space University, Strasbourg,
France*

Nigel P. Fillery
EADS Astrium, Portsmouth, UK

C. Richard Francis
*European Space Research and Technology
Centre (ESTEC), European Space Agency,
The Netherlands*

Geoffrey Hall
Moreton Hall Associates, Maidenhead, UK

John M. Houghton
EADS Astrium, Stevenage, UK

J. Barrie Moss
School of Engineering, Cranfield University, UK

Terry Ransome
EADS Astrium, Stevenage, UK (retired)

Ken M. Redford
British Aerospace, Bristol, UK

Chris J. Savage
*European Space Research and Technology
    Centre (ESTEC), European Space Agency,
    The Netherlands (retired)*

Ray E. Sheriff
*School of Engineering, Design and Technology,
    University of Bradford, UK*

David Stanton
Keltik Ltd, Hampton Hill, UK

Martin N. Sweeting
*Surrey Space Centre, University of Surrey,
    Guildford, Surrey, UK*

Adrian R. L. Tatnall
*Aeronautics and Astronautics, Faculty of
    Engineering and the Environment,
    University of Southampton, UK*

Craig I. Underwood
*Surrey Space Centre, University of Surrey,
    Guildford, Surrey, UK*

PREFACE TO THE FOURTH EDITION

When I was thinking about what to say in this foreword to the fourth edition, I had a look back over the previous editions to get a flavour of what was going on when they were published—the first two decades ago! Obviously a great deal has changed in that time, which is of course reflected in the current book's content. However, one aspect that has remained constant throughout that time is the influence that the US Space Shuttle has had as the work-horse of the West's human spaceflight programme.

The Shuttle's first launch three decades ago, in 1981, was for me one of those landmark events that somehow spurs the memory to recall exactly where you were and what you were doing at the time. For me, the 12th of April 1981 was a glorious spring day, during which my wife and I enjoyed the climb of a peak in the remote north-west Highlands of Scotland. However, sensing the historic character of the day's events, I do recall a resolve at the end of that glorious day to find out how the first historic flight of Shuttle Columbia had gone. The subsequent history of the Shuttle programme is well documented. Despite the high cost of operations, the programme has overall been hugely successful, but also overshadowed by the human cost of desperate tragedies. Coming full circle, this year sees the retirement of this remarkable machine, again an event with a personal dimension—the commencement and retirement of the Shuttle's space career have coincided closely with my own career in the space industry and academia! Consequently, like an old friend, it's always been there.

The Shuttle retirement has inevitably forced a rethink of the US human spaceflight programme. As a consequence, the Bush administration proposed the Constellation programme which centred on a new crewed spacecraft Orion. This was to be lifted to orbit by the Shuttle replacement—the man-rated Ares 1 launcher. The other significant component of the programme was a heavy-lift launch vehicle called Ares 5, which would independently orbit the massive payloads required for human exploration beyond Earth orbit. The main objectives of the programme were a return to the moon by 2020, and preparations for a crewed landing on Mars in the longer term. However, the incoming Obama administration has effectively overturned the 'Bush vision', throwing open the development of human access to orbit to private venture, abandoning the immediate prospect of human exploration beyond Earth orbit, and extending the lifetime of the International Space Station to 2020. In the short-term this has led to the rather bizarre situation of focusing US human spaceflight activities on Earth orbit, but without the independent means of US astronauts to reach it. At the time of writing, the future development of US human spaceflight is unclear, which raises the prospect that the next footprints on the moon's surface may be those of Chinese taikonauts. As far as the book content is concerned, this has not been a good time to attempt to write about this aspect of space activity. For example, reference to the Space Shuttle is minimal throughout the current edition, and the emphasis in the launch vehicles section (Chapter 7) is on the European Ariane launcher programme (although there is some discussion of the Ares launchers, which in the fullness of time may, or may not, be relevant).

The majority of the book content, however, focuses on the design and engineering of unmanned spacecraft, and around the turn of the millennium, the 'faster, better, cheaper' design philosophy was particularly influential in reducing the size and mass of science spacecraft in particular. However, this has been tempered somewhat by the occurrence of inopportune in-orbit failures, which have provided lessons that maybe faster and cheaper are not necessarily better. However, the explosion of interest in small, capable spacecraft continues unabated, and this is reflected in an updated Chapter 18 on small satellite engineering.

At the other end of the size range, there are a number of major robotic spacecraft programmes that will be making the headlines soon after this edition hits the book shelves. Perhaps, the most significant of these is the follow-on to the Hubble Space Telescope, which has been christened the James Webb Space Telescope. This is to be launched in around 2014 to the L_2 Sun-Earth Lagrange point, around 1.5 million km from Earth. With a mirror nearly three times the size of Hubble's, the scientists are looking forward to an explosion of new cosmological discoveries resulting from its operation. At the same time, the ESA comet probe Rosetta should be beginning its mission in orbit around comet 67P/Churyumov-Gerasimenko, and it is anticipated that the Rosetta data will provide a step function in our understanding of the origins of our local environment, the solar system. In the world of application satellites, a new global navigation satellite system called Galileo should become operational, also around 2014. This is a civil programme, funded by the European Union, involving the launch of a constellation of 30 satellites in Earth orbits at a height of around 23 000 km. It is hoped that the introduction of this non-military system will remove the reticence of civilian organisations to embrace the technology of satellite navigation in their operations. One significant development arising from this is the prospect of satellite navigation being fully utilized in the arena of civil air traffic control.

This fourth edition of *Spacecraft Systems Engineering* has been significantly revised and updated throughout, so that readers can master the many facets involved in an unmanned space vehicle project, like those mentioned above, from early system design through to in-orbit flight operations. There are also some 'all-new' features which are worthy of additional mention. Current trends in interplanetary missions have suggested that a new section on low-thrust trajectories would be helpful, and this has been added to the already-extensive Mission Analysis chapter (Chapter 5). The previous Chapter 14 on Ground Stations has been rewritten. The new version has been entitled 'Ground Segment' to emphasize that this area is not just about ground station aspects, but also includes many other activities such as flight operations. A new chapter (Chapter 17) has been added devoted to the important topic of Assembly, Integration and Verification, which focuses on the later stages of a spacecraft project when the whole system is brought together and tested prior to launch. The old chapter on Product Assurance has been completely rewritten (Chapter 19) to reflect the diverse aspects of PA, including that of software. This is particularly pertinent in the space sector, in which the manufactured 'products' often must survive many years in a hostile environment without the benefit of maintenance.

The final chapter on Spacecraft System Engineering (Chapter 20) has also been rewritten, changing the emphasis to discuss system design methods—in particular that of concurrent engineering design. System design in action is illustrated by discussion of the design development of the ESA Cryosat spacecraft, which is used as a case study.

Finally, the editors wish to thank the army of contributors who have given their time and effort to bring this edition to fruition—without them a new edition would not have been possible. We are also indebted to the team at Wiley, in particular to Nicky Skinner and Gill Whitley, whose assistance throughout the period of compilation of the manuscript was invaluable. As this stage was drawing to a close, and the production process was beginning, we were shocked and saddened by the sudden death of Nicky Skinner in March. My regret is that our working relationship was conducted purely by email as is often the case these days. Although I did not have an opportunity to meet up and consolidate that relationship, nevertheless I feel I got to know Nicky very well. I am thankful for her assistance throughout, and it is entirely appropriate that this edition is dedicated to her memory.

Graham Swinerd
Southampton, March 2011

PREFACE TO THE THIRD EDITION

Graham Swinerd, my friend and colleague, took over the running of the Space Technology short courses at Southampton University when I retired in 1989. Who would be a better choice than Graham to take over the role of principal editor for this new edition of Spacecraft Systems Engineering? I am sure that Graham will build on the reputation that the past editions have achieved, and I wish him success in his new role. Over to you, Graham...

Peter Fortescue, *Southampton, July 2002*

Since the publication of the previous edition, Dan Goldin's 'Faster, Better, Cheaper' space mission philosophy has had a major impact upon American activities. As a consequence, the last of the heavyweight interplanetary spacecraft, Cassini, was launched in October 1997 on its mission to Saturn. Programmes such as NEAR Shoemaker, which launched a relatively small but capable spacecraft in February 1996 to orbit and ultimately to land on a small body—the asteroid 433 Eros—have substituted this type of mission. These 'small missions' have significantly influenced current and proposed planetary exploration programmes.

In the same interim period, we have also seen the launch of constellations into low Earth orbit, for global mobile communications using handheld telephones—in particular, the Iridium constellation, the first satellites of which were lofted in May 1997. Although financial problems have impacted this programme, it nevertheless heralds large-scale use of constellation systems in many application areas. There are great benefits to the usage of these distributed systems, not only in communications and navigation applications but also in improving the temporal coverage of Earth observation. There is also an implicit trend here to use a number of small, but capable, spacecraft to do the job of one or two large satellites.

The principal driver for the development of small satellite technology is the reduction in cost associated with access to space. The elements contributing to this philosophy are low launch costs, a short design, build and test period, a less complex ground interface and operations, and the recognition of a means of testing new spacecraft technologies in a relatively low financial risk environment.

At the other end of the size spectrum, December 1998 saw the first elements of the International Space Station (ISS) being brought together in orbit. If development continues as originally planned, around 2005, the ISS will become the largest structure ($\sim$400 tonnes) ever to be deployed in Earth orbit, marking the beginning of permanent habitation in space.

These various developments have had a significant influence on the structure of the new edition of the book. The major changes involve the removal of the chapter on atmospheric re-entry, and the addition of a new chapter on small satellite engineering and applications. Much of the removed material has been redistributed in other chapters, however, for example, Earth atmosphere re-entry is included in Chapter 7 (Launch Vehicles), and sections on aero-manoeuvring have been included in Chapter 5 (Mission Analysis). The new Chapter 18 on Small Satellites has been contributed by Martin Sweeting and Craig Underwood of the Surrey Space Centre, based at the University of Surrey, UK. Both individuals are recognized internationally for their expertise in this field. The chapter, built on the huge expertise of the Surrey Space Centre, gives insights into small satellite systems engineering in general. Given the growing activity in this area, no textbook of this kind is complete without such a contribution.

Other chapters have been rewritten—in particular, Chapter 8 (Spacecraft Structures), Chapter 11 (Thermal Control), Chapter 16 (Electromagnetic Compatibility) and Chapter 19 (Spacecraft Systems Engineering)—and most of the others have been substantially revised, including a discussion of constellation design and small-body missions in Chapter 5 (Mission Analysis).

Some of the authors of the second edition have retired, and new names have appeared in the contributors list. The editors are grateful to all of them for their contributions. It is also sad to report that three of our previous authors have died in the interim—Howard Smith (Telecommunications), Les Woolliscroft (Spacecraft Electromagnetic Compatibility Engineering) and Mervyn Briscoe (Spacecraft Mechanisms). Each of them will be sadly missed.

The reader may have noticed the dedication at the front of the book to one of these authors, Mervyn Briscoe, who was actively involved in revising his chapter on Mechanisms when he died in 2001. Our thanks are also due to Guglielmo Aglietti who jumped into the hot seat to complete the revision of Mervyn's chapter as a co-author. Mervyn gave loyal service as a contributor to the short course activity at Southampton over many years, and we would like to acknowledge this by dedicating this edition to his memory.

Finally, it is appropriate to thank both Peter Fortescue and John Stark for their pioneering work in bringing the previous editions to fruition, and for their valued assistance with this one.

Graham Swinerd
Southampton, July 2002

PREFACE TO THE SECOND EDITION

This second edition comes in response to a phone call that we editors had been dreading. 'Had we thought of producing a second edition?' After much consideration our answer was 'Yes', and here it is.

Not only has it given our contributing authors a chance to update the material in their chapters—the technology is developing all the time and five years is a long time! It has also given us the opportunity to rectify some of the errors in the first edition (and possibly to introduce some new ones), and to respond to suggestions from readers about the content and to our inevitable 'second thoughts' on the matter. As a result there are two new chapters.

The first is on Mechanisms—important equipment on spacecraft. They are an essential part of many of the systems that are covered in the other chapters, but having their own requirements we have given them chapter status here. They are a specialist topic, involving the problem of moving one mechanical part relative to another. For an application that has a long life, no servicing, no disturbance to the structure, and ideally no single point failure as design objectives, mechanism designers are faced with a challenging task. Chapter 16 tells us how they have responded to it.

The second additional chapter addresses the subject of System Engineering. The first edition has no hyphens in its title. Those who read the title as meaning 'The Engineering of Spacecraft Systems' will probably have found that the content was much as they had expected. Indeed there have been enough satisfied readers to cause the dreaded question of a second edition to be raised. However, it could also be read as meaning 'Systems Engineering of Spacecraft', and those who interpreted it as such will no doubt have been disappointed to have found little on the discipline of System Engineering in the book.

So our response is to retain the same ambiguous title, and to retain the same thrust as in the first edition. But we have added a new chapter (No. 19), which focuses on the subject of Systems Engineering of spacecraft. It is written by authors within the spacecraft industry who have experience in that activity. We hope it will bring together the pieces of the jigsaw puzzle that are to be found among the other chapters, and will show how they can be fitted together harmoniously to form a viable whole—a spacecraft that meets its design objectives in a viable manner.

Since the first edition some of our authors have moved to new locations; some have retired. New names have appeared in our list of contributors. We editors are grateful to them all—new and old—and trust that this edition presents 'second thoughts' that are an improvement on the first.

PREFACE TO THE FIRST EDITION

This book has grown out of a set of course notes, which accompany a series of short courses given at Southampton University. These courses started in 1974 with a two week 'space technology' course, and they are aimed at the recent science or engineering graduate who wishes to become a spacecraft engineer. The courses are still thriving, now serving much of European industry, with one-week versions for experienced engineers, sometimes senior ones, who are specialists in their own fields.

On the courses, the attendees work in competing teams on a project that involves designing a spacecraft in response to an overall objective. Over the years, mission designs have been directed at all application areas: science, astronomy, communications and Earth observations. There is now a 'museum' of models that demonstrate vehicle layouts and support the attendees' presentations covering operation, subsystem specification and launch constraints. These models demonstrate system viability rather than detailed design. The projects are designed at 'system level', and their supervision has provided a basis for deciding the level of detail that should be included in this book.

The coverage in this book is therefore aimed at giving the breadth that is needed by system engineers, with an emphasis on the bus aspect rather than on the payload. The specialist engineer is well served with textbooks, which cover many of the subsystems in detail and in depth. He is unlikely to learn very much about his own specialist topic from this book. But he may well learn something about other specialists' disciplines, and, it is hoped, enough for him to appreciate the trade-offs that affect his own subsystem in relation to others.

Chapters 2 to 5 set the general scene for spacecraft, and particularly for satellites. They must operate in an environment that is generally hostile compared to that with which we are familiar on Earth, and the main features of this are described in Chapter 2. Chapters 3 and 4 address the dynamics of objects in space, where the vehicles will respond to forces and moments that are minute, and which would be discounted as of no significance if they occurred on Earth. Indeed, most of them do occur here, but we do not often operate in a fully free state, and our Earth-bound vehicles are subject to other, much larger forces.

Chapter 5 relates the motion of the spacecraft to Earth rather than to the inertially based reference system of celestial mechanics.

Chapters 6 to 15 address the main subsystems. Chapters 7 and 8 cover the subjects of getting off the ground and returning through the atmosphere. Chapters 6, 9 to 12 and 14 deal with the main subsystems on board the spacecraft, that include the on-board end of the telemetry and control link (Chapter 14) with ground control (Chapter 15). The communication link is covered in Chapter 13 in which the fundamentals of the subject are included together with their rather special application to spacecraft. This is relevant to the telemetry and control link and to a communications payload.

Chapter 16 introduces electromagnetic compatibility (EMC), one of the subjects that must be addressed by the systems engineer if the various subsystems are to work in harmony. Product assurance is of vital concern to spacecraft engineers. Their product(s) must survive a hostile launch environment and then must last many years without the luxury of any maintenance. It does great credit to the discipline they exercise, that so many of their products do so.

We editors would like to express our thanks to the authors who have contributed chapters in the book. Most of them have lectured on the courses mentioned above. Our task has been to whittle down the material they have provided since they have been very generous. We are grateful too for their patience. The conversion of course notes into a book was expected to be a short process. How wrong we were!

We would also like to thank colleagues Graham Swinerd and Adrian Tatnall, who read some of the texts and gave advice. And finally our thanks to Sally Mulford, who has converted some much-abused text into typescript with patience and good humour.

LIST OF ACRONYMS

AATSR	Advanced Along-Track Scanning Radiometer		AQAP	Allied Quality Assurance Publication
ABM	Apogee Boost Motor		ARQ	Automatic report queuing
AC	Alternating current		ASAP	Ariane Structure for Auxiliary Payloads
ACS	Attitude Control System			
ACU	Antenna Control Unit		ASAR	Advanced Synthetic Aperture Radar
A/D	Analogue to digital			
ADEOS	Advanced Earth Observing System		ASIC	Application-specific integrated circuit
ADR	Active Debris Removal		ASK	Amplitude-shift keying
AFT	Abbreviated Functional Test		ASS	Antenna Support Structure
AGC	Automatic gain control		ASW	Address and synchronization word
AIT	Assembly, integration and test		ATCS	Active Thermal Control System
AIV	Assembly, integration and verification		ATSR	Along-track Scanning Radiometer
AKM	Apogee Kick Motor			
AM	Amplitude modulation		AU	Astronomical Unit (mean distance from Earth to Sun)
AMI	Active Microwave Instrument			
AMOOS	Aero-Manoeuvring Orbit-to-Orbit Shuttle		BCDT	Binary code data transfer
			BCH	Bose-Chaudhuri-Hocquenchem
AO	Announcement of Opportunity		BCR	Battery Charge Regulator
AOCS	Attitude and Orbit Control System		BDR	Battery Discharge Regulator
AOP	Announcement of Opportunity Package		BER	Bit error rate
			BMDO	Ballistic Missile Defence Organization
AOS	Acquisition of signal, also Advanced Orbiting Systems		BMU	Battery Management Unit
			BOL	Beginning of life
AOTV	Aero-assisted Orbital Transfer Vehicle		BPF	Band-pass filter
			BPSK	Bi-phase-shift keying
APM	Antenna Pointing Mechanism, also Attached Pressurized Module		BRTS	Bilateration Ranging Transponder System
			BSF	Back-Surface Field
			BSR	Back-Surface Reflector
APS	Active Pixel Sensor		CA	Contingency Analysis

CAD	Computer Aided Design	COTS	Commercial off the shelf
CADU	Channel Access Data Unit	CPL	Capillary-pumped loop
CAM	Civil, Aircraft, Military	CPM	Coarse pointing mechanism
CCB	Configuration Control Board	CR	Corrosion resistance
CCD	Charge coupled device	CRP	Contingency Recovery Procedure
CCIR	Comité Consultatif International de Radiocommunication	CSG	Centre Spatial Guyanais
		CTM	Collapsible Tube Mast
		CVCM	Collected volatile condensable materials
CCITT	Comité Consultatif International de Téléphonie et de Télégraphie	CW	Continuous-wave
		DARPA	Defence Advanced Research Project Agency
CCSDS	Consultative Committee for Space Data Systems	DB	Data Base
		DC	Direct current
CCU	Central Communications Unit	DCP	Data Collection Platform
		DoD	Department of Defence (USA)
CDF	Concurrent Design Facility		
C&DH	Control and Data Handling	DLR	(German Aerospace Centre)
C&DM	Configuration and Data Management	DM	Development Model
		DOF	Degree of freedom
CDMA	Code-division multiple access	DORIS	Determination of Orbit and Radiopositioning Integrated by Satellite
CDR	Critical Design Review	DPL	Declared Parts List
CE	Concurrent Engineering		
CEV	Command Execution Verification	DPSK	Differential phase-shift keying
CFDP	CCSDS File Delivery Protocol	DRS	Data Relay Satellite
		DSBSC	Double side-band suppressed carrier modulation
CFRP	Carbon fibre reinforced plastic		
CHAMP	CHAllenging Minisatellite Payload	DS-CDMA	Direct Sequence Code Division Multiple Access
CHRIS	Compact High Resolution Imaging Spectrometer	DSN	Deep Space Network
		DSP	Digital signal processing, also Digital Signal Processor
CIS	Confederation of Independent States		
CLA	Coupled loads analysis	ECR	Engineering Change Request
CLTU	Command Link Transfer Unit		
		ECSS	European Cooperation for Space Standardization
CMG	Control moment gyroscope		
CMOS	Complementary metal oxide semiconductor	EDA	Electrically Despun Antenna
CNES	Centre National d'Etudes Spatiales (French National Space Agency)	EDAC	Error detection and correction
		EEE	Electrical, Electronic and Electromechanical
COMSAT	Communications Satellite		

EGSE	Electrical Ground Support Equipment	FEA	Finite element analysis
EIRP	Equivalent isotropic radiated power	FEEP	Field emission electric propulsion
ELV	Expendable Launch Vehicle	FEM	Flight Engineering Model
EM	Electrical Model, also Engineering Model	FET	Field effect transistor
		FFSK	Fast frequency-shift keying
EMC	Electromagnetic compatibility	FGSE	Fluids Ground Support Equipment
EMI	Electromagnetic interference	FIFO	First-In-First-Out
EMP	Electromagnetic pulse	FIRST	Far Infra-Red Space Telescope
EOEP	Earth Observation Envelope Programme	FITS	Failures per 10^9 hours
EOL	End of lifetime	FM	Flight Model, also Frequency Modulation
EOS	Earth Observing System	FMECA	Failure Mode Effects and Criticality Analysis
e.p.	Equivalent particle		
ER-MIL	Established Reliability-MIL	FOG	Fibre optic gyroscope
ERS	Earth Resources Satellite	FOP	Flight Operations Procedure
ERT	Earth Received Time		
ESA	European Space Agency	f.o.r.	Frame of reference
ESATAN	European Space Agency Network Analyser	FOV	Field of view
		FPGA	Field Programmable Gate Array
ESD	Electrostatic discharge		
ESOC	European Space Operations Centre	FPM	Fine Pointing Mechanism, also Fine-pointing mode
ESTEC	European Space Research and Technology Centre	FRB	Failure Review Board
		FRR	Flight Readiness Review
ESTL	European Space Tribology Laboratory	FRSI	Flexible reusable surface insulation
EUMET-SAT	European Meteorological Satellite Organization	FS	Fail safe
		FSK	Frequency-shift keying
EURECA	European Retrievable Carrier	FTA	Fault Tree Analysis
		GEM	Giotto Extended Mission
EUTEL-SAT	European Telecommunications Satellite Organization	GEO	Geostationary Earth orbit
		GLONASS	Global Navigation Satellite System
EVA	Extra-vehicular activity	GMT	Greenwich Mean Time
EWSK	East-West Station Keeping	GNSS	Global Navigation Satellite System
FAR	Flight Acceptance Review		
FBC	Faster, Better, Cheaper	GOCE	Gravity field and steady state Ocean Circulation Explorer
FCP	Flight Control Procedure		
FDIR	Fault detection, inspection and recovery		
		GOES	Geostationary Orbit Environmental Satellites
FDMA	Frequency-division multiple access	GOMOS	Global ozone monitoring by the occultation of stars
FE	Finite element	GPS	Global Positioning System

GRACE	Gravity Recovery And Climate Experiment	IPN	Interplanetary Internet
GRO	Gamma Ray Observation	IPNRG	IPN Research Group
GSE	Ground Support Equipment	IQ	In-phase and Quadrature
GSFC	Goddard Space Flight Center	IRAS	Infra-Red Astronomical Satellite
GSOC	German Space Operations Centre	IRIG	Inter-Range Instrumentation Group
G/T	Ground track	IRTF	Internet Research Task Force
GTO	Geostationary transfer orbit	ISC	Integrated System Check
HEO	Highly elliptical orbit	ISDN	Integrated Services Digital Network
HGA	High Gain Antenna		
HGAS	High Gain Antenna System	ISO	Infrared Space Observatory, also International Organization for Standardization
HOTOL	Horizontal take-off and landing		
HPA	High power amplifier		
HRG	Hemispherical Resonator Gyroscope	ISS	International Space Station
		IST	Integrated System Test
HRSI	High-temperature reusable surface insulation	IT	Information Technology
		ITU	International Telecommunications Union
HST	High Speed Telemetry, also Hubble Space Telescope		
		JAXA	Japan Aerospace Exploration Agency
IC	Integrated circuit		
ICBM	Inter-Continental Ballistic Missile	JGM	Joint Gravity Model
		JHUAPL	Johns Hopkins University Applied Physics Laboratory
ICU	Instrument Control Unit, also Intelligent Control Unit		
		JPL	Jet Propulsion Laboratory
IDE	Integrated Design Environment	KSA	K-band Steerable Antenna
		KSC	Kennedy Space Center
IDHT	Instrument data-handling and transmission	LAM	Liquid Apogee Motor
		LBR	Low bit-rate
IDM	Integrated Design Model	LDEF	Long Duration Exposure Facility
IF	Intermediate frequency		
IFOV	Instantaneous field of view	LEAF	Large European Acoustic Facility
IFR	Inertial frame of reference		
IFRB	International Frequency Registration Board	LED	Light emitting diode
		LEO	Low Earth Orbit
IM	Intermodulation	LEOP	Launch and Early Orbit Phase
INMAR-SAT	International Maritime Satellite Organization		
		LET	Linear energy transfer
INTELSAT	International Telecommunications Satellite Organization	LHP	Loop heat pipe
		LISA	Laser Interferometer Spaceborne Antenna
IOAR	In Orbit Acceptance Review	LISN	Line impedance stabilization network
IOT	In Orbit Test		
IP	Internet Protocol	LNA	Low noise amplifier

LO	Local oscillator		MOS	Metal oxide semiconductor
LOS	Loss of signal		MOSFET	Metal oxide semiconductor field effect transistor
LPF	Low-pass filter			
LRR	Laser retro-reflector		MPD	Magneto-plasma-dynamic
LRSI	Low-temperature reusable surface insulation		MPG	Multi Point Grounding
			MPS	Mission Planning System
LSS	Large Space Simulator		MRB	Materials Review Board
LST	Low speed telemetry		MSG	Meteosat Second Generation
LTP	Long-haul Transport Protocol		MSH	Micro Humidity Sounder
			MSS	Multishock shield
MA	Multiple access		MST	Micro systems technology
MAC	Medium Access Control		MTBF	Mean time between failures
Mbps	Mega bits per second		MTTF	Man-Tended Free Flyer
MCC	Mission Control Centre		MVS	Machine Vision System
MCPC	Multi-Channel Per Carrier		MW	Momentum wheel
			MWR	Microwave Radiometer
MCS	Monitoring and Control System		NASA	National Aeronautics and Space Administration
MCU	Mode Control Unit		NASP	National Aero-Space Plane
MDBS	Mesh double bumper shield		NBFM	Narrow-band frequency modulation
MEMS	Micro Electro Mechanical Systems			
			NCR	Non Conformance Report
MEO	Medium height Earth Orbit		NDT	Non-destructive testing
MERIS	Medium Resolution Imaging Spectrometer		NEAR	Near Earth Asteroid Rendezvous
MET	Mission Elapsed Time		NERVA	Nuclear engine for rocket vehicle applications
MFR	Multi-function receiver			
MGSE	Mechanical Ground Support Equipment		N-MOS	N-type metal oxide semiconductor
MHS	Micro Humidity Sensor		NSSK	North-South Station Keeping
MIL-STD	Military Standard			
MIPAS	Michelson interferometer for passive atmospheric sounding		OBC	On-Board Computer
			OBDH	On-board data handling
			OBS	On-board software
MIPS	Million instructions per second		OCC	Operations Control Centre
			OMS	Orbital Manoeuvring Subsystem
MLI	Multi-layered insulation			
MMBS	Multiple mesh bumper shield		OSCAR	Orbiting satellite carrying amateur radio
MMH	Mono-methylhydrazine		OSR	Optical Solar Reflector
MMIC	Monolithic microwave integrated circuit		OTA	Optical Telescope Assembly
MMS	Multi-mission Modular Spacecraft		OTV	Orbital Transfer Vehicle
			PA	Product Assurance
MOD	Ministry of Defence		PAEHT	Power-Augmented Hydrazine Thruster
MOP	Multiple On-line Peripheral, also Meteosat Operational Programme			
			P-MOS	P-type metal oxide semiconductor

PAM	Payload Assist Module	PSK	Phase-shift keying
PAM-A	Payload Assist Module—Atlas-sized	PT	Prototype
		PTCS	Passive Thermal Control System
PAM-D	Payload Assist Module—Delta-sized	PTFE	Polytetrafluoroethylene
PC	Personal computer	PU	Pattern Unit
PCB	Printed circuit board	PUS	Packet Utilization Standard
PCDU	Power Control and Distribution Unit	PVG	Piezo-electric Vibratory Gyroscope
PCM	Pulse Code Modulation, or Phase Change Material	QA	Quality Assurance
		QF	Quality Factor
PDC	Project Design Centre	QM	Qualification Model
PDF	Probability Density Function	QPL	Qualified Parts List
PDHT	Payload data handling and transmission	QPSK	Quadrature phase-shift keying
PDR	Preliminary Design Review	QR	Qualification Review
PDUS	Primary Data User Station	RA	Radar altimeter
PEB	Payload Equipment Bay	RAL	Rutherford Appleton Laboratory
PEEK	Polyether Ether Ketone		
PEM	Payload Electronics Module	RAM	Random Access Memory, or Radio-frequency Anechoic Material
PF	Protoflight		
PFM	Protoflight Model		
PID	Proportional, integral and differential	RAMS	Reliability, Availability, Maintainability and Safety
PIM	Passive intermodulation product	RARR	Range and range rate
PLC	Payload Carrier	RBI	Remote Bus Interfaces
PLM	Payload Module	RCC	Reinforced carbon-carbon
PM	Phase modulation, also Pulse modulation	RCE	Reaction Control Equipment
		RF	Radio Frequency
PMC	Payload Module Computer	RFA	Request for approval
PML	Preferred Materials List	RFC	Radio frequency compatibility
PMP	Parts, materials and processes	RFD	Request For Deviation
PN	Pseudo-random noise	RFW	Request For Waiver
POEM	Polar Orbiting Earth-observation Mission	RIG	Rate-Integrating Gyroscope
		RKA	Russian Federal Space Agency
PPF	Polar Platform	RLG	Ring Laser Gyroscope
PPHSS	Preservation, Packaging, Handling, Storage and Shipping	RLV	Reusable Launch Vehicle
		RMS	Remote Manipulator System
PPL	Preferred Parts List	ROM	Read-only Memory
PRARE	Precise Range and Range-rate Equipment	Rpm	Revolutions per minute
		RS	Reed-Solomon
PRK	Phase-reversal keying	RTG	Radioisotope Thermoelectric Generator
PRR	Preliminary Requirements Review		
		RTU	Remote Terminal Unit

RW	Reaction Wheel	SOP	Spacecraft overhead pass
SA	Single access	SPA	Solar Power Array
SAD	Solar Array Drive	SPE	Solid Polymer Electrolyte
SADM	Solar array drive mechanism	SPELDA	Structure Porteuse pour Lancement Double Ariane
SAO	Smithsonian Astrophysics Observatory	SPG	Single-point grounding
		SP-L/PM	Split Phase-Level/Phase Modulation
SAR	Synthetic Aperture Radar		
SAS	Solar Array System	SPOT	Satellite Pour l'Observation de la Terre
SAW	Surface Acoustic Wave		
SBE	S-Band Exciter	SPS	Satellite Power System
SCC	Stress corrosion cracking	S³R	Sequential switching shunt regulation
SCF	Satellite Control Facility		
SCIA-MACHY	Scanning Imaging Absorption Spectrometer for Atmospheric Cartography	SRE	Spacecraft Ranging Equipment
		SRP	Solar radiation pressure
		SRR	System Requirements Review
SCPC	Single channel per carrier		
SCPS	Space Communications Protocol Standards	SSA	S-band Steerable Antenna, also Space Situational Awareness
SCPS-FP	SCPS File Protocol		
SCPS-NP	SCPS Network Protocol	SSB	Single side-band
SCPS-SP	SCPS Security Protocol	SSLV	Standard Small Launch Vehicle
SCPS-TP	SCPS Transport Protocol		
SCRAM-JET	Supersonic Combustion RAMJET	SSM	Second Surface Mirror
		SSMA	Spread-spectrum multiple access
SDR	System Design Review		
SDUS	Secondary Data User Station	SSME	Space Shuttle Main Engine
		SSPA	Solid State Power Amplifier
SEE	Single event effect		
SEL	Single-event latch-up	SSTL	Surrey Satellites Technology Limited
SEU	Single event upset		
SFT	System Functional Test	SSTO	Single Stage To Orbit
SGL	Space-to-ground link	SSUS	Solid Spinning Upper Stage
SIRAL	SAR and Interferometric Radar Altimeter	STDN	Spaceflight Tracking and Data Network
SL	Safe life	STRV	Space Technology Research Vehicle
SLR	Satellite laser ranging		
SM	Structure Model	STS	Space Transportation System
SMM	Solar Maximum Mission		
SMOS	Soil Moisture and Ocean Salinity	SVM	Service Module
		SVT	System Validation Test
SNAP-19	System for Nuclear Auxiliary Power	SW	Software
		SYLDA	Système de Lancement Double Ariane
SNR	Signal-to-noise ratio		
SOHO	Solar and Heliospheric Observatory	TCP/IP	Transmission Control Protocol/Internet Protocol

TC&R	Telemetry, Command and Ranging	UOSAT	University of Surrey Satellite
TCS	Thermal Control System	UQPSK	Unbalanced quadrature phase-shift keying
TDM	Time-Division Multiplexed		
TDMA	Time-division multiple access	USAF	United States Air Force
		USB	Upper side-band
TDPS	Tracking and Data Processing Station	USNO	US Naval Observatory
		USSTRAT-COM	US Strategic Command
TDRS	Tracking and Data Relay Satellite	UT	Universal Time
TDRSS	Tracking and Data Relay Satellite System	UTC	Universal Time Co-ordinated
TIU	Time Interval Unit	VCDU	Virtual Channel Data Unit
TLE	Two Line Elements	VCHP	Variable-Conductance Heat Pipe
TLM	Telemetry		
TM	Thermal Model	VCO	Voltage-Controlled Oscillator
TML	Total Mass Loss		
TMM	Thermal Mathematical Model	VCXO	Voltage-Controlled Crystal Oscillator
TM/TC	Telemetry/Telecommand	VHDL	Very High level Design description Language
TPS	Thermal Protection Subsystem	VHF	Very high frequency
TRB	Test Review Board	VLSI	Very large scale integrated
TRL	Technology Readiness Level	WARC	World Administrative Radio Conference
TRR	Test Readiness Review		
TT&C	Tracking, Telemetry and Control	WBFM	Wide-Band Frequency Modulation
TTL	Transistor-transistor logic	WRC	World Radio Conference
TWT	Travelling Wave Tube	WSGT	White Sands Ground Terminal
TWTA	Travelling Wave Tube Amplifier		
		WTR	Western Test Range
UARS	Upper Atmosphere Research Spacecraft	WWW	World Wide Web
		XMM	X-ray Multi-mirror Mission
UART	Universal Asynchronous Receiver and Transmitter	XPD	Cross-Polar Discrimination
		YSM	Yaw-steering mode
UHF	Ultra High Frequency		

1 INTRODUCTION

John P. W. Stark[1], Graham G. Swinerd[2] and Adrian R. L. Tatnall[2]

[1] *School of Engineering and Material Science, Queen Mary, University of London*
[2] *Aeronautics and Astronautics, Faculty of Engineering and the Environment, University of Southampton*

Man has only had the ability to operate spacecraft successfully since 1957, when the Russian satellite Sputnik I was launched into orbit. At the time of writing (2010) the Space Age is just over half a century old. In that time technology has made great strides, and the Apollo human expedition to the Moon and back is now a rather distant memory. In little more than five decades, unmanned explorer spacecraft have flown past all the major bodies of the Solar System, apart from the 'dwarf planet' Pluto—this exception will soon be remedied, however, by the 'New Horizons' spacecraft that is due to fly through the Pluto-Charon system in 2015. Space vehicles have landed on the Moon and Venus, and in recent years Mars has seen a veritable armada of orbiters, landers and rovers in preparation for a hoped-for future human expedition to the red planet. The Galileo Jupiter orbiter successfully deployed a probe in 1995, which 'landed' on the gaseous 'surface' of Jupiter. The Cassini/Huygens spacecraft has been a stunning success, entering orbit around Saturn in 2004, and executing a perfect landing on Titan of the European built Huygens probe in 2005. Minor bodies in the Solar System have also received the attention of mission planners. The first landing on such a body was executed by the Near Earth Asteroid Rendezvous (NEAR) Shoemaker spacecraft, when it touched down on the Eros asteroid in February 2001. This was succeeded in 2005 by the attempted sampling of material from the Itokawa asteroid by the Japanese Hayabusa spacecraft. Although the sampling operation was unsuccessful, the spacecraft is now on a return journey to Earth in the hope that some remnants of asteroid material may be found in its sealed sampling chamber. Similarly, a prime objective of the ambitious European Rosetta programme is to place a lander on a cometary body in 2014. There is also a growing awareness of the impact threat posed by near-Earth asteroids and comets, which is driving research into effective means of diverting such a body from a collision course with Earth.

Since our brief sojourn to the Moon in 1969–1972, human spaceflight has been confined to Earth orbit, with the current focus on construction and utilization of the International Space Station (ISS). The United States, Europe, Russia and Japan are all involved in this

Spacecraft Systems Engineering, Fourth Edition.
Edited by Peter W. Fortescue, Graham G. Swinerd and John P. W. Stark.
© 2011 John Wiley & Sons, Ltd. Published 2011 by John Wiley & Sons, Ltd.

ambitious long-term programme. The ISS has been a major step for both the technology and politics of the space industry, and has been a useful exercise in learning to live and work in space—a necessary lesson for future human exploration of the Solar System. The 'work horse' of this activity has been the US Space Shuttle, which has been the United States' principal means of human access to orbit over almost three decades. However, 2011 sees the retirement of the Shuttle. This is a major event in NASA's space operations, and it has forced a radical rethink of the United States' human spaceflight programme. This led to the proposal of a less complex man-rated launch vehicle, Ares 1, which is part of the Constellation Programme. The objective of this programme is to produce a new human spaceflight infrastructure to allow a return of US astronauts to the Moon, and ultimately to Mars. However, the shuttle retirement coincides with a deep global financial recession, and the political commitment to the Constellation Programme appears to be very uncertain. This re-evaluation by the US will perhaps herald the reinvigoration of the drive towards the full commercialization of the space infrastructure.

There is no doubt, however, that the development of unmanned application spacecraft will continue unabated. Many countries now have the capability of putting spacecraft into orbit. Satellites have established a firm foothold as part of the infrastructure that underpins our technological society here on Earth. There is every expectation that they have much more to offer in the future.

Before the twentieth century, space travel was largely a flight of fantasy. Most authors during that time failed to understand the nature of a spacecraft's motion, and this resulted in the idea of 'lighter-than-air' travel for most would-be space-farers [1, 2]. At the turn of the twentieth century, however, a Russian teacher, K. E. Tsiolkovsky, laid the foundation stone for rocketry by providing insight into the nature of propulsive motion. In 1903, he published a paper in the *Moscow Technical Review* deriving what we now term the rocket equation, or Tsiolkovsky's equation (equation 3.20). Owing to the small circulation of this journal, the results of his work were largely unknown in the West prior to the work of Hermann Oberth, which was published in 1923.

These analyses provided an understanding of propulsive requirements, but they did not provide the technology. This eventually came, following work by R. H. Goddard in America and Wernher von Braun in Germany. The Germans demonstrated their achievements with the V-2 rocket, which they used towards the end of World War II. Their rockets were the first reliable propulsive systems, and while they were not capable of placing a vehicle into orbit, they could deliver a warhead of approximately 1000 kg over a range of 300 km. It was largely the work of these same German engineers that led to the first successful flight of Sputnik 1 on 4 October 1957, closely followed by the first American satellite, Explorer 1, on 31 January 1958.

Five decades have seen major advances in space technology. It has not always been smooth, as evidenced by the major impact that the Challenger (1986) and Columbia (2003) disasters had on the American space programme. Technological advances in many areas have, however, been achieved. Particularly notable are the developments in energy-conversion technologies, especially solar photovoltaics, fuel cells and batteries. Developments in heat-pipe technology have also occurred in the space arena, with ground-based application in the oil industry. Perhaps the most notable developments in this period, however, have been in electronic computers and software. Although these have not necessarily been driven by space technology, the new capabilities that they afford have been rapidly assimilated, and they have revolutionized the flexibility of spacecraft. In some cases they have even turned a potential mission failure into a grand success.

But the spacecraft has also presented a challenge to Man's ingenuity and understanding. Even something as fundamental as the unconstrained rotational motion of a body is now better understood as a consequence of placing a spacecraft's dynamics under close scrutiny. Man has been successful in devising designs for spacecraft that will withstand a hostile space environment, and he has found many solutions.

1.1   PAYLOADS AND MISSIONS

Payloads and missions for spacecraft are many and varied. Some have reached the stage of being economically viable, such as satellites for communications, weather and navigation purposes. Others monitor Earth for its resources, the health of its crops and pollution. Determination of the extent and nature of global warming is only possible using the global perspective provided by satellites. Other satellites serve the scientific community of today and perhaps the layman of tomorrow by adding to Man's knowledge of the Earth's environment, the solar system and the universe.

Each of these peaceful applications is paralleled by inevitable military ones. By means of global observations, the old 'superpowers' acquired knowledge of military activities on the surface of the planet and the deployment of aircraft. Communication satellites serve the military user, as do weather satellites. The Global Positioning System (GPS) navigational satellite constellation is now able to provide an infantryman, sailor or fighter pilot with his location to an accuracy of about a metre. These 'high ground' space technologies have become an integral part of military activity in the most recent terrestrial conflicts.

Table 1.1 presents a list of payloads/missions with an attempt at placing them into categories based upon the types of trajectory they may follow. The satellites may be categorized in a number of ways such as by orbit altitude, eccentricity or inclination.

It is important to note that the specific orbit adopted for a mission will have a strong impact on the design of the vehicle, as illustrated in the following paragraphs.

Table 1.1   Payload/mission types

Mission	Trajectory type
Communications	Geostationary for low latitudes, Molniya and Tundra for high latitudes (mainly Russian), Constellations of polar LEO satellites for global coverage
Earth resources	Polar LEO for global coverage
Weather	Polar LEO, or geostationary
Navigation	Inclined MEO for global coverage
Astronomy	LEO, HEO, GEO and 'orbits' around Lagrange points
Space environment	Various, including HEO
Military	Polar LEO for global coverage, but various
Space stations	LEO
Technology demonstration	Various

Note: GEO: Geostationary Earth orbit; HEO: Highly elliptical orbit; LEO: Low Earth Orbit; MEO: Medium height Earth Orbit.

Consider geostationary (GEO) missions; these are characterized by the vehicle having a fixed position relative to the features of the Earth. The propulsive requirement to achieve such an orbit is large, and thus the 'dry mass' (exclusive of propellant) is a modest fraction of the all-up 'wet mass' of the vehicle. With the cost per kilogram-in-orbit being as high as it currently is—of the order of $30 000 per kilogram in geostationary orbit—it usually becomes necessary to optimize the design to achieve minimum mass, and this leads to a large number of vehicle designs, each suitable only for a narrow range of payloads and missions.

Considering the communication between the vehicle and the ground, it is evident that the large distance involved means that the received power is many orders of magnitude less than the transmitted power. The vehicle is continuously visible at its ground control station, and this enables its health to be monitored continuously and reduces the need for it to be autonomous or to have a complex data handling/storage system.

Low Earth orbit (LEO) missions are altogether different. Communication with such craft is more complex as a result of the intermittent nature of ground station passes. This resulted in the development, in the early 1980s, of a new type of spacecraft—the tracking and data relay satellite system (TDRSS)—operating in GEO to provide a link between craft in LEO and a ground centre. This development was particularly important because the Shuttle in LEO required a continuous link with the ground. More generally, the proximity of LEO satellites to the ground does make them an attractive solution for the provision of mobile communications. The power can be reduced and the time delay caused by the finite speed of electromagnetic radiation does not produce the latency problems encountered using a geostationary satellite.

The power subsystem is also notably different when comparing LEO and GEO satellites. A dominant feature is the relative period spent in sunlight and eclipse in these orbits. LEO is characterized by a high fraction of the orbit being spent in eclipse, and hence a need for substantial oversizing of the solar array to meet battery-charging requirements. In GEO, on the other hand, a long time (up to 72 min) spent in eclipse at certain times of the year leads to deep discharge requirements on the battery, although the eclipse itself is only a small fraction of the total orbit period. Additional differences in the power system are also partly due to the changing solar aspect angle to the orbit plane during the course of the year. This may be offset, however, in the case of the sun-synchronous orbit (see Section 5.4 of Chapter 5), which maintains a near-constant aspect angle—this is not normally done for the benefit of the spacecraft bus designer, but rather because it enables instruments viewing the ground to make measurements at the same local time each day.

It soon becomes clear that changes of mission parameters of almost any type have potentially large effects upon the specifications for the subsystems that comprise and support a spacecraft.

1.2   A SYSTEM VIEW OF SPACECRAFT

This book is concerned with spacecraft systems. The variety of types and shapes of these systems is extremely wide. When considering spacecraft, it is convenient to subdivide them into functional elements or subsystems. But it is also important to recognize that the satellite itself is only an element within a larger system. There must be a supporting ground control system (Figure 1.1) that enables commands to be sent up to the vehicle and status and payload information to be returned to the ground. There must also be a launcher system that sets the vehicle on its way to its final orbit. Each of the elements of the overall system must interact with the other elements, and it is the job of the system designer to achieve an overall optimum in which the mission objectives are realized efficiently. It is, for example, usual for the final orbit of a geostationary satellite to be achieved by a combination of a launch vehicle and the boost motor of the satellite itself.

This starts us towards the overall process of systems engineering, which will be treated in detail in the final chapter of this book. Figure 1.1 shows the breakdown of the elements needed to form a satellite mission. Each of these may be considered to perform functions that will have functional requirements associated with them. We can thus have an overriding set of mission requirements that will arise from the objectives of the mission itself. In the process of systems engineering, we are addressing the way in which these functional requirements can best be met, in a methodical manner.

Chambers Science and Technology Dictionary provides the following very apt definition of the term 'system engineering' as used in the space field:

'A logical process of activities that transforms a set of requirements arising from a specific mission objective into a full description of a system which fulfils the objective in an optimum way. It ensures that all aspects of a project have been considered and integrated into a consistent whole.'

The 'system' in question here could comprise all the elements within both the space and the ground segments of a spacecraft project, including the interfaces between the major elements, as illustrated in Figure 1.1. Alternatively, the system approach could be applied on a more limited basis to an assembly within the space segment, such as an instrument within the payload. In the case of an instrument, the system breakdown would include antenna elements or optics and detectors as appropriate, and the instrument's mechanical and electrical subsystems.

The *mission objectives* are imposed on the system by the customer, or user of the data. They are statements of the aims of the mission, are qualitative in nature and should

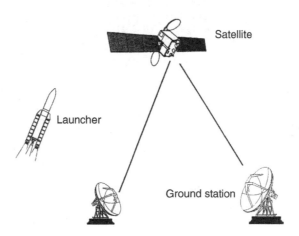

Figure 1.1   The total system — the combined space and ground segments

be general enough to remain virtually unchanged during the design process. It is these fundamental objectives that must be fulfilled as the design evolves.

For example, the mission objectives might be to provide secure and robust three-dimensional position and velocity determination to surface and airborne military users. The *Global Positioning System (GPS)* is a method adopted to meet these objectives.

An illustration of the range of methods and the subsequent requirements that can stem from mission objectives is given by the large number of different concepts that have been proposed to meet the objective of providing a worldwide mobile communication system. They range from an extension of the existing *Inmarsat* spacecraft system to schemes using highly eccentric and tundra orbits (see Chapter 5 for the definitions of these), to a variety of concepts based around a network of LEO satellites, such as The Globalstar or Iridium constellations.

This example demonstrates an underlying principle of system engineering, that is, that there is *never* only one solution to meet the objectives. There will be a diverse range of solutions, some better and some worse, based on an objective discriminating parameter such as cost, mass or some measure of system performance. The problem for the system engineer is to balance all these disparate assessments into a single solution.

The process that the system engineer first undertakes is to define, as a result of the mission objectives, the mission requirements. The subsequent requirements on the system and subsystems evolve from these initial objectives through the design process. This is illustrated in Figure 1.2, which shows how a hierarchy of requirements is established. In Chapter 20 this hierarchy is further explained and illustrated by considering a number

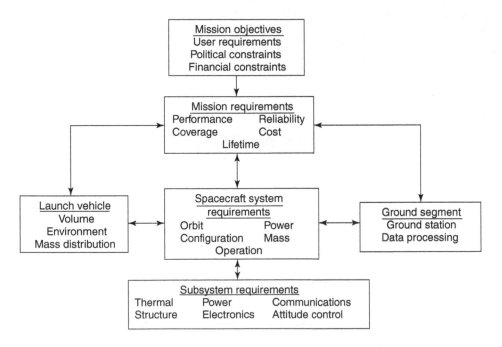

Figure 1.2 Objectives and requirements of a spacecraft mission

of specific spacecraft in detail. At this point, however, it is important to note the double-headed arrows in Figure 1.2. These indicate the feedback and iterative nature of system engineering.

We turn now to the spacecraft system itself. This may be divided conveniently into two principal elements, the payload and the bus (or service module). It is of course the payload that is the motivation for the mission itself. In order that this may function it requires certain resources that will be provided by the bus. In particular, it is possible to identify the functional requirements, which include:

1. The payload must be pointed in the correct direction.
2. The payload must be operable.
3. The data from the payload must be communicated to the ground.
4. The desired orbit for the mission must be maintained.
5. The payload must be held together, and on to the platform on which it is mounted.
6. The payload must operate and be reliable over some specified period.
7. An energy source must be provided to enable the above functions to be performed.

These requirements lead on to the breakdown into subsystems, which is shown in Figure 1.3. Inset in each of these is a number that relates it to the functions above.

The structure of this book recognizes this overall functional breakdown, shown in Figure 1.3. The individual subsystems are covered separately in the chapters. Thus, in Chapter 8 the structural subsystem is considered, and in Chapter 15, mechanism design is outlined. The power subsystem, including the various ways in which power can be raised on a spacecraft, is described in Chapter 10. The main elements of an attitude control subsystem are indicated principally in Chapter 9, although the underlying attitude motion of a free body such as a satellite is covered in Chapter 3. Telemetry and command subsystems may be conveniently considered alongside on-board data handling (OBDH); these

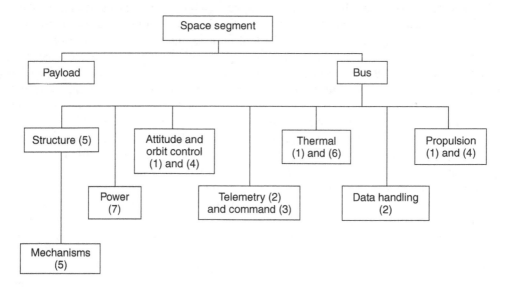

Figure 1.3   Spacecraft subsystems

topics are covered in Chapter 13, with the underlying principles and practice of space-craft communications in the previous chapter. The thermal control subsystem appears in Chapter 11. Propulsion, as it relates to on-board systems, is described in Chapter 6, while its application to launch systems is described in Chapter 7.

One facet of these subsystems is that the design of any one has impacts and resource implications on the others. A most important feature of spacecraft system design is to identify what aspects of the mission and what elements of the design provide the major influences on the type of satellite that may meet the specific mission requirements. This process is the identification of the *'design drivers'*. In some cases the drivers will affect major features of the spacecraft hardware. The varied mission requirements, coupled with the need to minimize mass and hence power, has thus led to a wide variety of individual design solutions being realized. However, the spacecraft industry is now evolving towards greater standardization—in the shape of the specific buses that may be used to provide the resources for a variety of missions (e.g., the SPOT bus, the Eurostar bus, Mars/Venus Express, etc.—see Chapter 20).

It is not simply the nature of its payload that determines the design that is selected for a given mission, although this will have a considerable influence. Commercial and political influences are strongly felt in spacecraft engineering. Individual companies have special-ist expertise; system engineering is dependent on the individual experience within this expertise. This was perhaps most notably demonstrated by the Hughes Company, which advanced the art of the spin-stabilized satellite through a series of *Intelsat* spacecraft. Spacecraft systems engineering is not all science—there is indeed an art to the discipline.

This leads to another major feature of spacecraft system design, namely, the impact of reliability. The majority of terrestrial systems may be maintained, and their reliability, while being important, is not generally critical to their survival. If a major component fails, the maintenance team can be called in. In space, this luxury is not afforded and while the Shuttle did provide in-orbit servicing for a limited number of satellites, this was an extremely expensive option. This requires that the system must be fault-tolerant, and when this tolerance is exceeded the system is no longer operable and the mission has ended.

There are two principal methods used to obtain high reliability. The first is to use a design that is well proven. This is true for both system and component selection.

The requirement to validate the environmental compatibility of components (Chapter 2) leads to relatively old types being used in mature technology, especially in electronic components. This tends to lead to a greater demand for power than the terrestrial 'state-of-the-art' technology. At system level a 'tried and tested' solution will minimize development risk, reducing system cost while also achieving high reliability.

The second method of achieving high reliability is *via* de-rating (Chapter 19). By reduc-ing the power of the many electronic components, for example, a greater life expectancy can be obtained. This leads to an overall increase in mass.

The net effect of designing for high reliability is that spacecraft design is conservative—'if it has been done before then so much the better'. Much of satellite design is thus not state-of-the-art technology. Design teams evolve a particular design solution to meet varied missions—because it is a design they understand—and hence system design is an art as well as a science.

In making the selection of subsystems for the spacecraft, the designer must have a good grasp of the way in which the subsystems work and the complex interactions between them, and they must recognize how the craft fits into the larger system. Further, the designer must be able to trade off advantages in one area with the disadvantages in another and achieve a balance in which the end result will work as a harmonious whole. While each subsystem will have its own performance criterion, its performance must nevertheless be subordinated to that of the system as a whole.

1.3 THE FUTURE

As we enter the second half century of the Space Age, we are approaching a new frontier in space. Up until now, we have been able to gain access to space, and demonstrate a competent exploitation of this environment principally in terms of the use of application satellites—however, our utilization of it is still limited. This limitation is mainly related to the very high cost of access to orbit, and this obstacle needs to be overcome in opening up the new frontier. Beyond the frontier, however, we will require to establish space infrastructure; including the prime elements of communications, and safe and reliable transportation, with a permanent human presence in space—initially on space stations, but then on the Moon and Mars. Over the last 40 years or so, space exploration has had to adapt to changes in world politics. Before that, in the Apollo era, the funding available was motivated by a political end—that of winning the 'space race' and so demonstrating the superiority of one political ideology over another.

Clearly, to set up the infrastructure there must be a space transportation system. The first step—getting off the ground—requires the development of next-generation launchers, which are truly reusable, having aircraft-like operational characteristics. This poses huge technological challenges principally for propulsion engineers and material scientists. However, the rewards for such a breakthrough would be enormous—the resulting reduction in cost to access to orbit would open up the new frontier, not only in terms of space applications and science, but also for human space exploration and space tourism.

The early part of the twenty-first century will see the completion and operation of the ISS. Although this has been a controversial and expensive programme, it will no doubt be superseded by future space stations, which will be used as staging posts, where 'a new team of horses' can be obtained. Surely such orbital staging posts will eventually become assembly and servicing posts as well, so that spacecraft venturing beyond Earth orbit do not have to be designed to withstand the full rigours of launch, when their subsequent stages of travel are relatively stress-free. Manufacturing in space also has significant potential, not only for exotic materials, but also for lightweight structural materials extruded in zero gravity, for use in zero gravity. A communications infrastructure is already in use. There will be a requirement for 'accommodation units' in orbit, fulfilling both scientific and space tourism needs. This will be aided by the store of knowledge being gathered each year of Man's ability to live in space. There is a requirement for a power generating and supply system ... and so the list goes on.

We are, however, at a crossroad in the way we develop our presence in space. Throughout the past 50 years there has been debate concerning the presence of humans in space: what role should we have and how should this be accomplished. Over the past 30 years there has also been the parallel debate about how space endeavours should be financed—whether by governmental funds or private capital. The philosophies of the past may of course reassert themselves, with the result that we see funding for space programmes once again dominated by tax payers money, but maybe not. This issue is central to the current controversy (of 2009/10) which has been stimulated by the Obama administration and the Augustine Commission, resulting in the cancellation of the Constellation programme.

The issues are drawing together in a way never before witnessed. Up until very recently, the advent of space tourism has only been the province of the dreamer and the science fiction aficionado. This is now changing, stimulated without doubt by the winning of the Ansari X-prize in October 2004 by SpaceShipOne, built by the company Scaled Composites. Their subsequent teaming with Virgin Galactic in 2005 means that it is now possible (in 2010) to reserve a seat online on the first commercial sub-orbital flights. There is, as a result, the potential for the commercial airline industry (which itself was initiated by the not dissimilar Orteig Prize, won by Charles Lindbergh in 1927) to address the issue of access to space. The enabling technologies are gradually emerging alongside the commercial realization that some people can afford to fly into space at a commercial price tag. This is all coming at a time when the prevailing political situation in the USA is pointing to a reassessment of the role NASA should play in future access to space. It seems quite probable that this nexus will indeed further stimulate the progress to the commercial utilization of space. There is a subtle but significant shift. Rather than attempting merely to commercialize the sale of a largely government-funded collection of data products from satellites, typified by the approaches adopted in the 1980s and 90s, a movement can be perceived towards the commercialization of the core process of access to space by individuals.

There is a whole new exciting arena waiting to be explored, occupied and used for the benefit of all mankind.

REFERENCES

[1] de Bergerac, Cyrano (1649) *Voyage dans la lune*.
[2] Winter, F. H. (1983) *Prelude to the Space Age*, Smithsonian.

2 THE SPACECRAFT ENVIRONMENT AND ITS EFFECT ON DESIGN

John P. W. Stark

School of Engineering and Materials Science, Queen Mary, University of London

2.1 INTRODUCTION

Spacecraft operation is characterized by its remoteness from the Earth and thus the loss of the Earth's protective shield, namely, the atmosphere. This atmosphere evidently provides a suitable stable environment in which the human species has been able to evolve. Coupled with the gravitational force of the Earth, 'the one-g environment', it provides familiarity in design, and its removal has significant and sometimes unexpected implications. The aim of this chapter is to introduce the reader to the nature of the spacecraft's environment and the implications that it has on spacecraft design, by considering both the way materials behave and the way in which systems as a whole are influenced. The final section looks very briefly at the implications for human space flight.

Before considering the environment in detail it should be noted that the different phases in the life of a space vehicle, namely, manufacture, pre-launch, launch and finally space operation, all have their own distinctive features. Although a space vehicle spends the majority of its life in space, it is evident that it must survive the other environments for complete success. Whilst the manufacturing phase is not specifically identified in the following section, it has an effect upon the reliability and the ability to meet design goals. Cleanliness, humidity and codes of practice are critical in the success of spacecraft missions, and these issues are addressed in Chapters 17 and 19.

2.2 PRE-OPERATIONAL SPACECRAFT ENVIRONMENTS

2.2.1 Pre-launch environment

The design, manufacture and assembly of a large spacecraft, and its final integration into a launch vehicle is a lengthy process, lasting typically 5–10 years. Components and

Spacecraft Systems Engineering, Fourth Edition.
Edited by Peter W. Fortescue, Graham G. Swinerd and John P. W. Stark.
© 2011 John Wiley & Sons, Ltd. Published 2011 by John Wiley & Sons, Ltd.

subsystems may be stored for months or even years prior to launch (e.g. the Galileo mission to Jupiter). Careful environmental control during such periods is essential if degradation of the spacecraft system as a whole is to be avoided.

2.2.2   The launch phase

From an observer's viewpoint, the launch of a space vehicle is evidently associated with gross noise levels. This impinges on the structure of a satellite contained within the launcher's shroud. The launch sequence entails high levels of vibration, associated both with the noise field and structural vibration, modest-to-high levels of acceleration during ascent, mechanical shock due to pyrotechnic device operation and stage separation, a thermal environment that differs from both laboratory and space environments, and for most launch vehicles, a rapidly declining ambient pressure. These features are described separately below.

The *severe acoustic/vibration environment* during launch is due to both the operation of the launch vehicle's main engines, and also the aerodynamic buffeting as the vehicle rises through the lower region of the Earth's atmosphere. Two peak levels occur.

The first peak occurs at the moment of lift-off. The overall build-up of the rocket motor firing and the exhaust products reflected from the ground reaches a peak at launcher release. During ascent, the contribution from ground reflection decreases, but a variety of mechanical components, such as liquid fuel turbopump operation, continue to excite the vehicle's structure. The vibration is not only directly transmitted to the spacecraft through structural components, but it also excites the launch shroud to generate a secondary acoustic field. For light, flexible components such as the solar array, the acoustic environment may be more severe than the mechanically induced vibration [1].

The second peak in the acoustic field occurs during transonic flight. The launch shroud is again excited, this time by the unsteady flow field around the vehicle.

Measurement of the field is generally made in dB with reference to a pressure of 2×10^{-5} Pa. The sound pressure level I is then measured in decibels, given by

$$I = 20 \log_{10} \left(\frac{F}{2 \times 10^{-5}} \right) \text{dB}$$

where F is the acoustic field intensity. The frequency spectrum of the noise field will be dependent on the launch vehicle and will change during the launch sequence. Data is provided by the launch agency for design purposes, thus enabling the designer to quantify the vibrational inputs that individual components and the whole space vehicle will experience. The design noise spectrum for Ariane 5 is compared with that of the now decommissioned Ariane 4 in Figure 2.1 [2, 3]. The random vibration is indicated in Figure 2.2 for Ariane 4. Data for the Ariane 5 launch vehicle shows a similar trend, but with the peak amplitude shifted to a lower frequency.

The *steady component of launch acceleration* must achieve a speed increase of about 9.5 km/s. Its time history is again dependent on the launch vehicle used. Low-mass payload vehicles such as Scout, sounding rockets, and air-launched systems such as Pegasus generally exhibit high peak acceleration, whereas those of large payload/crewed vehicles are smaller. For multi-stage vehicles, the acceleration increases during the burn of each stage and peaks at burn-out/stage-separation. This is demonstrated for Ariane 5 in Figure 2.3.

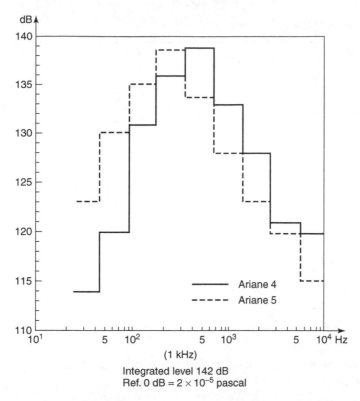

Figure 2.1 Ariane noise spectrum under the SPELTRA fairing, Ariane 4 [2] and SPELTRA/Ariane 5 [3]. (Reproduced by permission of Arianespace)

During this launch profile, event H1 is the solid rocket booster flame-out and separation, event FJ is the fairing jettisoning, event H2 is the main cryogenic stage engine shutdown and separation and event H3 is the upper stage shutdown. The air-launched Pegasus, a three-stage solid rocket vehicle, shows peak static acceleration at stage burnout, with the actual value being heavily dependent upon the payload mass. Thus for a 50 kg payload at stage 3 burn-out, the 3σ acceleration is $13\,g_0$ [4]. However, for the near maximum payload of 600 kg this falls to nearly $4.5\,g_0$ (g_0 is the surface gravitational acceleration, $9.8\,\text{m/s}^2$). Table 2.1 shows the design loads that a satellite must achieve associated with various mission events for the Ariane 5 launch vehicle (see also Section 8.4.2 for discussion of the impacts of these launch loads on structural design).

For human flight, it is necessary to place the astronaut in a suitable position to withstand peak acceleration levels; Figure 2.4 shows the typical maximum levels that may be withstood. Greatest protection is provided when the astronaut is on his or her back during launch.

Mechanical shock is experienced when devices such as latches or explosive bolts are used, or at ignition of rocket motor stages and their subsequent separation, launch vehicle/payload separation, or when docking or landing.

These instantaneous events can provide extremely high-acceleration levels lasting only a few milliseconds locally, or in some cases, extending to the complete system.

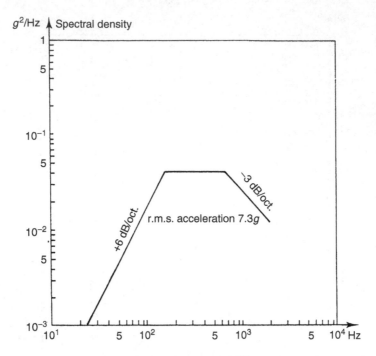

Figure 2.2   Ariane 4 random vibration spectrum [3]. (Reproduced by permission of Arianespace)

Table 2.1   Acceleration values for the Ariane 5 launch vehicle [3]. (Reproduced by permission of Arianespace)

Acceleration (g) Critical flight events	Longitudinal		Lateral	Additional line load (N/mm)
	Static	Dynamic	Static + Dynamic	
Lift-off	− 1.8	± 1.5	± 2	10
Maximum dynamic pressure	− 2.7	± 0.5	± 2	14
SRB end of flight	− 4.55	± 1.45	± 1	20
Main core thrust tail-off	− 0.2	± 1.4	± 0.25	0
Max. tension case: SRB jettisoning	+ 2.5		± 0.9	0

Their frequency spectrum is characterized by high-frequency components. In the case of Ariane 5, during payload separation the peak excitation at the spacecraft interface is some $2000\,g_0$ for frequencies in the range 1–10 kHz. For Pegasus, the highest shock loading is experienced above a frequency of 1.5 kHz where the design load is $3500\,g_0$.

The *thermal environment* experienced during launch is determined generally by the temperature reached by the launch shroud. Its high temperature arises from the

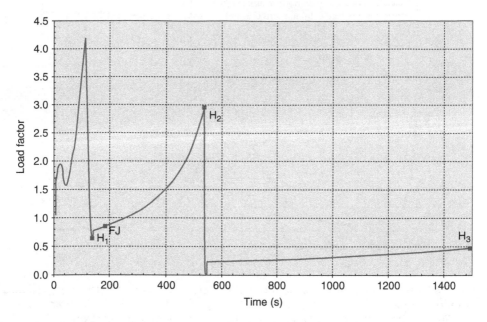

Figure 2.3 Ariane 5 static acceleration profile [3]. (Reproduced by permission of Arianespace)

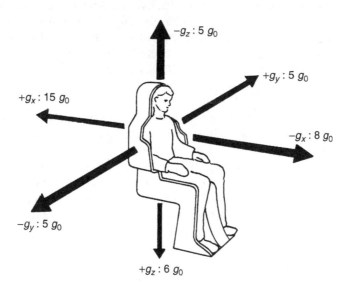

Figure 2.4 Typical tolerance to sustained linear acceleration as a function of the direction of the acceleration

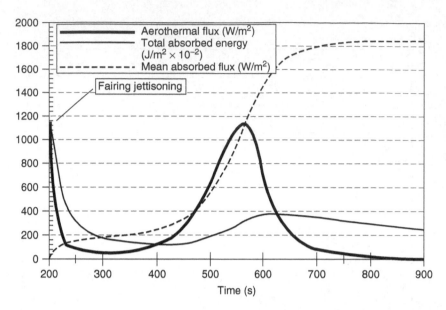

Figure 2.5   Aerothermal fluxes encountered on a standard Ariane 5 trajectory [3]. ——
Aerothermal flux (W/m^2). ___ Total absorbed energy (J/m^2 × 10^{-2}). - - -Mean absorbed
flux (W/m^2). Fairing jettison and second flux peak constrained at 1135 W/m^2.
(Reproduced by permission of Arianespace)

aerodynamic frictional forces of the vehicle moving at high velocity through the atmosphere. The temperature reached is determined by the specific heat of the shroud material and a balance between friction heating and radiative and convective heat losses. The subsequent temperature rise of the payload within the shroud is dominated by radiative and heat conduction paths between shroud and payload. Once the shroud has been jettisoned, payload heating arises directly as a result of frictional forces, but the low density of the atmosphere then results in only modest heat fluxes.

Figure 2.5 shows the heating experienced during the Ariane 5 launch ascent into geostationary transfer orbit (GTO) [3]. The altitude at which the fairing is jettisoned is governed by an aerothermal flux limit which ensures that the launcher payload is not exposed to a level beyond 1135 W/m^2 once the fairing has been ejected.

The *ambient atmospheric pressure* declines during launch. The rate at which depressurization occurs depends on the venting of the shroud volume. Generally, this is fixed by the inclusion of venting ports; for example, on Ariane 5 the static pressure declines at a rate of 2 kPa/s although there is a brief period during ascent where this may reach 4.5 kPa/s. Venting control is particularly important because of possible adverse static loads being placed on structural members. It should be noted that venting of electronic boxes is also generally required within the satellite payload.

Great care is required during payload integration to ensure that *electromagnetic interference* (EMI) does not present a hazard. Hazards may present themselves in a variety of forms but the most severe are cases in which EMI may result in the activation of part of the payload, which could lead to death of attendant personnel, perhaps via the ignition of an on-board propulsion system. Figure 2.6 shows the EMI environment associated with Ariane 5.

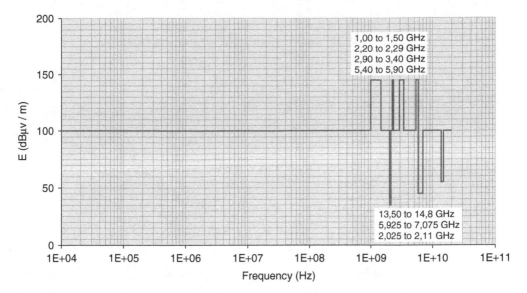

Figure 2.6 Electromagnetic interference for Ariane 5 [3]. (Reproduced by permission of Arianespace)

2.3 OPERATIONAL SPACECRAFT ENVIRONMENTS

2.3.1 Solar radiation

The Sun provides all the heat input to the solar system (excluding planetary radioactive decay processes), and its mass is 99.9% of the total. It is not surprising that it dominates the space environment of the whole Solar System, and it is therefore appropriate to outline its significant features as they affect the near-Earth environment.

The Sun itself is not a particularly significant body in the universe. It is a star of mass $\sim 2 \times 10^{30}$ kg, modest by stellar standards, and is one of $\sim 10^{11}$ stars that form our galaxy. It is classified as a G2V star, having a yellowish appearance because its radiated light peaks at ~ 460 nm, and it is termed *a yellow-dwarf star*. Its radius is 7×10^8 m.

After the Sun, the nearest star is 3.5 light years away (1 light year $= 9.46 \times 10^{12}$ km) and between the stars the gas density is low, with hydrogen as the dominant species. The density amounts to only 3 atoms/cm^3, in comparison to the nominal number density of our own atmosphere at sea level of $\sim 3 \times 10^{19}$ molecules/cm^3.

The Sun is fundamentally a giant thermonuclear fusion reactor, the surface temperature of which is ~ 5800 K. The *photosphere* is optically thick, and its spectrum approximates to that of a black body.

The *solar spectrum* is shown in Figure 2.7, with a 5900 K black body for comparison. It is evident that it departs from the black-body spectrum at some wavelengths, these discrepancies arising in the solar atmosphere. There are two primary regions of this. The lower, or *chromosphere*, extends to a few thousand kilometres above the photosphere and is a region of increasing temperature, peaking at $\sim 10\,000$ K; it is responsible for enhanced ultraviolet (UV) emission. The upper atmosphere, called the *corona*, becomes

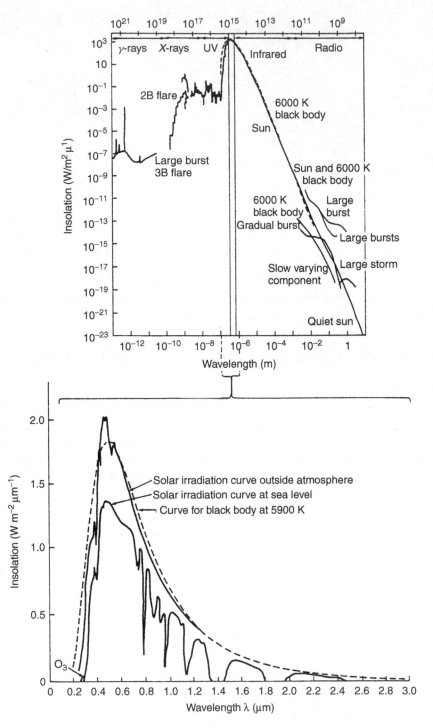

Figure 2.7   The solar spectrum. (Reproduced by permission from Hynek, J.A. (1951) *Astrophysics*, McGraw-Hill)

more tenuous and extends to several solar radii. Its nominal temperature is around 2×10^6 K, and it emits substantial amounts of X-rays. The nominal release of energy from the Sun is at a rate of 3.85×10^{26} W. In order to gain an insight to this power level, if the Earth's fossil fuel resources were to be burnt at this rate, then they would be exhausted in a mere 50 milliseconds.

The *solar wind* is another outward flux from the Sun. It is a flow of plasma expelled at high velocity. In reality it forms the outermost layer of the solar atmosphere, being continuously driven outward as a result of the Sun's radiation pressure. At Earth, the speed of the wind is ~450 km/s, its density is ~9 protons/cm^3 and its kinetic temperature is ~100 000 K.

Sunspots are an indication that there are significant disturbances taking place on the Sun's surface and through its atmosphere. These, first observed by Galileo, are regions of its disc which are cooler than the surrounding surface. They emit less radiation and thus appear as dark spots. Periods of high solar activity occur when there are a large number of sunspots and then enhanced emission of radiation occurs, most notably at radio wavelengths and at X-ray and γ-ray energies. This enhanced emission is generally associated with solar flares, which occur at sites near sunspots. They may last from a few minutes to several hours and occur as frequently as one every two hours during high solar activity.

The Zurich sunspot number R_z is used to quantify the overall number of sunspots on the Sun at any time. It is defined as $R_z = K(10g + f)$, where f is the number of sunspots that exhibit umbrae, and g is the number of groups into which these spots fall. K is a factor that relates to the observing instrument and is used as a normalization factor.

The detailed prediction of individual flares is not yet possible, but the general level of activity has a well-defined 11-year cycle as shown in Figure 2.8. However, owing to magnetic pole reversal of the Sun at peak solar activity, the real period is 22 years. It may

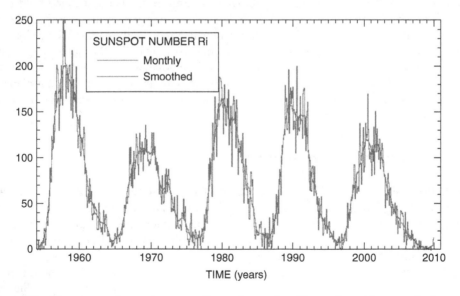

Figure 2.8 The sunspot number over five solar cycles. Reproduced by permission of the Solar Influences Data Center (http://sidc.oma.be)

Table 2.2   Solar variability [5]. (Reproduced by permission of NASA)

Spectral region	Wavelength	Flux $(J/(m^2 s \mu m))$	Variability
Radio	$\lambda > 1$ mm	10^{-11}–10^{17}	$\times 100$
Far infrared	1 mm $\geq \lambda > 10 \mu$m	10^{-5}	Uncertain
Infrared	10μm $\geq \lambda > 0.75 \mu$m	10^{-3}–10^2	Uncertain
Visible	0.75μm $\geq \lambda > 0.3 \mu$m	10^3	<1%
Ultraviolet	0.3μm $\geq \lambda > 0.12 \mu$m	10^{-1}–10^2	1–200%
Extreme ultraviolet	0.12μm $\geq \lambda > 0.01 \mu$m	10^{-1}	$\times 10$
Soft X-ray	0.01μm $\geq \lambda > 1$Å	10^{-1}–10^{-7}	$\times 100$
Hard X-ray	1Å $\geq \lambda$	10^{-7}–10^{-8}	$\times 10$– $\times 100$

be seen that the magnitude of the peak varies from cycle to cycle. The past five cycles are shown in Figure 2.8 and it is notable that the period of low solar activity in 2009/10 exhibited one of the lowest levels of activity on record. Regular updated sunspot activity may be readily accessed at a number of web sites including http://www.dxlc.com/solar, which provides daily data on a variety of forms of measuring solar activity, including sunspot number, the geomagnetic indices and maps of solar active regions. The site http://sidc.oma.be/index.php3 provides a long-term historical record of solar activity from 1700 when the first regular recording of sunspot activity took place.

Table 2.2 shows the typical intensity variability of the Sun at particular wavelengths where it has been established. There is good correlation between sunspot number and radiated power at some specific wavelengths and this is used to provide an additional measure of the overall level of solar activity. The most frequently used monitor is the solar flux at a wavelength of 10.7 cm. One source for such data is from NOAA, in the US.

An enhanced flux of radiation and high-energy particles is also associated with major solar flares. Two components may be distinguished when it arrives at the Earth. The first occurs approximately 20 min following the flare, when the electromagnetic emissions first increase, and then a more prolonged component arrives about a day later. These latter particles appear to be an enhanced component of the solar wind, travelling at velocities of $\sim 10^3$ km/s. Figure 2.9 shows the variability of particle fluxes in the interplanetary medium caused by solar activity. Detailed descriptions of ionizing radiations are given in the next section.

2.3.2   Earth orbit environment

The Earth orbits the Sun at a mean distance of one astronomical unit (AU), equal to 1.496×10^8 km. It is a nearly spherical body having a mass of only 3×10^{-6} times that of the Sun, but having a gravitationally bound atmosphere and a significant magnetic field. Each of these features is important in determining the near-Earth environment, in which the majority of space vehicles operate.

The *atmosphere* at sea level is predominantly molecular nitrogen (78%) and molecular oxygen (21%), with a variety of trace elements, the most significant being argon. Standard sea-level pressure is accepted to be 1.013×10^5 Pa [6]. The lower atmosphere, up to $\sim$86 km, is sufficiently turbulent to result in a homogeneous gas mixture, albeit with a

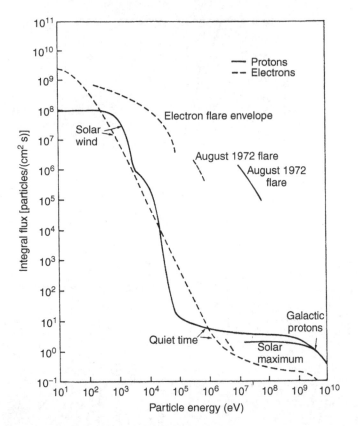

Figure 2.9 Electron and proton spectra. (From [1]: *Solar Cell Array Design Handbook* by Rauschenbach, Copyright ©1980 by Van Nostrand Reinhold. All rights reserved)

pressure decreasing exponentially. Above this height, photochemical processes disturb the homogeneity. Of particular importance is the absorption of UV radiation from the Sun, leading to the dissociation of oxygen in the upper atmosphere. The details of the processes are beyond the scope of this chapter and the interested reader is referred to Reference [6]; however, the resultant atmosphere above ~120 km is one in which each atmospheric constituent is decoupled from all the others. For each species, it is then possible to write down an equation of the diffusive equilibrium of the form

$$n_i \upsilon_i + D_i \left(\frac{\mathrm{d}n_i}{\mathrm{d}Z} + n_i \frac{(1 + \alpha_i)}{T} \frac{\mathrm{d}T}{\mathrm{d}Z} + \frac{gn_i}{R^*} \frac{M_i}{T} \right) = 0 \tag{2.1}$$

where n_i is the number density of species i, having a molecular weight M_i at altitude Z, υ_i is the vertical transport velocity of the species, and D_i and α_i are its molecular and thermal diffusion coefficients, T is the atmospheric temperature, R^* is the universal gas constant and g is the height-dependent acceleration due to gravity. If negligible vertical transport takes place, and for species in which thermal diffusion is negligible, equation (2.1) reduces to a hydrostatic equilibrium equation wherein the number density profile is driven by the atmospheric temperature. Figure 2.10 shows the variation of number density with height for different species.

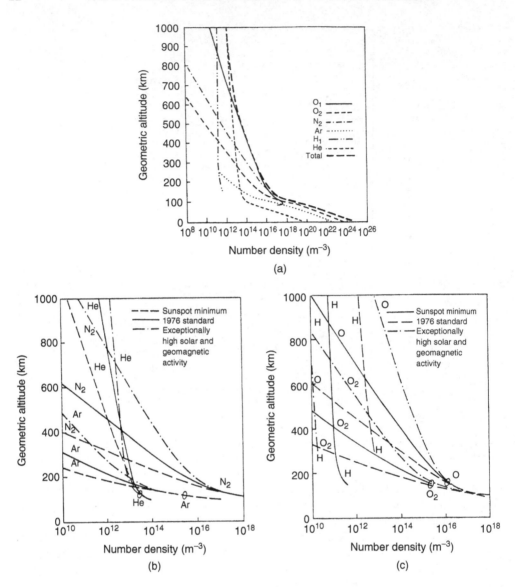

Figure 2.10   Species concentration (a) in the US standard atmosphere, and (b, c) under extreme conditions

For detailed calculation of the atmosphere, a variety of models are available (e.g. References [7, 8]). Each of these requires the specification of a temperature at some height, and then by using a pre-defined temperature profile, the density may be determined through equation (2.1). Figure 2.11 shows the general form of such profiles. It can be seen from this that at extreme altitudes the temperature tends to a limiting value, the so-called exospheric temperature, T_∞. Within the exosphere the atmosphere is effectively isothermal. T_∞ rises through increased solar activity; most models relate it to the flux of solar radiation at 10.7 cm ($F_{10.7}$) through an algorithm, and also include the effects of geomagnetic activity,

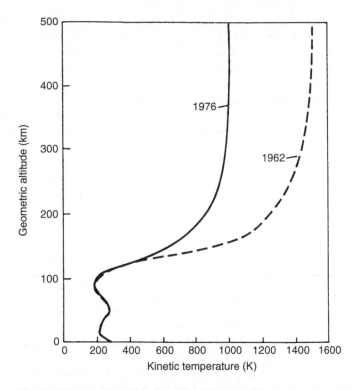

Figure 2.11   The temperature profile of the atmosphere with altitude

stimulated by the interaction of the solar wind and the Earth's magnetic environment or magnetosphere (see below). Since under conditions of hydrostatic equilibrium, the density falls from that at sea level, ρ_{SL}, at a rate

$$\rho = \rho_{SL} \exp\left(\frac{-gM_i}{R^*T}Z\right) \tag{2.2}$$

then for a given altitude, the atmospheric density will increase with solar activity and this will result in reduced lifetimes for low orbiting vehicles.

Equation (2.2) provides the simplest formulation for air density as a function of altitude, and other more sophisticated analytic descriptions have been developed to yield expressions for the rate of orbit decay for low Earth orbiting (LEO) vehicles. The pioneering work of King-Hele in this field [9] provides the most comprehensive of these analytic models, which have the potential to describe the variation of scale height in the atmosphere with altitude, and the effects of atmospheric oblateness and rotation. Whilst these expressions can become complex, for the case wherein only scale height variations around a single orbit are concerned, the equivalent density may be written:

$$\rho = \rho_p\left[1 + \frac{\mu_0}{H_p}(r - r_p)\right]^{-1/\mu_0} \tag{2.3}$$

where suffix p refers to conditions at perigee of the orbit, H is the scale height and μ_0, the altitude gradient of H, is a constant of order of the eccentricity of the orbit (see Chapter 4).

For circular orbits in which e is zero, clearly equation (2.3) provides difficulty. However, in such cases the altitude of the satellite remains constant, and it is then inappropriate to consider the variation in scale height.

The most recent US standard atmosphere 1976 [6] is shown in Figure 2.10. At altitudes typical of LEO vehicles (300–900 km) the density is low, but not insignificant. At geostationary altitude, ~36 000 km, the density of the neutral atmosphere is approximately the same as in the interplanetary medium, namely, 10^{-20} kg/m^3, and the pressure is ~10^{-15} Pa. The dominant species of the atmosphere in LEO is generally either atomic oxygen or helium. The effects of atomic oxygen on surfaces are considered in Section 2.4.1.

One of the most frequently used models of the thermosphere is the MSIS model [7]. This was originally produced in 1986 but has been revised in the MSISE-90 model and in 2001 as NRLMSISE-00. This includes both shuttle flight data and incoherent radar scattering data. It is now possible to access a variety of atmospheric models on the NASA FTP site at ftp://nssdcftp.gsfc.nasa.gov/models/atmospheric. This can be used to produce density profiles of the major atomic and molecular species together with temperature and total density information on a longitude, latitude specific basis.

Few *atomic/molecular collisions* take place between components of the atmosphere and spacecraft in either LEO or GEO. Table 2.3 shows that above 200 km altitude the mean free path is significantly greater than the dimensions of most space vehicles. This has two consequences.

Firstly, the ability to exchange heat energy with the environment is solely as a result of radiation. Thus the dominant radiative heat input is due to solar radiation, which in the near-Earth environment is 1371 ± 5 W/m^2. Secondary input occurs as a result of Earth albedo (the reflection of solar radiation from the top of the atmosphere) and Earth shine (the black-body radiation of the Earth), which has a magnitude of ~200 W/m^2. The neutral atmosphere at ~10^3 K and the solar wind at ~2×10^5 K provide negligible heating. The temperature that a space vehicle reaches is thus dependent upon a balance of radiative heat input and output (see Chapter 11).

Secondly, the aerodynamics of spacecraft at orbiting altitudes must be based upon free molecular flow as briefly described in Chapter 4. Since the density is low, the frictional heating forces are negligible, even though the relative velocity approaches 8 km/s.

The *ionosphere*, above ~86 km, is a region of increasing plasma density caused by photo-ionization, due to incident UV photons. The plasma has a significantly lower density

Table 2.3 Mean free path λ_0 as a function of altitude

Altitude (km)	λ_0 (m)	Altitude (km)	λ_0 (m)
100	0.142	300	2.6×10^3
120	3.31	400	16×10^3
140	18	500	77×10^3
160	53	600	280×10^3
180	120	700	730×10^3
200	240	800	1400×10^3

than the neutral density below an altitude of ~1000 km, even though its peak value occurs at 300–400 km. It has significant influences upon the propagation of radio waves; for an electron density n_e electrons cm^{-3}, frequencies below the plasma frequency, given by $f_p \sim 9000\sqrt{n_e}$ Hz, cannot propagate, and thus radio waves of lower frequency incident upon such a plasma will be reflected. If there is additionally a magnetic field present, as is the case in the near-Earth environment, then the polarization of any electromagnetic radiation propagating through the plasma will be rotated due to *Faraday rotation*; inefficiencies can then occur in communication systems if linearly polarized radio waves are used (see Chapter 12).

The *Earth's magnetic field* has two primary sources. The dominant one at its surface is due to currents circulating within its core, while at higher altitudes the currents caused by the differential motion of electrons and ions in the *magnetosphere* play a significant role. The solar wind plasma, carrying its own magnetic field, distorts the Earth's simple dipole field into the shape shown in Figure 2.12, with both open and closed magnetic field lines.

From Figure 2.12 it is apparent that at high altitude the magnetic field structure is complex; however, at lower altitude it is possible to make certain observations. Firstly, the overall strength of the magnetic field is not constant, but is decreasing at ~0.05% per year. This field is weakest on the equator, and Figure 2.13 shows its dependence on both latitude and altitude.

Ionizing radiations

In the near-Earth environment, there are several types of ionizing radiations that impact upon spacecraft systems.

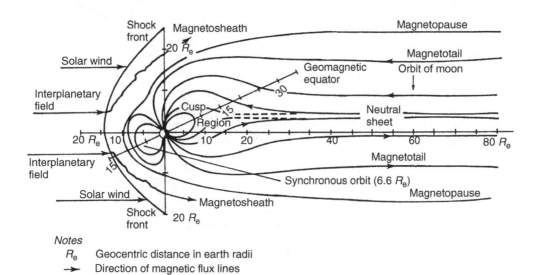

Figure 2.12   The Earth's magnetosphere. (From Wertz, R. *et al.* (1978) *Spacecraft attitude determination and control*, p. 858. Reproduced by permission of Kluwer Academic Publishers)

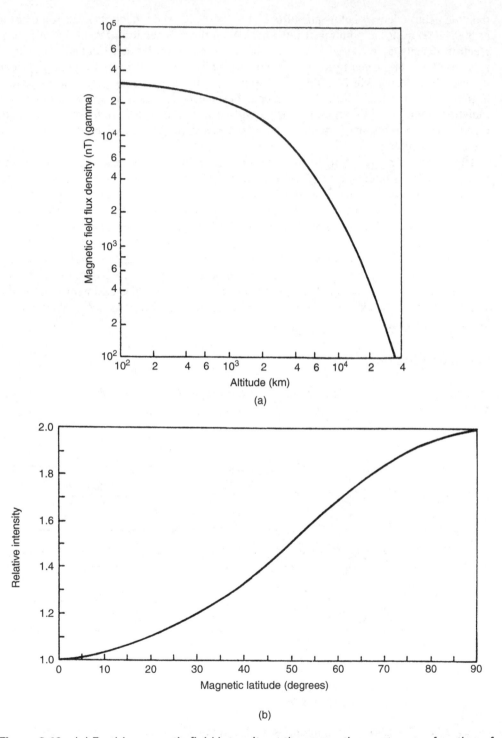

Figure 2.13 (a) Earth's magnetic field intensity at the magnetic equator as a function of altitude (Adapted from [11]) and (b) relative intensity of the Earth's magnetic field as a function of magnetic latitude. (From Wertz, R. *et al.* (1978) *Spacecraft attitude determination and control*, p. 858. Reproduced by permission of Kluwer Academic Publishers)

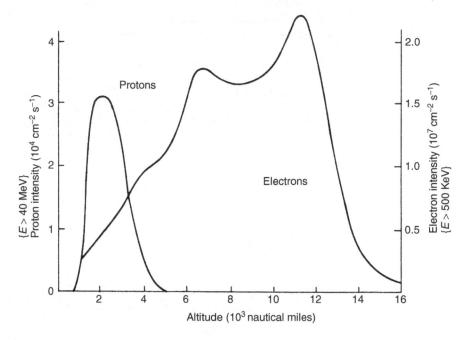

Figure 2.14   The structure of the Van Allen radiation belts (idealized)

The *Van Allen radiation belts* contain energetic protons and electrons that are trapped in the Earth's magnetic field and generally follow the magnetic field lines. There may also be significant fluxes of heavy ions such as of helium (He), nitrogen (N) and oxygen (O) whose atmospheric densities depend on solar and geomagnetic activity. The overall structure of the Van Allen belts can be seen from the highly simplified diagram in Figure 2.14.

Proton energies range from 0.01 to 400 MeV with fluxes in the range of 10^8 to 600/cm^2 s, respectively. Electron energies are in the range from 0.4 to 4.5 MeV with fluxes from 4×10^8 to 100/cm^2 s, respectively. One of the most notable features of the radiation belt particles is the large variation with both altitude and latitude. Figure 2.15 shows typical contour plots of the electron and proton fluxes at various locations [12].

To predict electron and proton fluxes for LEO missions, the NASA models AE8 [13] and AP8 [14] are used. Since these particle fluxes vary with the solar cycle, there are two variants for each model, the maximum and minimum, corresponding to the maximum and minimum of the solar cycle, respectively. However, it should be noted that the solar cycle variations of the fluxes are not well understood. Furthermore, the data on which these models are based are old (circa 1960s and 1970s), and the model should only really be used to predict integrated fluxes (fluences) for periods greater than about six months. More recent models are also available through NASA, as may be found at http://spdf.gsfc.nasa.gov/space_weather/Space_Weather_at_SSDOO.html.

At low altitudes and low inclinations, the dominant feature of the radiation environment is the region known as the *South Atlantic Anomaly*. Because of the offset and tilt of the geomagnetic axis relative to Earth's rotation axis, this is a region of enhanced radiation in which parts of the radiation belt are brought to lower altitudes (see Figure 2.16).

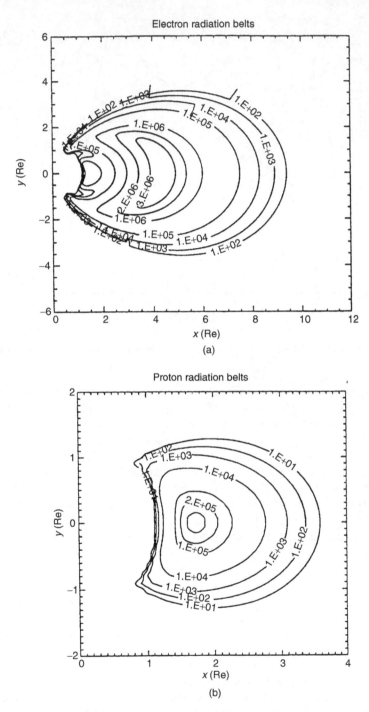

Figure 2.15 Radiation belt model particle fluxes. Contour plots in 'dipole space' in a plane containing the geo-dipole axis (y): (a) electron fluxes at greater than 1 MeV and (b) proton fluxes at greater than 10 MeV. (Axes are calibrated in Earth radii, R_e.) [12]. (Reproduced by permission of the International Astronautics Federation)

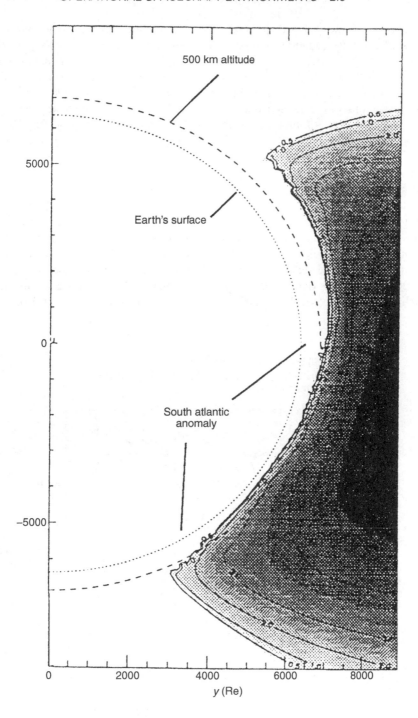

Figure 2.16 Proton radiation belt at low altitude, derived from the AP8 model. Flux contours at longitude 325 °E. Tilt and offset brings the contours below 500 km altitude. This is the South Atlantic anomaly [12]. (Reproduced by permission of the International Astronautical Federation)

The effects of the trapped particles are degradation of electronic parts due to accumulated dose, degradation of solar array performance due to displacement damage, single-event upsets (SEUs) and dielectric charging. While the detailed processes associated with these damage types is somewhat complex, the key phenomena that take place result from the impact of the high-energy particle upon both the energy structure and lattice structure of the semiconductor material. The damage influence of an individual massive particle, such as a proton, is significantly greater than that from an electron due to the higher momentum carried by the former. In particular, if, for example, a proton is stopped in the material, because of the momentum exchange a significant *displacement* of a lattice atom will occur. This displacement can result in local ionization and the disruption, again locally, in the energy structure, possibly with the creation of additional energy states between the conduction and valence bands. The ionization process results in an impulsive release of charge, leading to an SEU. Changes to the energy structure result in a reduction in the efficiency of solar cells converting sunlight to electricity (see Chapter 10). Both electrons and protons contribute to the total dose and the major contribution will depend on the particular orbit; in general, low altitude orbits (less than about 800 km) will be dominated by protons, whereas high altitude orbits including geosynchronous orbits, will be dominated by electrons (see Figure 2.17). Both types of

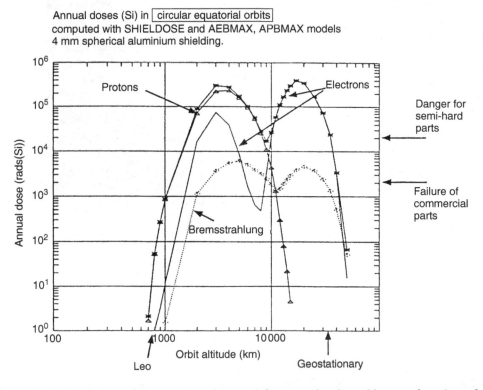

Figure 2.17 Variation of the computed annual dose on circular orbits as a function of altitude, computed with the AE8 and AP8 models, and SHIELDOSE [12]. (Reproduced by permission of the International Astronautical Federation)

particles will cause displacement damage, but protons are more effective. Because of the relative magnitudes of electron and proton fluxes, it is not easy to generalize about their contributions to the displacement damage (usually expressed as a 1 MeV equivalent fluence), but as with total dose, protons usually dominate at low altitude and electrons at high altitude.

Galactic cosmic radiation is composed of high-energy nuclei, believed to propagate throughout all space unoccupied by dense matter. Its origin is still a matter of scientific debate and may have both galactic and extragalactic sources.

Primary cosmic radiation is by definition that which is incident on the Earth's atmosphere. Cosmic rays propagating through the atmosphere undergo nuclear collisions and generate secondary rays consisting of all known nuclear and sub-nuclear species. The flux of galactic cosmic radiation is believed to be essentially isotropic outside the heliosphere; inside it, propagation effects result in an isotropy of approximately 1%.

High-energy cosmic ray particles have a large amount of kinetic energy and this can have a permanent effect upon the material through which they pass.

Primary cosmic radiation observed at the Earth's orbit consists of approximately 83% protons, 13% alpha particles, 1% nuclei of atomic number greater than two and 3% electrons. The composition extends over an energy range from a few hundred MeV to greater than 10^{20} MeV. Figure 2.18 [15] shows typical quiet-time spectra for several elements. The intensity of cosmic rays observed is dependent on the solar cycle, decreasing as the sunspot number cycle increases.

During their travel from their source regions to the vicinity of the Earth, cosmic rays interact with the interstellar medium. This causes fragmentation to occur, resulting in a depletion of the heavy charge primary rays and an increase in the lighter nuclei. The low-energy portion of the cosmic ray spectrum is quite variable, reflecting its dependence on solar modulation; from solar cycle minimum to maximum, the energy density of near-Earth primary galactic rays decreases by about 40%. The ray intensity observed at Earth does not change smoothly from maximum to minimum values, but decreases in a series of sharp drops followed by a partial recovery until the minimum intensity is reached near to the maximum in the solar activity.

Solar energetic particle events

Part of the energy in solar flares is in the form of nuclei accelerated to high energies and released into space. These are commonly referred to by descriptive names such as solar cosmic events, solar proton events, solar electron events, solar cap absorption events and ground level events. The most commonly measured components of these are the proton and electron events. Major solar particle events occur at random, with a frequency that varies from approximately one every two months to one every two years. Events initiated by solar flares may also contain a small apparently variable flux of heavier elements.

Solar proton events are statistical in nature, showing a wide variation in characteristic parameters such as integrated flux (fluence), peak flux and energy spectrum. One method for predicting proton fluence for space missions is that of Feynman *et al*. [16] and is based on a combination of observations made from the Earth's surface, from above the atmosphere between 1956 and 1963 and from spacecraft in the vicinity of Earth between 1963 and 1985. Analysis of these data shows that there are seven hazardous years during each solar cycle, beginning two years before the solar maximum year and extending to four

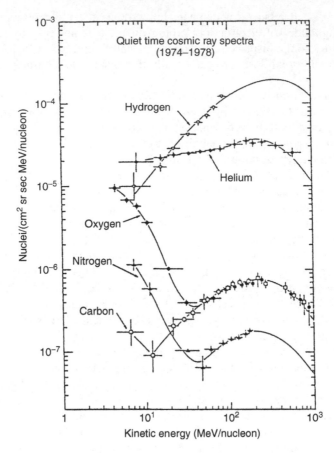

Figure 2.18 Quiet-time energy spectra for the elements H, He, C, N and O, measured at
1 AU over the solar minimum period from 1974 to 1978. Note the 'anomalous'
enhancements in the low-energy spectra of He, N and O. The data are from Caltech and
Chigago experiments on IMP-7 and IMP-8 [15]. (Reproduced by permission of the Jet
Propulsion Laboratory, California Institute of technology, Pasadena, California)

years after it; the remaining four years of the cycle contribute negligibly to the expected
mission fluence. More recent tools for prediction are available through the ESA. A suitable
link to this via http://www.esa.int/TEC/Space_Environment/SEMNW3SMTWE_0.html.

Figure 2.19 shows the probability of exceeding a given fluence level over the life of a
mission, assuming a constant heliocentric distance of 1 AU. If a mission is in space for
more than one solar cycle, the best method for finding the total expected fluence is to estimate the fluence per cycle from the seven-year curve and add the appropriate number of
additional yearly fluences to this line. The use of the one-year curve to estimate the additional fluence expected on a long mission is not recommended because it will lead to an
overestimate. Correction for the dependence on the distance from the Sun (r) may be made
by assuming that it varies as r^{-3} for r less than 1 AU, and as r^{-2} for r greater than 1 AU.

Solar protons are accelerated in solar active regions during solar flares, and X-ray, radio
and optical emissions are indicators that particle acceleration is occurring. The decaying

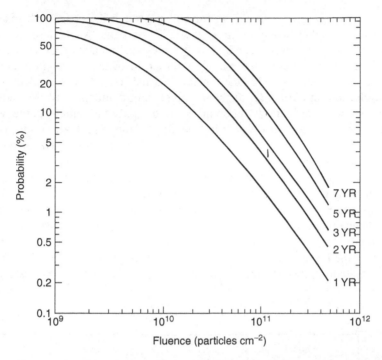

Figure 2.19   The probability of exceeding given levels of fluence energy >10 MeV versus active years of solar cycle. (Reproduced by permission of the Jet Propulsion Laboratory, California Institute of Technology, Pasadena, California)

portion of a flare is normally exponential in character, the flux decay typically having a time constant of two to three days, but there is a slight dependence on the energy, and there can be a large variability between events.

In addition to energetic protons, solar energetic particles produce heavy ions. Their fluxes and fluences can be found by extrapolation from those of protons, using elemental abundance ratios normalized to hydrogen. Typical values are shown in Table 2.4 [17].

Table 2.4   Normalized abundances of solar energetic particle events [17]. (Reproduced by permission of the Jet Propulsion Laboratory, California Institute of Technology, Pasadena, California)

		1 MeV	1–20 MeV	10 MeV	6.7–15 MeV
1	H	1.0	1.0	1.0	1.0
2	He	2.2 E-2	1.5 E-2		1.5 E-2
3	O	3.2 E-4	2.2 E-4	2.2 E-4	2.8 E-4
4	Al	3.5 E-6	3.5 E-6	3.1 E-6	3.3 E-6
5	Ca	2.3 E-6	2.6 E-6	3.1 E-6	3.2 E-6
6	Fe	4.1 E-5	3.3 E-5	3.4 E-5	

Electrostatic charging of a spacecraft travelling through the near-Earth space environment will occur, whether it is in or out of the radiation belts. As a consequence, currents will occur between the space vehicle and the plasma, imbalance of which will cause spacecraft to develop a charge that may be returned to balance through arcing. The two major sources of currents are [11] the ambient plasma itself and photoelectron emission due to sunlight, and in particular, the short wavelength component of this radiation. The latter phenomenon is particularly important when the spacecraft enters and leaves eclipses [18]. Severe problems arise if differential charging of the spacecraft surface occurs. The simplest method of preventing this is to use conductive surfaces wherever possible. One primary area in which this is not possible is on the solar array; an alternative solution is then to apply a near transparent coating of indium oxide to the cell cover glass material, which typically reduces the resistivity of the glass surface to less than 5000 Ω/cm^2 [19].

Meteoroids and micrometeoroids occur with a frequency that varies considerably with the type of space mission. These are solid objects whose mass and size vary over many orders of magnitude. Their mass spectrum is shown in Figure 2.20. Near large gravitational masses such as the Earth their fluxes tend to be enhanced. The asteroid belt is also a region of enhanced meteoroid density. Impact of micrometeoroids generally causes a degradation

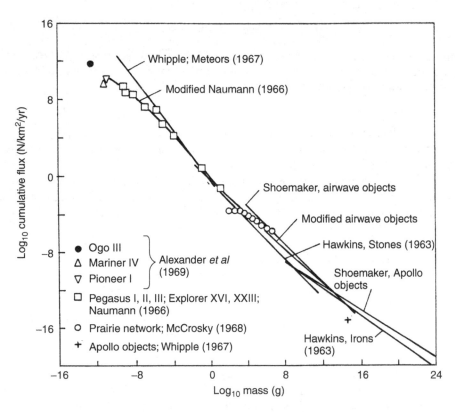

Figure 2.20 Terrestrial mass-influx rates of meteoroids. N is the flux of particles with mass greater than m [20]. (Reproduced by permission of NASA; see reference [5] for details)

of surface thermal properties, although the possibility of component failure clearly exists. The most dramatic evidence of particle impacts on a spacecraft is from the Giotto spacecraft and its passage near to Halley's comet during 1986. Particle impacts led to the failure of some experiments and a change in the attitude of the vehicle at closest encounter.

Man-made space debris, consisting of aluminium oxide dust particles (from solid rocket exhausts), instrument covers, nuts and bolts, rocket upper stages, and so on is in addition to the naturally occurring micrometeoroid environment. The size varies from 0.001 mm to 10 m in diameter. Space debris varies with the orbit's altitude, eccentricity and inclination, and exhibits a strong directional dependence. The average velocity (relative to the orbiting spacecraft) is about 11 km/s, but it is a function of altitude, ranging from zero to twice the orbital velocity. Since the majority of the particles originate from satellite and launcher components, the average density is assumed to be that of aluminium alloys—about 2.8 g/cm^3. The threat to space-stations such as the ISS due to the larger sizes of particles, greater than 1 mm, has caused a great deal of interest in this environment, both in modelling and in measuring it. The collision between Iridium 33 and Cosmos 2251 in February 2009 has further highlighted the dangers posed to operational spacecraft of collisions, together with the substantial increase in the population of debris objects following such collisions. Figure 2.21 provides data on the predicted growth in debris objects by one of the models (LEGEND) used for such simulations. Four test scenarios were analysed: non-mitigation ('business-as-usual') scenario, in order to compare with possible active debris removal (ADR) scenarios initiated in 2020, with annual debris removal rates of either 5 (2020/5), 10 (2020/10), and 20 (2020/20) objects per year. In 2009 there were eight avoidance manoeuvres required to avoid potential impact for the

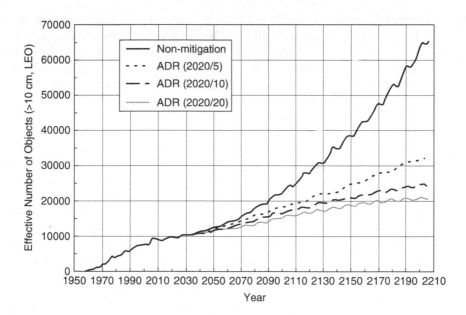

Figure 2.21 LEGEND-simulated LEO debris populations (objects 10 cm and larger) between 1957 and 2006 (historical), and between 2007 and 2206 (future projection). Each curve represented the average of 100 Monte Carlo runs [22]. (Reproduced by permission of Pergamon Press)

International Space Station, Space Shuttle operations and other smaller satellite systems [21]. As a result there is a growing realization that active removal of debris may become a requirement for sustained operation within the LEO environment [22].

The smaller particles, in the range from 10^{-3} to 10^{-9} g, have a flux that is high enough to erode surfaces and have enough energy to penetrate protective coatings. Of particular concern is their effect on large solar arrays, sensitive optical surfaces and detectors. The possibility of penetrating protective coatings necessitates the consideration of synergistic effects between the particulate and the atomic oxygen environments. See Section 2.4.1.

System requirements for meteoroid and debris protection amount generally to ensuring the safety of people for crewed spacecraft and the operational availability for unmanned craft. Usually, these requirements are expressed as the probability of no damage over a given time. For example, typical probability values for the Columbus Attached Laboratory range from 0.995 for endangering the crew or module's survivability for a 10-year life, to 0.999 for penetration or bursting over a one-year period. Having set the system requirement probabilities, the critical debris size can be calculated using the environment models and the spacecraft geometry. For ISS, the critical debris size ranges from about 0.5–1.0 cm [23]. One resource available for predicting the micrometeorite and debris impact on space vehicles is the ESABase software package, available through http://esabase2.net/.

Effective shielding can be achieved by using a double-walled bumper shield, in which the first wall fragments the impacting particle into a cloud of smaller, slower moving debris, which are then stopped by the second wall (see Chapter 8).

2.3.3   Solar System environments

The *Solar System* may be considered to be the region of space that is dominated gravitationally by the Sun. In Chapter 5, this will be defined in greater detail. However, at this point, we can consider the Solar System to be that region containing the eight major planets and their natural satellites, together with all the other minor bodies including the 'dwarf planet' Pluto, asteroids, comets and dust, and the residue of the solar wind. Following the detailed environment of the near-Earth region of space, in which the majority of space vehicles operate, it is appropriate to consider briefly the major features of those other solar system bodies, which have been the focus of space missions.

Table 2.5 lists the mass properties of the major Solar System bodies together with their orbital characteristics about the Sun. The definition of the orbital parameters may be found in Chapter 4, together with additional data on these objects. The key features discernible from the data presented in Table 2.5 are firstly, with the exception of Pluto, and to some extent Mercury, all the planetary bodies nearly lie in a plane. This can be seen from the angle of inclination i, being no more than a few degrees away from the ecliptic plane. Secondly, again with the exception of these two bodies, and evidenced by the eccentricity e, the planetary orbits are nearly circular. Whilst these two features provide evidence of the history of the solar system, from the consideration of space vehicle design they lead to significant design implications. Thus, the near planar nature of the solar system results in only relatively minor plane changes being required to send a probe to another solar system body. This has significant influence on the fuel load for the vehicle's propulsion system; details of this are provided in Chapter 5. The circularity of most planetary orbits means that following orbital insertion about a planet, the spacecraft will not be subject to any additional major changes in the thermal environment. Clearly the overall thermal

Table 2.5 Solar System data: mass characteristics and orbit

Object	Mass $(10^{24}$ kg)	$R_{equator}$(m) Equatorial radius	Density $(10^3$ kg/m$^3)$	Angular momentum $(10^{39}$ kg m^2/s)	Average speed (km/s)	a (AU) Semi-major axis of orbit[2]	e Orbital eccentricity	i (°) Orbital inclination[3]	Sidereal period (years)
Sun	1.99×10^6	0.696×10^9	1.409	170[1]	—	—	—	—	—
Mercury	0.33	2.44×10^6	5.46	0.906	47.9	0.3871	0.206	7.00	0.241
Venus	4.87	6.05×10^6	5.23	18.5	35.1	0.7233	0.007	3.39	0.615
Earth	5.97	6.38×10^6	5.52	26.7	29.8	1.0000	0.017	0.00	1.000
Mars	0.642	3.40×10^6	3.92	3.52	24.2	1.5237	0.093	1.85	1.881
Jupiter	1899	7.15×10^7	1.31	19 400	13.1	5.2028	0.048	1.31	11.862
Saturn	568	6.03×10^7	0.7	7840	9.64	9.5388	0.056	2.49	29.46
Uranus	87.2	2.56×10^7	1.3	1700	6.81	19.1914	0.046	0.77	84.01
Neptune	102	2.48×10^7	4.66	2500	5.44	30.0611	0.010	1.77	164.79
Pluto	0.66	1.20×10^6	4.9	17.9	4.75	39.5294	0.248	17.15	248.43

[1] Spin angular momentum of the Sun.
[2] 1 AU = 1.496×10^{11} m.
[3] Inclination of orbit plane relative to the ecliptic.
See also Tables 2.6, 2.7 and 4.1.

environment to a first approximation will scale as $(r_E/r_P)^2$, where r_E is the radial distance from the Earth to the Sun, and r_P is the distance of the planet from the Sun.

The majority of time, during most planetary space missions, is spent in interplanetary space. This environment is dominated by the solar wind, for which a description has already been provided. Both solar insolation and the density of the solar wind decrease with distance from the Sun, again with an approximate $1/r^2$ relationship. For missions beyond Mars, spacecraft need to traverse the asteroid belt, for which there is enhanced collision risk with sizeable objects.

The planetary environments themselves are very diverse. Table 2.6 provides data on those planets having an atmosphere, together with what is known about such atmospheres and any associated ionosphere. Only Mercury of the four inner planets does not have an atmosphere. The atmospheres of Venus, Earth and Mars are radically different, both in their composition and thermodynamic properties of pressure and temperature. It is particularly noteworthy that lander missions to Venus are required to operate both in the vacuum of space, and then on arrival at the surface of that planet survive pressures over 90 times that at the surface of the Earth, whilst the temperature is of order 750 K.

Titan is also included in Table 2.6, being the sole natural satellite of a planet that possesses a significant atmosphere. Our knowledge of Titan and its atmosphere has improved significantly following the successful landing by ESA's Huygens probe in January 2005, part of the joint ESA/NASA Cassini/Huygens mission to the Saturnian system. A complete review of the findings from this mission may be found in [24].

The four gas giants are listed here as having atmospheres. The surface properties noted are however for an assumed surface, on which the pressure of the atmosphere is the same as that on the Earth at sea level.

A necessary condition for a planet to have an ionosphere—a region of space that contains free electrons and ions—is that it also has an atmosphere. The major energy source for the ionization of the upper regions of an atmosphere is solar radiation through

Table 2.6 Solar System data: approximate atmospheric parameters for solar system bodies

Planet/ Moon	Composition %	Surface pressure (Bar)	Surface temperature (K)	Temperature @ 200 km (K)	Ionosphere (Electrons/ cm^3)
Mercury	None	—	—	—	
Venus	CO_2 (96); N_2 (3.5)	92	750	100–280	$\sim 10^6$
Earth	N_2 (77); O_2 (21); H_2 (1)	1	285	800–1100	$\sim 10^6$
Mars	CO_2 (95); Ar (1.6); N_2 (2.7)	0.006	220	310	$\sim 10^5$
Jupiter	H_2 (89); CH_4 (0.2); He (11)	Gaseous planet	165[1]		$\sim 10^5$
Saturn	H_2 (93); CH_4 (0.2); He (7)	Gaseous planet	130[1]		
Titan	N_2 (90–99); CH_2 (1–5); Ar (0–6)	1.5	95	150	$\sim 10^3$
Uranus	H_2 (85); $CH_4(< 1)$; He (15)	Gaseous planet	80[1]		
Neptune	H_2 (90); $CH_4(< 1)$; He (10)	Gaseous planet	70[1]		
Pluto	N_2 CH_4/CO (traces only)	—	40	—	—

[1]Temperature quoted where pressure is the same as Earth sea level (P = 1 Bar).
See also Tables 2.5, 2.7 and 4.1.

the process of photo-ionization. Typical peak electron densities in each of those bodies having an ionosphere is provided in Table 2.6. As might be expected, Venus and Earth, having the most significant atmospheres of the terrestrial planets, and also being relatively close to the Sun, have the highest ionospheric electron number density.

If a planet also possesses a magnetic field, then this will influence the properties of the ionosphere. As noted for the Earth, a magnetic field provides an opportunity to both trap and add energy to charged particles, which in general may originate from the planetary atmosphere/ionosphere or from the solar wind. The magnetic properties of the major solar system bodies are listed in Table 2.7. As we have already noted for the Earth, its magnetic field leads to the formation of the Van Allen radiation belts. The most significant magnetosphere in the solar system is however that of Jupiter, which acts as an energy source for the generation of relativistic particles that radiate by synchrotron emission. Whilst Jupiter's magnetosphere is far more complex than that of the Earth, its overall structure is similar to that shown in Figure 2.12. The magnetosphere for Jupiter is vast—the magneto-tail extends to the orbit of Saturn. This magnetic field co-rotates with Jupiter, which has a period of just less than 10 hours. This provides an additional source of energy to accelerate particles. Further, the satellite Io passes through the plasma sheet formed within the magnetosphere. Io contributes ions, principally from the dissociation products of SO_2, to this plasma sheet from intense volcanic activity on its surface. At this stage, it is not clear however whether Io has its own magnetic field. The overall interaction yields resultant particle fluxes that are several orders of magnitude greater than

Table 2.7 Solar System data

Planet/ Moon	Number of moons	Presence of rings	Magnetic field- Dipole moment (Relative to Earth)	Magnetopause height (units of $R_{equator}$)	Major missions
Mercury	—	—	0.0007	1.5	Mariner 10
Venus	—	—	<0.0004	—	Mariner 2/5/10 Venera 3–16 Pioneer Venus 1/2
Earth	1	—	1	10	—
Mars	2	—	<0.0002	—	Mariner 4/6/7/9 Mars 2/3 Viking 1/2 Mars Pathfinder
Jupiter	16	Yes	20 000	80	Pioneer 10/11 Voyager 1/2 Galileo
Saturn	18	Yes	600	20	Pioneer 11 Voyager 1/2 Cassini
Titan	N/A	—	None detected	—	Cassini/Huygens probe
Uranus	17	Yes	50	20	Voyager 2
Neptune	8	Yes	25	25	Voyager 2
Pluto	1	—	Unknown	Unknown	None

Notes: See also Tables 2.5, 2.6 and 4.1.

those at the peak flux of the Earth's own radiation belts. These particles in the Jovian system, coupled with the motion of Io's motion through the magnetosphere also provide an intense electromagnetic radiation source, extending from VLF $\sim$10 MHz through to 3 GHz. Not surprisingly therefore, all missions that transit the Jovian magnetosphere have shown signs of radiation damage in electronic circuits and the darkening of exposed optical systems. The radiation dose absorbed on the two grand-tour Pioneer spacecraft was a thousand times beyond the level of human tolerance [25]. As a result the Jovian system, having the additional feature common with the other gas giants of a ring system, proves to be one of the most hostile environments in the Solar System.

2.4 ENVIRONMENTAL EFFECTS ON DESIGN

2.4.1 Effects on materials

Outgassing or *sublimation* refers to the vaporization of surface atoms of a material when it is subjected to an ambient pressure that is comparable with its own vapour pressure. Such a pressure, 10^{-11}–10^{-15} Pa, occurs at spacecraft altitudes. This process occurs at an increasing rate as temperature rises. A summary of rates for different metals at a variety of temperatures is shown in Table 2.8 but they do not apply to alloy materials or those having a covering surface layer of a different material. Since the precise surface conditions of orbiting vehicles are difficult to determine, particularly for vehicles in LEO, exact figures for mass loss are not at present available. While structural problems arising because of outgassing are unlikely, the subsequent deposition of the material is hazardous to both optical and electrically sensitive surfaces. Thin plastic layers and oxide coatings are particularly sensitive to mass loss, especially if the material is used for its thermal properties; if mass loss is associated with specific constituents then modified emissive properties may result.

For plastic materials, high-vapour-pressure components evaporate rapidly, although initially mass loss is usually associated with the loss of adsorbed gases and water vapour.

Traditional lubricants used on the ground are clearly not appropriate to spacecraft operation. Generally, they have high vapour pressure and would thus outgas rapidly, but in many the lubricative action arises because of the presence of adsorbed gases and water vapour. While low-volatility oils are used, solid lubricant coatings such as MoS_2 are

Table 2.8 Temperature for a given sublimation rate ($^\circ$C)

Element	0.1 µm/yr	10 µm/yr	1 mm/yr
Cd	38	77	122
Zn	71	127	177
Mg	110	171	233
Au	660	800	950
Ti	920	1070	1250
Mo	1380	1630	1900
W	1870	2150	2480

frequently adopted. Reference [10] provides a detailed listing of ESA's materials database (but see also Chapter 15).

Atomic oxygen erosion

The atmospheric composition shown in Figure 2.10 indicates that atomic oxygen forms the major atmospheric species in LEO. Following the severe erosion noted for Kapton on STS3 [26], it became apparent that atomic oxygen provides an aggressive environment for materials used on space vehicles in LEO. This arises not only from its chemical activity, but most significantly from the fact that its atoms are travelling at $\sim$8 km/s relative to the vehicle (due to the vehicle's orbital motion). Silver is one of the few metals attacked by this environment, so due to its extensive use on solar arrays, it is important to avoid bare silver exposure.

The interactions between the oxygen atoms and the spacecraft surfaces are erosion, formation of stable oxide, scattering or reflection and chemiluminescent glow. When erosion takes place volatile products are formed, causing surface recession. Erosion rates are usually quoted in terms of yield in units of 10^{-24} cm^3/atom. Erosion yields range from 0.01 to 0.09 for materials like aluminium-coated kapton, FEP teflon and silicones at the low end, to between 2 and 4 for polyethylene, Kapton-H and Z-302 (glossy black) [27].

The net effect of this erosion interaction is to degrade the material properties (optical, thermal, mechanical and electrical) irreversibly, and so this must be taken into account when choosing materials for optical and thermal control coatings, structural material, thermal blankets, solar panels and optical components. The first major contribution to our knowledge of the space-based exposure to AO followed analysis of the NASA Long Duration Exposure Facility (LDEF), which spent 5.8 years in LEO and was retrieved in January 1990. Greatly improved understanding has been provided by the experiments on ISS, such as series MISSE [28]. As a result a detailed picture has been produced both of the erosion rates and methods for protection, and potential failure mechanisms in such protection [29].

Stable oxide formation can lead to significant property changes. For example, molybdenum disulfide, which is a lubricant, can oxidize to create an abrasive oxide [30]. Dimensional changes can also occur, leading to cracking in the formation of a surface oxide layer on silicone (due to contraction and expansion), and spalling in the case of oxidation of silver.

Indirect impingement caused by ram atoms scattered from adjacent surfaces can lead to erosion of surfaces 'shadowed' from direct attack. The chemiluminescent glow at visible wavelengths, observed on the Space Shuttle and around small spacecraft [31], is currently thought to be caused by surface mediated O + NO recombination, forming excited NO$_2$ molecules that then radiate [32].

Some work has been done on the use of protective coatings that are resistive to the attack of atomic oxygen [27], but the problems of either manufacture-induced pinholes or those created by debris/micrometeoroid impact must be overcome [33].

Simulation of the atomic oxygen environment has become particularly important. Further information on this critical feature of the environment may be found in References [34] and [35].

Material strength and fatigue life are also affected by a high-vacuum environment. Generally, fatigue life is improved. For many materials, it may be extended by more than an order of magnitude [36], although in some cases, for example, pure nickel and Inconel 550, the improvement is uncertain [36, 37]. The physical reasons for such changes in mechanical properties, and also changes in strength, are believed to arise because of one of two principal reasons. One of these is that gases absorbed into surface cracks either aid or hinder crack propagation. The second possible physical cause is that oxidation and gas diffusion absorbed into the material bulk influences material properties. A well-documented example of a material whose strength is improved in vacuum is that of glass, wherein a threefold improvement in strength has been noted at a pressure that is one thousandth that of atmospheric pressure.

Embrittlement is a form of material damage that is caused by exposure to UV radiation. Many polymers are particularly sensitive to such photons, whose energy is high enough to modify their chemical bonding structure.

Ultraviolet exposure also causes *electrical changes* in the form of resistivity modification, and *optical changes* affecting both thermal characteristics and opacity. A particularly UV-sensitive element is the solar array. More specifically the solar cell coverglass and its attendant adhesive are subject to darkening. This results in reduced cell illumination and an enhanced operating temperature, both being deleterious to cell operation (see Chapter 10).

Radiation damage affects all materials to some extent, and humans. High-energy corpuscular radiation is experienced most severely in the Van Allen radiation belts, but is also at a significant level in any space operation. Under even modest radiation doses, some metals such as cadmium and zinc may form metallic whiskers.

Semiconductor materials and biological tissue are particularly sensitive to damage caused by high-energy charged and neutral particles. This generally arises because of both the displacement of atoms from crystal lattice sites and the attendant local ionization, together with an ionization track caused by the penetrating radiation. Protection is a complex task, since screening material will itself generate secondary radiations due to the passage of a primary high-energy particle through it. It should be noted that 'heavy' particles such as protons and neutrons cause significantly more damage than light ones such as electrons. They cause a dense track of ionization through a material until they are stopped by an atomic collision. When this occurs, a significant displacement of the atom arises, leading to further ionization, which does not occur for electrons.

Radiation shielding analysis

To ensure that electronic components such as transistors, diodes and so on are capable of surviving the radiation environment, the total dose inside the spacecraft, in rads has to be calculated. This is done for the specific orbit in question by first calculating the fluence spectrum of the external (unshielded) particles—the number of particles/cm^2 at specific energies. The shielded environment is then determined by transporting the external fluence through the spacecraft mass. Usually, the first step is to calculate a one-dimensional dose—depth curve based on a spherical shell geometry of aluminium. The dose behind a thickness typical of the spacecraft (usually about 2.5–3.8 mm) is read from this curve. This dose is then used with some design margin, typically between 1.3 and 2, to set the rad hardness requirement for electronic parts. If parts to be used fall below this requirement, several options are available; a more detailed three-dimensional analysis can be carried out to determine the dose at the actual location of the 'soft' parts. This usually results in the lowering of the dose with the inclusion of more individual spacecraft components further reducing the dose values. If the dose is still too high, then spot shielding can be implemented (i.e. the placement of a shield of tantalum or tungsten at the location of the actual part), or a rad-hard version of the part can be sought.

Other radiation effects including single-event phenomena, noise in charge-coupled devices CCDs (and optical detectors), and radiobiological interactions also need to include the transport of the primary particles through the spacecraft structure. While, in general, shielding is beneficial and reduces the hazard caused by the radiation, in some cases large thicknesses of shielding can worsen the effects. For example, in considering single-event effects (SEEs), a heavy ion passing through a certain thickness of material will be slowed down to such an extent that its linear energy transfer (LET), and therefore its ability to produce ionization, is increased. Consideration must also be given to secondary particles produced as a result of the interaction of the primaries with the atoms of the shielding material. For example, secondary neutrons produced by proton fluxes can pose a problem for human space missions in orbits where the proton flux is high (trapped proton fluxes at altitudes greater than 500 km, and solar protons in high inclination orbits). Protons can also be responsible for SEEs in the South Atlantic Anomaly and in polar regions during solar flares, by producing secondary recoiling nuclei and light fragments [38].

The production of radioactive materials, which is called activation, can also be a problem. This may lead to dose problems on long, human space missions such as the International Space Station, and can also induce limiting background noise levels in low-level measurements made by radiation detectors.

In semiconductor materials, two damaging effects have been noted. Primarily, radiation damage reduces the effectiveness of semiconductor operation. Specifically in solar cells, it results in a reduction in the efficiency of conversion from sunlight to electrical energy. In order to quantify the effect of this, the spacecraft designer usually resorts to tables of fluence for particular orbits. The upper, exposed surface of the cell may be protected to some extent by the use of a coverglass. The lower surface is generally protected by the substrate on which it is mounted. Manufacturers' calibration data is generally available to enable the expected degradation in solar cell performance to be assessed during the mission lifetime. This topic will be returned to in greater detail in Chapter 10.

Single event effects (SEEs)

In addition to total dose effects, which are caused by the deposition of energy by many particles, there is a set of phenomena that are caused by single particles. A single-event upset (SEU) occurs when a heavy ion is incident on the sensitive area of an integrated circuit, producing sufficient charge in the form of electron—hole pairs to cause a change in the logic state of the device. This type of error is known as a *'soft' error* since it is reversible and causes no permanent damage. However, if it occurs in critical circuitry such as a control system or decision-making logic, then it can have serious consequences on the spacecraft operation—generating false commands such as thruster firings.

A single-event latch-up (SEL) occurs when the passage of a single charged particle leads to a latched low impedance state in parasitic PNPN devices in bulk complementary metal oxide semiconductor (CMOS) material, and it can result in burn-out. A more serious effect is single-event burn-out, which occurs when an incident ion produces a conducting path—in a metal oxide semiconductor field effect transistor (MOSFET), for example, that causes the device to latch-up; if this condition continues for a sufficiently long time then the device could be completely burnt out and destroyed.

Finally, single-event-induced dark current is caused by the passage of a particle which causes displacement damage in a single pixel.

Particles that cause SEUs include heavy ions, protons and neutrons. Both galactic cosmic rays and solar flares contain these. High-energy trapped protons can also cause SEUs, not by direct ionization but by the recoiling heavy reaction products. The parameter that characterizes a particle's ability to cause a SEE is its energy deposition rate, known as dE/dx, stopping power, or LET, and is measured in units of $MeV/g\,cm^2$ or $MeV/\mu m$.

SEU sensitivity is largely determined by the critical charge for a sensitive node, which is the quantity of charge needed to alter the device's memory status. The critical charge is a function of the feature size and roughly scales with the square of its dimension. SEU rates can be determined by combining the LET spectrum of the environment (i.e. the number of particles with a given energy loss) and the device dimensions.

In general, systems approaches to SEU-hardening can be divided into three categories: error toleration, error correction and error prevention. The first is costly and involves specifying tolerance levels for various parts of the system, determining maximum permissible error rates, and designing and constructing each part of the system within these constraints. Tests must be conducted at the subsystem and at the total system levels to ensure that the tolerance levels are met and that catastrophic failure is impossible. Error-correction techniques applied at the systems level include: redundant units, self-checking circuits, error-detecting and error-correcting codes, and serial calculation with error correction (hardware solution), concurrent programme execution and breakpoint reasonableness testing, checkpoint storage and roll-back for recovery (software solution), and repetitive execution and watchdog timers (time-related solutions) [39]. The last approach involves the choice of components that will not upset; however, there are only a limited number of SEU-hard devices for very large scale integrated (VLSI) levels of complexity and speed.

The potential for software errors is becoming more apparent as the processing power of on-board computers is increasing. This is a challenging area of current spacecraft design for which solutions are being sought.

2.4.2 Effects on humans

The removal of human beings from their natural environment or habitat necessarily introduces deleterious effects. A suitable life-support system will guard against the more obvious ones such as lack of atmosphere, thermal environment and possible high-acceleration loads. Details applicable to the design of a suitable life-support environment may be found in References [40] and [41]. Here we consider the environmental problems for which a life-support system offers no solution. It should be noted that the response of humans to extended exposure to the space environment still requires further improvement and understanding: even following a significant increase in the number of human-years in space, this exposure in reality is still modest. A bioastronautics road map has been produced with the aim of mitigating human risks [42].

In general terms, the response of human beings to spaceflight may be classified under one of four categories: transient effects lasting for short periods on initial exposure to, or removal from, the space environment; flight duration adaptation; cumulative effects dependent upon the length of the flight, which are reversible on return to the ground; cumulative effects that are irreversible.

The dominant effect of the space environment, which cannot be removed through life-support systems, is zero or microgravity. This causes a major disturbance to the human system, with effects that fall into each of the first three categories above. The most notable of these are the following:

1. *Blood volume redistribution.* On Earth the blood pressure of a person whilst standing decreases with height above the feet. Typically the pressure in the brain is only one third that in the feet. The immediate effect of a 'zero g' environment is to cause a redistribution of the blood volume, resulting in a 'puffy' face. Whilst on Earth, the body system is familiar with orientation changes (lying down, etc.), and a complex system of hormonal secretion results in a control system that adjusts heart rate to these conditions. In zero gravity, however, this hormonal action appears to result in a loss of sodium with consequent fluid loss. Typically stabilization occurs after $\sim$4 days when a loss of some 2–4 kg has occurred. Readaptation on return to Earth occurs over a short period of time, during which giddiness may be expected.

2. *Muscular atrophy.* Atrophy of all muscles occurs through long periods of inactivity. In zero g the heart itself is required to do less work since it is not pumping blood against gravity. Reduction in both heart muscle mass and heart rate occurs. It is believed that anaemia (a reduction in red blood cell count) is an additional side effect that has been noted in astronauts. To combat general atrophy of muscles astronauts spend much of their time in space undergoing physical exercise.

3. *Vestibular problems.* The human vestibular system is dependent upon both visual and inner ear sensors. In the inner ear the sensor has two orthogonal components, horizontal and vertical, which rely on gravity for their operation. Removal of gravity results in enhanced sensitivity to acceleration (including rotation) because of fluid motion in these sensors. Conflict between visual and inner ear sensors therefore arises during motion of the astronaut. The most notable effect is motion sickness, but astronauts also experience enhanced clumsiness. Adaptation generally takes three to four days.

4. *Locomotor system.* The major effect of zero gravity on the locomotor system is cumulative bone decalcification. This directly results in bone fragility and indirectly leads to problems of recalcification, external to the bones in, for example, the kidneys, forming kidney stones. High calcium diets do not appear to stop this increased calcium mobility and on long duration flights such as Salyut 6, a 2–8% loss of calcium was noted. Restabilization occurs on return to Earth.

The corpuscular radiation environment noted in preceding sections evidently provides a hazardous environment for humans. In the absence of specific solar particle events associated with eruptive phenomenon on the Sun, the most damaging health risk to astronauts is the galactic cosmic radiation [43]. Shielding materials must be chosen with care, as poor design can lead to a secondary radiation that is more hazardous than the primary radiation. It is notable that aluminium, a frequently used structural material is one of the worst materials to use in relation to shielding and thus the materials used tend to be composite materials, albeit encased within thin aluminium face sheets. The composites used are those containing elements smaller than carbon, since these reduce the production of harmful secondaries and fragmentation of the incident particle [44]. As is clear from

Figures 2.14 through 2.18, the averaged radiation dose is dependent upon the orbit altitude and inclination when in Earth orbit. The protective structure of and habitation regions of ISS result in a dose rate to astronauts of about 0.3 Sv per year [45, 46], approximately 10 times the dose rate a passenger would receive during a commercial airline flight.

In addition to the physiological effects, the opportunity for long duration missions to perhaps the Moon or even Mars has identified that research into psychological effects of remoteness must also be addressed. Studies directed towards this goal have commenced on ISS flights [42], and this is a clear area for further research.

ACKNOWLEDGEMENT

Thanks are due to Professor Stephen Gabriel for compiling material for this chapter.

REFERENCES

[1] Rauschenback, H. S. (1980) *Solar Cell Array Design Handbook*, Van Nostrand, New York.

[2] Ariane 4 Users Manual Issue 1 (1983) Arianespace.

[3] Ariane 5 Users Manual Issue 5 (2008) Arianespace.

[4] Pegasus User's Guide, Release 7.0, Orbital Sciences Corporation, April 2010.

[5] NASA TM 82478 (1982) Volume 1.

[6] US Standard Atmosphere (1976) NOAA, Washington, DC.

[7] Hedin, A. E. (1991) Extension of the MSIS thermospheric model into the middle and lower atmosphere, *J. Geophys. Res.*, **96**, 1159.

[8] Jacchia, L. G. (1977) SAO Special Report 375.

[9] King-Hele, D. G. (1987) *Satellite Orbits in an Atmosphere*, Blackie, London.

[10] ESA/ESTEC. http://esmat.esa.int/ (last accessed 17/09/ 2010).

[11] Garrett, H. B. (1979) *Rev. Geophys. Space Phys.*, **17**, 397.

[12] Daly, E. J., Adams, L., Zehnder, A. and Ljungfelt, S. L. (1992) IAG-92-0799, *43rd Congress of the International Astronautical Federation*, 28 Aug to 5 Sept, Washington, DC.

[13] Vette, J. I. (1991) The AE-8 Trapped Electron Model Environment, NSSDC-91-24, NASA-GSFC.

[14] Sawyer, D. M. and Vette, J. I. (1976) AP-8 Trapped Proton Environment for Solar Maximum and Solar Minimum, NSSDC-76-06, NASA-GSFC.

[15] Mewaldt, R. A., Spalding, J. D. and Stone, E. C. (1984) The isotopic composition of the anomalous low energy cosmic rays, *Astrophys. J.*, **283**, 450.

[16] Feynman, J. L., Armstrong, T. P., Dao-Gibner, L. and Silverman, S. M. (1990) Solar proton events during solar cycles 19, 20 and 21, *Solar Phys.*, **126**, 385.

[17] Feynman, J. and Gabriel, S. (eds) (1988) *Interplanetary Particle Environment: Proceedings of Conference*, JPL Publication 88-28, Jet Propulsion Laboratory, Pasadena, USA.

[18] Garrett, H. B. and Gaunt, D. M. (1980) *Prog. Astro. Aero.*, **71**, 227.

[19] Pilkington Space Technology (1984) Coverglass Specification PS 292.

[20] Gault, D. E. (1970) *Radio Sci.*, **5**, 273.

[21] Johnson NL (2010), International and NASA SSA and Safety of Flight Issues, JFCC Space/J3, Space Operations Conference January 2010, NTRS Document ID: 20100003476; Report Number: JSC-CN-19578.

[22] Liou, J. C. and Johnson, N. L. (2009) A sensitivity study of the effectiveness of active debris removal in LEO, *Acta Astronautica*, **64**, 236–243.

[23] Lambert, M. (1993) Shielding against natural and man-made space debris: a growing challenge, *ESA J.*, **17**, 31–42.

[24] Lebreton, J., Coustenis, A., Lunine, J., Raulin, F., Owen, T. and Strobel, D. (2009) Results from the Huygens probe on Titan, *Astron. Astrophys. Rev*. **17**, 149–179.

[25] Van Allen, J. A. and Bagenal, F. (1999) Planetary magnetospheres and the interplanetary medium, in *The New Solar System* (4th edn), ed. Beatty, J. K., Peterson, C. C. and Chaikin, A. L., Cambridge University Press, New York.

[26] Leger, L. J. (1983) AIAA Paper No. AIAA-83-0073.

[27] Tennyson, R. C. (1993) Atomic oxygen and its effect on materials, in *The Behaviour of Systems in the Space Environment*, ed. DeWitt, R. N., Duston, D. and Hyder, S. K., Kluwer Academic Publishers, Dordrecht, p. 233.

[28] NASA/TP–2009–213146 (2009).

[29] NASA/TM-2006-214482 (2006).

[30] Cross, J. B., Martin, J. A., Pope, L. E. and Koontz, S. L. (1989) Oxidation of MoS_2 by thermal to hyperthermal atomic oxygen, *3rd Int. SAMPE Electron. Conf*., **3**, 638.

[31] Garrett, H. B., Chutjian, A. and Gabriel, S. (1988) Space vehicle glow and its impact on space systems, *J. Spacecraft Rockets*, **25**, 321–340.

[32] Greer, W. A. D., Pratt, N. H. and Stark, J. P. W. (1993) Spacecraft glows and laboratory luminescence evidence for a common reaction mechanism, *Geophys. Res. Lett*., **20**, 731–734.

[33] NASA/TM-2008-215482 (2008).

[34] ESA-SP-616 Proceedings of the 10th ISMSE, 8th ICPMSE (2006).

[35] Proceedings of 11th ISMSE Conference (2009) http://esmat.esa.int//Materials_News/ISME09/cd_isme09.pdf (last accessed 17/09/2010).

[36] NASA TN-D 2563 (1965).

[37] NASA TN-D 2898 (1965).

[38] Harboe-Sorensen, R., Daly, E. J., Underwood, C., Ward, J. and Adams, L. (1990) The behaviour of measured SEU at low altitude during periods of high solar activity, *IEEE Trans. Nucl. Sci*., NS-37, 1938–1946.

[39] Nagle, H. T., Santago, P., Miller, T. K., McAllister, D. F. and Mehrotra, R. (1989) Fault tolerance for single events, *J. Rad. Effects Res. Eng*., **6**, 2.

[40] NASA SP-3006 (1973).

[41] Sharpe, M. R. (1969) *Living in Space*, Aldus Books 1, London.

[42] NASA/SP–2004–6113 (2005).

[43] Johnson, N. K. (2010) International and NASA SSA and Safety of Flight Issues, *JSFC Space Operations Conference*, 26–28 January 2010, available NTRS 20100003476.

[44] Sen, S., Schofield, E., O'Dell, J. S., Deka, L. and Pillay, S. (2009) The development of a multifunctional composite material for use in human space exploration beyond low-earth orbit, *Journal of the Minerals*, Metals and Materials Society, **61**, 23–31.

[45] Stanford, M. and Jones, J. A. (1999) Space radiation concepts for manned exploration, *Acta Astonautica* **45**, 39.

[46] National Council on Radiation Protection and Measurements (2000). *Radiation protection guidance for activities in low-earth orbit*. Report no. 132. ISBN 0-929600-65-7. The Council:Bethesda, MD.

3 DYNAMICS OF SPACECRAFT

Peter W. Fortescue and Graham G. Swinerd

Aeronautics and Astronautics, Faculty of Engineering and the Environment, University of Southampton

3.1 INTRODUCTION

This chapter serves as a general introduction to the subject of the dynamics of bodies and sets a framework for the subjects of celestial mechanics and attitude control (Chapters 4 and 9). For both of these, Newtonian dynamics will provide a sufficient means of forecasting and of understanding a spacecraft's behaviour. The summary presented here is chosen with a view to its relevance to spacecraft.

The approach adopted is to develop an understanding of dynamics in two stages. The first is to express the dynamics of both translation and rotation in terms of the appropriate form of momentum—linear or angular. Momentum becomes an important concept, in terms of which it is relatively easy to determine the consequences of forces or moments.

The second stage is to interpret the momenta in terms of the physical movement—the velocities, linear and angular. This is straightforward for linear momentum since momentum and velocity are in the same direction. The relationship between rotational movement and angular momentum is less intuitive.

3.1.1 Translation/rotation separation

One feature that is peculiar to spacecraft is that their translational (trajectory) motion is virtually independent of their rotational motion. This is due to the fact that the moments or torques that cause their rotation are not dependent upon their direction of travel, and the gravitational forces that determine their trajectory are not dependent on their attitude. Whilst this is not entirely true, it is approximately so, and spacecraft designers will normally aim to preserve this independence.

At a fundamental level it is convenient to think of a spacecraft as being a collection of particles and bodies that lie within a closed surface S (Figure 3.1). This concept allows the surface to be chosen at one's convenience, to embrace the complete craft or just part

Spacecraft Systems Engineering, Fourth Edition.
Edited by Peter W. Fortescue, Graham G. Swinerd and John P. W. Stark.
© 2011 John Wiley & Sons, Ltd. Published 2011 by John Wiley & Sons, Ltd.

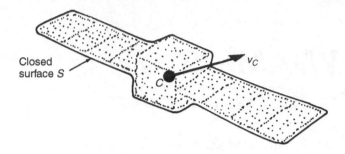

Figure 3.1 The closed surface S

of it or, when considering a docking manoeuvre or a tethered configuration, it may be chosen to embrace more than just the one spacecraft. But for the following, assume that it envelops just one spacecraft, unless stated otherwise.

The dynamics of the craft (or whatever is within the surface S) may now be described in terms of its momenta—its linear momentum $\mathbf{L}$ and its angular momentum $\mathbf{H}$—the former leading to equations that describe the trajectory and the latter leading to equations that describe the attitude motion.

In particular, it is the centre-of-mass C whose trajectory will be described. It may seem to be pedantic to pick on one specific point for this purpose, but the centre-of-mass has special properties that lead to it being chosen for the development of Newtonian dynamics, properties that make it sensible to separate the motion into

1. the motion of the centre-of-mass, C and
2. the motion relative to the centre-of-mass.

For highly specialized purposes, such as when using the orbit to improve man's knowledge of the geoid, the accuracy with which the orbit is determined may be less than the dimensions of the spacecraft, and then it is no longer pedantic to consider to which point of the vehicle its orbit refers.

3.1.2 The centre-of-mass, *C*

The centre-of-mass of the particles in S, relative to an arbitrary point O, is the point whose position vector $\mathbf{r}_{OC}$ obeys

$$M\mathbf{r}_{OC} = \sum (m\mathbf{r}_{OP}) \tag{3.1}$$

where
  $\mathbf{r}_{OP}$ is the position vector of a general particle P (see Figure 3.2),
  m is the mass of the general particle
and M is the total mass within S.

When the location of C is being determined for a spacecraft, then it is useful to note that equation (3.1) can be applied to objects such as items of equipment rather than to particles. In that case, $\mathbf{r}_{OP}$ refers to the centre-of-mass of the item and m is its mass.

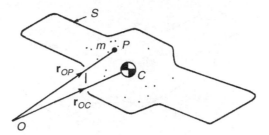

Figure 3.2 The centre-of-mass, C

When an object has a continuous mass distribution, the integral equivalent of equation (3.1) should be used, this being

$$M \mathbf{r}_{OC} = \int \mathbf{r}_{OP} \, dm \tag{3.2}$$

In the development of the theory of dynamics *the centre-of-mass C is useful as an origin or reference point* for the particles in S. Putting the origin there leads to $\mathbf{r}_{OC} = \mathbf{0}$, and so equations (3.1) and (3.2) become

$$\sum (m \mathbf{r}_{CP}) = \mathbf{0} \quad \text{and} \quad \int \mathbf{r}_{CP} \, dm = \mathbf{0} \tag{3.3}$$

Since this is true at all times, its derivatives are also zero, leading to

$$\sum (m \mathbf{v}_{CP}) = \mathbf{0} \quad \text{and} \quad \int \mathbf{v}_{CP} \, dm = \mathbf{0} \tag{3.4}$$

where $\mathbf{v}_{CP}$ is the velocity of a particle at P relative to the centre-of-mass C.

3.2 TRAJECTORY DYNAMICS

For trajectory purposes, it is convenient to treat a spacecraft as a particle, as is done in Chapters 4 and 5.

This *equivalent particle* (*e.p.*) has mass M equal to that of the spacecraft and is situated at its centre-of-mass C. Therefore, the e.p. moves with the velocity of C and its *momentum* $\mathbf{L}$ is the aggregate of all the particles and bodies that comprise the spacecraft (Figure 3.3),

$$\text{Momentum} \quad \mathbf{L} = M \mathbf{v}_C \tag{3.5}$$

This interpretation admits the possibility that the spacecraft may have moving parts or appendages. The surface S (Figure 3.1) may in fact be chosen to embrace any objects whose joint orbit(s) are of interest, although it will here refer to a single spacecraft or celestial body unless otherwise stated.

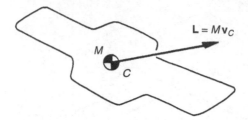

Figure 3.3 The equivalent particle at C

According to Newtonian mechanics—of a single particle—the only way of changing its momentum $\mathbf{L}$ is to apply a force. For a spacecraft—whose e.p. represents many particles—there are two ways:

1. *Application of external forces*. These include gravitational attraction towards the heavenly bodies, solar radiation pressure, aerodynamic forces and so on.

Internal forces, acting between any two particles or bodies *within S*, will not produce any change in the total momentum. It follows that objects moving within a spacecraft, such as fuel, astronauts, mechanisms or flexing structures and so on, will not cause the total linear momentum to change. Likewise, the total momentum of two docking spacecraft will not be changed by any forces between them.

This mechanism applies to a spacecraft that has no propulsion; its characteristic is that the mass M remains constant.

2. *Ejection of some of the particles from within S*. This occurs during rocket propulsion. The ejected particles take away both their mass and their momentum, leading to a change in the residual mass and momentum of the spacecraft.

In the above text it is important to recognize that the momentum must in all cases involve the velocity relative to inertial space rather than to the spacecraft.

It is convenient to deal separately with the two cases above, that is, with and without propulsion.

3.2.1 Translational motion with no propulsion

The effect of external forces $\mathbf{F}_{ext}$ on the momentum can be forecast from a diagram such as Figure 3.4. Their effect will depend upon not only the magnitude and direction of the forces but also upon the magnitude of the momentum $\mathbf{L}$ when the forces act. The additional momentum $\delta\mathbf{L}$ produced during an interval of time δt is the additional momentum

$$\delta\mathbf{L} = \mathbf{F}_{ext}\delta t \qquad (3.6)$$

This must be added vectorially as shown in Figure 3.4 in order to see the effect upon the momentum $\mathbf{L}$ during the time interval.

Equation (3.6) may be written as the Newtonian equation:

$$d\mathbf{L}/dt = d(M\mathbf{v}_C)/dt = \mathbf{F}_{ext} \qquad (3.7)$$

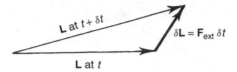

Figure 3.4 The incremental increase in momentum caused by an external force

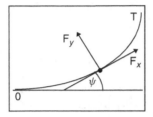

Figure 3.5 The force components along and normal to the trajectory

To establish the effect that an external force has upon the motion, it is best to resolve the force into two components—F_x along the trajectory and F_y at right angles to it—as shown in Figure 3.5. Then

1. F_x along the trajectory will change only the *magnitude* of the momentum, and hence only the speed—from v_0 to v_T say—where

$$M(v_T - v_0) = \int_0^T F_x \, dt \qquad (3.8)$$

2. F_y normal to the trajectory will change only the *direction* of the momentum. The rate at which the direction changes is $d\psi/dt = F_y/Mv$, and so a change of direction is best effected when the speed is least, such as at an apoapsis. If the speed remains constant ($F_x = 0$), then during a time interval from $t = 0$ to T, the trajectory will turn as in Figure 3.5, through an angle $(\psi_T - \psi_0)$, where

$$Mv(\psi_T - \psi_0) = \int_0^T F_y \, dt \qquad (3.9)$$

If F_y is constant, then the trajectory is an arc of a circle.

In Section 3.3.2, it will be seen that angular momentum **H** responds to a torque **T** in exactly the same way as linear momentum **L** responds to a force **F**; this also applies to the moment-of-momentum $M\mathbf{h}_I$ in response to a force whose moment is $\mathbf{M}_I$ (Section 3.2.3).

3.2.2 Moment-of-momentum *m*h and angular momentum H

Moment-of-momentum is a useful concept in celestial mechanics (Section 4.2 of Chapter 4). It is also a useful stepping stone towards angular momentum and the study of attitude motion.

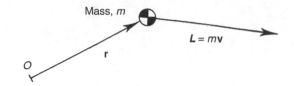

Figure 3.6   The situation leading to equation (3.10)

The *moment-of-momentum* $m\mathbf{h}_O$ referred to a point O may be defined as

$$m\mathbf{h}_O = \mathbf{r} \times m\mathbf{v} \tag{3.10}$$

$\mathbf{r}$ is the position vector from O to any point on the line of action of the momentum vector $m\mathbf{v}$ (Figure 3.6) and $\mathbf{v}$ is the velocity relative to the reference point O. This definition is equivalent to the product of the momentum with the perpendicular distance from O to the momentum vector.

For the orbit of (the e.p. of) a spacecraft whose mass is M, it is useful to take the centre of gravitational attraction—the centre of the Earth for Earth orbits, for example—as the reference point, O say. Then the moment-of-momentum vector $M\mathbf{h}_O$ will be normal to the orbit plane (Figure 3.7).

In the study of attitude dynamics, the aggregate of all the moments of the momenta of the particles comprising the spacecraft is a useful concept and it is defined as the *angular momentum*, $\mathbf{H}$. In this case the most useful reference point is the centre-of-mass, C. This subject is covered in Section 3.3.1.

Note that 'moment-of-momentum' $M\mathbf{h}_O$ is often loosely referred to as 'angular momentum' $\mathbf{H}_O$. This can lead to some confusion. Only in special cases are the two identical, such as when there is only one particle. In the more general case of a spacecraft, its angular momentum referred to O is the sum of the moment-of-momentum of its e.p. referred to O and its angular momentum referred to C:

$$\mathbf{H}_O = M\mathbf{h}_O + \mathbf{H}_C \tag{3.11}$$

In practice, $\mathbf{H}_C$ is very much less than $M\mathbf{h}_O$, and it is neglected in orbit studies.

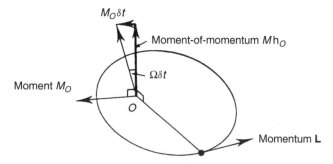

Figure 3.7   Precession of an orbit's plane due to a moment $\mathbf{M}_O$ lying in the plane. Its normal turns towards the direction of the moment vector

3.2.3 The rate of change of moment-of-momentum Mh_O

The moment-of-momentum of a particle $M\mathbf{h}_O$ referred to an inertially fixed point O is only changed if the forces on it have a moment $\mathbf{M}_O$ about O. The relationship may be expressed in the form of the Newtonian equation

$$\mathrm{d}(M\mathbf{h}_O)/\mathrm{d}t = \mathbf{M}_O \qquad (3.12)$$

This is also true for the e.p. of a spacecraft.

The forces on a spacecraft are conventionally separated into

1. the dominant central gravitational force and
2. the additional perturbative forces.

The former acts towards a point O that may be taken to be inertially fixed, such as the centre of the Earth for Earth orbiters, and so it contributes nothing towards the moment $\mathbf{M}_O$.

The similarity between equations (3.12) and (3.7) indicates that the moment-of-momentum responds to the moment $\mathbf{M}_O$ in the same way as linear momentum $\mathbf{L}$ responds to a force $\mathbf{F}_{\mathrm{ext}}$. When this is applied to an orbiting spacecraft, the consequences are

1. The component of $\mathbf{M}_O$ that is *normal* to the orbit's plane—in the same direction as $M\mathbf{h}_O$—will change the magnitude of the moment-of-momentum but not its direction, and so the orbit will remain in the same plane, but will change its shape. (See Section 9.4.7 of Chapter 4: Apsidal precession.)
2. The component of $\mathbf{M}_O$ *in* the orbit plane—at right angles to the moment-of-momentum $M\mathbf{h}_O$—will affect the direction of the normal to the plane, turning it towards $\mathbf{M}_O$ as shown in Figure 3.7. This represents a rotation of the orbital plane. The rate of turn Ω is given by

$$\Omega = M_O/Mh_O \qquad (3.13)$$

(See Section 4.4 in Chapter 4: Regression of the line of nodes, for example.)

3.2.4 The impulses

If a force $\mathbf{F}$ acts for only a limited duration—from $t = 0$ to τ, say—then its impulse $\mathbf{I}$ is defined as

$$\mathbf{I} = \int_0^\tau \mathbf{F}\,\mathrm{d}t \qquad (3.14)$$

From equation (3.6) it may be seen that $\mathbf{I} = \mathbf{L}_\tau - \mathbf{L}_0$, so that *an impulse is equal to the change in momentum that it causes*.

The impulse is used in situations in which there is an insignificant amount of movement during the time that the force acts. A collision is a common application. The firing of a rocket when in Earth orbit may also occur over a very small orbital arc, and if so then it may be treated as an impulse. A swing-by manoeuvre in heliocentric space is another application since the passage through the sphere of influence of a planet involves very little movement in the heliocentric orbit (see Chapter 5).

A torque impulse I_T is defined in terms of torque T in a manner similar to equation (3.14).

$$I_T = \int_0^\tau T \, dt \tag{3.15}$$

Since the relationship between torque T and angular momentum H (equation 3.28) is identical to that between force F and linear momentum L (equation 3.7), it follows that $I_T = H_\tau - H_0$, so that *a torque impulse is equal to the change in angular momentum that it causes*.

The restriction that there is little movement—angular in this case—during the torque impulse means that collisions and landing impacts are the main situations in which the torque may be deemed to be sudden.

In the case of a body in motion—linear and/or rotational—the impulses are a measure of what would cause the motion from an initially stationary state. It follows that the negative of the impulses will bring the body to rest.

3.2.5 Translational motion under propulsion

When a rocket motor is fired, there is an expulsion of particles out of the spacecraft—out of the surface S referred to above. Each particle of the exhaust gases takes away its contributions to two of the properties of the e.p., its mass M and its momentum $M v_C$ —the total momentum within S.

If the rocket's mass flow is σ, then this is the rate at which the mass of the e.p. decreases, that is,

$$dM / dt = -\sigma \tag{3.16}$$

Also, the absolute velocity of the exhaust is $(v_C + v_{ex})$, and so the rate at which the momentum of the e.p. changes is

$$d(M v_C)/dt = -\sigma (v_C + v_{ex}) \tag{3.17}$$

Here v_{ex} is the exhaust velocity relative to the centre-of-mass, taken to be positive in the direction of travel.

In the general case there will be an external force F_{ext} acting in addition, part of this being due to the back-pressure if the rocket operates in an atmosphere. Then the total rate of increase of momentum obeys the Newtonian equation

$$d(M v_C)/dt = F_{ext} - \sigma (v_C + v_{ex}) \tag{3.18}$$

From equations (3.16) and (3.18) it follows that the absolute acceleration a_C of the e.p. obeys

$$M a_C = F_{ext} - \sigma v_{ex} \tag{3.19}$$

Clearly when the rocket is restrained, as on a test bed, there will be no acceleration, and the restraining force F_{ext} will be equal to the thrust, σv_{ex}.

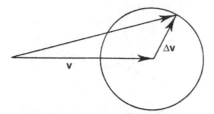

Figure 3.8   Possible changes of velocity due to a rocket's $\Delta\mathbf{v}$

The rocket equation is the integral of equation (3.19) for the case of no external forces, and the thruster points in a constant direction. The velocity increment $\Delta\mathbf{v}$ due to the burn is then

$$\Delta\mathbf{v} = -\mathbf{v}_{ex}\ln(M_0/M_1) \tag{3.20}$$

M_0/M_1 is the *mass ratio*, the ratio between the masses M_0 before and M_1 after the burn.

Equation (3.20) is frequently used as an approximation when the burn is short enough that it may be considered to be impulsive. Then, by pointing the rocket in different directions, it may be used to achieve an increase or decrease in speed, or a change of direction or any combination that is consistent with the vector diagram shown in Figure 3.8.

A *course change* $\Delta\psi$, for example, without any change of speed, will occur if the initial and final velocity vectors in Figure 3.8 form an isosceles triangle with $\Delta\mathbf{v}$ as base. The angle turned through will be

$$\Delta\psi = 2\arcsin\left(\tfrac{1}{2}|\Delta\mathbf{v}|/v\right) \tag{3.21}$$

and the thrust needs to be directed at an angle $(90 + \psi/2)$ degrees to the initial direction of travel. An assumption above is that the thrust vector passes through the centre-of-mass. If this is not so, then its moment about this point will produce a possible course change (see Section 3.3.2).

3.2.6   Translational kinetic energy

An energy equation may be obtained as the first integral of equation (3.7), and this relationship can be very useful in the subject of celestial mechanics.

The translational kinetic energy may be expressed as

$$\mathrm{KE} = \tfrac{1}{2}Mv_C^2 \quad\text{or } \tfrac{1}{2}M\,\mathbf{v}_C\cdot\mathbf{v}_C \quad\text{or } \tfrac{1}{2}\mathbf{L}\cdot\mathbf{v}_C \tag{3.22}$$

The change in this is equal to the work done by the external forces $\mathbf{F}_{ext}$, and it may be expressed as

$$\Delta\mathrm{KE} = \Delta\left(\tfrac{1}{2}M\,\mathbf{v}_C\cdot\mathbf{v}_C\right) = \int \mathbf{F}_{ext}\cdot d\mathbf{s} \tag{3.23}$$

where $d\mathbf{s}$ is the incremental change in position (see Figure 3.9).

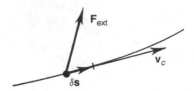

Figure 3.9   Vectors used in definition of translational kinetic energy

The conservative forces may be separated from the non-conservative ones since the former's contribution to the integral may be expressed in terms of potential energy (PE). The resulting energy equation becomes

$$\Delta KE + \Delta PE = \int \mathbf{F}_{nc} \cdot d\mathbf{s} \qquad (3.24)$$

where $\mathbf{F}_{nc}$ is the non-conservative contribution to the total external force $\mathbf{F}_{ext}$.

When the only force on the spacecraft is that due to gravitational attraction, a conservative force whose potential energy is $-\mu M/r$, then the orbital energy equation becomes

$$\tfrac{1}{2}M\mathbf{v}_C \cdot \mathbf{v}_C - \mu M/r = \text{constant} \qquad (3.25)$$

3.3   GENERAL ATTITUDE DYNAMICS

Trajectory dynamics supplies rules governing the motion of the centre-of-mass C relative to some inertially fixed frame of reference. Attitude dynamics on the other hand uses the centre-of-mass C as a reference point.

In terms of momentum, attitude dynamics is mathematically identical to trajectory dynamics. That is to say angular momentum $\mathbf{H}$ responds to a torque $\mathbf{T}$ in exactly the same way as linear momentum $\mathbf{L}$ does to a force $\mathbf{F}$. But the physical motions associated with the two types of momentum are quite different.

In order to establish the fundamental principles in terms of momentum, it is convenient to refer once more to the closed surface S (Figure 3.1), the boundary that separates particles that are of interest from those that are not. These results are then quite general, including spacecraft with fluids and moving parts as well as the important cases of a rigid body and multiple bodies; the rules for the physical motions of the latter are dealt with later.

3.3.1   Angular momentum H

The angular momentum $\mathbf{H}_O$ referred to a point O is defined as the aggregate of the moments of the momenta of all the particles within S. In mathematical terms, angular momentum referred to O is defined as

$$\mathbf{H}_O = \sum(\mathbf{r} \times m\mathbf{v}) \qquad (3.26)$$

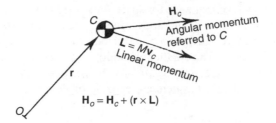

Figure 3.10   Transfer of reference point from C to O

where both **r** and **v** are relative to O. It is a product of rotation in much the same way as linear momentum is a product of translation.

In physical terms the angular momentum of a rigid body is a measure of the torque impulse that is needed to create its rotational motion. Thus the rotation of a body will be brought to rest by the application of a torque impulse that is equal and opposite to its angular momentum **H**.

The reference point that is most useful for attitude dynamics is the centre-of-mass C. For example, the rotational motion of a spacecraft is normally derived from equation (3.28), using its momentum $\mathbf{H}_C$ referred to its centre-of-mass.

The rule governing the transfer of reference point from C to some other point O is

$$\mathbf{H}_O = \mathbf{H}_C + M\mathbf{h}_O = \mathbf{H}_C + (\mathbf{r} \times M\dot{\mathbf{r}}) \tag{3.27}$$

Figure 3.10 illustrates the terms used. The transfer involves adding on the moment-of-momentum of the e.p., a process similar to the transfer of inertia from C to O (Appendix, equation 3.A9). Equation (3.27) could be used, for example, for expressing the contribution to a spacecraft's total angular momentum referred to *its* centre-of-mass, which arises from a momentum wheel (see also equation 3.35).

3.3.2   Rate of change of angular momentum, H

The following explanation applies only when the reference point is the centre-of-mass C or an inertially fixed point I. It does not apply for other points.

The angular momentum may be changed in two ways, the first being

- by applying an external couple or a force that has a moment about the reference point. The term 'external torque' (**T**) will be used as a general term to cover both of these. Its consequence is a rate of change of **H**, referred to C or I, described by the Newtonian equation

$$d(\mathbf{H})/dt = \mathbf{T} \tag{3.28}$$

Internal torques, acting between particles or bodies, will *not* change the total momentum. Thus mechanisms, fuel movement and so on will not change the total angular momentum of a spacecraft. Similarly, the forces between two docking spacecraft will not affect their combined angular momentum.

There is an important consequence for spacecraft designers since there will always be naturally occurring external disturbance torques (see Section 9.4 of Chapter 9). Their mean level will therefore cause a progressive build-up of the angular momentum over the lifetime of the craft. The rotational motion associated with this could be quite unacceptable. It follows that spacecraft *must* be fitted with means of controlling this build-up, and only external torquers are capable of doing so.

The second means of changing **H** is

- by the ejection from S of some particles whose momenta have moments about the reference point. This will occur during the firing of rockets, when their thrust vector does not pass precisely through the centre-of-mass. An analysis of this process will not be covered in this book. (A procedure similar to that used above in the development of equations (3.16), (3.17) and (3.20) can be followed if required.)

Equation (3.28) takes the same form as equation (3.7), and so the effect that a torque **T** has upon angular momentum **H** is the same as the effect that a force **F** has upon linear momentum **L**, stated as follows:

1. The component of a torque **T** that is in the same direction as the angular momentum **H** will change only the magnitude of the momentum.
2. The component T_N of the torque **T** that is at right angles to the angular momentum **H** will cause it to change direction towards the torque **T**.
   In this case, the torque impulse during a small interval of time δt is $T_N \delta t$ and is the change in the angular momentum during this time (see equation 3.15). This is shown in Figure 3.11.

The best-known physical illustration of this is the precession of a gyroscope. From Figure 3.11 it may be seen that the rate of precession is

$$d\psi/dt = T_N/H \tag{3.29}$$

As the angular momentum **H** becomes very large, the effect of a torque impulse becomes less, leading to the property known as *gyroscopic rigidity*.

This characteristic is made use of by spacecraft designers when they give their craft *momentum bias*, as a means of making the bias direction insensitive to disturbance torques. Making **H** large causes the precession rate to become small in response to a given torque.

The rate of change of the components of **H** depends upon the rotation of the axis system chosen, in addition to the change described by the Newtonian equation (3.28). When the

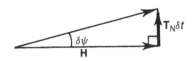

Figure 3.11 Precession due to an external torque

axis system has an angular velocity $\mathbf{\Omega}$ relative to inertial space, then equation (3.28) must be interpreted using the Coriolis theorem, as

$$d(\mathbf{H}_C)/dt = d/dt_{\text{compts}}(\mathbf{H}_C) + (\mathbf{\Omega} \times \mathbf{H}_C) = \mathbf{T} \qquad (3.30)$$

where $d/dt_{\text{compts}}(\ )$ means the rate of change of the components of $(\ )$.

This equation is used in the remainder of this chapter in order to develop the scalar dynamic equations for the rotational motion of spacecraft.

3.3.3   Angular momentum of rigid bodies

The angular momentum of a rigid body may be expressed in terms of its angular velocity ω. The angular momentum equations given above can then be used to develop equations in ω that describe the attitude and rotational motion of such a body, and of systems made up of rigid bodies.

The angular momentum $\mathbf{H}_C$ of a single rigid body referred to its centre-of-mass C may be expressed as

$$\mathbf{H}_C = [I_C]\omega \qquad (3.31)$$

where ω is its angular velocity relative to an inertial (non-rotating) frame of reference and $[I_C]$ is the inertia matrix based upon the centre-of-mass C.

In general, $[I_C]$ may be expressed as

$$[I_C] = \begin{bmatrix} I_{xx} & -I_{xy} & -I_{zx} \\ -I_{xy} & I_{yy} & -I_{yz} \\ -I_{zx} & -I_{yz} & I_{zz} \end{bmatrix} \qquad (3.32)$$

where I_{xx}, I_{yy} and I_{zz} are the moments of inertia.

I_{xy}, I_{yz} and I_{zx} are the products of inertia, broadly representing a measure of the lack of mass symmetry, leading to cross-coupled behaviour, as will be seen later.

The Appendix to this chapter contains a full treatment of the subject of the inertia matrix, including the dependency of the elements upon the choice of axes. It is sufficient to record here that every body has a set of orthogonal axes at each point, for which the products of inertia are zero. These are called *principal axes*, and there may well be more than one set of them, depending upon the mass symmetry of the body. Principal axes are eigenvectors of the inertia matrix.

It is evident from equation (3.31) that the components of the angular momentum may in general be expressed as

$$\mathbf{H}_C = \begin{bmatrix} (I_{xx}\omega_x - I_{xy}\omega_y - I_{zx}\omega_z) \\ (I_{yy}\omega_y - I_{yz}\omega_z - I_{xy}\omega_x) \\ (I_{zz}\omega_z - I_{zx}\omega_x - I_{yz}\omega_y) \end{bmatrix} \qquad (3.33)$$

When principal axes are used, then

$$\mathbf{H}_C = \{I_{xx}\omega_x, I_{yy}\omega_y, I_{zz}\omega_z\}^{\mathrm{T}} \qquad (3.34)$$

It is usual to develop the rotational equations for a body by choosing axes in which the inertias are constant. Axes fixed in the body will always achieve this, but there are other options if there is a mass symmetry. For example, a spinning object that has cylindrical mass symmetry about its spin axis, such as a typical wheel, will have constant inertias in non-spinning coordinate axes.

The angular momentum of a rigid body with spinning wheels, such as a spacecraft fitted with momentum or reaction wheels, can be expressed as the sum of the angular momentum of the rigid body containing the wheels in their non-spinning state (equations 3.33 or 3.34), together with the extra momentum due to the angular velocities of the wheels *relative to the body*.

For example, suppose a spacecraft is fitted with a wheel that is spinning with an angular velocity ω_{wh} relative to the craft. The vector ω_{wh} will be along the wheel's axis of symmetry, which will be in a fixed direction in the spacecraft's axes. If its moment of inertia about its own axis is I_{wh}, then an additional angular momentum equal to $I_{wh}\omega_{wh}$ must be added to that of the spacecraft's body. If there are several wheels and as a whole they contribute the additional momentum components $\{H_x, H_y, H_z\}^T$, say, then the total angular momentum of the body plus wheels becomes

$$\mathbf{H}_C = \begin{bmatrix} I_{xx}\omega_x - I_{xy}\omega_y - I_{zx}\omega_z + H_x \\ I_{yy}\omega_y - I_{yz}\omega_z - I_{xy}\omega_x + H_y \\ I_{zz}\omega_z - I_{zx}\omega_x - I_{yz}\omega_y + H_z \end{bmatrix} \tag{3.35}$$

If the body's principal axes are used, then this becomes

$$\mathbf{H}_C = \{(I_{xx}\omega_x + H_x), (I_{yy}\omega_y + H_y), (I_{zz}\omega_z + H_z)\}^T \tag{3.36}$$

Multiple rigid bodies of a more general nature need to be addressed in order to deal with the docking manoeuvre, ejecting payloads from a launcher's cargo bay, astronauts repairing spacecraft, tethered spacecraft and so on and indeed to the Earth/Moon combination. For these multi-body situations, each separate body will obey the dynamic equations that are stated elsewhere in this chapter. But it is worth noting that their combined momenta will not be changed by any forces or moments of interaction between them, and so will not be affected by a collision, separation or tethering. This applies to their absolute linear momentum, and also to the angular momentum about their combined centre-of-mass, C.

For example, consider two bodies with masses M_1, M_2, as in Figure 3.12, and total mass M. Assume that their centres-of-mass at C_1, C_2 have absolute velocities $\mathbf{v}_1$, $\mathbf{v}_2$ and that referring to these, they have angular momenta $\mathbf{H}_1$, $\mathbf{H}_2$.

Then, during a collision or separation or due to a tether

- their total absolute linear momentum remains constant, so

$$M\mathbf{v}_C = M_1\mathbf{v}_1 + M_2\mathbf{v}_2 \text{ remains constant} \tag{3.37}$$

- their angular momenta referred to C remains constant, so

$$\mathbf{H}_C = (M_1M_2/M)(\mathbf{r}_{12} \times \mathbf{v}_{12}) + \mathbf{H}_1 + \mathbf{H}_2 \text{ remains constant} \tag{3.38}$$

where $\mathbf{r}_{12}, \mathbf{v}_{12}$ are the position and velocity vectors of C_2 relative to C_1.

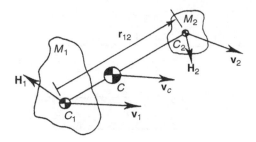

Figure 3.12 The combined momenta of two rigid bodies

These conservation of momentum laws also apply to the erection of solar arrays and so on, and when using pointing mechanisms.

3.3.4 Rotational kinetic energy

The rotational energy of a rigid body may be expressed as

$$E = \tfrac{1}{2}(\mathbf{H}_C \cdot \boldsymbol{\omega}) \text{ or } \tfrac{1}{2}([I_C]\boldsymbol{\omega} \cdot \boldsymbol{\omega}) \tag{3.39}$$

This is additional to the translational energy of the e.p. quoted in equation (3.22).

Work is done by a torque $\mathbf{T}$ at a rate $\mathbf{T} \cdot \boldsymbol{\omega}$, and the rate at which the rotational energy is increased by a torque $\mathbf{T}_C$ about C is $\mathbf{T}_C \cdot \boldsymbol{\omega}$. It follows that in the absence of any external torque $\mathbf{T}_C$ the rotational energy will remain constant.

When the body is not rigid, however, or when there are moving parts, the energy level can change in the absence of any external torque. In a spacecraft there will be internal dissipative mechanisms such as flexure, passive nutation dampers and so on that lead to the loss of kinetic energy, but at the same time the angular momentum $\mathbf{H}_C$ remains constant. A notable consequence of this is that the long-term tendency of a spinning body is towards spinning about its axis of maximum inertia, this being its minimum-energy state (see Section 3.4.2).

3.4 ATTITUDE MOTION OF SPECIFIC TYPES OF SPACECRAFT

The general theory in Section 3.3 may now be applied to specific types of spacecraft. For this purpose it will be assumed that all craft have rigid bodies, and have rigid moving parts unless otherwise stated. Whilst the response of their angular momentum to a torque is well ordered and straightforward, as explained above, the associated rotational motion is by no means necessarily so easily understood.

Spacecraft may be classified for convenience, as shown in Figure 3.13. The main subdivision depends upon whether the spacecraft has momentum bias or not. Bias means that the craft has a significant amount of angular momentum due to the spin of part or all of it. It then behaves like a gyroscope, with the associated characteristics of gyroscopic rigidity and a precessional type of response to a torque, as described in equation (3.29).

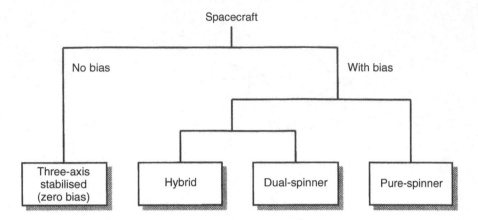

Figure 3.13 Categories of spacecraft

Some spacecraft may be spun-up for only a short time. It is common practice to do so prior to the firing of a high thrust rocket, for example. This is primarily to prevent any thrust offset from causing the craft to veer off course, the spin causing the mean path to be straight. Subsequently, the craft will be spun-down to the level of bias that is required for normal operation.

*Cross-coupling** is an essential feature of precessional response. The precession axis is at right angles to the axis of the torque that caused it. It is nevertheless a systematic and orderly effect. Normally, large cross-couplings are to be avoided so that control about each axis can proceed without interfering with the other axes, and vice versa. The rigid body dynamic equations developed in this section enable the sources of cross-couplings to be identified.

3.4.1 Three-axes-stabilized spacecraft (with no momentum bias)

Spacecraft in this class are usually large, with extensive solar arrays. They have a variety of different shapes such as the notional one shown in Figure 3.14. Their angular velocity is normally small, perhaps one revolution per orbit in order to maintain one face pointing towards Earth. The solar arrays may have even less angular movement, being required to point towards the Sun, and they may be mounted on a bearing to permit this.

The craft will be treated as a single rigid body. This is a reasonable approximation for the main structure, but the solar arrays are very flexible (see Section 3.5.2). Nevertheless, the presentation brings out some important points that are relevant to the 'rigid body' responses of these craft.

Using principal axes that are fixed in the body, their angular velocity Ω will be that of the body, ω. Inserting $\mathbf{H}_C$ from equation (3.34) into equation (3.30) leads to

---

* 'Cross-coupling' here means that a cause about one axis, such as a torque about the pitch axis, produces a response about another axis—the yaw axis, for example—and vice versa. (In electrical circuits it implies an undesired effect.)

$$\left.\begin{array}{l} I_{xx}\dot{\omega}_x - (I_{yy} - I_{zz})\omega_y\omega_z = T_x \\ I_{yy}\dot{\omega}_y - (I_{zz} - I_{xx})\omega_z\omega_x = T_y \\ I_{zz}\dot{\omega}_z - (I_{xx} - I_{yy})\omega_x\omega_y = T_z \end{array}\right\} \tag{3.40}$$

From these equations it may be seen that *cross-coupling* will potentially be present in attitude control systems (ACS) for a variety of reasons. The term is used when the acceleration in response to a torque does not take place solely about the torque axis.

For an initially stationary spacecraft a torque about a principal axis will produce a response about this axis, without any cross-coupling into the other axes. The response is an angular acceleration given by

$$\dot{\omega}_x = T_x/I_{xx} \tag{3.41}$$

Since torquers produce couples, their axes do not need to pass through the centre-of-mass, C.

In the more general case, when there are products of inertia, the initial acceleration from rest will be

$$\dot{\boldsymbol{\omega}} = [I_C]^{-1}\mathbf{T} \tag{3.42}$$

Cross-coupling will occur, and if a pure response $\dot{\omega}_x$ is required, then potentially all three torquers must be used.

Cross-coupled motion will also result if the x-component of torque is used when there is already an angular rate about the y- or z-axes. It must be assessed when a repointing manoeuvre is needed. It may be that the manoeuvre will best be carried out as a sequence of separate rotations about the principal axes, but there are a number of possible routes that may be taken in changing from one attitude to another.

In the absence of any external torque, a stationary spacecraft with no bias will remain in that condition indefinitely. Once it is rotating, its motion may become divergent. The stability of rotating motion is dealt with in Section 3.4.2.

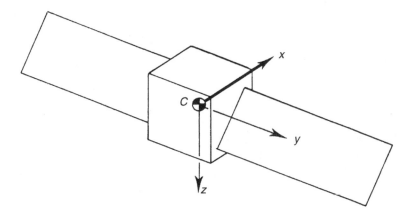

Figure 3.14   The three-axis-stabilized spacecraft

3.4.2 Pure-spin spacecraft

When the whole of the spacecraft structure spins, it has angular momentum. At very low values this leads to minor cross-couplings such as are mentioned above. But when the momentum becomes large and deliberate it is called *momentum bias*, and then the cross-couplings become so great that the behaviour needs to be looked at afresh. The behaviour of the spacecraft becomes like that of a gyroscope rotor, or usually its designer would like it to be so. This will only be achieved if he observes constraints upon the choice of spin axis and upon the mass distribution, as indicated below.

The *choice of spin axis* is important. From the mission designer's point of view this axis will be the direction of the momentum bias, gyroscopically rigid as described in Section 3.3.2, and so it is ideally orientated in a direction that is not required to change during the mission—normal to the orbit plane, for example. But the dynamics of the spacecraft requires that constraints on the mass distribution are observed.

If a body is spinning with angular velocity ω, then the spin axis will not in general remain in a fixed direction in space. It is the angular momentum $\mathbf{H}_C$ that has the fixed direction. If it is required that ω is also fixed, then it follows that it must be aligned with $\mathbf{H}_C$, that is, $[I_C]\omega$. This can only happen if the spin axis is along an eigenvector of the inertia matrix, which is to say that the spin must take place about a principal axis.

If rotation is initiated about some axis other than a principal axis, then the physical behaviour depends upon a number of factors. In general, the angular velocity vector varies in direction and magnitude. The motion may be anything from an untidy tumbling type of motion to an oscillatory coning motion, depending upon the magnitudes of the initial rotation, the departure of its axis from a principal axis and the disparity between the inertias. When faced with this type of behaviour, and also the instabilities referred to in the following paragraphs, it is difficult to believe that the angular momentum vector $\mathbf{H}_C$ is fixed in magnitude and direction.

The *stability of a spinning body* may be explored by using equation (3.40), with no torque components present. It will be assumed here that the spin is about the z-axis and has a rate $\omega_z = S$, consistent with Figure 3.15. Two of equations (3.40) then become

$$I_{xx}\dot{\omega}_x - (I_{yy} - I_{zz})S\omega_y = 0$$

and

$$\text{(3.43)}$$

$$I_{yy}\dot{\omega}_y - (I_{zz} - I_{xx})S\omega_x = 0$$

These are linear equations whose characteristic equation in terms of the Laplace operator, s, is

$$s^2 + (1 - I_{zz}/I_{xx})(1 - I_{zz}/I_{yy})S^2 = 0 \qquad \text{(3.44)}$$

It may be seen that for the spinning motion to be stable, I_{zz} must be the maximum or the least of the moments of inertia.

Bodies that are designed to spin will normally have a mass distribution that is axisymmetric, meaning that $I_{xx} = I_{yy}$. The moments of inertia about all radii through C then have this same value.

There are many examples of such axially symmetric spinning bodies. Gyroscope rotors, vehicle wheels, propellers and so on spin about their axes of maximum inertia. On the

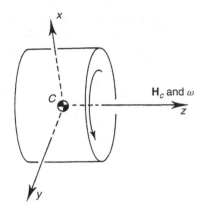

Figure 3.15   The pure-spin spacecraft

other hand rifle bullets, expendable launchers, guided missiles and so on spin about their axes of least inertia.

Long-term stability is important for spacecraft that must remain spinning for their lifetime. If they spin about their axis of least inertia, they will be stable in the short term only. Their long-term behaviour will be unstable if there is a loss of rotational energy brought about by internal dissipation rather than by an external torque. They will eventually adopt a cartwheeling type of motion—a spin motion about the axis of maximum inertia. During this process the angular momentum $\mathbf{H}_C$ remains constant, whilst the energy $\frac{1}{2}\mathbf{H}_C \cdot \boldsymbol{\omega}$ decreases, and it follows that the motion must move towards the minimum-energy state in which $\boldsymbol{\omega}$ is least. With $[I_C]\boldsymbol{\omega}$ remaining constant, this means that it moves towards rotation about an axis of maximum inertia.

Thus it follows that spacecraft that are pure-spinners will spin about their axis of maximum moment of inertia.

The effect of a torque $\mathbf{T}$ is to change the angular momentum $\mathbf{H}_C$, as explained in Section 3.3. For the physical response its effect upon the angular velocity $\boldsymbol{\omega}$ is required, and the appropriate equations may be obtained by inserting $\mathbf{H}_C$ in the form $[I_C]\boldsymbol{\omega}$ into equation (3.30).

In the following analysis it will be assumed that the spacecraft is initially spinning about its z-axis, a principal axis of maximum or least inertia. When this is the only angular motion, the craft is in a state of equilibrium, stable at least in the short term, as shown above.

Repointing of the spin axis will normally call for a rotation about an axis that is at right angles to the spin axis and that is fixed in space rather than fixed in the rotating structure. Coordinate axes that are convenient for the analysis are therefore non-spinning ones with the z-axis aligned with the spin axis. These axes have their angular velocity z-component equal to zero at all times, so that during a manoeuvre the angular velocity will be $\boldsymbol{\Omega} = \{\Omega_x, \Omega_y, 0\}^T$.

The spacecraft spins at a rate $\{0, 0, S\}^T$ relative to the coordinate axes, and so its angular velocity $\boldsymbol{\omega}$ is $\{\Omega_x, \Omega_y, S\}^T$.

The inertia matrix $[I_C]$ referred to the coordinate axes has elements that are in general changing. If the moments of inertia referred to the spacecraft's principal axes are

$\{I_{xx}, I_{yy}, I_{zz}\}$ and these axes are at an angle ψ to the coordinate axes, where $\dot{\psi} = S$, then the inertia matrix $[I_C]$ is obtainable from Appendix equation (3.A10):

$$[I_C] = \begin{bmatrix} (I_+ - I_-c) & I_-s & 0 \\ I_-s & (I_+ + I_-c) & 0 \\ 0 & 0 & I_{zz} \end{bmatrix} \tag{3.45}$$

where

$I_+ = \frac{1}{2}(I_{yy} + I_{xx})$, the mean inertia orthogonal to the spin axis,
$I_- = \frac{1}{2}(I_{yy} - I_{xx})$, a measure of lack of axial symmetry,
and $c, s = \cos 2\psi$ and $\sin 2\psi$, respectively.

The angular momentum $\mathbf{H}_C$ is then

$$\mathbf{H}_C = [I_C]\,\omega = \begin{bmatrix} \Omega_x(I_+ - I_-c) + \Omega_y I_-s \\ \Omega_x I_-s + \Omega_y(I_+ + I_-c) \\ SI_{zz} \end{bmatrix} \tag{3.46}$$

Noting that $dc/dt = -2Ss$ and $ds/dt = 2Sc$, then equation (3.30) leads to the following dynamic equations referred to non-spinning axes:

$$\left. \begin{aligned} I_+\dot{\Omega}_x + I_{zz}S\Omega_y + I_-\{-(c\Omega_x - s\Omega_y) + 2S(s\Omega_x + c\Omega_y)\} &= T_x \\ I_+\dot{\Omega}_y - I_{zz}S\Omega_x + I_-\{(s\Omega_x + c\Omega_y) + 2S(c\Omega_x - s\Omega_y)\} &= T_y \\ I_{zz}\dot{S} + I_-\{s\Omega_x^2 + 2c\Omega_x\Omega_y - s\Omega_y^2\} &= T_z \end{aligned} \right\} \tag{3.47}$$

A number of conclusions can be drawn:

1. *A torque T_z about the z-axis* causes a simple acceleration $\dot{S}$ about that axis when the spacecraft is in its equilibrium state. Note that in this state Ω_x and Ω_y are zero, so

$$\dot{S} = T_z/I_{zz} \tag{3.48}$$

2. *Repointing the spin axis* by means of a constant precession rate Ω_y about the y-axis, say, will require a torque T_x about the x-axis. Only if the moments of inertia I_{xx} and I_{yy} are equal, that is, $(I_- = 0)$, will the torque be constant. Under this condition,

$$\Omega_y = T_x/I_{zz}S \tag{3.49}$$

For example, a spacecraft with a moment of inertia of $400\,\text{kg}\,\text{m}^2$, spinning at 10 rpm, will precess at about 0.5 arcsec per second for each newton millimetre of torque.

This represents complete cross-coupling between x- and y-axes, which is a characteristic of a gyroscope's precessional behaviour. But in addition it should be noted that a torque produces an angular *velocity* rather than the acceleration produced by T_z. Figure 3.16(a) shows how the angular momentum $\mathbf{H}_C$ is rotated towards the torque vector $\mathbf{T}$ and Figure 3.16(b) shows the total angular velocity, with contributions from ω and Ω. If I_{xx} and I_{yy} differ, then additional oscillating torques will be needed, that is,

$$T_x = 2cSI_-\Omega_y \text{ and } T_y = 2s\,SI_-\Omega_y \tag{3.50}$$

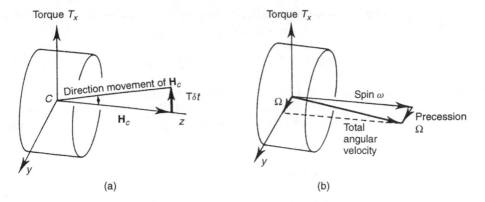

Figure 3.16 Precession of a pure-spinner due to a torque

Furthermore, if a constant rate of spin S is to be maintained, then a torque will be needed, which is equal to

$$T_z = -s\,I_-\Omega_y^2 \tag{3.51}$$

Failure to provide these oscillating torques will lead to an unsteady precession in response to a constant torque T_x. Most objects that are designed to spin will be given axial mass symmetry, that is, $I_- = 0$, or $I_{xx} = I_{yy}$, and this applies to spacecraft too.

It should be noted that the components T_x and T_y are in non-spinning axes and that torquers fixed in the spacecraft will need to apply resolved versions of T_x, T_y, that is,

$$
\begin{matrix} \text{Torque components} \\ \text{in spacecraft axes} \end{matrix} =
\begin{bmatrix} \cos\psi & \sin\psi & 0 \\ -\sin\psi & \cos\psi & 0 \\ 0 & 0 & 1 \end{bmatrix}
\begin{bmatrix} T_x \\ T_y \\ T_z \end{bmatrix} \tag{3.52}
$$

3. *There is an oscillatory 'nutation' mode*. This appears as a coning type of motion when the spacecraft is disturbed from its equilibrium state. Whenever precessional torques are applied, the mode will be activated; additional activation will occur when the torque ceases. If the duration of the precession is equal to an exact number of nutation periods, then the torque cessation will cancel the nutation, but engineered damping may be necessary (see Section 3.5.1).

It may already be seen from the characteristic equation (3.43) that an oscillatory mode exists, the frequency of which is

$$\omega_{\text{nut}} = S\sqrt{(1 - I_{zz}/I_{xx})(1 - I_{zz}/I_{yy})} \tag{3.53}$$

This is the frequency that may be observed on traces from rate gyroscopes that are mounted on the structure, that is, when it is observed in the spinning axes.

In the axisymmetric case, I_{zz}/I_{xx} lies in the range $1 \le I_{zz}/I_{xx} < 2$. The *observed* nutation frequency will depend upon the rate at which the observer is spinning.

When viewed from the (spinning) spacecraft frame of reference, as measured by instruments that are fixed to the spacecraft's structure, for example,

$$\omega_{nut} = S((I_{zz}/I_{xx}) - 1), \text{ and so } 0 \le \omega_{nut} < S \quad (3.54)$$

When viewed from a non-spinning (inertial) frame of reference,

$$\omega_{nut} = S(I_{zz}/I_{xx}), \text{ and so } S \le \omega_{nut} < 2S \quad (3.55)$$

3.4.3 Hybrid and dual-spin spacecraft

Two classes of spacecraft bring together the advantages of having momentum bias and yet provide a non-spinning base for accommodation of payload elements and torquers. These are the dual-spin spacecraft and the hybrid spacecraft. In each case momentum bias is provided by mounting a rotating body—or more than one—on the non-spinning part; it is part of the structure in the case of the dual-spin spacecraft, but consists of high-speed purpose-built wheels—momentum wheels—in the case of the hybrid spacecraft.

The spinning parts of these types of spacecraft have dynamic properties similar to those of the pure-spinner covered above. They will have axially symmetric mass properties for the reasons given. So their centres-of-mass lie on the spin axis, and their moments of inertia will be equal about axes normal to this. Their contribution to the total inertia $[I_C]$ of the spacecraft will then be independent of their rotation. Their rate of spin will remain constant unless there is a torque about their axis.

The non-spinning structure will behave like the three-axes-stabilized craft described in Section 3.4.1. Some of the torques on it will be applied via the bearings of the rotating parts in which there will be built-in torque motors.

The angular momentum of the hybrid and dual-spin spacecraft is then given by the expressions in equation (3.35) or (3.36). These may be substituted into equation (3.30) in order to obtain equations obeyed by the components of the angular velocity of the non-spinning structure, in any coordinate axis.

If the momentum bias direction is along the z-axis, a principal axis, then

$$\mathbf{H}_C = \{I_{xx}\omega_x, I_{yy}\omega_y, (I_{zz}\omega_z + H_z)\}^T \quad (3.56)$$

The dynamic equations for the components of angular velocity along principal axes are

$$\left.\begin{array}{l} I_{xx}\dot{\omega}_x + \omega_y\omega_z(I_{zz} - I_{yy}) + \omega_y H_z = T_x \\ I_{yy}\dot{\omega}_y + \omega_z\omega_x(I_{xx} - I_{zz}) - \omega_x H_z = T_y \\ I_{zz}\dot{\omega}_z + \dot{H}_z + \omega_x\omega_y(I_{yy} - I_{xx}) = T_z \end{array}\right\} \quad (3.57)$$

A number of observations may be made:

1. *Stability*. The momentum bias axis may be the axis of intermediate inertia without causing instability. The constraint imposed on the pure spinner is not necessary here.

The long-term stability cannot be forecast from the above equations. The dual-spinner will, however, be stable provided that energy dissipation in the non-spinning part exceeds that in the spinning part. The bias direction may then be along the axis of least inertia. Passive nutation dampers, for example, will be placed in the non-spinning part of the spacecraft.

2. There is a *nutation mode* whose frequency, when observed in non-spinning axes, is

$$\omega_{\text{nut}} = H_z / \sqrt{(I_{xx} I_{yy})} \tag{3.58}$$

For example, a spacecraft having both moments of inertia I_{xx} and I_{yy} equal to $400\,\text{kg m}^2$, and carrying a momentum wheel giving a bias of 40 N m s, will have a nutation mode with a period of about one minute.

3. *The response to a torque*. The inclusion of momentum bias makes the whole structure behave as a gyrostat, with a response that is very similar to that of a spinner. A constant torque about the x-axis will cause a constant rate of precession about the y-axis, for example, the magnitude of which is

$$\omega_y = T_x / H_z \tag{3.59}$$

Thus a torque of 1 N mm will produce a rate of precession of about 5 s of arc per second when the bias is 40 N m s. On the other hand a torque about the bias direction will produce an angular acceleration about that direction if it is parallel to a principal axis. In other cases there will be cross-coupling.

3.5 OSCILLATORY MODES

All vehicles have a tendency to oscillate and vibrate, and spacecraft are no exception. In their case the damping of the oscillatory modes is less than that experienced by Earthbound vehicles, and as a consequence any oscillation, once started, may last for a very long time—a characteristic of modes that are very close to being unstable.

It follows that it is important that these modes are identified by the designer and that artificial damping is introduced by the Attitude and/or Orbit Control system if possible; certainly the modes must not be destabilized. Care must also be taken to avoid exciting those modes whose ultimate damping remains very low.

The main oscillatory tendencies occur as either 'rigid body modes' or 'flexure modes' and are addressed below. Fuel movement inside tanks can also have an oscillatory tendency but this is normally controlled by means of baffles.

3.5.1 Rigid body modes

There are potentially two rigid body modes, nutation and libration.

The nutation mode is present in spacecraft that have momentum bias, whether this is achieved by spinning the complete structure or part of it or by using a momentum wheel. It is easiest to describe it as follows, when there is just one spinning part.

In the absence of any torque, the equilibrium state occurs when the axis of the spinning part is aligned with the angular momentum vector and they point at the same fixed direction in space. A small torque impulse will cause them to become misaligned, the momentum vector being turned to a new fixed direction. The ensuing nutation oscillation then has a pattern of motion in which the spin axis adopts a 'coning' type of behaviour around this fixed angular momentum direction.

At the start of a constant torque causing precession of the momentum bias vector, the step change in the torque will excite the nutation mode and the oscillation will be superimposed on the precession as shown in Figure 9.7. Further excitation will occur when the torque ceases; this may be timed so as to cancel the nutation.

A hand-held bicycle wheel may be used to demonstrate nutation. The wheel must be carefully balanced and fitted with a handle that is an extension of its axle. When the wheel is spinning rapidly with its axle horizontal, it may be supported by just one finger under the handle, whereupon it will rotate about a vertical axis at a constant rate. This is precession, obeying equation (3.29). If during this motion a sharp impulse is delivered to the handle, then an oscillatory motion will take place (in addition to the precession); this is nutation, the observed frequency obeying equation (3.55).

The libration mode is caused by the gravity gradient—the reduction in the gravitational field strength as the distance from Earth's centre increases.

A rigid body is in stable equilibrium in this field when its axis of minimum moment of inertia lies along the local vertical. The Moon is an example of a body that is roughly in this state. When the body in equilibrium is disturbed by a small amount, it will then oscillate like a conical pendulum if it is free to do so. The frequency of oscillation is given by equation (9.21) of Chapter 9; at the Earth's surface its period is about 48 min, about half the period of a satellite in Low Earth Orbit (LEO).

A few spacecraft use this phenomenon as a means of achieving an Earth-pointing face, and they must then incorporate damping for this mode in their ACS algorithms. (The Earth's tidal system provides damping for the Moon.) In other spacecraft the torques of this mode appear as disturbance torques (see Section 9.4.3 of Chapter 9).

3.5.2   Flexure modes

All structures are elastic and will vibrate at their modal frequencies when they are appropriately excited. The appendages of spacecraft in particular can be very flimsy structures compared with any Earthbound equivalents, and their fundamental frequency—their lowest modal frequency—can be very low indeed. The solar arrays of the Hubble telescope have a mode with frequency 0.11 Hz for example. Theory indicates that there are an infinity of modes with frequencies higher than the fundamental, each with its corresponding modal shape.

The solar arrays shown in Figure 9.3 of Chapter 9, the form in which they appear on many spacecraft, are cantilevered outwards from the central body and each will have bending and torsional modes. The frequencies of the bending modes of a single, rigidly attached uniform cantilever are given by

$$f = (K^2\pi/8L^2)\sqrt{(EI/\rho A)}\text{Hz} \tag{3.60}$$

where
$K \approx 1, 2, 3, 5, 7, \ldots$ for the different modes,
L = array length (m)
E = Young's modulus (N/m^2)
I = second moment of area (m^4)

ρ = density (kg/m^3)

A = cross-sectional area (m^2)

However, on a spacecraft it is unlikely that the attachment will be rigid, and the actual fundamental frequency may be only about 50% of the value given by this equation. A large number of the overtones' frequencies fall within the passband of the ACS.

Any oscillation of the solar arrays will be transmitted to the main structure of the spacecraft by the bending moments and shear forces at their roots. In consequence, the payload will tend to oscillate in sympathy with the flexure modes. This occurred on the Hubble telescope, the oscillation being initiated by the thermal shock that took place when it moved from being in eclipse to being in sunlight.

This simple example of an interaction is compounded by interactions between different flexure modes, affecting their modal frequencies. Forecasting the modal properties of the spacecraft as a whole is a complicated process for which specialist software packages are used (see also Chapter 8).

It should be noted that flexural oscillations do not directly involve external forces and moments on the spacecraft and so they do not directly affect the momenta, linear or angular.

3.6 IN CONCLUSION

The material in this chapter has been aimed primarily at the dynamics of the main categories of spacecraft. The benefits of adopting certain mass distributions and torque axes have been shown, together with the consequence of doing otherwise.

There are likely to be increasingly many occasions in which the designer has to deal with non-optimum mass distributions, as, for example, when reusable launch vehicles part with their payload and when large space structures are assembled or constructed in orbit. The fundamental laws still apply but the control systems will have to adjust to the changes in the mass distribution.

The methods contained in the chapter are not confined to the conventional configurations. Any configurations that may be treated as being made up of rigid bodies may be dealt with in the manner shown. Their linear momenta, moments-of-momentum and angular momenta will obey the Newtonian equations (3.7) or (3.18), (3.12) and (3.28), respectively. The components of their angular momentum $\mathbf{H}_C$ may be found by using equations (3.31) or (3.35), having first obtained their inertia matrix using the Appendix. All the Newtonian equations should be developed by using the Coriolis theorem as illustrated in equation (3.30), enabling the motion equations to be expressed in terms of the components of their velocities, linear and angular, along any desired axes.

APPENDIX: THE INERTIA MATRIX

A1 Introduction

The inertia matrix $[I_O]$ referred to a point O is a property of the mass distribution about O. Its elements depend upon the directions of a right-handed orthogonal set of axes

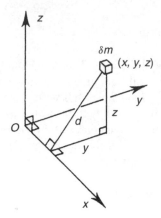

Figure 3.A1   An incremental mass δm

x, y, z through O, and in particular it contains the moments and products of inertia associated with these axes. The matrix at the centre-of-mass C plays an important part in the rotational behaviour of a spacecraft, and it must be evaluated and controlled during its design.

This appendix defines terms that are associated with mass distributions and presents formulae that are useful for the evaluation of the inertia matrix.

A2   Definitions

- *Moments of inertia I_{xx}, I_{yy}, I_{zz}*
  A moment of inertia is the second moment of mass about an axis. The contribution of an increment of mass δm at a distance d from the axis is $d^2 \delta m$ (see Figure 3.A1). The moment of inertia about the x-axis is, for example,

$$I_{xx} = \int (y^2 + z^2)\, dm \tag{3.A1}$$

  where the integral extends over the whole mass distribution.
- *Products of inertia I_{xy}, I_{yz}, I_{zx}*
  The product of inertia associated with the x-axis is

$$I_{yz} = \int yz\, dm \tag{3.A2}$$

Products of inertia are measures of the lack of symmetry in a mass distribution.

If there is a plane of symmetry, then the product of inertia associated with all axes in that plane will be zero. For example, an aircraft whose xz-plane is a plane of symmetry will have I_{xy} and I_{yz} equal to zero.

If two of the coordinate planes are planes of symmetry, then all three of the products of inertia will be zero. This applies to axially symmetric bodies such as many expendable launchers.

- *Principal axes* are sets of orthogonal axes for which all three products of inertia are zero. There is always one such set at each point.
- *The inertia matrix* $[I_O]$ referred to $O(x, y, z)$ is defined as

$$[I_O] = \begin{bmatrix} I_{xx} & -I_{xy} & -I_{zx} \\ -I_{xy} & I_{yy} & -I_{yz} \\ -I_{zx} & -I_{yz} & I_{zz} \end{bmatrix} \tag{3.A3}$$

For a single particle with mass m at (x,y,z) this becomes $[I_{Om}]$, say, where

$$[I_{Om}] = \begin{bmatrix} m(y^2 + z^2) & -mxy & -mzx \\ -mxy & m(z^2 + x^2) & -myz \\ -mzx & -myz & m(x^2 + y^2) \end{bmatrix} \tag{3.A4}$$

- *Rotation matrix* $[R]$

  If the components of a vector **V** in one set of axes are expressed as the terms in a (3×1) column matrix $\mathbf{V}_1$, say, and $\mathbf{V}_2$ consists of its components in a second set that is rotated relative to the first, then $\mathbf{V}_2$ may be expressed as $\mathbf{V}_2 = [R]\mathbf{V}_1$. Then, $[R]$ is known as a rotation matrix.

If Euler angles are used to describe the rotation as a sequence of separate rotations about the coordinate axes, then it is convenient to use a notation $[X(\phi)], [Y(\theta)]$ and $[Z(\psi)]$ for

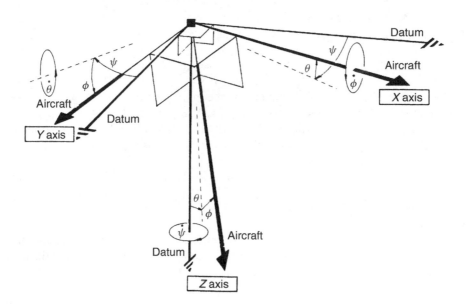

Figure 3.A2   The standard Euler angles used to specify an aircraft's attitude

the separate rotations. $[X(\phi)]$ is the rotation matrix for a clockwise rotation through an angle ϕ about the x-axis, and similarly for $[Y(\theta)]$ and $[Z(\psi)]$. These matrices are

$$[X(\phi)] = \begin{bmatrix} 1 & 0 & 0 \\ 0 & \cos\phi & -\sin\phi \\ 0 & \sin\phi & \cos\phi \end{bmatrix} \quad [Y(\theta)] = \begin{bmatrix} \cos\theta & 0 & \sin\theta \\ 0 & 1 & 0 \\ -\sin\theta & 0 & \cos\theta \end{bmatrix}$$

(3.A5)

$$[Z(\psi)] = \begin{bmatrix} \cos\psi & -\sin\psi & 0 \\ \sin\psi & \cos\psi & 0 \\ 0 & 0 & 1 \end{bmatrix}$$

For an aircraft whose attitude is defined in the conventional manner as the sequence of rotations yaw ψ, followed by pitch θ, followed by roll ϕ, as in Figure 3.A2, the matrix $[R]$ for converting the components of its velocity $\mathbf{V}$, say, from its own axes $\mathbf{V}_{a/c}$, back to the datum axes $\mathbf{V}_{datum}$, is

$$\mathbf{V}_{datum} = [R]\mathbf{V}_{a/c}, \text{ where } [R] = [Z(\psi)][Y(\theta)][X(\phi)] \tag{3.A6}$$

A3 Useful Formulae

- *Inertia invariant*

$$I_{xx} + I_{yy} + I_{zz} = 2\int r^2 \, dm = 2I_O, \text{ say} \tag{3.A7}$$

where I_O is the second moment of mass about the origin O and is independent of the direction of axes.

- *Perpendicular axis theorem* — for laminas only.
  For a lamina lying entirely in the yz-plane,

$$I_{yy} + I_{zz} = I_{xx} \tag{3.A8}$$

- *Transfer of reference point* (parallel axis theorem) (see Figure 3.A3).
  If an object whose centre-of-mass G is at (X, Y, Z) has an inertia matrix $[I_G]$ referred to G, then add on the inertia matrix of its e.p. referred to O, in order to obtain the inertia matrix $[I_O]$ referred to parallel axes at O, that is,

$$[I_O] = [I_{OM}] + [I_G] \tag{3.A9}$$

where equation (3.A4) may be used for $[I_{OM}]$.

- *Rotated axes theorem*
  If the rotation matrix for transforming a vector's components from axes number 1 to axes number 2 is $[R]$ (see rotation matrices above), then the inertia matrix $[I]$ may be transformed between the same sets of axes by using

$$[I_2] = [R][I_1][R]^{-1} \tag{3.A10}$$

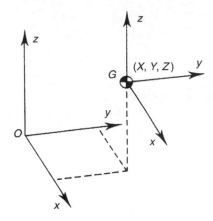

Figure 3.A3 Parallel axes used in transferring inertia matrix between points

A4 Contribution of a Piece of Equipment to $[I_C]$

Suppose that a piece of equipment with mass M is installed with its centre-of-mass G at (x, y, z) relative to the spacecraft's axes at C. Suppose that $[I_{eq}]$ is its inertia matrix referred to its own natural axes and that these axes are rotated through an angle ϕ about the spacecraft's x-axis when it is installed.

Then its contribution to the inertia matrix $[I_C]$ of the spacecraft will be $[X(\phi)] [I_{eq}] [X(-\phi)]$ plus the inertia matrix of its e.p. referred to C (equation 3.A4). In general,

$$[I_C] = [R][I_{eq}][R]^{-1} + [I_{CM}] \qquad (3.A11)$$

4 CELESTIAL MECHANICS

John P. W. Stark[1], Graham G. Swinerd[2] and Peter W. Fortescue[2]

[1] *School of Engineering and Material Science, Queen Mary, University of London*
[2] *Aeronautics and Astronautics, Faculty of Engineering and the Environment, University of Southampton*

4.1 INTRODUCTION

The theory of celestial mechanics underlies all the dynamical aspects of the orbital motion of spacecraft. The central feature is the mutual gravitational force of attraction that acts between any two bodies. This was first described by Newton, and together with his laws of motion (see Chapter 3), it provides us with the theoretical framework for celestial mechanics. The orbits that it forecasts will be relative to an Inertial Frame of Reference (IFR) that is fixed with respect to the stars; the consequential motion relative to the ground will also be covered in Chapter 5.

The simplified case in which the gravitational force acts between two pointlike objects gives a good approximation to orbital motion for most spacecraft situations. It may easily be shown that if a body has a uniform mass distribution within a spherical surface, then outside it the gravitational force from the body does indeed appear to emanate from a pointlike source. This so-called two-body problem has a solution—a Keplerian orbit.

Kepler, whose major works were published during the first 20 years of the seventeenth century, consolidated the observations of planetary motion into three simple laws, which are illustrated in Figure 4.1. These are

1. the orbit of each planet is an ellipse with the Sun occupying one focus;
2. the line joining the Sun to a planet sweeps out equal areas in equal intervals of time;
3. a planet's orbital period is proportional to the mean distance between Sun and the planet, raised to the power 3/2.

Newton's theory of gravity, predicting the above results, was not developed until more than 65 years later.

Spacecraft Systems Engineering, Fourth Edition.
Edited by Peter W. Fortescue, Graham G. Swinerd and John P. W. Stark.
© 2011 John Wiley & Sons, Ltd. Published 2011 by John Wiley & Sons, Ltd.

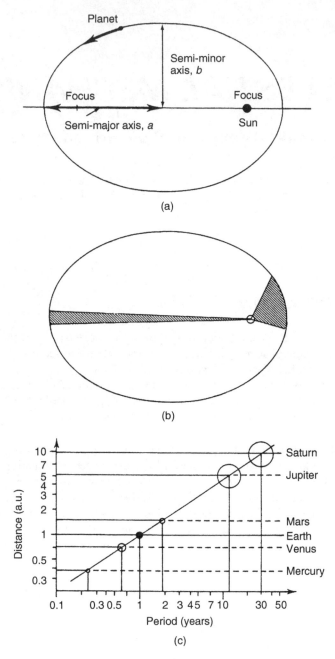

Figure 4.1 (a) Kepler's first law: orbit shape; (b) Kepler's second law: equal areas in equal time and (c) Kepler's third law: period $\tau \propto a^{1.5}$

Various perturbing forces must be included in order to produce a better approximation to a spacecraft's orbit. As the altitude of a spacecraft's orbit decreases, the departure from spherical symmetry of the Earth's mass and shape becomes more important, together with aerodynamic forces. At high altitudes such as in geostationary orbits ($\sim$36 000 km) the gravitational pull of the other celestial bodies such as the Moon and Sun become significant, together with radiation pressure from the Sun.

The theory of celestial mechanics is required in order that the motion of a spacecraft may be predicted. In most situations the accuracy required for the vehicle's position is relatively modest—typically a few kilometres. However, there are significant deviations from this, one important case being when the space vehicle must perform a manoeuvre to change its orbit (this topic is left to the next chapter). It should be noted that for some vehicles, particularly those that employ active remote sensing instrumentation, precise orbit determination is required. An accuracy of the order of 1 m was required for Seasat, and values of 10 cm or less are a requirement for the European Envisat spacecraft. Such precision in orbit determination is difficult because our knowledge of the Earth's gravity field remains limited, and more importantly because of the uncertainties associated with surface forces, in particular, aerodynamic drag. The former problem will be alleviated when dedicated spacecraft missions to determine the Earth's gravity field are flown. The solution to the latter problem of quantifying drag is much more elusive, since difficulties are associated with adequately modelling the atmospheric density, atmospheric winds, the response of the atmosphere to solar activity and the magnitude of lift and drag coefficients relevant to spacecraft. Consequently, the precise orbit can only be determined retrospectively. Fortunately, the accuracy level required for orbit prediction, derived for tracking and operational activities, is much less stringent.

4.2 THE TWO-BODY PROBLEM – PARTICLE DYNAMICS

The simplest problem in celestial mechanics concerns the predicted motion of two co-rotating point masses. It is assumed in this analysis that the sole force acting between them is their mutual gravitational attraction. Newton demonstrated that the gravitational field from any pointlike body is solely dependent upon its mass. The field's potential U at a distance R from a particle of mass M is simply

$$U = -\frac{GM}{R} \tag{4.1}$$

where G is the universal constant of gravitation, having a numerical value in SI units of 6.670×10^{-11} N m^2/kg^2.

If two masses m_1 and m_2 are a distance r apart, as shown in Figure 4.2, then the equation governing their motion due to their mutual gravitational attraction is easily obtained by combining Newton's Laws of gravity and motion. Newton's Law of gravity gives the forces $\mathbf{F}_1$ and $\mathbf{F}_2$ acting on the masses as

$$\mathbf{F}_1 = \frac{Gm_1m_2}{r^2}\left(\frac{\mathbf{r}}{r}\right),\ \mathbf{F}_2 = \frac{Gm_1m_2}{r^2}\left(\frac{-\mathbf{r}}{r}\right) \tag{4.2}$$

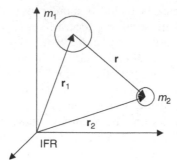

Figure 4.2   The two-body problem geometry in an inertial frame of reference (IFR)

The motion of each mass with respect to the IFR is expressed by Newton's second law of motion,

$$m_1 \ddot{\mathbf{r}}_1 = \mathbf{F}_1, \quad m_2 \ddot{\mathbf{r}}_2 = \mathbf{F}_2 \tag{4.3}$$

By noting that $\mathbf{r} = \mathbf{r}_2 - \mathbf{r}_1$, we may combine equations (4.2) and (4.3) to give

$$\ddot{\mathbf{r}} + \frac{G(m_1 + m_2)}{r^2} \left(\frac{\mathbf{r}}{r}\right) = \mathbf{0}$$

which is the equation of unperturbed motion due to gravity of mass m_2 with respect to mass m_1.

The restricted two-body problem assumes that the mass of one body far exceeds that of the other, $m_1 \gg m_2$ say. This applies to a spacecraft of mass m, say about 10^3 to 10^4 kg, compared with a heavenly body of mass M; Earth's mass, for example, is 6×10^{24} kg. In this case the equation above reduces to:

$$\ddot{\mathbf{r}} + \frac{(GM)}{r^2} \frac{\mathbf{r}}{r} = \mathbf{0} \tag{4.4}$$

Equation (4.4) is also a good approximation for the planets and the Sun, since even the largest one, Jupiter ($\sim 2 \times 10^{27}$ kg), is dominated by the Sun ($\sim 2 \times 10^{30}$ kg).

The parameter GM, here referred to as μ, is the gravitational parameter for the body about which the motion is taking place. Table 4.1 lists its values for a selection of solar system bodies.

The orbit constants

Equation (4.4) may be solved and will lead to the orbit equation if the initial conditions are known. The constants associated with a particular orbit may be thought of as being constants of integration of equation (4.4) or they may be approached in physical terms *via* conservation laws, as in Chapter 3.

Moment-of-momentum conservation follows from the fact that the only force acting has no moment about the centre of the primary body. The moment of the momentum vector $m\,\mathbf{h}$ is therefore constant, in both magnitude and direction.

Table 4.1 Some physical properties of the major bodies in the solar system, including Earth's Moon. See also Tables 2.5, 2.6 and 2.7 of Chapter 2. The sphere of influence (see Section 5.8.1 of Chapter 5) is expressed with respect to the Sun as the disturbing body

Parameter\ Body	μ (m^3/s^2) Gravitational parameter	Equatorial surface gravity (m/s^2)	Surface escape velocity (m/s)	Oblateness J_2	Sphere of influence (10^6 km)	Axial rotation period (sidereal)
Sun	1.327×10^{20}	273.98	6.18×10^5	—	—	~27 days
Mercury	2.203×10^{13}	3.70	4250	—	0.09–0.14	58.646 days
Venus	3.249×10^{14}	8.87	10 360	2.7×10^{-5}	0.61–0.62	243.019 days
Earth	3.986×10^{14}	9.81	11 180	0.001083	0.91–0.94	$23^h 56^m 22.7^s$
Mars	4.283×10^{13}	3.71	5020	0.001964	0.52–0.63	$24^h 37^m 22.6^s$
Jupiter	1.267×10^{17}	23.12	59 530	0.01475	45.9–50.5	~$9^h 50^m$
Saturn	3.794×10^{16}	9.05	35 560	0.01645	51.6–57.5	~$10^h 15^m$
Uranus	5.780×10^{15}	7.77	21 250	0.012	49.4–54.1	~$17^h 50^m$
Neptune	6.871×10^{15}	11.00	23 540	0.004	85.7–87.6	~$19^h 10^m$
Pluto	1.021×10^{12}	0.40	1300	—	11.4–18.8	6.387 days
Moon	4.903×10^{12}	1.62	2380	0.0002027	0.157–0.162	27.322 days

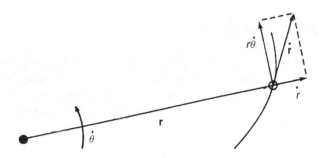

Figure 4.3 The particle's position and velocity vectors

By definition $\mathbf{h} = \mathbf{r} \times \dot{\mathbf{r}}$, and so its direction, perpendicular to both the position vector and the velocity vector, is normal to the orbit plane and is constant. The fact that the magnitude of $\mathbf{h}$ is constant, is consistent with Kepler's second law. The value of this is $r^2 \dot{\theta}$ (see Figure 4.3), which is twice the areal velocity $\dot{A}$.

Energy conservation follows from the fact that the only external force on the spacecraft is due to the gravitational field, a conservative force whose potential energy per unit mass is $-\mu/r$.

The total orbital energy per unit mass ε (the sum of the kinetic and gravitational energies) remains constant, and so

$$\frac{1}{2}V^2 - \frac{\mu}{r} = \varepsilon \qquad (4.5)$$

This is known as the energy equation or *vis-viva* integral.

Solution to the two-body problem

The vector solution to equation (4.4) may be obtained by first taking the cross-product with the constant **h** and integrating once with respect to time. This yields

$$\dot{\mathbf{r}} \times \mathbf{h} = \mu \left(\frac{\mathbf{r}}{r} + \mathbf{e} \right) \tag{4.6}$$

where **e** is the vector constant of integration called the eccentricity vector, which lies in the plane of the orbit as shown in Figure 4.4

The final solution to equation (4.4), obtained by taking the dot product of equation (4.6) with **r**, is

$$r = \frac{h^2/\mu}{1 + e\cos\theta} \tag{4.7}$$

where θ is the angle between **r** and **e**. This is the equation of a conic section and demonstrates the first of Kepler's laws.

The *eccentricity e* determines the type of conic. It is a circle when $e = 0$, an ellipse when $0 < e < 1$, a parabola when $e = 1$ and a hyperbola when $e > 1$. These shapes are shown in Figure 4.4. The dominant mass is at one of the two foci.

Elliptic and circular orbits ($0 \le e < 1$)

The *ellipse* is the general form of a closed orbit and is shown in Figure 4.5. The major axis is called the *line of apsides*, recalling that half an ellipse is shaped like the apse of a church. The angle θ is termed *the true anomaly*. The dimensions are related and may be expressed in terms of the physical constants h and ε. For example,

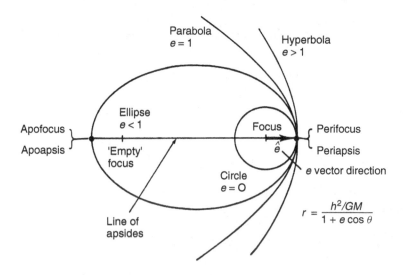

Figure 4.4 Conic sections

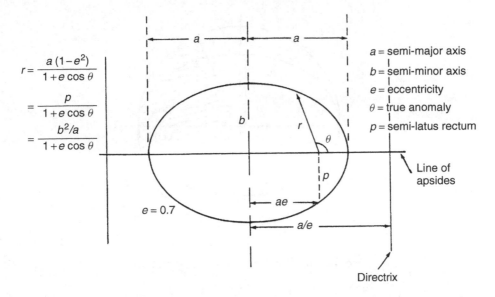

$$r = \frac{a(1-e^2)}{1+e\cos\theta}$$

$$= \frac{p}{1+e\cos\theta}$$

$$= \frac{b^2/a}{1+e\cos\theta}$$

$e = 0.7$

$a =$ semi-major axis
$b =$ semi-minor axis
$e =$ eccentricity
$\theta =$ true anomaly
$p =$ semi-latus rectum

Line of apsides

Directrix

Figure 4.5   Ellipses

- The semi-latus rectum p and semi-major axis a obey:

$$p = a(1 - e^2) = h^2/\mu \tag{4.8}$$

- The point of closest approach to the main body is termed the periapsis or perifocus and its distance is

$$r_p = p/(1 + e) = a(1 - e) \tag{4.9}$$

- The most distant point, the apoapsis or apofocus, is at a distance

$$r_a = p/(1 - e) = a(1 + e) \tag{4.10}$$

Clearly $r_p + r_a = 2a$.

The apses normally carry the name of the main body; so periapsis is called the perigee if the orbit is round the Earth, perihelion if round the Sun and so on. Equating the energy and the moments of momenta at the apses leads to

$$\varepsilon = -\mu/2a \tag{4.11}$$

Kepler's third law is simply obtained, since the orbit period is the time taken for the complete area of the ellipse to be swept out by the radius vector. Since $\dot{A} = h/2$ and the area of an ellipse is $\pi a^2 \sqrt{(1 - e^2)}$, then by making use of the relationships above it follows that the orbit period τ is given by

$$\tau = 2\pi \sqrt{(a^3/\mu)} \tag{4.12}$$

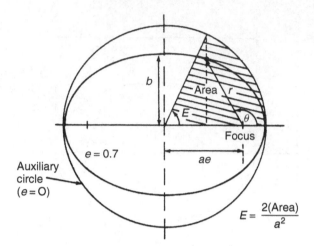

Figure 4.6 Eccentric anomaly definition

The *position versus time relationship* will be required for ground station passes, manoeuvres and other mission activities.

For *elliptical motion* $(0 < e < 1)$, a drawing of the trajectory may be achieved by constructing a circumscribing circle of radius a about the ellipse, as shown in Figure 4.6. This is called the *auxiliary circle*, and it is used to transform from the variable angle θ to a new variable E, called the *eccentric anomaly*.

Some useful expressions can be obtained from the geometry:

- The position on the ellipse may be written in terms of E as

$$r = a(1 - e \cos E) \qquad (4.13)$$

- The relationship between θ and E is

$$\tan \theta/2 = \tan E/2 \sqrt{\left[\frac{1+e}{1-e}\right]} \qquad (4.14)$$

To find a position *versus* time relationship, differentiation of equation (4.13) with respect to time gives

$$\frac{\mathrm{d}r}{\mathrm{d}t} = ae \sin E \frac{\mathrm{d}E}{\mathrm{d}t}$$

If we differentiate equation (4.7) to find $\mathrm{d}r/\mathrm{d}t$ and note from the geometry in Figure 4.6 that $r \sin \theta = a\sqrt{1 - e^2} \sin E$, we obtain

$$\frac{\mathrm{d}E}{\mathrm{d}t} = \frac{1}{r}\sqrt{(\mu/a)}$$

Separating the variables this gives

$$a(1 - e \cos E) \, dE = dt \sqrt{(\mu/a)} \tag{4.15}$$

which on integration yields

$$E - e \sin E = \sqrt{(\mu/a^3)}(t - t_p) \tag{4.16}$$

where t_p is the time of perifocal passage. Also, note that the *mean motion n* is defined by

$$n = 2\pi/\tau = \sqrt{(\mu/a^3)} \tag{4.17}$$

Therefore, if we suppose that the spacecraft passes through perifocus at time zero (i.e. $t_p = 0$), then the time of flight t from perifocus to the current position is given by

$$E - e \sin E = M \tag{4.18}$$

where the *mean anomaly* is defined by

$$M = nt \tag{4.19}$$

Equation (4.18) is known as *Kepler's equation*. It provides a simple relationship between time and position, which can be used in conjunction with equations (4.13) and (4.14) to provide a complete specification of position with time. The problem of the determination of position given the time of flight from perifocus can be represented symbolically by the following chain of operations,

Time t—equation (4.19) → M—equation (4.18) → E—equation (4.14) → Position θ.
Conversely, to determine the time to reach a given position the above chain is reversed, Position θ—equation (4.14) → E—equation (4.18) → M—equation (4.19) → Time t.

It is important to note that the latter chain of operations can be performed analytically. However, the $M \to E$ step of the former chain requires a computational solution of Kepler's equation (4.18). This problem has attracted a great deal of attention since Kepler first wrote down his famous equation, and over a 100 ways of solving it have been presented in the literature. Newton's method of successive approximations is well suited to the task. Given the values of M and e, successive estimates of E can be obtained iteratively using

$$E_{i+1} = E_i - \left(\frac{E_i - e \sin E_i - M}{1 - e \cos E_i} \right), \quad E_0 = M \tag{4.20}$$

Example: Estimate the time for a spacecraft to move through the 60° of true anomaly centred around the apogee point, for an orbit with $a = 24\,000$ km and $e = 0.7$.

Solution: Let perigee be point P_0, the point at true anomaly $150°$ be point P_1 and the point at true anomaly $210°$ be point P_2. We need to find the time of flight from point P_1 to point P_2. Using the chain of operations above for $\theta \to t$, we find the time from perigee to P_1. At point P_1 we have $\theta_1 = 2.6180$ rads, $E_1 = 2.0060$ rads and $M_1 = 1.3713$ rads. This gives $t(P_0 \to P_1) = 8075.61$ s, which by symmetry is identical to $t(P_2 \to P_0)$. Hence the time to traverse the apogee region from P_1 to P_2 is

$$t(P_1 \longrightarrow P_2) = \tau - t(P_2 \longrightarrow P_1) = 37\,002.25\,\text{s} - (2 \times 8075.61\text{s})$$
$$= 20851.03\,\text{s} \approx 5.8\,\text{h}.$$

A similar calculation shows that the spacecraft traverses a similar $60°$ interval of true anomaly centred on perigee in 807.57 s ≈ 0.22 h, which illustrates the significantly different velocity of the vehicle at perigee compared to apogee, for this highly eccentric orbit.

The *velocity relationships* for elliptic orbits may be obtained primarily from equation (4.7) and the *vis-viva* integral (equations 4.5 and 4.11), namely,

$$\frac{1}{2}V^2 - \frac{\mu}{r} = \varepsilon = -\frac{\mu}{2a} \tag{4.21}$$

This equation is also the basis for velocity calculations for all conic section trajectories.

For a *circular orbit* $(e = 0)$, the circular velocity is given by

$$V_{\text{circ}} = \sqrt{(\mu/r)} \tag{4.22}$$

Using this, one can note that for Earth-orbiting space vehicles having a 24-h period, the orbit radius will be 4.2×10^4 km, so the velocity is 3.1 km/s.

For a spacecraft to orbit the Earth at its surface (equatorial radius 6378 km), the velocity needed is 7.91 km/s. This is substantially larger than the Earth's equatorial velocity due to rotation of $\sim$0.46 km/s. Thus this increment in velocity, 7.45 km/s, will be required just to orbit the Earth at zero altitude. This feature of orbit attainment will be developed further in Chapter 7, which deals with launch vehicles.

For *elliptical and circular orbits* $(0 \le e < 1)$, the magnitude of the potential energy per unit mass is greater than the kinetic energy. The total energy ε per unit mass is negative, a consequence of the potential energy zero datum being at infinity. This is a necessary condition for a closed orbit, and may be used to determine planetary capture requirements for interplanetary vehicles.

Parabolic trajectories ($e = 1$)

From equation (4.8) it is apparent that as e tends to unity, the semi-major axis becomes infinitely large; the parabolic trajectory is open to infinity (see Figure 4.7). The distance of its perifocus from the centre of the primary body is obtained from equation (4.7),

$$r_p = p/2 \tag{4.23}$$

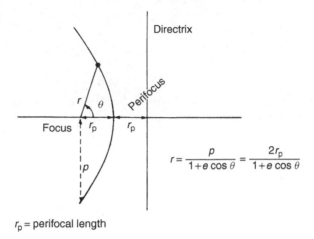

Figure 4.7   Parabola ($e = 1$)

A *position versus time relationship* may be found by noting that $h = r^2\dot{\theta}$ which, together with equation (4.7) yields

$$\dot{\theta} = \sqrt{(\mu/p^3)}(1 + e \cos\theta)^2$$

Whilst this equation is generally valid for all values of eccentricity, direct integration is only straightforward in the case $e = 1$, leading to Barker's equation

$$2\sqrt{(\mu/p^3)}(t - t_p) = 2M = \tan\frac{\theta}{2} + \frac{1}{3}\tan^3\frac{\theta}{2} \qquad (4.24)$$

where the time of perifocal passage t_p is usually taken as zero.

The *energy of a parabolic trajectory* is zero, and as such it represents the boundary case between a captive, orbiting satellite and an escaping one. It is easy to see from equation (4.5) that the speed approaches zero at very large distances from the primary body. Consequently the parabola can be regarded as the minimum energy escape trajectory. Indeed, with $\varepsilon = 0$ in equation (4.5), the minimum speed required for escape is shown to be

$$V_{\text{esc}} = \sqrt{\left(\frac{2\mu}{r}\right)} \qquad (4.25)$$

Hyperbolic trajectories ($e > 1$)

It can be seen for equation (4.8) that the hyperbolic trajectory has a negative semi-major axis, a curious attribute that influences consideration of the hyperbolic motion and geometry. The hyperbola is an open 'orbit' corresponding physically to planetary fly-by trajectories or to escape trajectories when the spacecraft retains some residual speed at a great distance from the primary body.

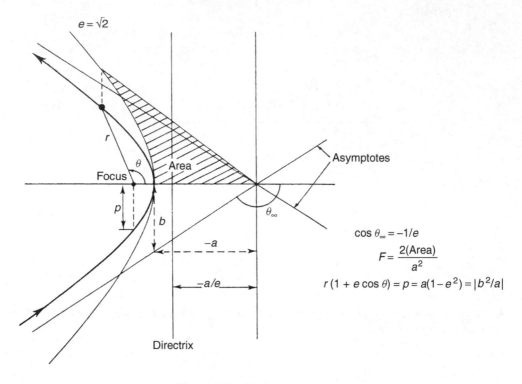

Figure 4.8 Hyperbola $(e > 1)$

To obtain a *position versus time relationship* a method analogous to that applied in the elliptic motion case may be used. A hyperbolic eccentric anomaly F is introduced, the geometrical realization of which is given in terms of the shaded area shown in Figure 4.8. The position on the hyperbola is given in terms of focal distance

$$r = a(1 - e \cosh F),\tag{4.26}$$

and true anomaly θ, which is related to F by

$$\tan \frac{\theta}{2} = \tanh \left(\frac{F}{2} \right) \sqrt{\left(\frac{e + 1}{e - 1} \right)}\tag{4.27}$$

The required relationship between position and time is then found to be analogous to Kepler's equation,

$$M = e \sinh F - F\tag{4.28}$$

where in this case

$$M = n(t - t_p)$$

Here the time of perifocal passage is usually taken as zero (i.e. $t_p = 0$), and the hyperbolic mean motion is defined by

$$n = \sqrt{\mu/(-a)^3}$$

Example: As a result of a swing-by manoeuvre (see Section 5.8.2 of Chapter 5) of Jupiter, a spacecraft is injected into a hyperbolic trajectory (relative to the Sun) to Pluto, with parameters $a = -3.5 \times 10^8$ km and $e = 3.0$. If the heliocentric distances of Jupiter and Pluto are 7.8×10^8 km and 59.0×10^8 km, respectively, produce a first estimate of the transfer time from Jupiter to Pluto, neglecting the effects of the gravitational field of each planet.

Solution: From equations (4.7) and (4.8), the true anomaly values on the hyperbola at Jupiter and Pluto are $\theta_J = 0.5291$ rads and $\theta_P = 1.7468$ rads, respectively. From equation (4.27), the corresponding values of the hyperbolic eccentric anomaly are $F_J = 0.3879$ and $F_P = 2.4698$. Using the hyperbolic analogy of Kepler's equation (4.28), the transit time can be estimated from

$$T_{\text{transit}} = t_P - t_J = \frac{(e \sinh F_P - F_P) - (e \sinh F_J - F_J)}{\sqrt{\mu/(-a)^3}} = \frac{(15.1335) - (0.8053)}{5.5633 \times 10^{-8}}$$

$$= 2.5755 \times 10^8 \, \text{s} \approx 8.2 \, \text{years}.$$

For hyperbolic trajectories, the energy ε is positive, and so the motion is dominated by kinetic energy. As r becomes very large ($r \to \infty$), equation (4.21) shows that the speed becomes

$$V_\infty = \sqrt{\left(\frac{-\mu}{a}\right)} \tag{4.29}$$

V_∞ is called the *hyperbolic excess velocity*.

The asymptotic direction along which the spacecraft escapes is obtained from equation (4.7) and Figure 4.8 as

$$\theta_\infty = \cos^{-1}\left(-\frac{1}{e}\right) \tag{4.30}$$

and the angle δ through which the trajectory is deflected by the encounter is given by

$$\delta = 2 \sin^{-1}\left(\frac{1}{e}\right) \tag{4.31}$$

To determine these geometrical aspects, the eccentricity is required. This may be related to the physical motion [1] by combining equations (4.9) and (4.29). This gives

$$e = 1 + \frac{r_p V_\infty^2}{\mu} \tag{4.32}$$

4.3 SPECIFYING THE ORBIT

The orbit equation (4.4) is equivalent to a sixth-order scalar differential equation that requires six initial conditions in order to determine the six constants of integration. At orbit injection these will be the three components of position and of velocity. A standard way of specifying an orbit is to use *orbital elements*, which refer the orbit to a frame of reference that is fixed relative to the stars. This is used both for astronomy and for satellites.

The *frame of reference* that is commonly used can be defined in terms of X-, Y-, and Z-axes (see Figure 4.9). The X- and Y-axes lie in Earth's equatorial plane and the Earth spins about the Z-axis.

The X-axis is in a direction from Earth to the Sun at the vernal equinox (~21 March). The direction thus indicated is termed the *first point of Aries*. Some 2000 years ago this direction did point towards the constellation of Aries, but at present it points to Aquarius, and is moving along the equator at about 0.8′ per year.

The Z-axis is along the Earth's spin axis, in the northerly direction. It is at an angle of 23°27′8″ to the normal of the ecliptic plane, changing slowly with respect to the stars, with a period of about 25 725 years. The overall precessional rate is ~0.12″ per year and may be neglected for most spacecraft applications.

The Y-axis makes up a 'right-handed' orthogonal set with the X-, Z-axes.

The *celestial sphere* is the name given to a sphere with infinite radius centred on the Earth. The Z-axis meets it at a point known as the North celestial pole. The motion of the Sun on the celestial sphere is indicated in Figure 4.9. It moves in the ecliptic plane, tilted about the X-axis through 23°27′8″, known as the obliquity of the ecliptic, ε.

The location of a point P on the celestial sphere, strictly a direction OP, can be expressed in terms of the two angles α and δ. The great circle through P and the North

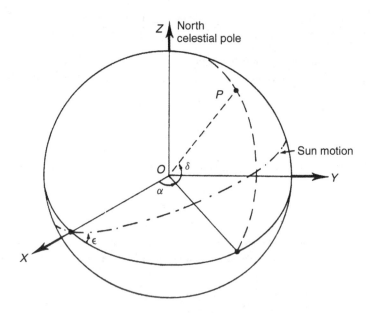

Figure 4.9 Celestial sphere

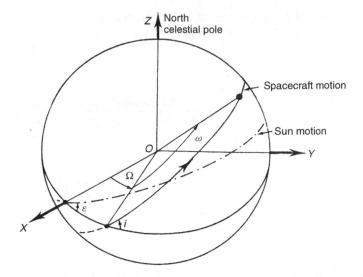

Figure 4.10 Spacecraft orbital elements

celestial pole is called the hour circle of P. The angle α between OX and the equatorial radius to the hour circle is called the right ascension of P. Its declination δ is the angle between OP and the equatorial plane (see Figure 4.9).

The *six orbital elements* of a spacecraft's orbit round the Earth are chosen to represent different features. The orbital plane in Figure 4.10 intersects the equatorial plane in a line called the line of nodes; the ascending node is the point on the equator at which the spacecraft moves from the southern into the northern hemisphere. The right ascension of this node together with the inclination angle i define the plane of the orbit. The orientation within the plane is defined by the angle ω, known as the argument of periapsis, which is the angle measured in the direction of motion between the line of nodes and the vector **e**, which points to the periapsis. The shape of the orbit is defined by its eccentricity e, and its size by its semi-major axis a. The sixth element defines the position of the spacecraft in its orbit. The time (or epoch) of last passage through the periapsis may be used, or the mean anomaly M (equation 4.19).

4.4 ORBIT PERTURBATIONS

At the distances of orbiting spacecraft from the Earth its asphericity and non-uniform mass distribution result in its gravitational potential departing from the simple $1/r$ function, which was assumed in Section 4.2. Equation (4.1) is not valid in this situation and the equation of motion (4.4) must be modified to take account of the Earth's gravitational field.

There are additional forces that act on space vehicles, which were not included in the Keplerian formulation. Some are from additional masses that provide secondary gravitational fields; for Earth orbit, the Moon and the Sun provide such forces. Also, at low altitudes (typically at less than 1000 km altitude) the Earth's atmosphere imposes a drag force. Table 4.2 lists the major perturbing forces, and their relative importance, for space vehicle orbital analysis.

Table 4.2   Magnitude of disturbing accelerations
acting on a space vehicle whose area-to-mass ratio is
A/M. Note that A is the projected area perpendicular to
the direction of motion for air drag, and perpendicular
to the Sun for radiation pressure

Source	Acceleration (m/s^2)	
	500 km	Geostationary orbit
Air drag*	$6 \times 10^{-5}A/M$	–
Radiation pressure	$4.7 \times 10^{-6}A/M$	$4.7 \times 10^{-6}A/M$
Sun (mean)	5.6×10^{-7}	3.5×10^{-6}
Moon (mean)	1.2×10^{-6}	7.3×10^{-6}
Jupiter (max.)	8.5×10^{-12}	5.2×10^{-11}

*Dependent on the level of solar activity

The equation of motion for a space vehicle about a body taking into account perturbative influences may be written in the form

$$\ddot{\mathbf{r}} = -\nabla U + \mathbf{b} \tag{4.33}$$

where U is the gravitational potential field and $\mathbf{b}$ is the force vector per unit mass due to other sources of perturbation, to which the vehicle is subject. A general closed solution is not possible, but there are a variety of solution methods that are appropriate for spacecraft dynamics. The 'variation of orbital elements' method is described here. Other methods such as those first proposed by Cowell and Crommelin [2] and Encke [3] are summarized by Cornelisse et al. [4].

The method of the variation of orbital elements may be considered in the following way. The elements referred to in the preceding section are constants for a Keplerian orbit, as derived in Section 4.2. When perturbative forces exist, they are no longer constant but for small forces they will change slowly.

A simple, physical model will serve to demonstrate this. Consider a spacecraft in circular orbit about a spherically symmetrical planet possessing an atmosphere of density ρ kg/m^3. If it is assumed that the perturbative drag force is small, then it is to be expected that the orbit will remain near circular. Now the velocity in a circular Keplerian orbit is given by $\sqrt{(\mu/r)}$. If the spacecraft's projected area in the direction of flight is S, then the work performed by the atmosphere as the vehicle moves round the orbit is given by $\sim -\pi r \rho S C_D \mu/r$, where C_D is an appropriate drag coefficient for the vehicle. Since this reduces the energy of the system, it is apparent that the energy constant ε in equation (4.21) will decrease and the orbital element a must do so too.

The variation of orbital elements method assumes that the actual orbit of a body, at any given instant, may be considered to have instantaneous values of Keplerian orbital elements. These are defined so that if the perturbing forces are removed at that instant, thus leaving only a central gravitational field whose potential is proportional to $1/r$, then the orbit will follow the Keplerian orbit that has the instantaneous orbital elements. These are called the *osculating elements*. It must be emphasized that the method is only appropriate for perturbing forces having a magnitude significantly smaller than μ/r^2.

The normal method for so describing an orbit is with recourse to Lagrange's planetary equations (see e.g. Reference [5]). One Gaussian form of these is the following [6]:

$$
\left.
\begin{aligned}
\frac{da}{d\theta} &= \frac{2pr^2}{\mu(1-e^2)^2}\left\{e\sin\theta S + \frac{p}{r}T\right\} \\
\frac{de}{d\theta} &= \frac{r^2}{\mu}\left\{\sin\theta S + \left(1+\frac{r}{p}\right)\cos\theta T + e\frac{r}{p}T\right\} \\
\frac{di}{d\theta} &= \frac{r^3}{\mu p}\cos(\theta+\omega)W \\
\frac{d\Omega}{d\theta} &= \frac{r^3\sin(\theta+\omega)}{\mu p\sin i}W \\
\frac{d\omega}{d\theta} &= \frac{r^2}{\mu e}\left\{-\cos\theta S + \left(1+\frac{r}{p}\right)\sin\theta T\right\} - \cos i\frac{d\Omega}{d\theta} \\
\frac{dt}{d\theta} &= \frac{r^2}{\sqrt{(\mu p)}}\left\{1+\frac{r^2}{\mu e}[\cos\theta S - \left(1+\frac{r}{p}\right)\sin\theta T]\right\}
\end{aligned}
\right\}
\qquad (4.34)
$$

where S, T, W form a triad of perturbing accelerations in a spacecraft-centred coordinate reference frame, S acting radially, T transverse to S in the orbital plane and directed positively in the sense of the spacecraft motion, and W normal to the orbit plane giving a right-handed system of accelerations.

It should be noted that the particular form of Lagrange's equations adopted depends on the type of orbit under investigation. For example, the set defined in equations (4.34) fails for circular ($e = 0$) and equatorial ($i = 0$) orbits. These may be dealt with by introducing parameter transformations of the type given, for example, by Roy [7].

Solutions of equations (4.34) must in general be performed using numerical techniques. This requires the formulation of functions for U and **b**, and these are given below for certain cases of particular interest.

4.4.1 Gravitational potential of the Earth

The most convenient method for describing Earth's gravitational field outside its surface is to use a spherical harmonic expansion [8], given by

$$
U(r, \Phi, \Lambda) = \frac{\mu}{r}\left\{-1 + \sum_{n=2}^{\infty}\left[\left(\frac{R_E}{r}\right)^n J_n P_{n0}(\cos\Phi)\right.\right.
$$

$$
\left.\left. + \sum_{m=1}^{n}\left(\frac{R_E}{r}\right)^n (C_{nm}\cos m\Lambda + S_{nm}\sin m\Lambda)P_{nm}(\cos\Phi)\right]\right\}
\qquad (4.35)
$$

where $U(r, \Phi, \Lambda)$ is the gravitational potential at a distance r from the centre of the Earth and Φ, Λ are the latitude and longitude. P_{nm} are Legendre polynomials. J_n, C_{nm} and S_{nm} are numerical values dependent on the mass distribution of the body, in this case the Earth. Terms of the form J_n are called *zonal harmonic coefficients*; they reflect the mass distribution of the Earth independently of longitude. C_{nm} and S_{nm} are the Earth's *tesseral harmonic coefficients* for $n \neq m$ and the *sectoral harmonic coefficients* for $n = m$.

CELESTIAL MECHANICS

These coefficients have mainly been determined from the motion of Earth-orbiting spacecraft. Whilst the lower-order terms were determined during the early 1960s, determination of the Earth's gravitational field continues to be an area of active research. Consequently there is a plethora of 'standard' global gravity field models, for example, the Joint Gravity Model (JGM) series [9], of which the JGM-3 model is an example. This gives the harmonic coefficient values to degree and order 70.

One of the major problems in determination of the higher-order terms is due to their rapid decrease in magnitude with altitude; from equation (4.35) terms decrease with $(R_E/r)^n$. However, at low altitudes there are also difficulties, since the gravity effects are difficult to separate from other perturbations, in particular, those due to variable air drag. This situation is greatly improved by the launch of dedicated spacecraft missions to determine the higher-order harmonic coefficients in the Earth's gravity field. Promising proposals, such as the US GRAVSAT and the European ARISTOTELES missions, were stalled in the 1990s because of fiscal problems. However, dedicated in-orbit gravity field determination got underway with the launch of the CHAMP spacecraft in July 2000, and the twin GRACE satellites in March 2002. Further significant advancement in gravity field modelling is in prospect with the launch of the ESA GOCE spacecraft in March 2009. This has been launched into a very low 250 km altitude circular orbit to enhance the sensitivity of the measurement of the Earth's field. Premature re-entry of the spacecraft is avoided by the use of ion propulsion to compensate for the atmospheric drag perturbations.

Table 4.3 gives the magnitude of some of the lower-order coefficients taken from a current geoid model [9]. From this it is apparent that the term J_2 is some three orders of magnitude larger than the others, and to a first approximation it dominates the gravitational perturbative influences of the Earth. This term represents the polar flattening of the Earth (or equatorial bulge), and its magnitude can be summarized by noting that the Earth's poles are approximately 21 km closer to the Earth's centre than the equator. This 'excess' gravitational mass in the equatorial regions influences the motion of an orbiting spacecraft principally in two ways—*the regression of the line of nodes and the precession of the line of apsides*. Before discussing these, it is worth commenting briefly on the effects of the smaller tesseral harmonic coefficients.

As can be seen from equation (4.35), the C_{nm} and S_{nm} coefficients are associated with the longitudinal variation in the geopotential surface. These variations are generally smaller than those due to the dominant zonal harmonic terms, producing variations of the order of a few tens of metres in the height of the geopotential surface as measured above a reference spheroid. These small longitudinal variations tend to be 'averaged-out' by the effect of the Earth's rotation beneath the orbital plane of a low Earth-orbiting satellite. The resulting perturbing effect on the satellite's motion is small and periodic, generally producing a negligible net effect on each orbit. Their short-periodic effect must, however,

Table 4.3 Magnitude of low-order J, C and S values for Earth

J_2	1082.6×10^{-6}	C_{21}	0	S_{21}	0
J_3	-2.53×10^{-6}	C_{22}	1.57×10^{-6}	S_{22}	-0.90×10^{-6}
J_4	-1.62×10^{-6}	C_{31}	2.19×10^{-6}	S_{31}	0.27×10^{-6}
J_5	-0.23×10^{-6}	C_{32}	0.31×10^{-6}	S_{32}	-0.21×10^{-6}
J_6	0.54×10^{-6}	C_{33}	0.10×10^{-6}	S_{33}	0.20×10^{-6}

be taken into account for spacecraft missions requiring very precise orbit determination (e.g. geodetic missions).

Tesseral harmonics can, however, produce moderate changes in the satellite's orbit if the orbit is in a *resonant state* with respect to the Earth's gravitational field. This occurs when the orbit period is such that the ground track of the satellite repeats precisely after a number of orbits and days. The orbit perturbations in this case are small but secular, causing a build-up of orbit changes while the satellite remains in the resonant state. It should be emphasized, though, that the resonant state is the exception rather than the rule, and generally the effect of the tesseral harmonic component is small for low Earth-orbiting satellites. A spacecraft in a Geostationary Earth orbit (GEO) encounters a form of resonance with respect to the lowest-order sectoral harmonic coefficients ($n = m = 2$), and this is often referred to as *triaxiality*. This is also discussed below.

Regression of the line of nodes

The equatorial bulge produces a torque that rotates the angular momentum vector. For prograde orbits ($i < 90°$), the orbit rotates in a westerly direction, leading to a regression of the line of nodes as shown pictorially in Figure 4.11. Neglecting all harmonic coefficients other than J_2, the rate of nodal regression may be written [7] to the first order in J_2 as

$$\overline{\Omega} = \Omega_0 - \frac{3}{2}\frac{J_2 R_E^2}{p^2}\overline{n}t \cos i + O[J_2^2] \tag{4.36}$$

where $\overline{n}$ is the mean angular velocity, $\sqrt{(\mu/a^3)}$. Thus, the secular rate of nodal regression per orbit is

$$\Delta\Omega = -\frac{3\pi J_2 R_E^2}{p^2} \cos i \quad \text{rad/rev} \tag{4.37}$$

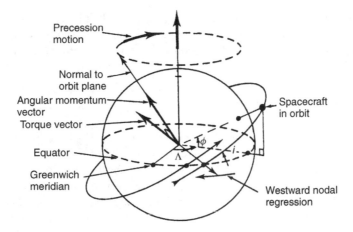

Figure 4.11 Nodal regression

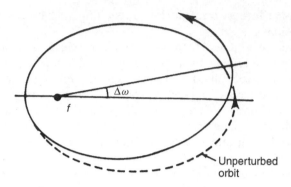

Figure 4.12 Apsidal precession

Precession of the line of apsides

The second influence of the equatorial bulge may be considered in the following way. Since the mass 'seen' by the spacecraft crossing the equator is greater than the mean mass, the orbit tends to curve more rapidly. Since the gravitational field of the Earth is conservative, however, this leads to an overall rotation of the orbit, *within* the orbit plane, as demonstrated in Figure 4.12. This motion implies rotation of the semi-major axis and is termed precession of the line of apsides. The secular effect is given by

$$\bar{\omega} = \omega_0 + \frac{3}{2}\frac{J_2 R_E^2}{p^2}\bar{n}\left(2 - \frac{5}{2}\sin^2 i\right)t + O[J_2^2] \tag{4.38}$$

or per orbit

$$\Delta\omega = 3\pi\frac{J_2 R_E^2}{p^2}\left(2 - \frac{5}{2}\sin^2 i\right) \quad \text{rad/rev} \tag{4.39}$$

At an inclination of $\sim$63.4° the precession is zero. In the 1960s, the Soviet Union devised a highly eccentric orbit with this inclination, called a *Molniya* orbit. The resulting 'frozen apogee' condition was used to good effect for high-latitude communications, as discussed in Chapter 5.

Differential evolutionary rates

Each of the rates of orbital evolution noted in equations (4.36) to (4.39) indicate the absolute changes in the orbital elements relative to an inertial frame. As such, it is clear from the numerical values given in Table 4.3 why it is that the J_2 term dominates the evolution of an orbit (see also Figure 4.15). There are, however, several important situations when it is the *differential rate of evolution* of two or more orbits that should be considered. Cases to note are, for example, a satellite constellation requiring orbiting vehicles to maintain a fixed relative position, or where satellites are required to co-orbit in formation. A third example is the differential evolution of a space debris cloud [10].

Expressions for the evolution of an orbit due to higher harmonic terms can be found in Reference [11]. These may be partially differentiated to find the relative evolutionary rate of two orbits as a function of changes in each of the orbit elements, the semi-major axis, eccentricity and so on. If the semi-major axis is varied, then the differential rates of evolution in right ascension of the ascending node and the argument of perifocus may be written in the form

$$\frac{\partial \Omega_n}{\partial a} = \Delta \Omega_n \quad \text{and} \quad \frac{\partial \omega_n}{\partial a} = \Delta \omega_n$$

where for the second, third and fourth harmonic terms we have

$$\frac{\partial \Omega_2}{\partial t} = -5.25 a^{-9/2} (1 - e^2)^{-2} \mu^{1/2} J_2 R_E^2 \cos i$$

$$\frac{\partial \Delta \Omega_3}{\partial t} = -1.6875 a^{-11/2} e (1 - e^2)^{-3} \mu^{1/2} J_3 R_E^3 \cot i \, (15 \sin^2 i - 4) \sin \omega$$

$$\frac{\partial \Delta \Omega_4}{\partial t} = 1.1786 a^{-13/2} (1 - e^2)^{-4} \mu^{1/2} J_4 R_E^4$$

$$\times \cos i [(1 + 1.5 e^2)(7 \cos^2 i - 3) - (7 \cos^2 i - 4) e^2 \cos 2\omega]$$

$$\frac{\partial \Delta \omega_2}{\partial t} = -5.25 a^{-9/2} (1 - e^2)^{-2} \mu^{1/2} J_2 R_E^2 (2 - 2.5 \sin^2 i)$$

$$\frac{\partial \Delta \omega_3}{\partial t} = 1.6875 a^{-11/2} e^{-1} (1 - e^2)^{-3} \mu^{1/2} J_3 R_E^3$$

$$\times \sin i [5 \cos^2 i - 1 + e^2 (35 \cos^2 i - 4 \operatorname{cosec}^2 i)] \sin \omega$$

$$\frac{\partial \Delta \omega_4}{\partial t} = -2.5781 a^{-13/2} (1 - e^2)^{-4} \mu^{1/2} J_4 R_E^4$$

$$\times [\{(49 \cos^4 i - 36 \cos^2 i + 3) + 0.75 e^2 (63 \cos^4 i - 42 \cos^2 i + 3)\}$$

$$+ \{\sin^2 i (7 \cos^2 i - 1) - 0.5 e^2 (63 \cos^4 i - 56 \cos^2 i + 5)\} \cos 2\omega]$$

Using these relationships it has been found that under some circumstances the differential evolution of the argument of perifocus due to J_3 can be more significant than that due to J_2.

Triaxiality perturbation

The terms representing the longitudinal variation of the Earth's gravitational field have their most significant influence on geostationary satellites, since non-synchronous orbits will average them out. The term J_{22} equal to $(C_{22}^2 + S_{22}^2)^{1/2}$ has the greatest influence due to the $(R_E/r)^n$ weighting of coefficients. This term represents the slight ellipticity of the Earth's equatorial cross-section, having its major axis aligned approximately along $15\,°\text{W} - 165\,°\text{E}$.

The term 'triaxiality' is derived from the geometrical form of the Earth resulting from the combination of the oblateness and J_{22} characteristics. To lowest order, the Earth may be modelled as an oblate spheroid with an elliptical equatorial cross-section. This form is

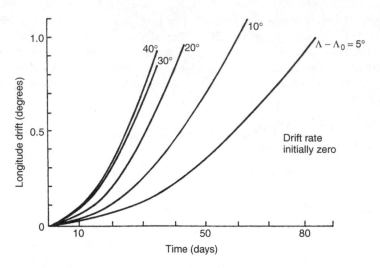

Figure 4.13 Longitudinal drift acquired over a period in geostationary orbit, as a function of the difference ($\Lambda - \Lambda_0$) in initial operating longitude and minor axis (stable) longitude

generated by three mutually orthogonal axes (tri-axes) centred on the Earth's centre, each corresponding to a different Earth radius. A satellite will be stable if it is at a minimum of the Earth's potential field, that is, on the minor axis of the equator. Departure from these two longitude positions provides an increasing perturbation. Agrawal [12] gives the following expression for the longitudinal acceleration:

$$\ddot{\Lambda} = k^2 \sin 2(\Lambda - \Lambda_0) \tag{4.40}$$

where $k^2 = -18J_{22}\omega_e^2(R_E^2/a_s^2) \approx -1.7 \times 10^{-3}$ deg/day^2 and $\Lambda_0 = 75°$ E or $255°$ E (stable longitudes). Here ω_e is the sidereal rotation rate of the Earth and a_s is the semi-major axis of the synchronous orbit.

Longitude drift over time is shown in Figure 4.13.

Triaxiality can be seen to provide an East/West station-keeping problem for a geostationary satellite. This is referred to in Chapter 5.

4.4.2 Atmospheric drag

For low Earth-orbiting spacecraft, the perturbation due to the atmosphere cannot be neglected. Atmospheric effects lead to a force that may be expressed as two orthogonal components—drag along the direction of travel and lift at right angles to it. The simplest form in which to write the drag force is

$$\mathbf{F}_D = \frac{1}{2}\rho S C_D V_r^2 \left(\frac{-\mathbf{V}_r}{|V_r|}\right) \tag{4.41}$$

where $\mathbf{V}_r$ is the velocity vector of the spacecraft relative to the atmosphere, ρ is the atmospheric density, S a reference area for the vehicle and C_D the vehicle's coefficient of drag referred to the reference area.

The lifting force is obtained by replacing C_D with the coefficient of lift C_L, in equation (4.41), and by noting that the lift force acts perpendicular to the velocity vector $\mathbf{V}_r$. For most spacecraft the effects of lift are negligible compared with drag, but when precise orbit determination is required ($\sim$ few metres accuracy) its influence must be included [13].

The drag coefficient for space vehicles is somewhat complex due to the type of flow regime within which the motion takes place. Because of the large mean free path, the flow cannot be treated in a continuum manner. Indeed, the most suitable description is that of free molecular flow wherein the molecular description is such that molecules reflected from the spacecraft surfaces do not interact further with the flow field; no shock wave is therefore formed about a body moving through the atmosphere at orbit altitudes. The drag force experienced is dependent on the gas—surface interaction, for which there is only sparse experimental data at typical incident velocities. Stalder and Zurick [14], Schaff and Chambre [15] and Schamberg [16] provide the basic theory for rarefied flow, and Cook [17] provides height-related C_D data adopting the Schamberg model. Typically values of $C_D \sim 2.5$ are predicted by these theories, when the reference area is taken as the spacecraft's cross-sectional area projected normal to the velocity vector.

With reference to equation (4.41), drag is most significant at perigee where both the velocity and density are greatest. King-Hele [5] provides detailed analysis of the secular changes caused by drag. To first order, these approximate to an impulsive, negative velocity increment occurring at perigee. This will lead to a reduction in the semi-major axis of an elliptical orbit. For a circular orbit, drag will occur continuously around the orbit. Assuming that the change in the radius of such an orbit is small, the $\Delta\tau$ in orbit period for a circular orbit of radius r will be given by

$$\frac{\Delta\tau}{\tau} \approx -3\pi\rho r(SC_D/M) \tag{4.42}$$

where ρ is the density at r (measured from the centre of the Earth) and M/SC_D is called the vehicle ballistic parameter; M is the vehicle mass, and S the projected area.

The *dominant influences of drag* are thus to cause orbit contraction and circularization, with eventual re-entry. Since the Earth's atmosphere rotates approximately synchronously with the Earth, in general the drag force has a component perpendicular to the orbit plane. This results in a change in inclination of the orbit. King-Hele [5] gives expressions for the radial, transverse and normal acceleration components for a vehicle moving through an atmosphere rotating at an angular rate α rad/s. These are

$$S = -\frac{1}{2}\rho\upsilon\delta\left(\frac{\mu}{pF}\right)^{1/2} e\,\sin\theta$$

$$T = -\frac{1}{2}\rho\upsilon\delta\left(\frac{\mu}{pF}\right)^{1/2}\left(1 + e\,\cos\theta - r\alpha\left(\frac{p}{\mu}\right)^{1/2}\cos i\right) \tag{4.43}$$

$$W = -\frac{1}{2}\rho\upsilon\delta\frac{r\alpha}{\sqrt{F}}\,\sin i\,\cos(\theta + \omega)$$

where

$$F = \left(1 - \frac{r_p\alpha}{\upsilon_p}\cos i\right)^2$$

and

$$p = a(1 - e^2)$$

Here, r_p, υ_p represent conditions at perigee, υ is the absolute velocity of the vehicle and δ is a modified ballistic parameter given by $\delta = FSC_D/M$.

4.4.3 Additional gravitational fields

Luni-solar perturbations

Other bodies in the solar system impose additional gravitational forces on spacecraft orbiting the Earth. The proximity and mass of the Moon provides the most significant influence. The Sun is a great deal further away but its larger mass generates an influence that is of a similar order of magnitude. These perturbations are collectively termed *luni-solar perturbations*. Since in general these bodies will not lie in the same plane as the vehicle orbit, their most significant influence will be to change the inclination of the orbit with respect to the equator. The formulation of this three-body interaction does not admit a general closed-form solution, and numerical techniques must be employed.

The disturbing acceleration a_d of a satellite due to a disturbing body having a mass M_d and gravitational parameter μ_d is given by

$$a_d = \mu_d \sqrt{(\mathbf{R} \cdot \mathbf{R})} \tag{4.44}$$

where

$$\mathbf{R} = \frac{\mathbf{r}_{sd}}{r_{sd}^3} - \frac{\mathbf{r}_d}{r_d^3}$$

and $\mathbf{r}_{sd}$ and $\mathbf{r}_d$ are defined in Figure 4.14.

It can then be shown [4] that the maximum value of the ratio of the disturbing acceleration a_d to the central acceleration a_c is given by

$$\frac{a_d}{a_c} = \frac{M_d}{M_c} \left(\frac{r_s}{r_d}\right)^3 \sqrt{(1 + 3\cos^2 \beta)} \tag{4.45}$$

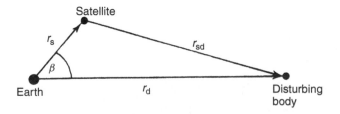

Figure 4.14 Disturbing body and satellite positions

or

$$\frac{a_d}{a_c} \leq 2 \frac{M_d}{M_c} \left(\frac{r_s}{r_d}\right)^3$$

The angle β is shown in Figure 4.14; M_c is the mass of the central body about which the vehicle is in orbit.

At geostationary orbit this ratio provides values of 3.3×10^{-5} and 1.6×10^{-5} for the Moon and Sun, respectively. The effect of Jupiter is some five orders of magnitude lower than these.

Expressions that provide approximate average rates of change of orbital elements for a single disturbing body, given by Cook [18], are

$$\left.\begin{array}{l}
\dfrac{da}{dt} \approx 0 \\[2mm]
\dfrac{de}{dt} \approx \dfrac{-15}{2}\dfrac{K}{n}e(1-e^2)^{1/2}\left[AB\,\cos 2\omega - \dfrac{1}{2}(A^2 - B^2)\sin 2\omega\right] \\[3mm]
\dfrac{d\Omega}{dt} \approx \dfrac{3KC}{4n(1-e^2)^{1/2}\sin i}[5Ae^2\sin 2\omega + B(2 + 3e^2 - 5e^2\cos 2\omega)] \\[3mm]
\dfrac{d\omega}{dt} + \dfrac{d\Omega}{dt}\cos i = \dfrac{3}{2}\dfrac{K(1-e^2)^{1/2}}{n}\left[5\left\{AB\sin 2\omega + \dfrac{1}{2}(A^2 - B^2)\cos 2\omega\right\}\right. \\[3mm]
\qquad \left. -1 + \dfrac{3}{2}(A^2 + B^2) + \dfrac{5a}{2er_d}\left\{1 - \dfrac{5}{4}(A^2 + B^2)\right\}(A\cos\omega + B\sin\omega)\right] \\[3mm]
\dfrac{di}{dt} = \dfrac{3KC}{4n(1-e^2)^{1/2}}[A(2 + 3e^2 + 5e^2\cos 2\omega) + 5Be^2\sin 2\omega]
\end{array}\right\} \quad (4.46)$$

where

$$K = \frac{GM_d}{r_d^3}$$

and

$$A = \cos(\Omega - \Omega_d)\cos u_d + \cos i_d \sin u_d \sin(\Omega - \Omega_d)$$
$$B = \cos i[-\sin(\Omega - \Omega_d)\cos u_d + \cos i_d \sin u_d \cos(\Omega - \Omega_d)] + \sin i \sin i_d \sin u_d$$
$$C = \sin i[\cos u_d \sin(\Omega - \Omega_d) - \cos i_d \sin u_d \cos(\Omega - \Omega_d)] + \cos i \sin i_d \sin u_d$$

Angles $\Omega_d, u_d = (\theta_d + \omega_d)$ and i_d are the orbital elements of the disturbing body referred to the equatorial-based system of Section 4.3.

4.4.4 Solar radiation pressure (SRP) perturbations

A spacecraft moving within the solar system will experience, to some degree, a perturbation to its trajectory owing to the incidence of solar radiation upon its illuminated surfaces.

Electromagnetic radiation carries momentum, and the reflection of incident radiation at a surface represents an exchange of momentum. Consequently the radiation exerts a small but measurable pressure on the spacecraft of mean magnitude

$$P = F_\odot/c$$

where $F_\odot$ is the solar energy flux at the spacecraft, and c is the speed of light. For a spacecraft in near-Earth orbit, where $F_\odot = 1400\,\text{W/m}^2$, the resulting mean SRP is approximately $P_E = 4.7 \times 10^{-6}\text{N/m}^2$. The effect of this pressure as a force acting on the vehicle is dependent upon the reflective characteristics of the spacecraft surfaces.

An Earth-orbiting spacecraft is also immersed in other electromagnetic radiation fields, such as those due to Earth albedo and infra-red emission, which similarly produce a perturbing force on the vehicle. However, these are generally subordinate in magnitude to the dominate solar radiation effect.

The perturbing effect of SRP on a spacecraft's trajectory is directly dependent upon the vehicle's area-to-mass ratio A/m, and inversely proportional to the square of its distance from the Sun. Hence, the magnitude of the SRP force per unit spacecraft mass (the disturbing acceleration) along the Sun—spacecraft line may be expressed as

$$f_{\text{SRP}} = s\frac{A}{m}P_E \left(\frac{a_\odot}{r_\odot}\right)^2 \tag{4.47}$$

where $r_\odot$ and $a_\odot$ are the spacecraft's distance and the Earth's mean distance from the Sun, respectively; s is a constant whose value, between 0 and 2, depends upon the reflective properties of the spacecraft's surface.

To estimate how the orbit elements are perturbed by SRP, Lagrange's planetary equations (4.34) may be used, with the triad of disturbing accelerations given by [19]:

$$S = f_{\text{SRP}}S_0(\theta), \quad T = f_{\text{SRP}}T_0(\theta), \quad W = f_{\text{SRP}}W_0 \tag{4.48}$$

where $S_0(\theta), T_0(\theta)$ and W_0 are direction cosines,

$$
\begin{aligned}
\begin{Bmatrix} S_0(\theta) \\ T_0(\theta) \end{Bmatrix} =&\; -\cos^2\frac{i}{2}\cos^2\frac{\varepsilon}{2}\begin{Bmatrix} \cos \\ \sin \end{Bmatrix}(\lambda_\odot - u - \Omega) \\
&- \sin^2\frac{i}{2}\sin^2\frac{\varepsilon}{2}\begin{Bmatrix} \cos \\ \sin \end{Bmatrix}(\lambda_\odot - u + \Omega) \\
&- \frac{1}{2}\sin i \sin\varepsilon\left[\begin{Bmatrix} \cos \\ \sin \end{Bmatrix}(\lambda_\odot - u) - \begin{Bmatrix} \cos \\ \sin \end{Bmatrix}(-\lambda_\odot - u)\right] \\
&- \sin^2\frac{i}{2}\cos^2\frac{\varepsilon}{2}\begin{Bmatrix} \cos \\ \sin \end{Bmatrix}(-\lambda_\odot - u + \Omega) \\
&- \cos^2\frac{i}{2}\sin^2\frac{\varepsilon}{2}\begin{Bmatrix} \cos \\ \sin \end{Bmatrix}(-\lambda_\odot - u - \Omega), \tag{4.49}
\end{aligned}
$$

$$
\begin{aligned}
W_0 =&\; \sin i \cos^2\frac{\varepsilon}{2}\sin(\lambda_\odot - \Omega) - \sin i \sin^2\frac{\varepsilon}{2}\sin(\lambda_\odot + \Omega) \\
&- \cos i \sin\varepsilon \sin\lambda_\odot. \tag{4.50}
\end{aligned}
$$

Here $u = \theta + \omega, \varepsilon$ is the obliquity of the ecliptic, and $\lambda_\odot$ is the ecliptic longitude of the Sun.

A method of integrating these equations was developed by Harwood and Swinerd [20], who applied the theory to the satellites Explorer 19 and Lageos [21].

Communication spacecraft in GEO, many of which have large solar array surfaces, experience SRP perturbations to their orbits. Generally, this leads to an increase in the eccentricity of the orbit, which has implications for station-keeping activities (see Chapter 5). For Low Earth Orbit (LEO) spacecraft, below around 600 km in altitude, the effects of air drag, however, dominate those of radiation pressure.

4.4.5 Concluding remarks

Figure 4.15 illustrates the relative magnitude of the main sources of perturbation acting upon an Earth-orbiting spacecraft. For each effect, the logarithm of the disturbing acceleration, normalized to 1 g, is shown as a function of altitude.

From the diagram it is apparent that the primary inverse square law gravity field of the Earth is very much the dominant influence. The only effect to compete with this is aerodynamic drag, the drag curve reaching the 1 g level at low altitude around 80 km. In this case the spacecraft encounters an atmospheric re-entry situation, when the magnitude of inertial accelerations due to aerodynamic effects can equal and exceed the gravitational acceleration (see Chapter 5).

After primary gravity, the zonal harmonic potential terms begin to play a role. The J_2, or Earth oblateness, effects are around three orders of magnitude smaller than primary gravity, but nevertheless lead to significant perturbations in the orbit perigee and nodal positions as we have seen in Section 4.4.1. These effects must be modelled accurately

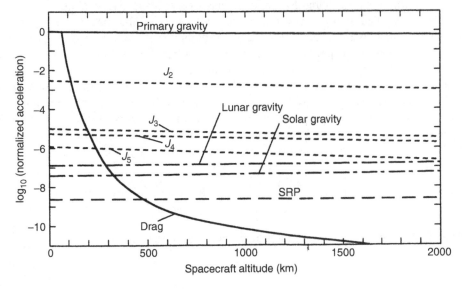

Figure 4.15 Comparisons of the disturbing accelerations for the main sources of perturbation

when planning LEO operations, for example, rendezvous strategies. The higher-order zonal harmonics are of lesser importance, but still lead to significant long-period perturbations in the orbit elements. These higher-order terms decrease rapidly in magnitude with altitude, and are comparable to the third-body forces of lunar and solar gravity.

The surface forces, drag and SRP, are dependent upon the spacecraft area-to-mass ratio, and a representative value of $A/m = 0.005$ m^2/kg was used for the purpose of compiling Figure 4.15. Furthermore, the drag curve is dependent upon the level of solar activity. Here a moderate level was chosen, although it is worth noting that at, say, 500 km altitude the drag acceleration can be an order of magnitude higher at solar maximum than at solar minimum.

The SRP curve is effectively independent of altitude, since it is a function of distance from the Sun, rather than from the Earth. The heliocentric distance of an Earth-orbiting spacecraft obviously does vary, but the effect on the disturbing acceleration is essentially second order. It is interesting to note the altitude at which the SRP and drag curves intersect. Figure 4.15 suggests that above 600 to 700 km altitude SRP dominates. To try to estimate more precisely the altitude at which this transition occurs is difficult since it is dependent upon the level of solar activity.

4.5 RESTRICTED THREE-BODY PROBLEM

The *three-body problem* refers to the problem of predicting the motion of three bodies, each influenced by the gravitational fields of the others. There is no general solution to the equations governing this. Of particular relevance to spacecraft is the *circular restricted* form of this problem, in which two of the bodies have masses that far outweigh the mass of the third, and are in circular orbits about each other.

An important example of this is when a spacecraft moves within the region of space that is dominated by both Earth and Moon; their masses are, respectively, about 6×10^{24} and 7×10^{22} kg. Another example is a rendezvous with a comet that has a mass distribution like a dumbbell. In both of these cases, the two dominant masses move in approximately circular orbits about their joint centre-of-mass—their barycentre—both lying on the same diameter, which rotates at a constant angular rate Ω.

Using a frame of reference (f.o.r.) that is centred on the barycentre and such that the two dominant masses are at fixed positions in it, then this f.o.r. is not an inertial one; it is rotating at a fixed rate Ω. A small mass m, at position $\mathbf{r}$ relative to the barycentre, and at $\mathbf{r}_1$ and $\mathbf{r}_2$ relative to M_1 and M_2 as shown in Figure 4.16, will experience gravitational forces $\mathbf{F}_1, \mathbf{F}_2$, and possibly a disturbance or thrusting force $\mathbf{F}_d$, where

$$\mathbf{F}_1 = -\frac{mGM_1}{r_1^3}\mathbf{r}_1 \quad \text{and} \quad \mathbf{F}_2 = -\frac{mGM_2}{r_2^3}\mathbf{r}_2$$

The equation governing the motion of the mass m within this f.o.r. is:

$$\ddot{\mathbf{r}} + 2(\mathbf{\Omega} \times \dot{\mathbf{r}}) + (\mathbf{\Omega}\cdot\mathbf{r})\mathbf{\Omega} - \Omega^2\mathbf{r} = -G\left(\frac{M_1}{r_1^3}\mathbf{r}_1 + \frac{M_2}{r_2^3}\mathbf{r}_2\right) + \frac{1}{m}\mathbf{F}_d \qquad (4.51)$$

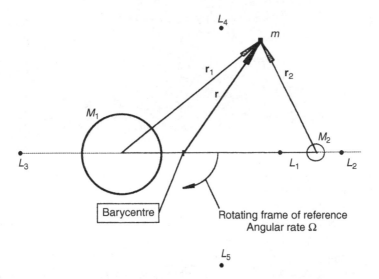

Figure 4.16   The restricted three-body situation, showing the Libration points L_1 to L_5

It should be noted that $\dot{\mathbf{r}}$ is not the *absolute* velocity $\mathbf{v}$; it has components that are the rates of change of the components of $\mathbf{r}$, along axes that are fixed in the (rotating) f.o.r. It is thus the velocity as it would be seen by an observer at the barycentre in Figure 4.16 and who is rotating with the same angular velocity as the f.o.r. described above. $\ddot{\mathbf{r}}$, defined in similar fashion in terms of the rate of change of the components of $\dot{\mathbf{r}}$, is the acceleration of the mass m as seen by the same observer.

When the disturbance or thrust force $\mathbf{F}_d$ is zero, equation (4.51) may be integrated once, to give the *Jacobi Integral*, the equivalent of the *vis-viva* equation (4.5) for this three-body situation:

$$\frac{1}{2}\dot{\mathbf{r}}^2 - \frac{1}{2}(\mathbf{r} \times \boldsymbol{\Omega})^2 - G\left(\frac{M_1}{r_1} + \frac{M_2}{r_2}\right) = J \qquad (4.52)$$

J is a constant of integration having the dimensions of energy per unit mass, whose value depends on the initial conditions.

Equation (4.52) may be used in ways such as the following:

- A trajectory in the three-body region cannot be obtained analytically. J remains constant along a trajectory, however, and so equation (4.52) may be used as a 'check' when a numerical integration of equation (4.51) is used to predict a trajectory.
- *Escape criterion*: The requirement for a body to escape from the two dominant bodies is for J to be at least zero. The reasoning is similar to that used in Section 4.2, 'Parabolic trajectories'.

For each value of J less than zero, a surface surrounding the two dominant bodies may be determined when $\dot{\mathbf{r}} = \mathbf{0}$. This represents a boundary beyond which a spacecraft with this amount of J cannot go. The intersection of this zero-velocity surface with the $x - y$ plane of the rotating frame of reference is shown in Figure 4.17(a) to (f) for

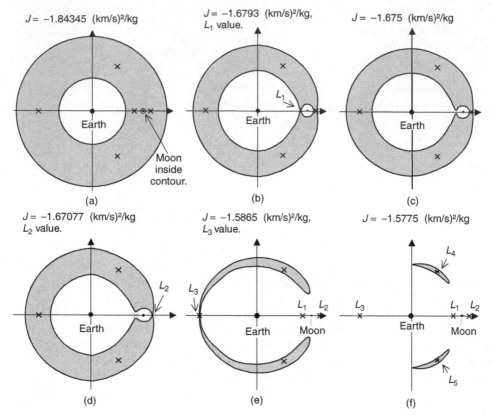

Figure 4.17 The intersection of the zero-velocity surface with the Earth–Moon $x - y$ plane as the energy level J changes

different values of J. The example shown is the Earth–Moon system. Forbidden regions are shaded. In (a) a spacecraft with a low energy level ($J < -1.84345$ (km/s)^2/kg) will be restricted to within the region around the Earth that is shown as approximately circular in this plane. If it is setting out from the Moon, it is restricted to a smaller circular region there. On the other hand, if an object is outside the outer circle then it cannot reach the vicinity of either the Earth or the Moon.

Increasing energy progressively reduces the size of the restrictions, as shown in (a) to (f), with transfer between Earth and Moon becoming possible from (b) onwards ($J > -1.6793$ (km/s)^2/kg), and journeys beyond the Moon to outer space from (d) onwards ($J > -1.67077$ (km/s)^2/kg). 'Gateways' open at the points L_1, L_2 and L_3 when the appropriate energy levels are reached, and all restricted areas shrink to zero when the energy level $J = -1.5738$ (km/s)^2/kg of L_4 and L_5 is exceeded. These special points are the *Lagrangian Points*. It is worth noting that in the Earth–Moon system the energy levels differ very little between these points, with only 0.5% difference between L_1 and L_2, for example.

- *Lagrangian, or Libration Points:* These are five points fixed within the rotating frame of reference at which a stationary body will be in equilibrium. Denoted by L_1 to

L_5, they all lie in the plane defined by the orbits of the primary bodies around each other (see Figures 4.16 and 4.17). Three of these, L_1, L_2 and L_3, lie on the line joining the primary bodies, and correspond to unstable equilibrium positions. The L_1 point, between the Earth and the Sun in this case, is a good position for solar observatories, and the ESA SOHO spacecraft has been 'orbiting' this point since its launch in 1995. The outer Lagrange point L_2 is becoming a popular operational location for astronomical observatories, such as the ESA Herschel infra-red telescope launched in 2009, and the James Webb Space Telescope which is due for launch in 2014. Because of the unstable nature of the equilibrium about these points, station-keeping activity is required to stabilize the 'orbits' of such spacecraft. Conversely, it can be shown that the off-axis points L_4 and L_5 are stable with respect to small displacements. An impressive example of this stability is the presence of a number of large asteroids, called the Trojans, oscillating about the L_4 and L_5 points with respect to the Jupiter–Sun system.

- *Effect of* $\mathbf{F_d}$: The inclusion of the extra force $\mathbf{F_d}$ into the Jacobi equation (4.52) leads to the right-hand side becoming

$$J + \frac{1}{m} \int \mathbf{F_d} \cdot \dot{\mathbf{r}} \, dt \qquad (4.53)$$

From this it may be seen that the best direction for a thrust force aimed at changing the value of J is along the trajectory in this (non-inertial) frame of reference.

REFERENCES

[1] Brown, C. D. (1998) *Spacecraft Mission Design* (2nd edn), AIAA, Reston.
[2] Cowell, P. H. and Crommelin, A. C. D. (1910) *Investigation of the Motion of Halley's Comet from 1759–1910*, Appendix to Greenwich Observatory.
[3] Encke, J. F. (1852) *Astron. Nachr.*, **33**, 377–398.
[4] Cornelisse, J. W., Schoyer, H. F. R. and Wakker, K. F. (1979) *Rocket Propulsion and Spaceflight Dynamics*, Pitman, London.
[5] King-Hele, D. G. (1987) *Satellite Orbits in an Atmosphere: Theory and Applications*, Blackie, London.
[6] Brouwer, D. and Clemence, G. M. (1961) *Methods of Celestial Mechanics*, Academic Press, London.
[7] Roy, A. E. (2004) *Orbital Motion* (4th edn), Taylor and Francis, London.
[8] Wertz, J. R. (1978) *Spacecraft Attitude Determination and Control*, Reidel, Dordrecht.
[9] Tapley, B. D. et al. (1996) The Joint Gravity Model 3, *J. Geophys. Res.*, **101**, 28029–28049.
[10] Stark, J. P. W. (2001) The evolution of debris clouds to microscopically chaotic motion, *J. Spacecraft Rockets*, **38**, 554–562.
[11] Groves, G. V. (1959) Motion of a satellite in the Earth's gravitational field, *Proc. R. Soc. London* **254**, 48.
[12] Agrawal, B. N. (1986) *Design of Geosynchronous Spacecraft*, Prentice-Hall, Englewood Cliffs.
[13] Stark, J. P. W. (1986) ESA SP 255, 239–246.
[14] Stalder, J. R. and Zurick, V. J. (1951) NACA Tech. Note 2423.
[15] Schaff, S. A. and Chambre, P. L. (1958) in *Fundamentals of Gas Dynamics*, ed. Emmons, H., Oxford University Press, Oxford.
[16] Schamberg, R. (1959) Rand Corp. R.M. 2313.

[17] Cook, G. E. (1965) *Planet. Space Sci.*, **13**, 929.
[18] Cook, G. E. (1962) *Geophys. J.*, **6**, 271.
[19] Aksnes, K. (1976) *Celestial Mech. Dynam. Astron.*, **13**, 89.
[20] Harwood, N. M. and Swinerd, G. G. (1995) *Celestial Mech. Dynam. Astron.*, **62**, 71.
[21] Harwood, N. M. and Swinerd, G. G. (1995) *Celestial Mech. Dynam. Astron.*, **62**, 81.

5 MISSION ANALYSIS

John P. W. Stark[1] and Graham G. Swinerd[2]

[1] *School of Engineering and Material Science, Queen Mary, University of London*
[2] *Aeronautics and Astronautics, Faculty of Engineering and the Environment, University of Southampton*

5.1 INTRODUCTION

Whereas celestial mechanics deals with a spacecraft's dynamics and orbit relative to the stars, one aspect of mission analysis is to relate both orbit and attitude to the ground, the Sun and so on. This ensures that operational aspects such as spacecraft–ground communications, power raising and spacecraft tracking can be achieved, which is of course crucial to mission success. However, beyond these operational issues, the mission design process is also a systems level activity, performed early in the feasibility study stage, which has a significant influence upon the spacecraft design. Generally, a careful analysis of the mission objectives, and consequently the payload operation and orbital aspects, is required. This, in turn places *design requirements* upon the spacecraft subsystem elements supporting the payload. The mission design then becomes an iterative process, wherein the impacts on the spacecraft system design may compromise an optimal solution for the mission design, and *vice versa*. These system aspects of mission analysis are touched upon briefly in this chapter, but a more detailed discussion is given in Chapter 20.

For study purposes, it is convenient to place missions into categories based upon their orbits. The following cover most applications:

- Low Earth Orbit (LEO)
- Medium height Earth orbit (MEO)
- Geostationary Earth orbit (GEO)
- Highly elliptical orbits (HEO)
- Non-geocentric orbits (e.g. Lagrange points or interplanetary)

There is also a potentially increasing category embracing space stations, on-orbit servicing and rescue. However, this chapter will focus on the five categories mentioned above since they illustrate most of the aspects that link orbits to missions.

Spacecraft Systems Engineering, Fourth Edition.
Edited by Peter W. Fortescue, Graham G. Swinerd and John P. W. Stark.
© 2011 John Wiley & Sons, Ltd. Published 2011 by John Wiley & Sons, Ltd.

A history of each mission may be viewed as a sequence of events starting at pre-launch and extending to end-of-life. The *pre-launch phase* includes all those operations that are required in order to effect launch vehicle ignition and separation from the umbilical cable between space system (spacecraft plus launcher) and the ground launch facilities.

The *launch phase* involves a sequence of events, many of which are pre-programmed and automatic. Careful planning is required for placing the spacecraft into an appropriate intermediate orbit from which the operational one may be attained. One evident constraint during the launch phase is the requirement for continuous communications and tracking. These facilities may be required for a ground-centre override to abort the mission, but also there is generally a requirement to monitor the performance of a launch vehicle. Figure 5.1 shows the ground-centre network used for a typical Ariane 5 launch from the launch site in French Guiana.

Orbit transfer is the next major event. This involves transferring the spacecraft from the orbit into which the launch vehicle *actually* places it, to the one from which the operational phase of the mission can commence. There is *a priori* uncertainty in the orbit at launcher burn-out as given by the performance envelope of the launch vehicle (see Chapter 7); but there is *a posteriori* uncertainty because of the orbit determination process that inevitably includes some errors in the range and range-rate measurements of the spacecraft trajectory (see Chapter 4). Transfer between these orbits requires propellant, and it is the task of the mission planners to determine how much is required in order to attain the desired orbit from any given initial one, with a given level of confidence. This is particularly important for geostationary spacecraft, since the final operations needed to acquire their orbit call for the use of a secondary propulsion system that may subsequently be used for attitude and orbit control. Excessive use or under-budgeting of fuel will therefore affect the available payload mass and reduce the operational life of the space system as a whole.

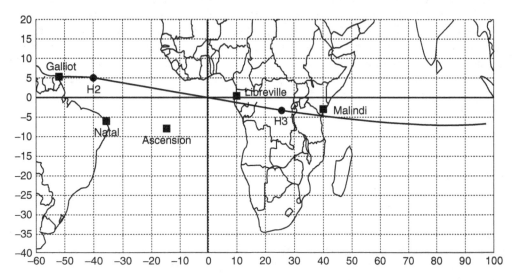

Figure 5.1 Typical Ariane 5 ground track, showing launch monitoring ground stations. H2 indicates location of first stage engine shutdown and separation, and H3 indicates upper stage shutdown and injection point. (Reproduced by permission of Arianespace)

On-station operations include various mission events. Fuel budgeting for station-keeping and attitude control will be necessary.

For sky survey observatory missions, the scientific goals may require continuous changing of the orientation of the spacecraft in such a way that a wide variety of directions or even complete coverage of the celestial sphere may be achieved. Careful planning is required in order to optimize the observing programme; both the rate at which the spacecraft is reorientated and the angular distance through which its attitude is changed between observations will influence fuel consumption. Repointing operations are normally performed using reaction wheels, with thrusters (propellant) being used to periodically dump angular momentum from the wheels (see Chapter 9). Whilst contingency planning is made for unscheduled repointing, many scientific missions inevitably cease only when the fuel has been exhausted. Other constraints that will be of concern in deriving mission profiles for scientific observatories include minimum sun/telescope angle, earth limb/telescope angle, passage through the South Atlantic Anomaly in the Earth's magnetosphere and thermal balance of the spacecraft and its sensors. Analysis and planning activities associated with a number of specific types of mission during their operational phases are detailed in the following sections.

Decommissioning the spacecraft is the final event associated with a space mission. In GEO, the demand for longitude slots is increasing, and thus an uncontrolled satellite in this orbit is wasteful and also constitutes a collision hazard [1]. It has therefore become common practice to remove an obsolete spacecraft from GEO into a higher orbit by the use of residual propellant in the secondary propulsion system at the end of its life. This manoeuvre is frequently and appropriately called a 'graveyard burn'. It is also becoming the practice in LEO missions to provide a controlled re-entry into the Earth's atmosphere, or at least place it into an orbit with a reduced lifetime. The reason for this is that uncontrolled re-entry can lead to the vehicle breaking up, providing a hazard on the ground and adding to the problem of space debris.

As mentioned earlier, a key aspect of mission analysis is to identify critical features of the mission, which have an impact upon system and subsystem design. Every space vehicle is required to meet specific mission objectives and these, thus provide design requirements.

The analysis of these, in the context of a particular spacecraft mission, usually leads to a subset of design requirements that are influential in shaping the whole configuration of the vehicle. These requirements are often referred to as *design drivers*. For example, the thermal control system design may be routinely achieved, relatively speaking, in the case of a GEO communications satellite. However, for an infrared astronomical observatory spacecraft, it may become very influential in terms of governing the spacecraft configuration and operation (e.g. Herschel).

To illustrate briefly how mission design influences spacecraft system design, consider the operation of a LEO spacecraft compared to one in GEO. At top level, the most obvious difference is the requirement for the GEO spacecraft to have primary propulsion (with the consequent impact upon the vehicle's mass budget) to achieve transfer from the initial orbit to the operational orbit, whereas the LEO vehicle often has no need for this. The LEO spacecraft will experience an eclipse each orbit period, whereas the GEO spacecraft remains in sunlight for the majority of its orbital lifetime. The rapid cycling of solar input experienced by the LEO spacecraft leads to differing design requirements, principally in the power and thermal control subsystem areas, compared to the GEO vehicles. The requirements in the communications subsystem areas (payload data, telemetry and command) are dissimilar for the two categories of mission in terms of both ground station coverage and communications link budget. These simple examples illustrate well the impact of the mission specification on the overall design of space vehicles.

A further example, this time of an interplanetary mission, is the ESA/NASA spacecraft Ulysses [2]. The mission objective of this vehicle, launched by the Space Shuttle in October 1990, was to explore the interplanetary medium away from the plane of the ecliptic. The spacecraft was finally decommissioned in July 2009, after nearly 19 years of operation. Despite the longevity of the mission, it remains an excellent example in demonstrating the interaction between the spacecraft mission and system design. To achieve the mission objective, the probe needed to be inserted into a heliocentric orbit, the plane of which was near-normal to that of the Earth's orbit about the Sun. To acquire such an orbit a velocity increment (ΔV) of the order of 30 km/s was required after Shuttle deployment. This mission requirement became a principal design driver for the spacecraft. To achieve such a large ΔV, two mission design options were available. The velocity would be acquired either by using an electrical propulsion system, or by using the gravitational field of a major planet to swing the plane of the orbit out of the ecliptic (see Section 5.8.2 for details of swing-by trajectories). The latter option was chosen and the vehicle was injected into a trajectory, which it took to Jupiter. A swing-by manoeuvre was then performed (in February 1992) over the Jovian pole to insert it into the required mission orbit. As a consequence of this choice of mission, the design of the spacecraft's subsystems in nearly every respect (communications, thermal control, power raising, radiation hardening, etc.) was profoundly influenced by the necessity for the visit to the relatively cold and distant Jovian environment.

It is important to appreciate that the attitude history of a space vehicle is inextricably linked with the mission analysis. For example, an astronomical observatory spacecraft requires to point the payload to specified regions of the celestial sphere, the communications antenna to the ground, the solar array to the Sun, and possibly thermal radiators to 'deep space' (e.g. Hubble space telescope). Clearly for some spacecraft, pointing requirements can be very influential in shaping the vehicle configuration. The control of attitude behaviour will be discussed in Chapter 9.

5.2   KEPLERIAN ORBIT TRANSFERS

The situation frequently arises where a space vehicle must be transferred from one orbit to another. The detailed optimization of these manoeuvres is beyond the scope of this text. (For detailed analysis, see for example References [3, 4].) However, it is appropriate to consider the nature of simple impulsive manoeuvres here, applied to Keplerian orbits.

If a single impulsive manoeuvre is performed, then the initial and final orbits intersect at the location of the manoeuvre; thus, a single manoeuvre may only transfer a vehicle between intersecting orbits. *At least* two manoeuvres will be required to transfer a vehicle between two non-intersecting orbits.

A transfer from orbit A to an intersecting orbit B will occur if at some point, say r_1, the velocity vector is instantaneously changed from its value on orbit A to that which it would be on orbit B. The simplest case is that of a coplanar transfer from a circular orbit to an elliptical orbit. If, on the circular orbit the velocity is increased, then the semi-major axis will be expanded.

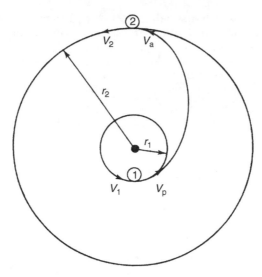

Figure 5.2   Transfer velocity definitions

A *Hohmann transfer* makes use of two such manoeuvres in order to transfer from one circular orbit to a larger coplanar one, as shown in Figure 5.2. It is the minimum energy, two-manoeuvre transfer, and is optimal (minimum velocity increment ΔV) if $r_2/r_1 < 11.8$.

The velocity increments needed for the two manoeuvres may be assessed by using the *vis-viva* integral equations (4.5) and (4.11). The first one requires an increase from the circular orbit speed $\sqrt{(\mu/r_1)}$ to the speed V_p at the periapsis of the transfer ellipse, where

$$V_p^2 = 2\mu \left\{ \frac{1}{r_1} - \frac{1}{r_1 + r_2} \right\}$$

This represents a ΔV given by

$$\Delta V = \sqrt{\left(\frac{\mu}{r_1} \right)} \left\{ \sqrt{\left(\frac{2r_2}{r_1 + r_2} \right)} - 1 \right\} \tag{5.1}$$

A second, similar manoeuvre will transfer the vehicle into a circular orbit at apoapsis, with a radius of r_2. The ΔV required is

$$\Delta V = \sqrt{\left(\frac{\mu}{r_2} \right)} \left\{ 1 - \sqrt{\frac{2r_1}{r_1 + r_2}} \right\} \tag{5.2}$$

Table 5.1 provides details of Hohmann transfers within the solar system. The transfer time is taken as half of the transfer orbit period. The velocity increments are for Earth departure only, from an assumed 300 km altitude circular Earth orbit. The Δv for the arrival phase is strongly dependent upon the insertion orbit parameters at the destination planet.

Table 5.1 Hohmann transfer data from the Earth to the planets. Δv requirements are for the Earth departure phase only, from an assumed 300 km altitude circular Earth orbit

Planet	Transfer orbit semi-major axis (AU)	$T_{transfer}$ transfer time (years)	Δv for Earth departure (km/s)
Mercury	0.694	0.289	5.55
Venus	0.862	0.400	3.48
Mars	1.262	0.709	3.59
Jupiter	3.101	2.731	6.30
Saturn	5.269	6.049	7.28
Uranus	10.096	16.040	7.98
Neptune	15.531	30.605	8.25
Pluto	20.265	45.616	8.37

Plane rotation manoeuvres and rotation of the line of apsides may be treated similarly to the above analysis. They involve a change of direction for which the ΔV requirement is given in Chapter 3, equation (3.21). This can involve appreciable use of fuel. If no change of speed is entailed, then the velocity increment required is proportional to the velocity at the time of the manoeuvre, and these transfers should, if possible, be performed when the velocity is at a minimum. For example, to execute large plane changes of an Earth orbit it is generally more efficient, from a propellant usage point of view, to perform the manoeuvre at a large geocentric distance where the spacecraft is moving slowly. This point will be developed later in this chapter.

5.3 MISSION ANALYSIS

As mentioned earlier, *mission analysis* is a 'front-end' systems level activity that takes aspects such as payload operational requirements and spacecraft system constraints as inputs, and generates as an output the mission specification. Clearly, a central feature of this process is the selection of the orbital parameters of the final *mission orbit* (as well as intermediate orbits during the early orbit-acquisition phase) that satisfy these requirements. It is important to note, however, that the mission specification also leads to design requirements on the spacecraft systems and subsystems. For example, the attitude control subsystem engineer can take the orbital characteristics and compute the on-orbit disturbance torques produced for a given spacecraft configuration, which in turn determines the angular momentum profile (see Chapter 9). It is these aspects that aid the engineer in producing design specifications for the attitude actuators and sensors.

As can be seen, mission analysis is not just concerned with the celestial mechanics discussed in Chapter 4, but also involves aspects governing operational and design issues, such as launch vehicle selection (see Chapter 7), ground station visibility, spacecraft ground coverage (see Section 5.5), eclipse duration (which impacts particularly on the thermal control and power subsystem design) and orbital lifetime. Summarizing this discussion, we can conclude that the principal spacecraft design inputs come from the payload interface with the spacecraft and its operational requirements, the spacecraft interface with the launch vehicle, and the orbit. Clearly, the mission analyst has a major role to play in this process. In this section, we will address some of these operational and system aspects related to the orbit.

5.3.1 Ground station visibility

A satellite is visible at all points on the Earth's surface within a circle that is centred on the sub-satellite point, and whose diameter increases with satellite altitude (see Table 5.2). However, signals from satellites at the horizon limit are considerably attenuated by the atmosphere, and so for practical purposes the surface coverage is restricted to the region in which the *satellite elevation* above the horizon is greater than ~5°. Figure 5.3 shows the geometry associated with a satellite S in a circular orbit of height h, passing immediately over a ground station at G. If it is visible down to elevations equal to ε, typically 5° to 10°, then the geocentric semi-angle ϕ over which it is visible is given by

$$\phi = -\varepsilon + \cos^{-1}\left(\frac{R_E}{R_E + h}\cos\varepsilon\right) \tag{5.3}$$

Table 5.2 Relationship between satellite altitude, geocentric semi-angle ϕ and slant range s

Altitude h (km)	Geocentric semi-angle ϕ (deg)			Slant range s (km)		
	$\varepsilon = 0°$	$\varepsilon = 5°$	$\varepsilon = 10°$	$\varepsilon = 0°$	$\varepsilon = 5°$	$\varepsilon = 10°$
100	10.08	6.24	4.16	1134	707	477
500	21.98	17.52	14.05	2574	2078	1696
1000	30.18	25.55	21.64	3709	3194	2763
5000	55.91	51.05	46.49	9423	8882	8379
35 786	81.30	76.33	71.43	41 679	41 126	40 585

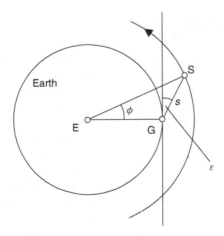

Figure 5.3 Geometry of a satellite passage directly overhead a ground station

The range of the satellite from the ground station, referred to as the *slant range s* is given by

$$s = (R_E + h) \sin \phi / \cos \varepsilon \qquad (5.4)$$

The duration of an overhead pass with semi-angle ϕ is

$$\tau = 2\phi / \omega_{ES} \qquad (5.5)$$

where ω_{ES} is the orbital angular velocity of the satellite relative to the Earth. ω_{ES} may be obtained from the Earth's angular rate ω_E (7.3×10^{-5} rad/s) and the satellite's orbital rate ω, using

$$\omega_{ES}^2 = \omega_E^2 + \omega^2 - 2\omega_E \omega \cos i \qquad (5.6)$$

For LEO, the error produced by using ω in place of ω_{ES} in equation (5.5) (around 5 to 10%) may be acceptable. For a circular orbit, ω in rad/s is given by $631(R_E + h)^{-3/2}$, where R_E and h are in km. For example, a satellite in a 500 km altitude orbit will have an elevation $>5°$ for 9.2 min and $>10°$ for 7.4 min during an overhead pass. Most passes will be shorter than this.

For non-overhead passes, equation (5.5) may be used with ψ in place of ϕ, where

$$\psi = \sin^{-1}(\sin^2 \phi - \tan^2 \alpha \cos^2 \phi)^{1/2}$$

ϕ is now the geocentric semi-angle of the cone within which the spacecraft is visible from the ground site, defined by the minimum elevation ε—as given by equation (5.3). Here the non-overhead pass orbit plane makes an angle α—measured at the Earth's centre—to the directly overhead one. Often non-overhead passes are characterized in terms of the maximum elevation ε_{max} of the satellite. In this case, α and ε_{max} are related by

$$\alpha = \cos^{-1}\left[\left(\frac{R_E}{R_E + h} \right) \cos \varepsilon_{max} \right] - \varepsilon_{max}, \qquad \varepsilon_{max} = \tan^{-1}\left[\cot \alpha - \left(\frac{R_E}{R_E + h} \right) \csc \alpha \right]$$

5.3.2 Eclipse duration

A spacecraft in an Earth orbit will generally encounter an *eclipse period*. This is particularly the case for orbits at low altitudes. The frequency and duration of eclipse periods is strongly dependent on orbital inclination and altitude. For example, in a low-altitude, equatorial orbit, the satellite resides in the Earth's shadow for approximately 40% of every orbit. Conversely, for dawn–dusk sun-synchronous orbits (see Section 5.4), even at low altitude, several months of wholly sunlit operation may be obtained. Spacecraft in highly elliptic orbits (see Section 5.7) with relatively low perigee altitudes will generally encounter eclipse periods when near the Earth. These are relatively brief however, compared to the orbit period, because of the spacecraft's high speed in the perigee region. For example, the eclipse period in a Molniya orbit varies from zero to a maximum of around 50 min (7% of orbit period), depending upon perigee height and season. In Geostationary Earth orbit (GEO) (see Section 5.6), a satellite spends most of its time in sunlight. The eclipse periods in this case are brief and depend on the season. In the period around the solstices, the operation is eclipse-free, whereas near the Spring and Autumn equinox the satellite-Sun vector is near the orbit plane and eclipses are encountered with durations up to a maximum of 72 minutes. This is still, however, a small fraction of the total orbit period of 24 h.

The calculation of the duration of eclipse is important from the point of view of the spacecraft design. For example, if the spacecraft's primary power source is solar arrays, backed up by a battery storage system, then the sizing of the power subsystem is strongly influenced by the length of the eclipse period (see Chapter 10). Similarly, the thermal input to the spacecraft from the Sun is governed by the eclipse period and so influences the design of the thermal control subsystem (see Chapter 11). For some orbits, such as circular orbits, which contain the Earth–Sun vector, the eclipse duration calculation is simple. However, in more general cases the process is more complicated and a computational technique is required.

The following outlines such a technique, and allows the computation of the eclipse period for a given orbit (defined generally by its Keplerian orbit elements) on any day of the year. This implied seasonal dependence is illustrated in Figure 4.9, which shows the motion of the Earth–Sun vector throughout the year. The unit dimensionless vector defining the direction from the Earth to the Sun is given by

$$\mathbf{r}_{sun} = \cos L_{\odot} \mathbf{i} + \sin L_{\odot} \cos \varepsilon \mathbf{j} + \sin L_{\odot} \sin \varepsilon \mathbf{k} \tag{5.7}$$

where $L_{\odot}$ is the Sun's ecliptic longitude, measured East along the ecliptic from the vernal equinox ($\mathbf{i}$ axis), ε is the obliquity of the ecliptic (23.45°), and ($\mathbf{i}$, $\mathbf{j}$, $\mathbf{k}$) defines the geocentric inertial frame. The unit vector defining the direction to the satellite from the Earth's centre, in the same frame of reference, is given by

$$\mathbf{r}_{sat} = [\cos \Omega \cos(\theta + \omega) - \sin \Omega \cos i \sin(\theta + \omega)]\mathbf{i} \tag{5.8}$$

$$+ [\sin \Omega \cos(\theta + \omega) + \cos \Omega \cos i \sin(\theta + \omega)]\mathbf{j} + \sin i \sin(\theta + \omega)\mathbf{k}$$

where the Keplerian elements have their usual notation. The criterion to establish whether the satellite is in eclipse is illustrated in Figure 5.4. The angle β that defines the entry

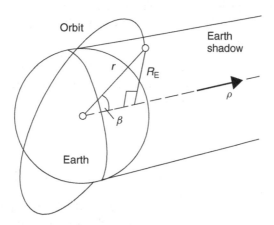

Figure 5.4   Geometry illustrating orbit's intersection with the Earth's shadow

and exit positions is given by

$$\beta = \sin^{-1}(R_E/r) \tag{5.9}$$

where R_E is the Earth radius, and r is the geocentric distance of the spacecraft given by equations (4.7) and (4.8). Note that we have made simplifying assumptions (e.g. the Earth's shadow is cylindrical and the Earth is spherical). The satellite is in eclipse if the angle between $\mathbf{r}_{sat}$ and the vector defining the anti-Sun direction $\rho = -\mathbf{r}_{sun}$ is less than or equal to β. This is expressed as

$$\cos^{-1}(\rho \cdot \mathbf{r}_{sat}) \leq \beta \tag{5.10}$$

The computational method to determine the eclipse period on a particular day of the year is based on the following steps using equations (5.7) to (5.10):

1. Evaluate $\rho = -\mathbf{r}_{sun}$ from equation (5.7). An approximate value of $L_\odot$ can be calculated in degrees from

$$L_\odot \approx \{(D - D_0)/365\}360° \tag{5.11}$$

   where D is the day number, assuming $D = 0$ is 0.0 h on 1 January (e.g. midday on 2 January gives $D = 1.5$), and $D_0 = 79.0$ (this assumes that the Northern Hemisphere Spring Equinox occurs at 0.0 h on 21 March). Clearly, equation (5.11) is approximate, and ignores complications such as leap years. However, it is adequate for most design calculations. More precise expressions can be obtained, if required, from published solar ephemeris data.
2. Cycle through values of true anomaly θ from 0° to 360° in steps of, say, 1° to calculate β and $\mathbf{r}_{sat}$ at each step, and use the condition (5.10) to determine the values of θ at eclipse entry and exit. To save on computational effort, an initial search with a larger θ step (10° or 20° say) may be performed to find the eclipse region (bearing in mind that the eclipse duration may be small or zero). If the entry and exit values are required to a greater accuracy than 1°, the values of θ can be further subdivided in the computation in the entry and exit regions. This is recommended for spacecraft entering eclipse when at large geocentric distance, since it will take a relatively long time period to traverse a 1° step in θ.
3. Use equations (4.14), (4.18) and (4.19) to convert true anomaly values at entry and exit to eclipse duration.

The eclipse period of a typical Space Shuttle orbit (200 km altitude circular orbit, inclined at 28°) when the orbit plane contains the Earth–Sun vector is 37.3 min, and the above method gives 36.9 min for a 1° step in θ.

5.3.3 Launch windows

The time of launch of a spacecraft is often constrained by dynamical aspects related to reaching the mission orbit, or by system requirements. Since the orbital plane is essentially inertially fixed, the satellite can be launched only as the launch site rotates through the orbital plane. The limited interval of time during which launch can be effected is referred to as the *launch window*.

The dynamical constraints are emphasized for launch into a LEO sun-synchronous orbit (see Section 5.4), for example, when the orbit's line of nodes must be set at a specified angle to the Earth-Sun vector. In this case, the launch window amounts to only a few minutes twice each day. Further, if there is a constraint on *launch azimuth* at the site, only one of these opportunities may be possible. The execution of a launch through such a small window is a non-trivial exercise. For missions launched to *rendezvous* with an orbiting object [e.g. International Space Station (ISS) transit missions], the dynamical issues are again paramount, since there are not just geometrical constraints to consider, but also temporal issues to do with phasing the launch to reach the target with minimum propellant.

The time of launch to acquire a GEO, on the other hand, is additionally constrained by systems issues which usually lead to two launch opportunities each day around the launch site midday and midnight. The launch window is constructed by combining constraints related to power raising, thermal control, sun-blinding of sensors, and so on, in the transfer orbit. More detail is given in Section 5.6.1.

For interplanetary missions, the *planetary alignment* at launch becomes an issue in defining the launch window. To illustrate this, consider a mission to Mars for which the heliocentric trajectory is a Hohmann transfer (see Section 5.2). Figure 5.5 shows the outbound transfer orbit. If the mission leaves Earth at the point E_{depart}, then clearly Mars must be at point M_{arrival} when the spacecraft has traversed the transfer orbit. For this to be the case, the position of Mars at the Earth departure time M_{depart} is such that Mars moves from M_{depart} to M_{arrival} in the same time as the spacecraft takes to execute half a transfer orbit period (~259 days). This defines the illustrated planetary geometry at Earth departure time, with a heliocentric angle of ~45° between Earth and Mars. The

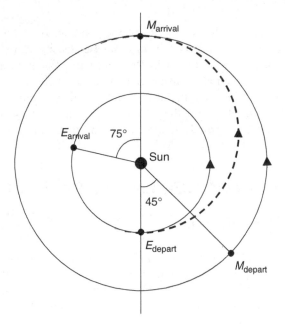

Figure 5.5   Planetary geometry at Earth departure and Mars arrival for a Hohmann transfer mission

opportunity to launch into a Hohmann trajectory is therefore limited by this planetary alignment, which recurs every 780 days (the Martian synodic period ~ 2.14 years). The planetary alignment defines a launch window of approximately two weeks duration centred around the ideal departure time, and Earth rotation during this period will impose a further restriction allowing launch for a relatively short period ($\sim$ half hour to one hour) each day. The position of Earth E_{arrival} at Mars arrival is determined in the same way, and allows an estimate of the communications distance. If the mission were to return on a Hohmann transfer, the wait for the return launch window to open would mean that a stay at Mars of approximately 16 months would be required.

5.3.4 Orbit lifetime

For LEO vehicles, aerodynamic drag will eventually result in re-entry as described in Chapter 4, Section 4.4.2. To the accuracy of a few per cent the lifetime of an uncontrolled space vehicle will be

$$\tau \sim \frac{e_0^2}{2B} \left(1 - \frac{11}{6}e_0 + \frac{29}{16}e_0^2 + \frac{7}{8}\frac{H}{a_0} \right) \tag{5.12}$$

where e_0 and a_0 are the initial values of eccentricity and semi-major axis once control has ceased, H is the scale height of the atmosphere near perigee and B is given by

$$B \sim \sqrt{\left(\frac{\mu}{a_0^3} \right) \frac{AC_D}{M} \rho_{p0} a_0 e_0 I_1 \left(\frac{a_0 e_0}{H} \right) \exp\left(-e_0 \left(1 + \frac{a_0}{H} \right) \right)} \tag{5.13}$$

This expression neglects the effects of atmospheric rotation. I_1 is the Bessel function of the first kind and order 1 and ρ_{p0} is the atmospheric density at the initial perigee. A more elaborate treatment may be found in Reference [5]. For large space vehicles and for military vehicles, re-entry may cause problems as a result of both the geographical position of the re-entry corridor and the size of individual spacecraft elements on ground impact. Most notably, the effect on the public of Skylab re-entry over Australia, and Cosmos 943B over Canada, provide historical evidence of some of the problems that may be encountered in re-entry.

5.4   POLAR LEO/REMOTE-SENSING SATELLITES

5.4.1   Mission orbit design

The choice of orbit for a LEO remote sensing spacecraft is governed by the mission objectives and payload operational requirements. To achieve near-global coverage, the plane of the orbit must be inclined at about 90° to the Earth's equator. In this near-polar orbit, the spacecraft ground track will span a latitude coverage of $\pm i$ for prograde orbits, where i is the orbital inclination.

In terms of orbit altitude, this is principally established by a trade-off between instrument resolution and the fuel required to maintain the orbit in the presence of aerodynamic

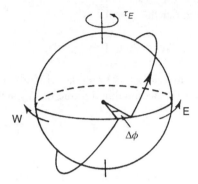

Figure 5.6   Orbit plane motion

drag. For civil remote sensing systems, which are required to operate for an extended period, an altitude typically of the order of 600 to 900 km results. It is worth noting, however, that the mission requirements for military surveillance spacecraft may be different, leading to more complex operational scenarios.

Other payload-derived requirements often demand that the mission orbit be *Earth-synchronous*, or *Sun-synchronous*, and sometimes both (e.g. SPOT). These further constrain the choice of orbit inclination and height in a manner that is discussed in more detail below.

Consequently, for viewing all parts of the Earth's surface at close quarters it is necessary to adopt a low-altitude polar orbit. With the orbit fixed in space and the Earth rotating underneath it, the result is that the ground tracks of successive orbits cross the equator at points that move westward as shown in Figure 5.6.

An Earth-synchronous orbit results when the sub-satellite point follows a ground track identical to some previous orbit after a certain period of time. This need occurs as a result of a payload-derived requirement to revisit ground target sites or regions. The repetition occurs on a regular basis, and it can be achieved in a variety of ways. (A geosynchronous orbit is of course another example of an Earth-synchronous orbit.) Between successive orbits, the sub-satellite point on the equator will change in longitude by $\Delta\phi$ radians, this angle being determined by two effects. The first of these is as a result of the rotation of the Earth beneath the orbit, and the second is caused by nodal regression. It will be assumed that a positive $\Delta\phi$ means a move towards the east.

The Earth rotates through one revolution in its *sidereal* period of τ_E, where $\tau_E = 86\,164.1$ s, neglecting small secular variations. This sidereal period of 23 h 56 min is with respect to the stars. The rotation period with respect to the Sun is of course the mean solar day of 24 h. If the satellite's nodal period is τ, then the contributions to $\Delta\phi$, which is caused by the Earth's rotation, will be given by

$$\Delta\phi_1 = -2\pi \frac{\tau}{\tau_E} \quad \text{rad/orbit} \tag{5.14}$$

The regression of the line of nodes (equation 4.37) contributes

$$\Delta\phi_2 = -\frac{3\pi J_2 R_E^2 \cos i}{a^2(1 - e^2)^2} \text{ rad/orbit} \tag{5.15}$$

The total increase in longitude at the equator is

$$\Delta\phi = \Delta\phi_1 + \Delta\phi_2 \text{ rad/orbit} \tag{5.16}$$

Clearly, if we wish to have an Earth-synchronous orbit, then we will require that some integer number of orbits later the accumulated value of $\Delta\phi$ will equal 2π. In general, we may, therefore, write

$$n|\Delta\phi| = m2\pi \tag{5.17}$$

where n is the total number of orbits performed and m is the number of Earth revolutions (equivalent to days) before an identical ground track will occur.

Sun-synchronism occurs when the orbit plane rotates in space at the same rate as the Earth moves round the Sun—at one revolution per year, or roughly one degree per day eastwards. Figure 5.7 illustrates this over a period of about three months, during which the orbit clearly needs to rotate through 90° in order to be synchronous.

The required rotation rate is

$$\Delta\phi_2 = 2\pi \frac{\tau_E}{\tau_{ES}} \frac{\tau}{\tau_E} \text{ rad/orbit} \tag{5.18}$$

where $\tau_{ES} = 3.155\,815 \times 10^7$ s is the orbital period of Earth round the Sun.

From this discussion, it can be seen that when an orbit is Sun-synchronous, the relative geometry of the Earth–Sun vector and the orbit plane remains approximately constant throughout the year. This implies that spacecraft in such orbits pass over particular ground sites at the same local solar times each day (one pass in daylight and one at night). The need for Sun-synchronism for some remote sensing missions is a payload-derived requirement to use this attribute. The operator of such a vehicle will often focus interest on a particular geographical region. By an appropriate choice of orbit-node position (equator crossing) with respect to the Earth–Sun vector, the solar aspect on the ground can be chosen optimally to fulfil the remote sensing mission objectives. This advantage of Sun-synchronous orbits is illustrated in Figure 5.8.

In general, configuration (a) is unattractive for Earth viewing since the Sun is always low on the horizon at the sub-satellite point, resulting in long shadows and low illumination. Indeed, during winter in the northern hemisphere the ground-track is in darkness while

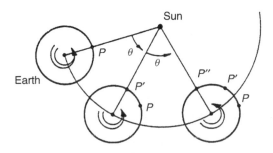

Figure 5.7   Sidereal and solar motion

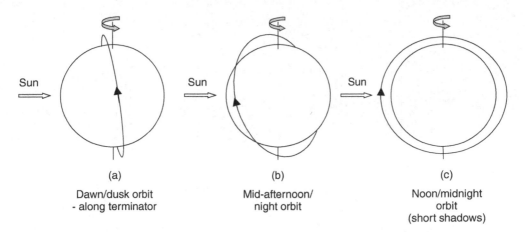

(a)	(b)	(c)
Dawn/dusk orbit - along terminator	Mid-afternoon/ night orbit	Noon/midnight orbit (short shadows)

Figure 5.8 Implications of changing local time coverage

the spacecraft is north of the equator. However, the orbit is attractive for the power subsystem as it provides lengthy eclipse-free periods, making energy storage minimal, and the array may be fixed relative to the body of the vehicle. Further, since the array is aligned along the direction of flight, a minimum projected area to the velocity vector of the satellite is obtained, thus reducing the influence of drag and increasing satellite lifetime. For the noon–midnight orbit (c), remotely sensed images have low definition due to short shadows on the ground, and the payload sensors may experience specular reflection from sea surfaces. The choice of orbit plane is usually a compromise, as shown in Figure 5.8(b).

The other feature to note about Sun-synchronous orbits is that since the orbit plane is fixed relative to the solar vector it is possible to cant the array relative to the spacecraft body, providing optimal illumination conditions for the solar array.

To achieve Sun-synchronism, the Earth-oblateness perturbation governing the regression of the line of nodes may be used without the use of fuel. Equation (5.15) indicates that an inclination in excess of 90° will be needed.

For Sun and Earth synchronism equations (5.16) to (5.18) apply, and these lead to the condition

$$n\tau\left(1 - \frac{\tau_E}{\tau_{ES}}\right) = m\tau_E \qquad (5.19)$$

The angular displacement between successive orbits in a westward direction is given by

$$\Delta\phi = 2\pi\tau\left(\frac{1}{\tau_{ES}} - \frac{1}{\tau_E}\right) = -7.27 \times 10^{-5}\tau\,\text{rad} \qquad (5.20)$$

From equation (4.12) we have

$$\tau = 9.952 \times 10^{-3}a^{3/2}\,\text{seconds} \qquad (5.21)$$

(where a is in kilometres). Since for remote sensing satellites in LEO the altitude is approximately 600 to 900 km, then $\tau \sim 96$ to 103 min, and so $|\Delta\phi| \sim 4.3 \times 10^{-1}$ rad.

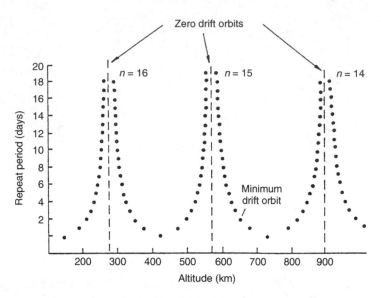

Figure 5.9 Locus of points for minimum drift orbits as a function of repeat period in days (m) and altitude

At the equator, this translates into a distance of $\sim$2800 km between ground tracks, although this decreases at higher latitudes. It is clear from this that wide instrument swath-widths will be required if complete Earth coverage is to be obtained. By extending the period between repetitions for a given set of ground tracks, a more densely packed set may be achieved. For example, if the requirement is for a daily repeat ($m = 1$) of a set of ($n =$) 14, 15 or 16 tracks, then from equations (5.19) and (5.21), it may be deduced that the corresponding orbit altitudes are 894, 567 and 275 km, respectively. More precisely, if account is taken of the higher order effects of Earth oblateness on the spacecraft's nodal period, then these altitudes become 888, 561 and 268 km, respectively. This type of orbit is known as a *zero drift* orbit, and has no infilling between ground tracks on subsequent days. A greater density of tracks may be achieved by infilling the tracks of Day 1 on subsequent days. The cycle repeats itself over a number of days in excess of one ($m > 1$), with a number (n) of tracks, which is not a multiple of m.

If the condition $(n \pm 1)/m = k$ is satisfied, where k is an integer, then a *minimum drift* orbit is obtained. In this situation, two successive tracks of a given day are infilled sequentially on subsequent days. Non-minimum drift orbits, $(n \pm 1)/m \neq k$, infill in a non-sequential manner.

Since τ is a function of altitude (equation 5.21), it is possible to plot the repeat period of Earth synchronization *versus* altitude. This, for minimum drift orbits, is shown in Figure 5.9. As an example, SPOT5 was launched into a 26 day repeat period, non-minimum drift orbit in May 2002. The SPOT (Satellite Pour l'Observation de la Terre) spacecraft series are the space segment of a French national remote sensing programme. The characteristics of the orbit are

- Sun- and Earth-synchronous
- Ground track repeat cycle $(m, n) = (26, 369)$

- Orbit parameters.
  semi-major axis $a = 7200$ km (height ~ 822 km, period $\tau = 101.3$ min)
  eccentricity e ≈ 0, inclination $i = 98.7°$
  Node time (descending) $= 10.30$ local solar time (LST)

At the equator the separation between successive orbits is 2838 km. However, over the 26-day repeat period, this separation is filled-in (non-sequentially) so that the distance between adjacent ground tracks is reduced to $2\pi R_E/n \sim 108.6$ km. This ground repetition geometry was chosen to ensure that the SPOT imaging system, with a nominal swath-width of 117 km, can achieve global coverage.

5.5 SATELLITE CONSTELLATIONS

5.5.1 Introduction

The use of *constellations of satellites* to achieve a mission objective is by no means a new phenomenon. For example, the United States Department of Defence (DoD) deployed the *TRANSIT system* in the 1960s, which comprised a constellation of spacecraft in polar orbits. This was used principally by US Naval ships to determine position. Because of the way the fix was determined (tracking the range rate of a single satellite), the system was unsuitable for navigation of highly dynamic vehicles, such as aircraft or missiles. This constraint was removed, however, with the implementation of another US military navigation system, the *Navstar Global Positioning System (GPS)* constellation. This has had a major impact on both military and civilian sectors. To satisfy global coverage requirements, the nominal GPS constellation comprises 24 satellites deployed in 6 orbit planes, inclined at 55° to the equator. Each orbit plane contains 4 spacecraft, equally spaced in true anomaly, and the node of each plane is spaced equally in right ascension around the equator. Each orbit is near-circular, and has an orbit period of half a sidereal day (height $\sim 20\,200$ km). A similar system—GLObal NAvigation Satellite System (*GLONASS*)—was deployed concurrently by the former Soviet Union, primarily for military use. This also comprises 24 satellites, but this time with 8 satellites in each of 3 planes with an inclination of 65°.

The prime stimulus of interest in LEO constellations, particularly during the last decade of the twentieth century, has been their use for global data and communication services. Satellite communications were formerly based principally on GEO systems (see Section 5.6), relaying data between relatively large, fixed ground stations. However, the perceived market for *personal* communications led to the consideration of satellites deployed in LEO to reduce the slant range, and therefore the required radiated power of the user terminal (phone) to physiologically acceptable levels. This simple driver gave rise to a large number of proposals for *personal global communication systems*, on the basis of LEO constellation configurations. Some of these proposals have become a reality, for example *ORBCOMM*, *IRIDIUM* and *GLOBALSTAR*, and others are to be built imminently (at the time of writing). For example, the IRIDIUM constellation comprised 66 satellites (plus one spare per plane) deployed at 780 km altitude in 6 equally spaced orbit planes, inclined at 86.4° to the equator.

This activity was dampened somewhat by the unfortunate financial demise of the *IRIDIUM* system around 1999. However, these problems were overcome, and the system operation was resumed soon afterwards. The overall trend is now one of growth, and the mood one of optimism. Although communication systems have led the trend, other applications for LEO constellation systems are being proposed and implemented. These are civilian and military remote sensing systems. As sensor technology progresses resulting in greater spatial resolution, the next step is the improvement of temporal resolution through the use of constellations. One aspect that inhibits this development is, of course, the cost involved in launching multiple satellites. This, however, is offset to some degree by the development in recent years of capable, small satellites, and their use in constellations. This theme is developed further in Chapter 18.

At the mission design phase of a project, there is the need to determine whether the objectives can be achieved with a single satellite or a constellation. The main advantage of a single spacecraft is clear—the cost benefit of a single build, launch and operations activity. On the other hand, a collection of satellites in a constellation will offer considerable improvements in *coverage*, in *reliability*—the system will degrade more gracefully in the event of a failure, and *survivability* in the case of a military system. In some instances, the requirements make the choice of a constellation mandatory—for example, the Navstar GPS user must be able to see at least four spacecraft with a diverse line-of-sight geometry to acquire a good navigational fix.

Once the need for a constellation is clear, the design of its geometry is a non-trivial problem, since there are a large number of possibilities (see Section 5.5.3). Since coverage is the prime benefit derived from a constellation, the choice normally involves a trade-off using coverage (as a measure of performance) against the number of spacecraft (as a measure of cost). However, even this process is not straightforward. For example, a constellation with $t + 1$ members may exhibit a step function in improved performance over one with t members, but the larger constellation may be deployed at a lower height or inclination, so reducing the cost of launch.

5.5.2 Ground coverage

Figure 5.10 shows a typical ground coverage geometry, with the satellite at height h. The 'viewing angle' of the payload is 2α. This may be the field of view of an imaging system or the beamwidth of a communications antenna. This defines a *circle of coverage* on the ground of geocentric semi-angle ϕ and diameter D km. The satellite elevation and slant range at the edge of coverage (useful concepts for communications missions) are ε and ρ, respectively.

From Figure 5.10, we have

$$\varepsilon = \pi/2 - \phi - \alpha \tag{5.22}$$

and (as seen in Section 5.3.1),

$$\phi = -\varepsilon + \cos^{-1}\left(\frac{R_E}{R_E + h}\cos\varepsilon\right) \tag{5.23}$$

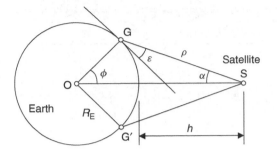

Distance over the ground GG′ = D = coverage diameter.

Figure 5.10 Typical ground coverage geometry

The coverage diameter is then

$$D = 2\phi R_{\mathrm{E}} \tag{5.24}$$

with ϕ in radians, and the slant range is given by

$$\rho = R_{\mathrm{E}} \left(\frac{\sin \phi}{\sin(\pi/2 - \varepsilon - \phi)} \right) \tag{5.25}$$

From these geometrical considerations, it is possible to relate coverage aspects to orbit height and payload characteristics.

A further issue is the degree of coverage required. For example, for communications missions, the user may require to 'see' just one satellite at any time to enable the service. However, as mentioned above, the Navstar GPS user requires *multiple coverage*—in this case four-fold coverage. Line of sight to at least four spacecraft is required at any time, to acquire the information needed to solve four unknowns—the three components of the user's position, plus the offset between the GPS satellite time and the user's clock. Figure 5.11 illustrates the concept of multiple coverage. The central horizontal line represents one orbit plane in the constellation and the dots denote the s satellites in that plane. The integers give the instantaneous levels of multiple coverage within the intersecting circles of coverage. The figure also illustrates the concept of *streets of coverage*, introduced in Reference [6] and developed by Adams and Rider [7], which is useful in analysing constellation systems requiring multiple levels of coverage. The street width giving j-fold coverage is given by

$$d_j = 2 \cos^{-1}(\cos \phi / \cos(j\pi/s)), \quad j = 1, 2, 3, \ldots, \quad s \geq 3 \tag{5.26}$$

where s is the number of satellites in each orbit plane. It should be noted that multiple coverage can also be provided by the overlapping of streets of coverage from neighbouring orbit planes, particularly at high latitudes where orbit planes converge.

Coverage is often the crucial issue when deciding upon a constellation geometry. Coverage statistics can be accumulated by numerical simulation of the candidate constellation designs and an appropriate *coverage figure of Merit* devised to distinguish between them.

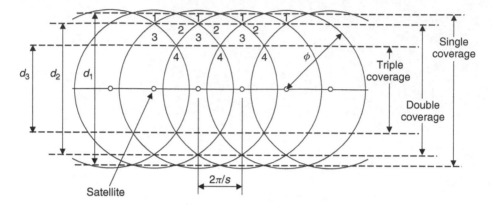

Figure 5.11 Constellation orbit plane geometry, illustrating multiple coverage and 'streets of coverage'

5.5.3 Constellation geometries

As seen below, there are a large number of possible constellation configurations that may satisfy a particular mission requirement. One factor that is common to the majority of these designs, however, is the invariance of the constellation geometry when subjected to orbital perturbations. In other words, it is usually the case that the orbit of each member satellite is similar in terms of height, eccentricity and inclination, so that each orbit is perturbed by the effects described in Chapter 4, Section 4.4 in the same manner (to first order). In this way, the constellation geometry can be preserved without recourse to an excessive station-keeping requirement. Generally, the only acceptable exception to this is the combination of inclined and equatorial orbit planes in a constellation design.

A further factor that should be considered in the design is the use of appropriate phasing of the member satellites within each orbit plane to ensure that collisions are avoided at orbit plane intersections. This is dealt with in some detail in Reference [7]. The most commonly considered constellation designs are described below.

Walker Delta pattern constellation. This geometry, first proposed by Walker [8, 9], can be uniquely specified by a small number of parameters; namely, the total number of satellites t, the number of orbit planes p, an integer f that determines the relative spacing between satellites in adjacent planes, and finally the orbit inclination i. Therefore, the Delta pattern denoted by

$$i : t/p/f, \qquad 0 \le f \le (p - 1)$$

comprises a design with p orbit planes, the ascending nodes of which are equally spaced in right ascension. All orbits are nominally circular, and of the same height and inclination. Each orbit plane contains $s(= t/p)$ satellites, equally spaced in true anomaly. To define the phasing of satellites within the pattern, Walker introduced a 'pattern unit' $PU = 360°/t$. The relative phasing is defined by the condition that, if a satellite is at its ascending node, then the satellite in the next most-Easterly plane will be $f \times PU$ past its node. An example, specified by $55°:25/5/1$, is illustrated in Figure 5.12, in which the $PU = 360°/25 = 14.4°$. In the figure, satellite one is positioned at its ascending node, with satellite six, therefore, one PU beyond its nodal position in the next most-Easterly plane.

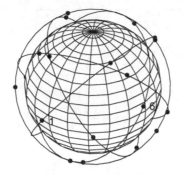

Figure 5.12 A Walker Delta pattern constellation, defined by 55°: 25 / 5 / 1

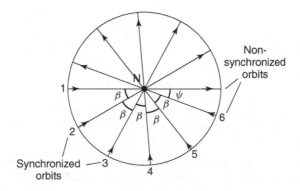

Figure 5.13 Illustration of a six-plane polar constellation

n-plane polar constellations. As the name suggests, this design comprises *n* orbit planes, each of which is of polar (or near-polar) inclination. Distributed within each of the *n* planes is an equal number of satellites. Walker [8] originally referred to this type of geometry as a (Walker) 'Star pattern' constellation. Figure 5.13 shows, as an example, the geometry of a six-plane polar constellation, as viewed when looking down on the North Pole of the Earth. As can be seen, most adjacent orbits are 'synchronized' in that the satellites are co-rotating. However, by necessity two adjacent planes must be 'non-synchronized', in which the satellites are counter-rotating. The manner in which the coverage is achieved in these two situations is dissimilar, and generally leads to different nodal spacing of the orbit planes.

Consequently, the *n*-plane polar constellation is not an example of a Walker Delta pattern. Also, given the general convergence of the orbit planes at the poles, issues of in-plane phasing become critical to avoid collisions, which in turn can further affect coverage. A good example of the development of an *n*-plane polar constellation for a personal communication satellite system with global coverage is given by Benedicto *et al.* [10]. This demonstrates well, the interactions between coverage and satellite/orbit-node separation. Figure 5.14 illustrates their solutions, specifying the total number of satellites required as orbit height varies, for minimum 'edge of coverage' elevations of 10°, 30°

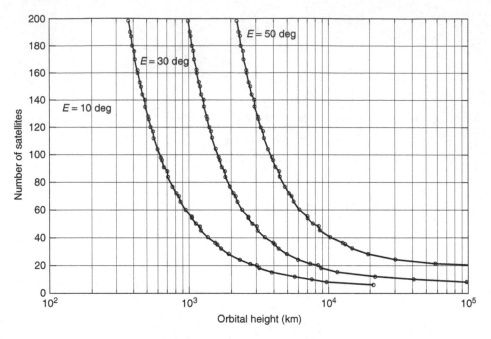

Figure 5.14 Number of satellites as a function of orbit height and minimum elevation. Reproduced by permission of ESA (authors: J. Benedicto *et al.*) (from Reference [10])

and 50°. This clearly demonstrates the strong dependence of the total number of satellites t on orbital altitude h, which is a general feature of constellations.

Geometries for regional coverage. The geometries discussed so far are useful candidates if global coverage is a requirement. However, if regional coverage is needed, a simpler constellation geometry should be sought, to gain the cost benefit of deploying fewer spacecraft. It is easy to see, for example, that coverage required within say 20° or 30° latitude of the equator, can be fulfilled by a number of satellites phased in true anomaly in an equatorial orbit. Similarly, regional coverage of the polar caps may be achieved by satellites optimally distributed in one (or two) near-polar orbits. To optimize coverage at mid-latitudes (30° to 60°), a region containing Europe, North America, the CIS, S. Australia and Japan, recourse to a number of orbit planes inclined at ∼55° to the equator may suffice. This utilizes the useful tendency for neighbouring orbits to converge at their Northern and Southern apexes. Consideration can also be given to the use of eccentric orbits when seeking solutions to regional coverage problems, bearing in mind the need to preserve the geometry when subjected to natural perturbations.

5.5.4 Design issues

In terms of constellation design, there are unfortunately no general rules to aid the designer in solving what is often a complex problem. The only semblance of a rule is perhaps the above-mentioned invariance of the geometry with respect to perturbations. However, even this was not considered a constraint by the designers of the ORBCOMM constellation, which remarkably exhibits a variety of orbit heights and inclination.

An observation that may assist in the design process is that there is often an advantage in deploying more satellites in fewer orbit planes. Underlying this statement is the relative ease with which spacecraft can be moved within an orbit plane, compared to the difficulty of moving them from one plane to another. The former involves a small, in-plane phasing manoeuvre, whereas the latter requires a prohibitively expensive plane change manoeuvre. In a constellation with p orbit planes, *performance plateaux* occur with the launch of the first satellite, then with one satellite in each plane, and so on. The major plateaux therefore occur with the launch of $1, p, 2p, 3p, \ldots$ satellites. Clearly, these plateaux occur more frequently, and with fewer spacecraft when the number of planes p is small, giving a cost benefit. Similarly, when considering satellite failures within a constellation, there is a greater degree of *graceful degradation* in performance in systems with fewer orbit planes. For example, in a system with t satellites, equally distributed in p planes, there are $s = t/p$ satellites in each plane. If a satellite was to fail, then the whole constellation could be re-phased with little cost, to a configuration with $s - 1$ satellites in each plane, corresponding to the $p(s - 1)$ performance plateau. Lower values of p again give benefits when considering degradation. However, despite these comments, it is often the coverage requirements that dictate the number of orbit planes, so that low values of p may not be possible.

5.6 GEOSTATIONARY EARTH ORBITS (GEO)

The utility of the geostationary orbit for providing global communications was first noted by Arthur C. Clarke [11]. Its primary attribute is that the sub-satellite point is fixed at a selected longitude, with $0°$ latitude. It does not have dynamic tracking problems. GEO spacecraft may therefore provide fixed-point to fixed-point communications to any site within the beam of their antennas. Figure 5.15(a) shows the horizon as viewed from GEO, and the region over which the satellite appears with an elevation in excess of $10°$; Figure 5.15(b) demonstrates that only three satellites are required to provide almost a global communications network.

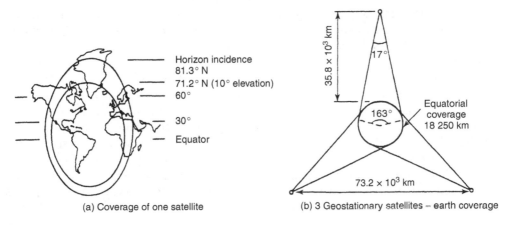

(a) Coverage of one satellite

(b) 3 Geostationary satellites – earth coverage

Figure 5.15 Geostationary views of the Earth (a) coverage of one satellite and (b) 'global' coverage using three satellites

The 24-hour geostationary orbit clearly offers unique advantages, providing almost complete global coverage (except for the immediate polar regions) from merely three satellites, and with no need for the ground antenna to switch between satellites. A potential disadvantage of GEO communication systems is the time delay for a transmission to reach its destination. This is clearly less of a problem for data services (e.g. internet, e-mail, etc.) than for live speech.

For a service of high quality and reliability, it is necessary to consider the control and replacement policy for a GEO spacecraft. The failure of a spacecraft would cause substantial financial penalties to the system operator, and since the ability to replace it is dependent on launcher availability, with substantial time being needed, the philosophy of having an in-orbit spare is frequently adopted. This spare is at a slightly different longitude, and offers the advantage of extra capacity should it be required.

Maintaining a spacecraft's orbit is an essential requirement for maintaining a communication link. The capability and method of achieving GEO and maintaining a specific location will now be discussed and fuel requirements for station-keeping will be indicated.

5.6.1   Geostationary orbit acquisition

The final stage of the launch vehicle will place the satellite in a nominal orbit. During the 1980s, the US Space Shuttle was used to launch commercial geostationary communication spacecraft. In order to make up the launcher's payload complement, typically two or three such spacecraft were launched at a time. In more recent times, the Shuttle has been used to launch large science, communications or military payloads that warrant its large lift capability, some of which were destined for GEO. However, with the anticipated retirement of the Space Shuttle system in 2011, *conventional expendable launch vehicles* (ELVs) will take over its role for the foreseeable future.

An Ariane 5 launch scenario, for example, is typical although the details may differ depending upon which ELV is considered. The final stage of the Ariane 5 launcher places the satellite directly into a *geostationary transfer orbit* (GTO), after a powered flight of the order of ∼27 min. The GTO is an elliptical orbit with an apogee near geostationary altitude (∼35 786 km), a perigee height between 200 and 650 km altitude, and an inclination of 7° to the equator. Further, the apogee of the GTO must occur at an orbit node (equator crossing) so that subsequently a single firing of an *apogee boost motor* (ABM) may both circularize and change the plane to become equatorial.

Several transfer orbit revolutions occur before injection of the satellite into a near-circular, near-GEO orbit. This period is essential for attitude manoeuvres and determination, and for tracking the satellite and determining its orbit before the ABM is fired. This motor increases the velocity of the satellite from ∼1.6 km/s to ∼3 km/s at apogee. Some satellites may be spin-stabilized during the entire transfer and orbit-acquisition phase, although vehicles using liquid apogee engines (low thrust) may be three-axis-stabilized. The GTO attitude and orbit control activities, and consequently the launch window (see Section 5.3.3), are very much dependent on the particular spacecraft and its system constraints. When specifying the time of launch, however, the constraints considered generally include

— the duration of eclipse periods, and the exposure of sufficient solar array surface to the Sun (particularly for spin-stabilized spacecraft) to ensure adequate power

— the Sun-relative attitude, and eclipse durations, to ensure thermal control can be maintained
— the relative Sun-Earth-spacecraft geometry, to ensure that the vehicle's sensors can provide information of adequate accuracy for attitude determination
— Sun blinding of sensors and so on.

For spinners, the orientation of the spin axis, and its control during motor firings, is particularly crucial not only for reasons of orbit attainment but also for power raising and thermal control. These latter constraints typically require a local midday or local midnight apogee firing to take place since the spin axis will be oriented along the thrust vector (see Figure 5.16).

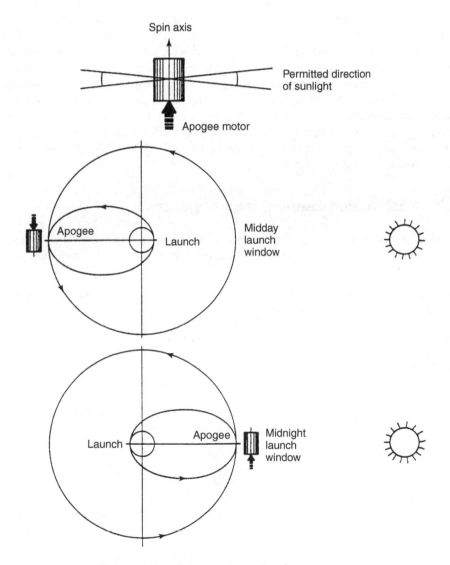

Figure 5.16 Vehicle orientation for injection

For vehicles that utilize a *liquid apogee motor* (LAM), a single firing at apogee is insufficient to transfer the vehicle into the desired near-GEO orbit. The low thrust of such motors ($\sim 10^2$ N rather than $\sim 10^4$ N for a solid rocket) would result in the need to rotate the thrust axis during firing in order to deliver the required impulse. Instead, typically number of firings of the LAM are utilized during successive apogee transits. As mentioned above, these apogee manoeuvres not only circularize the transfer orbit, but also rotate it into the equatorial plane. The cost of this plane change is directly proportional to the GTO inclination, so that the low inclination of the Ariane transfer orbit, for example, is preferable in this respect. To further reduce the Δv of this composite apogee manoeuvre, a *supersynchronous transfer orbit* is often used, which has an apogee in excess of the GEO altitude. Since the spacecraft moves more slowly at this higher altitude, the cost of the plane change is reduced. An additional in-plane burn is then required to lower the apogee, but the overall propellant mass can be reduced using this strategy.

If the one-impulse solid propellant ABM is used, there is a requirement to place the satellite into a slightly elliptical orbit, termed a drift orbit, whose apogee is at GEO altitude. Its period is less than the GEO period, the precise value being chosen to minimize the ΔV requirements, whilst maintaining a reasonable GEO attainment strategy. As a consequence, the satellite progressively drifts eastward. Through a sequence of manoeuvres in both latitude and longitude, it is eventually brought to rest at the required GEO location. These manoeuvres are similar to those described below for station-keeping. Final *station acquisition* may be up to a month after satellite launch. Table 5.3 summarizes some of the attributes of the LAM [12].

5.6.2 GEO station-keeping requirements

The primary disturbances on a satellite orbit were described in Chapter 4. The dominant effects for GEO are luni-solar perturbations, Earth triaxiality and solar radiation pressure.

Table 5.3 Characteristics of the liquid apogee motor for GEO acquisition ($\sqrt{}$ denotes advantage, × denotes disadvantage)

Attribute	Comments
Multiple engine restarts possible	Burn can be segmented $\sqrt{}$
Longer duration motor firings	Lower acceleration $\sqrt{}$
	Reduced efficiency ×
Burn duration controllable	Improved reaction to launcher errors $\sqrt{}$
	Improved reaction to execution errors $\sqrt{}$
	Manoeuvres can be calibrated $\sqrt{}$
Drift orbit not required	Satellite on-station sooner $\sqrt{}$
	Reduced propellant requirement $\sqrt{}$
Common fuel supply for primary and secondary propulsion	Fuel savings provide increased life on-station $\sqrt{}$
Extended transfer orbit phase	Increased cost of ground station support ×
	Power and thermal problems ×
More complex propulsion system	Cost ×
	Control problems induced by 'fuel slosh' ×

Solar radiation perturbation is complex to model and depends critically upon both the ratio of projected area to mass of the vehicle and its surface characteristics. Its predominant influence is upon the orbit eccentricity vector **e**. Generally, an effective area-to-mass ratio, which includes the reflectivity coefficient to solar illumination, is used. When this is less than $\sim 0.005\,\mathrm{m^2/kg}$, the effects of radiation pressure are significantly less than other perturbations. However, for larger values ($>0.01\,\mathrm{m^2/kg}$) it can cause significant perturbations.

The generalized techniques for examining the influence of perturbations, which were noted in Chapter 4, are not necessarily the optimum methods for analysis of GEO. This is particularly so since the orbit is ideally circular, and hence ω is undefined; the inclination is also ideally zero, resulting in Ω being undefined. A method that is applicable is to linearize the equations of motion for small perturbations of magnitude δa in the semi-major axis a. For small values of δa, i and e, a linearized solution to Kepler's equation yields [13]

$$r = A + \delta a - Ae \cos\left((t - t_0)\sqrt{\frac{\mu}{A^3}}\right) \tag{5.27}$$

$$\lambda = \Omega + \omega - t_0\sqrt{\frac{\mu}{A^3}} - \frac{3}{2}\frac{\delta a}{A}(t - t_0)\sqrt{\frac{\mu}{A^3}} + 2e\sin\left((t - t_0)\sqrt{\frac{\mu}{A^3}}\right) \tag{5.28}$$

$$\theta = i\sin\left(\omega + (t - t_0)\sqrt{\frac{\mu}{A^3}}\right) \tag{5.29}$$

where A is the semi-major axis of a truly geostationary orbit ($42\,164.5\,\mathrm{km}$), λ is the satellite longitude and θ its latitude.

The evolution of r, λ and θ with time, as shown in Figure 5.17, is easily described with the aid of these equations. The geocentric distance r (equation 5.27) oscillates simply about a mean distance $(A + \delta a)$ with an amplitude of Ae. Similarly, the latitudinal variation, from equation (5.29), is a simple oscillation about the equatorial plane with an amplitude equal to the orbital inclination. The longitude, however, manifests an oscillation of amplitude $2e$, but also a drift rate. It is worth noting that for circular orbits whose semi-major axes are given by A and whose inclinations are non-zero, the ground track will be a figure of eight of the form shown in Figure 5.20.

Earth triaxiality perturbation is due to the dominant J_{22} Earth tesseral harmonic that produces a force on the spacecraft whose direction is shown in Figure 5.18 [13]. From Section 5.2, it is evident that a positive ΔV will expand an orbit, changing both the semi-major axis and the eccentricity. Since an expanded orbit relative to GEO has a longer period, a positive ΔV (i.e. an eastward impulse) will lead to westward drift in satellite longitude. Hence, referring to Figure 5.18, the longitude drift rates are opposed to the acting force direction, and it is apparent that $105.3°\mathrm{W}$ and $75.1°\mathrm{E}$ are stable equilibria against equatorial drift. The magnitude of the acceleration as a function of longitude is shown in Table 5.4.

Luni-solar perturbations mainly cause out-of-plane forces acting on the spacecraft, leading to changes of inclination of the orbit. However, since a component of this force necessarily lies in the equatorial plane, then a, e and hence λ are also influenced. This equatorial component is nearly cyclic with the Earth orbit period, with a magnitude less than that caused by the J_{22} term. It therefore does not lead to a significant secular evolution of the orbit. More details may be found in Reference [13].

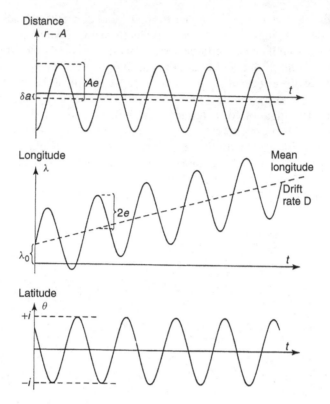

Figure 5.17 Linearized unperturbed spacecraft motion, given as distance (r), longitude (λ) and latitude (θ)

Returning to the out-of-plane components due to Moon and Sun, the periodic change of 18.6 years of the lunar orbit plane, plus precession of the Earth's spin axis, results in a 54-year period for the evolution of an uncontrolled geostationary orbit. The net force on the orbit plane evolution, shown in Figure 5.19, must clearly be opposed by an opposite ΔV. The magnitude of the ΔV arising from the luni-solar perturbation is typically ~50 m/s/year and thus the propulsion requirements for the control of orbit inclination (i.e. north/south station-keeping) is at least a factor of 20 larger than that required to overcome triaxiality (i.e. east/west station-keeping) effects (see Table 5.4). This results in the need for particularly careful control and planning of manoeuvres in order to avoid disadvantageous coupling between north/south and east/west control. This interaction is discussed in the following two sections.

In summary, the nature of station-keeping is to locate the spacecraft under the influence of the perturbations described within a specified range of longitude and latitude. This situation is shown in Figure 5.20. Whilst the apparent position of the spacecraft projected on the celestial sphere provides information to the ground control centre indicating whether the spacecraft is within the correct longitude/latitude region, the range and range rate are also required so that the actual orbit may be determined. This nominal orbit may then be used to schedule station-keeping manoeuvres to preclude departure from the required longitude/latitude location box. The size of this box necessarily impacts on the

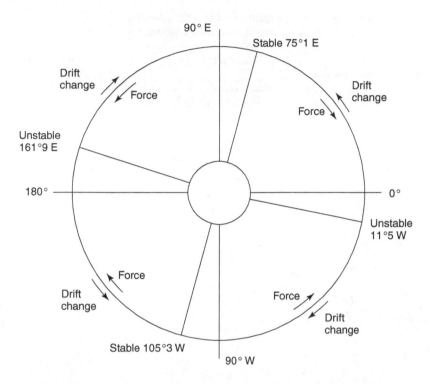

Figure 5.18 Force direction arising from J_{22} on a GEO vehicle

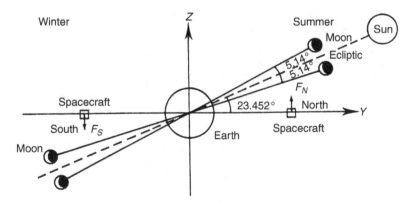

Figure 5.19 Out-of-plane forces F_N, F_S on a geostationary orbit. The Y-axis lies in the equatorial plane and the Z-axis is the Earth's spin axis. The diagram shows positions of the Sun and Moon in summer and winter, relative to the GEO spacecraft

Table 5.4 Acceleration and station-keeping requirements for geostationary vehicles

Longitude (degrees east)	Acceleration (m/s^2)	Fuel: ΔV (m/s/year)
−160	−5.3 × 10^{-8}	1.67
−140	−4.75 × 10^{-8}	1.50
−120	−2.30 × 10^{-8}	0.73
−100	+8.10 × 10^{-9}	0.26
−80	+3.39 × 10^{-8}	1.07
−60	+4.61 × 10^{-8}	1.45
−40	+3.99 × 10^{-8}	1.26
−20	+1.48 × 10^{-8}	0.47
0	−2.10 × 10^{-8}	0.66
20	−5.09 × 10^{-8}	1.61
40	−5.73 × 10^{-8}	1.81
60	−3.30 × 10^{-8}	1.04
80	+1.17 × 10^{-8}	0.37
100	+5.21 × 10^{-8}	1.64
120	+6.49 × 10^{-8}	2.05
140	+4.44 × 10^{-8}	1.40
160	+4.08 × 10^{-9}	0.13
180	−3.40 × 10^{-8}	1.07

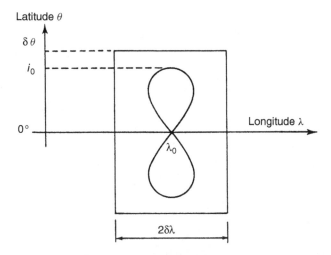

Figure 5.20 Dead band of inclined geosynchronous vehicle. The figure of eight demonstrates the apparent ground track of a synchronized GEO satellite inclined at an angle i_0 to the equator

station-keeping schedule: the tighter the requirements the greater the impact. Requirements for ~0.1° are typical, but this will become less when allowance is made for sensor errors.

5.6.3 Longitude station-keeping

The linearized approach noted above gives the subsequent change of longitude of a spacecraft following a small impulsive velocity increment ΔV as

$$\lambda = \lambda_0 + \frac{\Delta V}{r}\left\{-3t + 4\sqrt{\frac{A^3}{\mu}}\sin\left(t\sqrt{\frac{\mu}{A^3}}\right)\right\} \tag{5.30}$$

where t is the time since the impulsive burn. This is shown in Figure 5.21.

Longitudinal control can be effected by using an impulsive along-track burn when the spacecraft reaches one extreme of the required error box. The subsequent longitude drift due to J_{22} (see Figure 5.18) brings the spacecraft back to the same side of the box as shown in Figure 5.22, and the process is repeated as a limit cycle. This neglects the short period fluctuations caused by luni-solar perturbations.

If the ΔV is sized so that the spacecraft just reaches the opposite side of the box following a burn, and a constant acceleration f is assumed (as listed in Table 5.4), then the ΔV and the corresponding period T of the limit cycle are

$$\Delta V = 4\sqrt{\frac{rf\lambda_{max}}{3}} \quad \text{and} \quad T = 4\sqrt{\frac{r\lambda_{max}}{3f}} \tag{5.31}$$

where r is the orbit radius ($\sim 42 \times 10^6$ m) and the $\pm\lambda_{max}$ represents the box size.

Thus, for example, a satellite at a nominal longitude of $-160°$ ($f = -5.3 \times 10^{-8}$ m/s^2) requiring an error box of $\lambda_{max} = 0.1°$ (remember to convert to radians) would require $\Delta V = 0.14$ m/s and the period of the limit cycle would be $T = 31$ days.

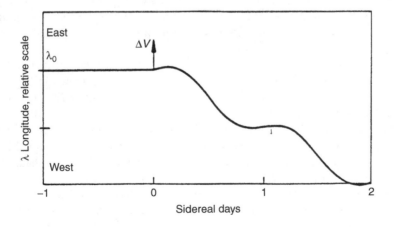

Figure 5.21 Longitude evolution following an east burn

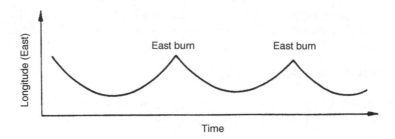

Figure 5.22 Parabolic shape of mean longitude free drift with station-keeping by east burns. In this example, the satellite is stationed where the triaxiality drift is Easterly, and the burns induce a Westerly drift

This technique could be referred to as the 'ping-pong method of East–West station-keeping', since a little thought shows that it is entirely analogous to repeatedly tapping a ping-pong ball on a bat. The impact on the bat corresponds to the impulsive burns, the constant gravitational acceleration represents the constant longitudinal acceleration within the dead band, and the constant height amplitude of the ball's motion reflects the size of the longitude dead band. This rather whimsical observation may aid understanding of this method of East–West station-keeping, since the two situations are physically equivalent.

5.6.4 Latitude station-keeping

Latitude drift may usefully be approached by introducing the two-dimensional vectors **i** and **e** given by Reference [13]:

$$\mathbf{e} = \begin{bmatrix} e_x \\ e_y \end{bmatrix} = \begin{bmatrix} e\cos(\Omega + \omega) \\ e\sin(\Omega + \omega) \end{bmatrix}$$

$$\mathbf{i} = \begin{bmatrix} i_x \\ i_y \end{bmatrix} = \begin{bmatrix} i\sin\Omega \\ -i\cos\Omega \end{bmatrix}$$

The natural evolution of inclination during a five-year period is shown in Figure 5.23, corresponding to a number of initial values of Ω. This wavy drift shows the half-yearly effect caused by the motion of the Earth about the Sun. The lunar periodicity of 14 days is smoothed out in this representation.

The objective of latitudinal station-keeping is to maintain the trajectory of the inclination vector within a specified region of the i_x, i_y plane. In the case shown in Figure 5.23, for example, the inclination is to be constrained within a circle of radius 2.5°, which represents a maximum orbit plane inclination of 2.5°.

The philosophy of inclination control is similar to that of longitude, a correcting nodal burn taking place just before the inclination drifts to the maximum permitted by the error box, $\pm i_{max}$, say. If the spacecraft's speed is V, then the directional change needed is $2i_{max}$, and this will require a ΔV of $2Vi_{max}$ (see equation 3.21).

For example, if $i_{max} = 0.1°$ and $V = 3075\,\text{m/s}$, then $\Delta V = 10.7\,\text{m/s}$.

Clearly, the ΔV impulses for controlling inclination are appreciably greater than those needed for longitude, and their errors due to cold starts will therefore represent a smaller percentage error.

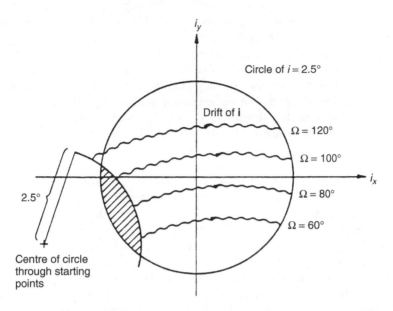

Figure 5.23   Uncontrolled evolution of inclination vector **i** in component form i_x, i_y

Table 5.5   GEO burns direction

	North burn	South burn
Spring	Morning	Evening
Summer	Midnight	Noon
Autumn	Evening	Morning
Winter	Noon	Midnight

Latitude manoeuvres imply a change of the orbit plane. Such manoeuvres must be performed on the line of nodes, using north/south burns at times of day indicated by Table 5.5. Ideally, the thrust vector should be perpendicular to the plane bisecting the initial and required orbit planes (see equation 3.21), and should produce no change in the spacecraft's speed. The thrust direction should be accurate since a directional error ε will lead to a speed change of approximately $2Vi_{max}\varepsilon$, and this would influence the longitude drift rate; errors of only a few degrees could swamp the natural longitude drift rates noted above.

5.7 HIGHLY ELLIPTIC ORBITS

5.7.1 Space-borne observatories

Spacecraft in elliptic orbits move more rapidly at perigee than at apogee (equation 4.21). This offers the prospect of a pass of increased duration over a ground station if the apogee is situated above it.

This type of orbit has been used to good effect in the past for astronomical observatories such as ISO and the X-ray Multi-mirror Mission (XMM Newton). These types of spacecraft are often used in *observatory mode*, which means the spacecraft instruments are operated as if they were located in a room adjacent to the astronomer's workstation. This requires that for extended periods of time the payload can be pointed to desired astrophysical targets whilst uninterrupted contact with a ground station is maintained. By placing such a spacecraft into highly eccentric orbits with a low perigee and an apogee substantially higher than the GEO radius, the observatory will spend the majority of its mission time in the apogee region. By optimizing its orbit period with respect to the ground station coverage, operation in observatory mode can be achieved.

Generally, there will be an interruption of observational time while the spacecraft passes through the perigee region. This also corresponds, however, with a traverse of the Van Allen radiation belt (see Chapter 2), which precludes the operation of certain types of payload, such as γ-ray, X-ray and UV detectors, which are particularly sensitive. Observation duty cycles in excess of 90% can be achieved in these orbits.

The alternative of placing this type of spacecraft into GEO has many obvious benefits such as ground station cover and long sensor integration times, but GEO is generally a higher energy orbit than a highly elliptical one, requiring greater ΔV to place it there. Moreover, the radiation environment in GEO may also preclude the operation of certain types of payload.

5.7.2 Communications spacecraft

In order to maintain the apogee above a ground station for a maximum period, the Earth-oblateness perturbation, which causes the line of apsides to precess can be used (Chapter 4, Section 4.4.1). From equation (4.39), the major axis of the orbit will rotate in the plane of the orbit by an amount, to first order, given by

$$\Delta\omega \approx \frac{3\pi J_2 R_E^2}{p^2} \left(2 - \frac{5}{2} \sin^2 i \right) \quad \text{rad/rev} \tag{5.32}$$

This will cause the apogee of the orbit to move away from above the region where the service is required.

However, there are two solutions of equation (5.32) for which $\Delta\omega = 0$, namely, $i = 63.4°$ (prograde) and $i = 116.6°$ (retrograde). At these inclinations, the line of apsides does not rotate and the apogee is stabilized at its initial position. Using this characteristic, spacecraft in highly elliptic orbits inclined at 63.4° have been used to provide communication links between high-latitude ground sites. Spacecraft in GEO cannot provide this service at latitudes in excess of 70°, North or South, since their elevation at ground sites is then less than 10°.

Molniya spacecraft

The use of highly elliptic inclined orbits was adopted by the former Soviet Union in April 1965 with the launch of the first Molniya (lightning) spacecraft.

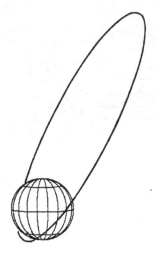

Figure 5.24   The Molniya orbit

The *Molniya orbit*, shown in Figure 5.24, is a highly elliptic one with a 12-hour period (strictly half a sidereal day, 11 h 58 min), inclined at 63.4° to the equator. The initial apogee, at a height of around 39 000 km, is placed above the Northern hemisphere, where it remains for the mission lifetime as a result of the choice of orbit inclination. Because the eccentricity is high, typically about 0.7, the spacecraft moves relatively slowly in the apogee region, and it consequently remains at high elevation over the high-latitude sites for an extended period. Each such spacecraft may then be utilized for communications purposes for typically 8 h of its 12-h orbit period.

To provide 24-h regional services, at least three Molniya spacecraft are needed. One of these resides in each of the three planes whose ascending nodes are spaced 120° apart. Many Molniya spacecraft have been launched; communications traffic in the CIS has grown to warrant the operation of typically eight Molniya I and eight Molniya III spacecraft in eight orbit planes spaced at 45° intervals.

A nominal set of Molniya orbital elements is given in Table 5.6. The ground track is shown in Figure 5.25(a). Since the spacecraft is at near-GEO altitude at apogee, its angular rate is approximately that of the Earth, and so the ground track over the required region closely follows a meridian as shown in Figure 5.25(a).

Interest in Molniya orbits is not confined to the CIS. Other space agencies are considering their use for high-latitude mobile communications and navigation programmes.

Table 5.6   Molniya orbit elements

$a = 26\,560$ km (12 h orbit period)
$e = 0.722$ ($h_\mathrm{p} = 1000$ km, $h_\mathrm{a} = 39\,360$ km)
$i = 63.4°$
$\omega = 270°$ (perigee in southern hemisphere)
Ω arbitrary (dependent upon regional coverage required)

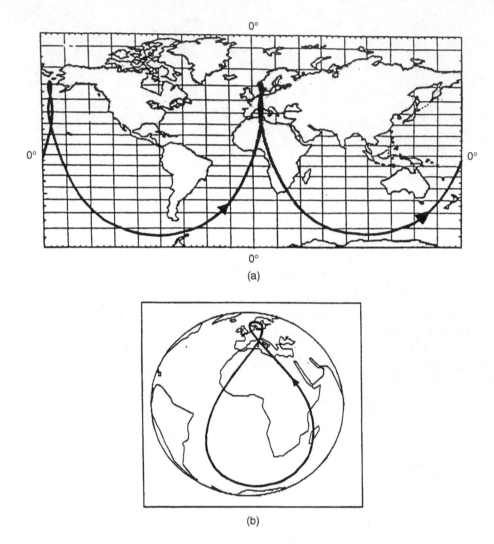

Figure 5.25 Typical global ground tracks for: (a) Molniya orbit and (b) Tundra orbit

Tundra orbit

The tundra orbit is a derivative of the Molniya orbit in that it is elliptical and has an inclination of 63.4°. The principal difference is that its period is one sidereal day—23 h 56 min. An example of a global ground track for a Tundra orbit with an eccentricity of 0.24 is shown in Figure 5.25(b). Its main advantages are that it can provide 24-h coverage with a minimum of only two spacecraft, and that the orbital parameters can be chosen so that the spacecraft does not traverse the Earth's radiation belts. Its main disadvantage is that a higher ΔV is needed for orbit acquisition.

Table 5.7 summarizes the advantages and disadvantages for communications purposes of the inclined highly elliptic orbit compared with GEO.

Table 5.7   Principal attributes of highly inclined communications orbits

Advantages	Comments
• Satellite at high elevation at high latitude ground sites • No eclipse during communications operation	This is an overriding benefit for mobile communication services at high latitude

Disadvantages	Comments
• Ground stations must track spacecraft • More than one satellite required for 24-hour regional coverage • Satellite switching protocol required • Variation in satellite range and range-rate	This has a number of impacts upon the communication payload design: — variation in time propagation — frequency variation due to Doppler effect — variation in received signal power — change of ground coverage pattern during each orbit
• Passage through Van Allen radiation belts each orbit	Accelerated degradation of power and electronic systems
• Orbit perturbations	Third-body forces may perturb the perigee height, causing atmospheric re-entry

5.8   INTERPLANETARY MISSIONS

5.8.1   Introduction

It is apparent that in the analysis of interplanetary trajectories, due account must be taken of perturbations caused by the gravitational influence of the Sun and planetary bodies within the solar system, and to a lesser extent to surface forces arising from solar radiation pressure.

For interplanetary spacecraft, the task of trajectory optimization and mission design is one of considerable complexity, requiring software support. Mission specialists at the NASA Jet Propulsion Laboratory in Pasadena have developed such computational tools to analyse complex missions such as that of the Jovian Galileo spacecraft, the Cassini mission to explore the Saturnian system, or the near-Earth asteroid rendezvous (NEAR) spacecraft to orbit a small body. Such a treatment is beyond the scope of this book. However, there is a simplified method, referred to as the *patched conics method*, which is widely used in feasibility studies of interplanetary flight.

The patched conics method assumes, in its simplest form, that at any one time only one body, a central body, is acting on the spacecraft. The region in which a particular body is dominant is called its *sphere of influence*, and in this region the spacecraft is assumed to follow a Keplerian orbit. As it leaves one region, it enters the sphere of influence of another body and its trajectory becomes a new conic with the new body at its focus. Thus, the entire trajectory is formed by patching together the various conic sections. Examples are shown in Section 5.8.2.

Spheres of influence are determined as in the following example of the Earth, which is subject to a disturbing gravitational field caused by the Sun. That is to say, a spacecraft at a distance r_s from the Earth, and subject to its dominant gravitational force, is also subject to a weaker force from the Sun, from which its distance is r_d (see Figure 4.14, Chapter 4). Equation (4.45) gives the ratio of the accelerations caused by the Earth as the central body and the Sun as the disturbing one. When this ratio is equal to that which is given by reversing roles (i.e. the Sun is the central body and the Earth is the disturbing one), then the spacecraft is, by definition, situated on the sphere of influence of the Earth. Equation (4.45) can be used again, for the reverse role.

For the Earth–Sun, the acceleration ratios are equal when the Earth–spacecraft distance is

$$r_s = r_d \left(\frac{m_\oplus}{m_\odot}\right)^{2/5} (1 + 3\cos^2 \beta)^{-1/10} \tag{5.33}$$

where $m_\oplus$ and $m_\odot$ are the masses of Earth and Sun, respectively.

This surface approximates to a sphere since $(1 + 3\cos^2 \beta)^{-1/10} \approx 1$. Equation (5.33) then leads to the radius R_{SI} of the sphere of influence about any planetary body of mass m_p with respect to a disturbing body of mass m_d being

$$R_{SI} \approx r_d \left(\frac{m_p}{m_d}\right)^{2/5} \tag{5.34}$$

where now r_d is the distance of the disturbing body from the central planetary body.

This equation shows that the Earth's R_{SI} is about 0.93×10^6 km, assuming the Sun to be the disturbing body. (This compares with the radius of the Moon's orbit of about 0.38×10^6 km.) Reference [14] provides a more detailed analysis.

Table 4.1 shows the radii of the spheres of influence of the planets relative to the Sun.

5.8.2   Patched conic examples

The following three examples show how the initial planning of interplanetary missions may be conducted using the patched conic technique. They focus on the estimation of ΔV. The techniques for calculating a first estimate of transfer times on various trajectories can be found in Chapter 4, Section 4.2. Each phase of the total mission is assumed to have just one central gravitational force leading, for that phase, to a trajectory, which is a conic with the central body at its focus. During each separate phase, the position and velocity of the spacecraft are relative to the phase's central body, and at the handover from one to another they have to be recalculated.

Example 1: interplanetary transfer

The transfer of a spacecraft from an orbit around the Earth to an orbit around Jupiter is considered using a Hohmann minimum-energy transfer (see Section 5.2). The orbits of the Earth and Jupiter round the Sun are taken to be circular and coplanar. Their radii and other relevant data are taken from Tables 2.5 and 4.1.

The sequence of events can be defined by the following mission phases:

- Phase 1: Geocentric—the boost from Earth orbit to a hyperbolic escape trajectory.
- Phase 2: Heliocentric—the cruise in transfer orbit from Earth to Jupiter.
- Phase 3: Jupiter centred—the powered deceleration from hyperbolic approach trajectory to Jupiter orbit.

First step: heliocentric phase

Consider the Hohmann transfer within the Sun's sphere of influence (Figure 5.2). In the case of this interplanetary transfer, equation (5.1) gives the required Earth-relative speed of the spacecraft as it exits the Earth's sphere of influence, as

$$V_\infty^E = \sqrt{\frac{\mu_S}{r_E}} \left(\sqrt{\frac{2r_J}{r_E + r_J}} - 1 \right) = 8.792 \, \text{km/s} \tag{5.35}$$

Parameter values corresponding to the Sun, Earth and Jupiter are denoted by sub- or superscripts S, E or J, respectively.

Similarly, the apoapsis equation (5.2) gives the speed of the spacecraft relative to Jupiter as it enters the Jovian sphere of influence as

$$V_\infty^J = \sqrt{\frac{\mu_S}{r_J}} \left(1 - \sqrt{\frac{2r_E}{r_E + r_J}} \right) = 5.643 \, \text{km/s} \tag{5.36}$$

The transfer time is estimated as half the orbital period of the transfer ellipse, 2.731 years in this case.

Second step: geocentric phase

This phase covers the spacecraft's departure from the Earth, while it is still within its sphere of influence. The geometry of the trajectory is shown in Figure 5.26(a). The spacecraft is initially assumed to reside in a circular orbit of height $h = 300$ km. An impulsive velocity increment ΔV_1 is provided at point A, in order to inject the vehicle into a hyperbolic escape trajectory having a speed V_∞^E at a great distance from Earth, as given in equation (5.35).

Since the transfer is to be of the Hohmann type, the hyperbola's asymptote must be in the same direction as the Earth's orbital velocity vector. The escape hyperbola is then patched to the Hohmann transfer ellipse at the edge of the Earth's sphere of influence.

The required velocity increment ΔV_1 is calculated by noting that the spacecraft's speed before the manoeuvre at A, given by equation (4.22), is

$$V_{\text{circ}} = \sqrt{\frac{\mu_E}{R_E + h}} = 7.726 \, \text{km/s}$$

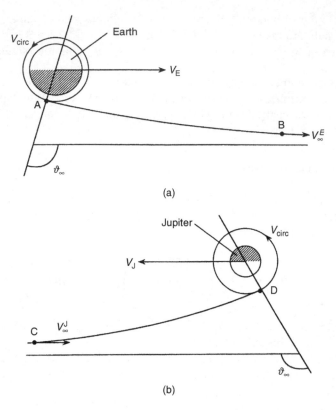

Figure 5.26 Geometry of planetocentric trajectories, showing: (a) departure from Earth, with rocket burn at A, and hyperbolic path to exit the sphere of influence at B; (b) arrival at Jupiter, entering its sphere of influence at C, hyperbolic path to D, and with a circularizing burn at D

After the manoeuvre, the speed is determined by noting that the hyperbolic path between A and B is at constant energy, and at B the potential energy is zero, so from equation (4.5):

$$\frac{1}{2}V_A^2 - \frac{\mu_E}{R_E + h} = \frac{1}{2}(V_\infty^E)^2$$

from which $V_A = 14.024$ km/s. So

$$\Delta V_1 = V_A - V_{circ} = 6.298 \text{ km/s}$$

The location of point A may be identified in terms of the angle θ_∞ (Figure 5.26a). Equation (4.32) may first be used, to find that the eccentricity $e = 2.2950$, and then equation (4.30) shows that $\theta_\infty = 115.8°$.

The final step: arrival at Jupiter

Within the Jovian sphere of influence the trajectory will appear as shown in Figure 5.26(b). The spacecraft enters the sphere at C with speed V_∞^J and follows a hyperbolic path until its conversion into a circular orbit of radius $6R_J$ at D, by means of an impulsive burn with velocity increment ΔV_2. It is assumed that D is the point of closest approach to Jupiter on the hyperbola.

On arrival at C, at the aphelion of the Hohmann ellipse, the spacecraft's heliocentric speed will be less than that of Jupiter; consequently the direction of its arrival will be opposite to Jupiter's direction of travel round the Sun.

The required velocity increment ΔV_2 at D may be calculated by using equation (4.5) to determine the speed V_D before the manoeuvre:

$$\frac{1}{2}V_D^2 - \frac{\mu_J}{6R_J} = \frac{1}{2}(V_\infty^J)^2$$

which leads to $V_D = 24.934$ km/s, and equation (4.22) to determine the circular speed V_{circ} afterwards ($V_{circ} = 17.173$ km/s). Thus

$$\Delta V_2 = V_D - V_{circ} = 7.761 \text{ km/s}$$

For this example, the eccentricity e is 1.1080 and θ_∞ is 154.5°.

Elliptic capture option

It must be noted that ΔV_2 can be reduced by specifying an elliptic capture orbit. For example, such an orbit with a perijove of $6R_J$ and an apojove of $20R_J$ will require a higher post-manoeuvre speed at D, that is, 21.301 km/s, and so ΔV_2 is reduced substantially, to 3.633 km/s.

Using a typical chemical propellant with a specific impulse of 300 seconds, the Tsiolkovsky equation (6.5) shows that the circular capture orbit above requires about 90% of the mass of the spacecraft that arrives at Jupiter to be fuel. For the elliptical capture case, this reduces to about 70%. The choice of capture orbit clearly has a major impact upon the spacecraft's mass budget and upon the mission's feasibility.

Example 2: planetary swing-by manoeuvres

A swing-by manoeuvre is a close pass of a planetary body by a spacecraft. Relative to the planet the trajectory is hyperbolic. If the geometry of the inbound and outbound asymptotes is constructed appropriately with respect to the orbital motion of the planet, then the spacecraft may increase or decrease its heliocentric velocity without expenditure of fuel. This makes the incorporation of these manoeuvres into interplanetary mission design a very attractive and powerful tool.

The Voyager spacecraft missions, launched in 1977 to explore the outer solar system, were of short enough duration to make the mission viable only through the use of this technique. Indeed, the Voyager 2 trajectory design included swing-bys of Jupiter, Saturn, Uranus and Neptune, spanning 12 years. This is to be compared with 30 years for a direct Hohmann-type transfer.

The mission designs for the Galileo Jupiter probe and the Cassini Saturn explorer both incorporate numerous swing-by manoeuvres in the capture and exploration mission phases.

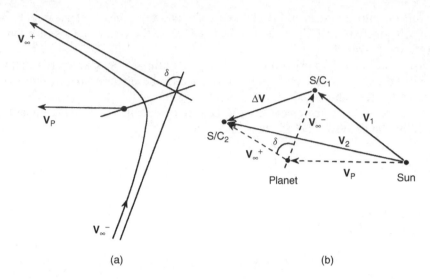

Figure 5.27 Swing-by passage *behind* a planet, showing: (a) hyperbolic trajectory in planet's sphere of influence and (b) relative velocity diagram (all vectors are coplanar)

The patched conic method can again be used, in combination with a relative velocity vector diagram, to provide a simple means of analysing spacecraft swing-bys. Figure 5.27(a) shows the swing-by geometry when the spacecraft passes 'behind' the planet (relative to the planet's forward motion), in a planet centred coordinate system. As the spacecraft enters the sphere of influence, the velocity relative to the planet is V_∞^-. If the hyperbola is a constant energy trajectory, then equation (4.5) shows that its outgoing velocity V_∞^+ is of equal magnitude, but its direction is deflected through an angle δ. Relative to the Sun, the planet's velocity is V_p. These vectors can be transferred to a relative velocity diagram shown in Figure 5.27(b), where they appear as dashed-line vectors. Nodes S/C$_1$ and S/C$_2$ represent the spacecraft before and after the encounter. Completion of the diagram as shown gives the Sun-relative velocities V_1 and V_2 before and after the swing-by. The change in the heliocentric velocity is ΔV.

It is clear that V_2 is greater than V_1, and so the spacecraft's speed has been increased by this passage *behind* the planet. A similar construction for a passage *in front of* a planet, shown in Figure 5.28, shows that this produces a reduction in speed. In both cases, the change in speed is of magnitude.

$$\Delta V = 2V_\infty \sin(\delta/2) \tag{5.37}$$

Example 3: Ulysses-type swing-by

The ESA/NASA probe Ulysses performed a swing-by of the planet Jupiter in February 1992, the purpose being to change the direction of Ulysses, so as to place it into an orbit in a plane that is approximately normal to the ecliptic (see Section 5.1).

After deployment from the Space Shuttle in October 1990, the spacecraft was accelerated to a record 15.4 km/s by a stack of upper stages. This gave the probe a heliocentric speed of 41.2 km/s when it left the Earth's sphere of influence, and this became

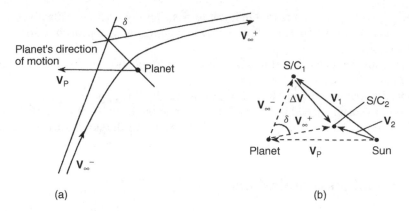

(a)   (b)

Figure 5.28 Swing-by passage *in front of* a planet, showing: (a) hyperbolic trajectory in planet's sphere of influence and (b) relative velocity diagram (all vectors are coplanar)

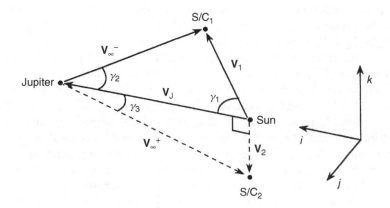

Figure 5.29 Relative velocity diagram for Ulysses-type swing-by. Solid-line vectors lie in the plane of the ecliptic $\mathbf{i} - \mathbf{j}$; dashed-line vectors lie in the orthogonal $\mathbf{i} - \mathbf{k}$ plane

$V_1 = 15.89$ km/s at the end of its journey to Jupiter, compared with Jupiter's $V_J = 12.6$ km/s. It's direction of travel was at an angle $\gamma_1 = 61.2°$ to Jupiter's. The geometry is shown in Figure 5.29.

Solving the ecliptic plane triangle leads to the conclusion that the probe's velocity relative to Jupiter as it entered its sphere of influence was $\mathbf{V}_\infty^- = 14.79$ km/s, with $\gamma_2 = 70.4°$.

The operational orbit for Ulysses was chosen to maximize the time spent over the polar regions, and for this its post-swing-by velocity $\mathbf{V}_2$ was not precisely normal to the ecliptic. However, for simplicity it is here assumed that it *is* normal, in the direction of $-\mathbf{k}$ in Figure 5.29. Noting that $\mathbf{V}_\infty^+ = \mathbf{V}_\infty^-$, this then yields $V_2 = 7.74$ km/s, and $\gamma_3 = 31.6°$. The deflection angle δ can be determined from the equation

$$\cos \delta = \frac{\mathbf{V}_\infty^- \cdot \mathbf{V}_\infty^+}{V_\infty^2}$$

when the $\mathbf{V}_\infty$ vectors are expressed in $\mathbf{i}, \mathbf{j}, \mathbf{k}$ form using Figure 5.29. This gives $\delta = 73.4°$, and equations (4.31) and (4.32) give e $= 1.673$ and the closest approach to Jupiter's centre is equal to $5.4R_J$.

The resulting heliocentric orbit is an ellipse with aphelion at Jupiter (5.2 AU) and a perihelion at 1.20 AU.

The mission parameters that have been evaluated by this approximate method are very similar to those of the actual event, and demonstrate the usefulness of a relatively simple method. A more exact solution would be given by using the equations of the restricted three-body problem (Chapter 4, Section 4.5), but these do not have an analytic solution.

5.8.3   Small object missions

Introduction

The majority of interplanetary missions during the twentieth century focussed on exploration of the major planets. There is, however, currently a surge of interest in investigating the properties of minor bodies in the solar system, mainly asteroids and comets, which has led to the proposal of many new spacecraft missions. The principal driver for this activity is the hypothesis that these small objects, particularly comets, hold the key to understanding the composition of the genesis solar nebula, and possibly to the origin of life. The term 'small' in this context has a wide interpretation however, 'small objects' ranging from the satellites of the major planets (some of which are larger than the planets Mercury and Pluto), to asteroidal chunks of rock perhaps a kilometre across. Missions to bodies of the former type are firmly on the agenda—for example, Jupiter's moon Europa has stirred interest, with the prospect of the discovery of life in a sub-ice ocean fuelled by tidal vulcanism. Objects such as Europa are relatively large, and near-spherical in shape, and methods discussed elsewhere in this chapter can be used to assess a mission profile. For example, a spacecraft orbiting Europa in combination with Jupiter, can be thought of as part of a circular, restricted three-body problem (see Chapter 4, Section 4.5) to a good approximation. A principal concern of a mission designer in this case would be the longevity of the mission orbit around the moon, when subjected to the powerful perturbing influence of the parent planet Jupiter.

The focus of the discussion in this section will be *asteroids* and *cometary bodies*, characterized by an irregular shape, and an overall dimension of up to a few tens of kilometres. Missions to orbit, and to land on such objects pose some interesting challenges to the mission designer. The first such mission, launched in February 1996, was the *NEAR Shoemaker* spacecraft, which was placed into orbit about the asteroid 433 Eros in February 2000. Figure 5.30 is a NEAR image of Eros, showing it to be an irregularly shaped object with a maximum dimension of around 35 km.

Near-body environment

The environment around small objects (particularly comets) is of great scientific interest in itself, but we will focus on those environments that influence the mission and spacecraft system design. Looking at Figure 5.30, perhaps the most obvious of these is that the

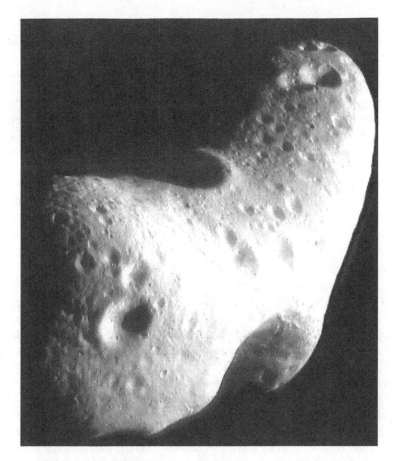

Figure 5.30 Image of asteroid 433 Eros, acquired by the NEAR Shoemaker spacecraft in 2000. (Reproduced by permission of NASA/JHUAPL)

gravity field does not approximate to that of a sphere, and a method of simulating orbits in such a field is discussed below.

If 433 Eros is to be regarded as typical, the near-body environment of asteroids appears to be 'clean', with minimal dust or debris posing a hazard to spacecraft (although asteroids have been observed to have natural satellites — the *Galileo* probe found the first, christened Dactyl, around the asteroid 243 Ida in 1993).

Cometary bodies are similarly thought to be irregularly shaped, but comprised mainly of ice and dust. As a consequence, they can have a very dynamic near-body environment, particularly when closer than around 3 AU to the Sun, when their surface layers become warm enough to trigger the sublimation of ices. A *nucleus* observed at close quarters is that of Halley's comet, when the European Space Agency (ESA) probe *Giotto* passed within 596 km in 1986. The resulting images revealed a very active nucleus about 16 km by 8 km in size, with bright jets of gas and dust emanating from the sunlit 'hemisphere'. Evidence suggests, however, that cometary bodies more typically have a dimension of roughly 1 or 2 km. When active, the nucleus is the source of all cometary phenomena, which comprise the visible *coma* (extending up to a million km in radius), the *hydrogen cloud* (spanning many millions of km) and of course, the tail. Comets may exhibit *dust tails* up to 10 million km in length, and *plasma tails* extending up to 10 times further. The interaction between the solar environment and these cometary features is complex, and the subject of much research, a review of which can be found in Reference [15]. In terms of mission and spacecraft design, the cometary environment poses greater challenges (and unknowns) than that of asteroids. For example, a near-comet dust and debris environment will cause orbit and attitude perturbations, which must be accounted for in the design. There is an inevitable conflict between the scientists and the engineers, the resolution of which is difficult. The former group would wish to encounter the environment fully, to maximize the scientific return, while the latter group would wish to avoid it to minimize risks to the spacecraft. The difficulty facing the designer, however, is that this environment is not known *a priori*, and therefore a range of environments need to be accommodated in the design.

Motion around small, irregularly shaped bodies

The motion of a spacecraft around a rotating, irregularly shaped body will approximate to the Keplerian orbits discussed in Chapter 4 when the vehicle is at sufficient distance. However, for close orbits where the orbit radius is of the same order as the body's size, the trajectory shape no longer approximates to a conic section, and the resulting equations of motion are sufficiently complex to make the use of a computational solution mandatory.

The gravity field of the body can be modelled using a variety of methods. In terms of an operational scenario, the objective is to determine the gravitational field of the body by observing the close proximity motion of an orbiting spacecraft. On approach, payload imagery will provide quantitative information regarding the size, shape and rotational state of the body. On the basis of the shape assessment, the gravitational potential can be expressed as a series of spherical harmonics (see Chapter 4, Section 4.4.1) or ellipsoidal harmonics. Subsequent precise tracking of the motion of the orbiter will allow the coefficients in these series expansions to be evaluated, so determining the gravitational potential of the object. This fairly complex process provides a general characterization of the body's gravity field, which takes into account its shape and any localized variations in body density. The small sample of evidence currently available suggests that body shapes can best be approximated by triaxial ellipsoids. For example, the asteroid 433 Eros can be approximated by an ellipsoid with semi-axes 17.9 km, 9.2 km and 7.9 km. For such a body, Garmier and Barriot [16] have demonstrated that the performance of the ellipsoidal harmonic expansion is superior to a spherical harmonic representation. Their investigation of lander trajectories showed that an ellipsoidal expansion up to degree 5 gave a landing position error of the order of 1 m, compared to 10 s of m for the spherical harmonic representation.

The above operational methodology concerns itself with determining the gravity field from the analysis of the orbiter's motion. Conversely, given the gravity field of the body, the motion of the orbiter may be determined. An approximate method of determining the orbiter motion, which is adequate for initial study work, is presented here. This is sometimes referred to as the 'mascon' method, and uses the approach of building the body's shape utilizing a collection of spherical masses of uniform size and density. These spherical 'building blocks' can be placed at the vertices of a three-dimensional grid, so

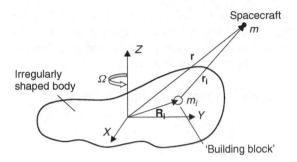

Figure 5.31 Geometry of the spacecraft in orbit around the irregularly shaped body

that they touch without overlap. The shape of the model of the irregularly shaped body can then be made arbitrarily close to the observed shape of an asteroid or comet, by allowing the size of these mass elements to decrease (within the bounds of computational effort). The overall mass and density of the body must be estimated bearing in mind that ~48% of the model is vacant. The gravitational potential at the spacecraft can then be found by summing the contribution from each element (or equivalently the gravitational force contribution from each may be summed vectorially). The situation is shown in Figure 5.31. The irregularly shaped body is shown rotating with angular velocity $\mathbf{\Omega}$ about the Z axis of an inertial co-ordinate frame (X, Y, Z), the origin of which is at the centre-of-mass of the body. The axis of rotation corresponds to the principal axis of maximum inertia of the body. From equation (4.2), the gravitational force acting on a spacecraft of mass m exerted by the body, comprised of n mass elements each of mass m_i, is

$$\mathbf{F} = \sum_{i=1}^{n} \left(\frac{Gmm_i}{r_i^2} \right) (-\mathbf{r_i}/r_i) \tag{5.38}$$

Noting that

$$\mathbf{r_i} = \mathbf{r} - \mathbf{R_i} = (x - X_i)\mathbf{i} + (y - Y_i)\mathbf{j} + (z - Z_i)\mathbf{k}$$

the equations of motion of the spacecraft may be written as

$$\ddot{x} = Ax + B_x, \ddot{y} = Ay + B_y, \ddot{z} = Az + B_z \tag{5.39}$$

where

$$A = -\sum_{i=1}^{n} \frac{Gm_i}{r_i^3}, B_x = \sum_{i=1}^{n} \left(\frac{Gm_i}{r_i^3} \right) X_i, B_y = \sum_{i=1}^{n} \left(\frac{Gm_i}{r_i^3} \right) Y_i, B_z = \sum_{i=1}^{n} \left(\frac{Gm_i}{r_i^3} \right) Z_i$$

The equations (5.39) may be easily transformed to a set of six coupled first-order differential equations given by

$$\dot{x} = u, \dot{y} = v, \dot{z} = w, \dot{u} = Ax + B_x, \dot{v} = Ay + B_y, \dot{w} = Az + B_z \tag{5.40}$$

which may be solved numerically, given appropriate initial values of the spacecraft's position and velocity. To compute the values of the A and $B_{x,y,z}$ functions, the position vector of the elemental masses m_i needs to be known as a function of time. Using the Taylor series for a vector function, we may write

$$\mathbf{R_i}(t + \delta t) = \mathbf{R_i}(t) + \delta t \dot{\mathbf{R_i}}(t) + \tfrac{1}{2}\delta t^2 \ddot{\mathbf{R_i}}(t) + \cdots$$

By noting that the vector $\mathbf{R_i}$ is constant in the co-ordinate frame rotating at angular velocity $\mathbf{\Omega}$ with the asteroid, it is easily shown that

$$\mathbf{R_i}(t + \delta t) = \mathbf{R_i}(t) + \delta t \{\mathbf{\Omega} \times \mathbf{R_i}(t)\} + \tfrac{1}{2}\delta t^2 \{\mathbf{\Omega} \times (\mathbf{\Omega} \times \mathbf{R_i}(t))\} + \cdots \qquad (5.41)$$

If using equation (5.41) to update $\mathbf{R_i}(t)$, it should be ensured that the combination of the magnitudes of δt, $\mathbf{R_i}$ and $\mathbf{\Omega}$ provide a convergent series. The above approximate method may be used to simulate close orbits around small irregular bodies. However, the overall accuracy of the method increases with the number of spherical building blocks used (so increasing computational effort), and perturbations due to density anomalies are not modelled.

Mission impact upon orbiter and lander system design

In this section, we discuss briefly the interaction between mission design and system design for a small body orbiter/lander mission. The focus will be on a comet rendezvous mission, since the challenges posed are greater than that of an asteroid mission. Many of the features are common to both. Such a cometary mission, called *Rosetta*, has been implemented by the European Space Agency. This was launched in March 2004, with a planned rendezvous with comet 67P/Churyumov-Gerasimenko in the early part of 2014 when the comet is about 5 AU from the Sun. A landing will be attempted in November 2014, and the spacecraft will continue to orbit the comet as it approaches the Sun to a perihelion distance of about 1.2 AU, when it is anticipated there will be vigorous solar-induced activity.

To plan such a mission, particularly the close orbit and landing phases, the design is very much dependent upon the physical properties of the comet and its environment. Since these are generally unknown prior to the mission, a useful course of action is to propose 'engineering models' of the target comet, which encompass the range of properties expected. This allows the designer to accommodate a range of possibilities in the design. Table 5.8 shows the characteristics of a small, typical and large comet, similar to those considered in the Rosetta mission design. It is assumed that the comet models are spherical during this design phase, to simplify trajectory calculations.

A first examination of the table shows that the operating environment for an orbiter/lander is an extremely 'low-energy' one. Spacecraft generally orbit at a comfortable walking pace—orbital speed, orbital period and escape velocity can be estimated as a function of orbital height from equations (4.22), (4.12) and (4.25), respectively. Landers take a long time to descend to the surface. Care is required when making trajectory manoeuvres not to exceed escape velocity. Landers need to be secured to the surface on touchdown. As we will see, these characteristics have an impact on the design of orbiter,

Table 5.8 'Engineering models' of a small, typical and large comet. The upper part of the table shows their physical characteristics. The lower part gives some dynamical parameters for an orbit radius of 20 km

	Comet size		
	Small	Typical	Large
Radius (m)	1000	2500	5000
Density (kg/m^3)	200	1000	1500
Gravity constant, $\mu = Gm$ (m^3/s^2)	56	4366	52 386
Surface g (m/s^2) (neglecting rotation)	5.59×10^{-5}	6.98×10^{-4}	2.10×10^{-3}
Rotation period (hours)	10	10	250
Equatorial surface speed due to rotation (m/s)	0.17	0.44	0.03
Surface escape velocity (m/s)	0.33	1.87	4.42
Surface weight of 50 kg lander (N)	2.8×10^{-3}	3.5×10^{-2}	1.0×10^{-1}
Orbit speed (m/s)	0.05	0.47	1.62
Escape velocity from orbit (m/s)	0.07	0.66	2.29
Orbit period (hours)	660.4	74.7	21.6

and in particular, lander operations. To highlight some of these impacts, we consider the deployment of a comet lander from an orbiting vehicle.

The first issue is release of the lander from the orbiter, which may be done by a mechanism, possibly in combination with thrusters. The difference between slowing the lander for descent and attaining escape velocity is only a matter of a few cm/s for small comets, so careful thought is needed to ensure the release mechanism is adequately calibrated. If thrusters are used to acquire the correct descent speed, then sizing these is also an issue for the same reason. These manoeuvres require the lander to have knowledge and control of its attitude. Initially, a gyro platform may be established from orbiter data. However, given that the descent time is likely to be large (see below), this may need to be updated using reference sensors (see Chapter 9). To minimize the mass of the lander, payload sensors may be utilized for this attitude-sensing task. The near-body environment of the comet will impact directly on the choice of sensor—an imaging device may not be able to discriminate between intended targets, such as a stellar image, and a particle of debris.

The time to descend to the surface and the surface impact speed is easily estimated if it is assumed that the lander release sequence arrests its orbital motion, so that it free-falls under gravity along a radial path. The time is given by

$$t = -\sqrt{\frac{r_0^3}{2\mu}} \left(\alpha - \frac{\pi}{2} - \frac{1}{2} \sin 2\alpha \right), \quad \alpha = \sin^{-1} \sqrt{\frac{r}{r_0}} \qquad (5.42)$$

where r_0 is the initial distance from the comet's centre, and r is the radius of the comet. The impact speed, estimated from the energy equation (4.5), is given by

$$V_{\text{imp}} = \sqrt{2\mu \left(\frac{1}{r} - \frac{1}{r_0} \right)} \qquad (5.43)$$

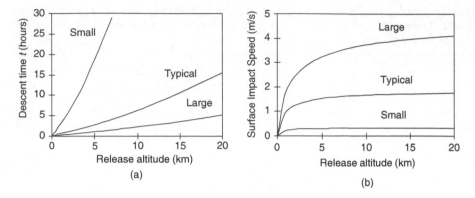

Figure 5.32 (a) descent time t and (b) surface impact speed V_{imp} as a function of release altitude for the three sizes of comet specified in Table 5.8

This 'impact speed' is a useful parameter as it is an estimate of the Δv required near the surface to halt the lander for a soft touchdown. Figure 5.32(a) and 5.32(b) display values of descent time t and V_{imp} as a function of release altitude, for the three comet types given in Table 5.8. The descent time for small comets is long, which raises issues concerning the attitude stability of the lander during this phase, especially if there are disturbance torques induced by a severe near-body environment. If this disturbance is to be controlled by a thruster limit cycle, then the duration of the descent will govern the amount of propellant used. The use of angular momentum bias may be considered, either in the form of a momentum wheel or by rotation of the lander itself, in order to reduce the sensitivity of the attitude to the disturbances. A sophisticated trade-off analysis is required, involving mass and complexity minimization, to establish the optimum solution. Given these interactions between system and mission design, a constraint on the descent time may be imposed (e.g. $t < 3\,\text{h}$), which in turn will dictate the release altitude. Very low release altitudes may be dangerous with respect to the near-body environment.

The next major event in the sequence is the soft landing. The vehicle may be arrested a few tens of metres above the surface, and may even be required to 'hover' under thruster power. The values of V_{imp} given in Figure 5.32(b) give a measure of the Δv that needs to be imparted by the 'main engine'. The fuel mass required may be approximated by

$$M_{fuel} = M_0\{1 - \exp[-(\Delta v + gt)/v_{ex}]\} \qquad (5.44)$$

where M_0 is the mass of the vehicle before the burn, g is the surface gravitational acceleration and v_{ex} is the effective exhaust velocity of the engine. The thruster sizing again is critical. Very low thrust levels will give good control, with respect to avoiding inadvertently acquiring escape velocity. On the other hand, the gravity loss 'gt' term in equation (5.44), and therefore the propellant mass, is larger for this type of engine. Pollution of the surface by the exhaust products may also be an issue from the point of view of the science objectives. The final phase is touchdown, and in particular, some form of hold-down manoeuvre, which may use an 'upward' firing thruster while the vehicle attaches itself mechanically to the surface.

This brief description of a typical lander mission highlights some of the system issues that pose a challenge to the design engineer.

5.8.4 Missions using low-thrust trajectories

In Section 5.8.2, the patched conic method was introduced to provide approximate estimates of important mission parameters, such as ΔV and transfer time. In doing so, an assumption was made that the engine firings were carried out using a high thrust chemical propulsion system, so that the manoeuvres were effectively impulsive. In this section, the consequences of abandoning this assumption are investigated by considering the effect on the trajectory of a continuous low-thrust. This is characteristic of an electric propulsion system, such as an ion thruster as discussed in Chapter 6. Low accelerations in the order of $10^{-4} - 10^{-5}$ m/s^2 may currently be anticipated, for Earth-orbiting or interplanetary spacecraft propelled by this type of system. Although this seems insignificant, nevertheless the electric propulsion system can operate for very long periods, so that ΔV values can be accumulated which are much larger than typical chemical systems can achieve. Also, this level of propulsive acceleration will influence the trajectory so that it is no longer Keplerian.

Such low-thrust propulsion systems are becoming more commonly used for orbit control activities in GEO, drag compensation in LEO (e.g. GOCE), lunar orbit missions (e.g. SMART-1) and missions to comets and asteroids (e.g. Deep Space 1). The European Space Agency BepiColombo programme is proposing to launch a mission in 2013 to orbit the planet Mercury. This will involve the use of a complex mission profile utilizing ion propulsion in combination with multiple gravity-assist manoeuvres. The spacecraft configuration is shown in Figure 5.33.

Figure 5.33 The current configuration of the ESA BepiColombo spacecraft, showing the ion propulsion system. (Reproduced by permission of ESA. Image by C. Carreau)

Low-thrust trajectories in a central force field

In Section 4.2 we looked at the restricted two-body problem, and derived equation (4.4) describing the unperturbed motion of a spacecraft around a central body possessing an inverse square law gravity field. When the motion of the spacecraft is additionally influenced by a continuous low-thrust, this equation is modified to give

$$\ddot{\mathbf{r}} + \frac{GM}{r^2}\frac{\mathbf{r}}{r} = \mathbf{a}, \quad \mathbf{a} = \frac{P\hat{\mathbf{u}}}{m} \tag{5.45}$$

where $\mathbf{a}$ is the perturbing thrust acceleration. Here M and m are the masses of the central body and the spacecraft respectively, P is the thrust magnitude, and $\hat{\mathbf{u}}$ is a dimensionless unit vector defining the direction of the thrust. The main rationale behind the analysis of low-thrust trajectories is to determine a steering law for the thrust vector, so that a particular objective can be achieved. For example, this may be the evaluation of a time profile for the thrust magnitude P and for the direction of the thrust vector $\hat{\mathbf{u}}$, so that a specified orbit transfer is achieved with minimum propellant, or minimum transfer time.

Lagrange's planetary equations

Some useful aspects of the steering law problem can be illustrated by examining the rates of change of the osculating elements using Lagrange's planetary equations. Following the method in Section 4.4, we can use the Gaussian form of these equations (4.34). By multiplying through by appropriate factors of $d\theta/dt \approx h/r^2$, we can express the time rates of change of the elements as

$$\left.\begin{aligned}
\frac{da}{dt} &= \tfrac{2a^2}{h}\left(e\sin\theta S + \tfrac{p}{r}T\right) \\[4pt]
\frac{de}{dt} &= \tfrac{1}{h}\{p\sin\theta S + [(p+r)\cos\theta + re]T\} \\[4pt]
\frac{di}{dt} &= \tfrac{r}{h}\cos(\theta + \omega)\,W \\[4pt]
\frac{d\Omega}{dt} &= \tfrac{r}{h}\frac{\sin(\theta+\omega)}{\sin i}\,W \\[4pt]
\frac{d\omega}{dt} &= \tfrac{1}{he}\{-p\cos\theta S + (p+r)\sin\theta T\} - \tfrac{r}{h}\frac{\sin(\theta+\omega)\cos i}{\sin i}\,W \\[4pt]
\frac{d\theta}{dt} &= \tfrac{h}{r^2} + \tfrac{1}{he}\{p\cos\theta S - (p+r)\sin\theta T\}
\end{aligned}\right\} \tag{5.46}$$

where S, T and W are a triad of accelerations as defined in Section 4.4, and $h = \sqrt{\mu a(1 - e^2)}$. Note that there are singularities (as in many other forms of these equations) in this formulation when e and i are zero. In this case, equations (5.46) may be expressed in terms of equinoctial elements—see for example Reference [17].

The thrust vector direction $\hat{\mathbf{u}}$ may be defined in terms of an azimuth angle α_2 and an elevation angle ε_2, as shown in Figure 5.34. Then the acceleration components S, T

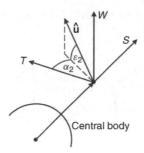

Figure 5.34 The thrust vector direction, expressed as a combination of azimuth and elevation angles

and W can be expressed as

$$S = (P/m)\sin\alpha_2\cos\varepsilon_2, \quad T = (P/m)\cos\alpha_2\cos\varepsilon_2, \quad W = (P/m)\sin\varepsilon_2 \quad (5.47)$$

The substitution of (5.47) into (5.46) gives the rates of the orbit elements as a function of these steering angles,

$$
\left.
\begin{aligned}
\frac{da}{dt} &= \frac{2a^2}{h}\frac{P}{m}\cos\varepsilon_2\{e\sin\theta\sin\alpha_2 + \cos\alpha_2(1 + e\cos\theta)\} \\[4pt]
\frac{de}{dt} &= \sqrt{\frac{p}{\mu}}\frac{P}{m}\cos\varepsilon_2\{\cos(\alpha_2 - \theta) + \cos\alpha_2\cos E\} \\[4pt]
\frac{di}{dt} &= \frac{r}{h}\frac{P}{m}\cos(\theta + \omega)\sin\varepsilon_2 \\[4pt]
\frac{d\Omega}{dt} &= \frac{r}{h}\frac{P}{m}\frac{\sin(\theta+\omega)}{\sin i}\sin\varepsilon_2 \\[4pt]
\frac{d\omega}{dt} &= \frac{1}{he}\frac{P}{m}\cos\varepsilon_2\{-p\cos\theta\sin\alpha_2 + (p + r)\sin\theta\cos\alpha_2\} \\[4pt]
&\quad -\frac{r}{h}\frac{P}{m}\frac{\sin(\theta+\omega)\cos i}{\sin i}\sin\varepsilon_2 \\[4pt]
\frac{d\theta}{dt} &= \frac{h}{r^2} + \frac{1}{he}\frac{P}{m}\cos\varepsilon_2\{p\cos\theta\sin\alpha_2 - (p + r)\sin\theta\cos\alpha_2\}
\end{aligned}
\right\} \quad (5.48)
$$

Steering laws

We can obtain some simple, yet important solutions to the steering angle problem using the equations (5.48). In this discussion we consider the Earth as the central body, but clearly the techniques can be applied with respect to any such body—e.g. interplanetary trajectories about the Sun.

- *Secular rates* of the orbit elements.
  For fixed values of the steering angles α_2 and ε_2, the secular change in the orbit elements can be estimated by integrating the osculating element rates over an orbit

period, and dividing by the orbit period. For example, for the semi-major axis a we find

$$\frac{da}{d\theta} = \frac{da}{dt}\frac{dt}{d\theta} \approx \frac{da}{dt}\frac{r^2}{h} = \frac{2a^3}{\mu}(1 - e^2)$$

$$\times \frac{P}{m}\cos\varepsilon_2 \left\{ e\sin\alpha_2 \frac{\sin\theta}{(1 + e\cos\theta)^2} + \frac{\cos\alpha_2}{1 + e\cos\theta} \right\}.$$

so that the change in a over one orbit period is given by

$$\Delta a = \int_0^{2\pi} \frac{da}{d\theta} d\theta = \frac{4\pi a^3}{\mu}\sqrt{1 - e^2}\frac{P}{m}\cos\alpha_2\cos\varepsilon_2$$

This gives the secular rate of change of a as

$$\left\langle \frac{da}{dt} \right\rangle = \frac{\Delta a}{\tau} = 2\sqrt{\frac{a^3}{\mu}}\sqrt{1 - e^2}\frac{P}{m}\cos\alpha_2\cos\varepsilon_2. \tag{5.49}$$

A similar analysis may be performed for the other orbital elements.

- Maximum rate of change of orbital energy.
  A steering law to maximize the change in orbital energy is useful when considering low-thrust orbit transfer manoeuvres, or planetary escape and capture strategies. In-plane accelerations are most effective in changing the orbital energy, so that we set the thrust elevation angle $\varepsilon_2 = 0$. From equation (4.11) the orbital energy is given by $\varepsilon = -\mu/(2a)$, so that $d\varepsilon/dt = (\mu/2a^2)(da/dt)$. From this we see that the maximum rate of change of energy occurs when the rate of change of a is maximized. To find the maximum or minimum value, we can substitute the expression for da/dt from equations (5.48) into the standard condition $\dfrac{\partial}{\partial\alpha_2}\left(\dfrac{da}{dt}\right) = 0$, which gives the following expression for the optimum value of α_2,

$$\tan\alpha_2 = \frac{e\sin\theta}{1 + e\cos\theta}. \tag{5.50}$$

The spacecraft's flight path angle γ has the same functional dependence, so that we may conclude that the rate of change of orbital energy is maximized or minimized if $\alpha_2 = \gamma$. That is, the thrust vector is aligned with the velocity vector, as indeed is the intuitive solution which underlies the impulsive transfers considered previously.

- Maximum rate of change of orbital inclination.
  This can be found from the condition,

$$\frac{\partial}{\partial\varepsilon_2}\left(\frac{di}{dt}\right) = \frac{r}{h}\frac{P}{m}\cos(\omega + \theta)\cos\varepsilon_2 = 0$$

indicating a thrust vector elevation angle of $\varepsilon_2 = \pm 90°$, with a direction that switches at the orbit apexes $\omega + \theta = \pm 90°$. This gives a maximum $|di/dt|$ at the orbit nodes

$\omega + \theta = 0°$ and 180°. A variable thrust magnitude P introduces an additional complication, as the out-of-plane acceleration is most effective in changing the orbital inclination when the spacecraft is at a greater distance from the central body, and therefore moving relatively slowly. This is particularly important in eccentric orbits, as illustrated in Figure 5.35, when the thrust magnitude can be reduced at the periapsis, and increased near the apoapsis.

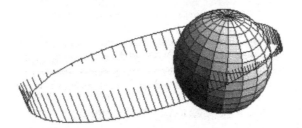

Figure 5.35 A typical plane changing strategy, showing direction and magnitude of thrust with orbital position for an eccentric orbit. (This figure is taken from NASA/TM-2002-211871 [18] and is used with permission of NASA)

Interplanetary missions using low-thrust

As discussed in Section 5.8.1, interplanetary missions are usually partitioned into three phases: Earth escape, heliocentric transfer and planetary capture.

Low-thrust Earth escape. The application of continuous low-thrust, applied along the spacecraft's velocity vector, will give the maximum rate of increase of orbital energy as discussed above. The resulting trajectory can be determined by numerically integrating the equations of motion (5.45), noting that the mass of the spacecraft decreases at each time step Δt,

$$m_{t+\Delta t} = m_t + \Delta m, \text{ where } \Delta m = -\frac{P}{g_0 I_{sp}} \Delta t. \qquad (5.51)$$

This solution, the classic spiral trajectory, is constrained by an initial boundary condition, for example departure from a circular orbit at a particular height, and by an escape condition usually characterized by a parabolic state (zero orbital energy, with an eccentricity of unity). The main mission parameters of interest are the resulting ΔV and time to escape, and these are shown in Figure 5.36 from Kemble [19] as a function of the altitude of the initial orbit, which is assumed circular. The ΔV plot shows that there is significant penalty in the use of low-thrust systems—for example, the ΔV for escape from a 200 km altitude circular orbit is in excess of 7 km/s, compared to 3.2 km/s for an impulsive manoeuvre. This results from the gravity loss term in the general Tsiolkovsky equation (see equation 7.6), as the flight path angle γ becomes significant for long periods of time as the spiral opens out near the escape condition. Also the time to escape is generally long, for example of the order of 2.5 years for a typical thrust acceleration of 10^{-4} m/s^2.

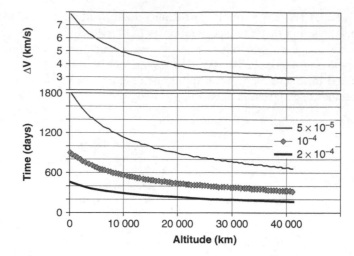

Figure 5.36 Escape ΔV and time from an initial circular orbit altitude. Constant acceleration, given in m/s², is assumed in making the estimates of escape time. (From Kemble (2006), with kind permission of Springer Science + Business Media)

A low-thrust Earth escape manoeuvre from an initially elliptical orbit is more efficient in increasing orbital energy if the thrust is applied only over an arc centred on perigee, although the time to escape is then increased. In the case of a spacecraft with a thrust acceleration of 10^{-4} m/s² departing from an initial GTO (see Section 5.6.1), the change in semi-major axis is ∼30 km for each m/s of ΔV applied near perigee, compared to ∼5 km change for each m/s in the apogee region [19].

Low-thrust planetary capture. To achieve capture around the destination planet, the reverse procedure to the escape strategy may be implemented. In this case the initial condition is a planet-relative (zero energy) parabolic approach state, followed by a low-thrust spiral arc to the final state of a specified orbit. Achieving the final arrival orbit in this process poses some difficulty, without the use of formal optimization methods, but this can be alleviated by a numerical integration of a time-reversed trajectory starting from the final orbit, remembering that the spacecraft mass must be increased at each time step [17]. Having acquired the planet-relative escape condition, further propagation of the trajectory is required in order to match the Sun-relative state of the spacecraft at the end of its heliocentric transfer. However this pragmatic approach is not without its difficulties—an iterative process is required to estimate the final spacecraft mass in orbit around the target planet (which is of course the initial mass in the time-reversed propagation).

Overall, the use of low-thrust propulsion for the escape and capture manoeuvres adds significantly to the total mission transfer time. In addition, the overall dose from the Earth's trapped radiation belts (and any contribution from the target planet) is appreciably increased. Both of these factors adversely impact spacecraft reliability. For missions where transfer time is already an issue, the escape phase can most sensibly be performed using high-thrust chemical propulsion of the launcher's upper stage for Earth departure. The implementation of a high-thrust chemical system onboard the spacecraft for the capture phase would be similarly beneficial, leaving the spacecraft's low-thrust system to perform the Sun-centred transfer phase.

Sun-centred transfer. If we adopt the above strategy of using high thrust chemical propulsion for the escape and capture manoeuvres, then the low-thrust system can be used in the heliocentric transfer phase. During this phase of the mission, in general the objective is to determine the time evolution of the control parameters, in this case the thrust magnitude and direction, in order to find a trajectory between initial and final boundary conditions which, for example, maximizes the spacecraft mass on arrival at the destination planet. This requires the use of trajectory optimization techniques, the description of which is beyond the scope of this text. For more information the reader is referred to the review by Betts [20] and the treatments of Kemble [19] and Chobotov [21].

However, there are some analytical techniques which can be used as a first approximation. The first of these, attributed to Kemble [19], provides a first estimate of the fuel mass for such an orbit transfer. In this formulation, it is assumed that the spacecraft is equipped with both chemical and electric propulsion systems. The vehicle, with initial mass M_0, departs from the perigee of an elliptic orbit using an impulsive burn from the chemical propulsion system which gives it an escape delta-V of ΔV_1. The perigee and apogee radii are r_{p1} and r_{a1}, respectively, and the resulting escape excess velocity is given by $V_{\infty,E}$. Similarly the capture manoeuvre ΔV_2 at the destination planet is also performed using the chemical propulsion system, with the spacecraft entering an elliptical orbit with periapsis and apoapsis radii of r_{p2} and r_{a2} from an approach excess velocity of $V_{\infty,C}$. As a first approximation, Kemble then assumes that the delta-V performed by the electric propulsion system during the heliocentric transfer ΔV_T is given by

$$\Delta V_T = (V_{\infty 1} - V_{\infty,E}) + (V_{\infty 2} - V_{\infty,C}), \tag{5.52}$$

where $V_{\infty 1}$ is the excess velocity required to enter a Hohmann transfer orbit after the escape manoeuvre. Similarly, $V_{\infty 2}$ is the excess velocity that would be required at the destination planet if the vehicle approached on the Hohmann trajectory. The total mass ratio for the mission, given by

$$\frac{M_{final}}{M_0} = \exp\left(\frac{-\Delta V_1}{g_0 I_{sp,Ch}}\right) \exp\left(\frac{-\Delta V_T}{g_0 I_{sp,El}}\right) \exp\left(\frac{-\Delta V_2}{g_0 I_{sp,Ch}}\right) \tag{5.53}$$

is then minimized analytically with respect to $V_{\infty,E}$ and $V_{\infty,C}$ to obtain the following estimates

$$V_{\infty,E}^2 = \frac{2\mu_1 I_{sp,Ch}^2}{r_{p1}(I_{sp,El}^2 - I_{sp,Ch}^2)}, \quad V_{\infty,C}^2 = \frac{2\mu_2 I_{sp,Ch}^2}{r_{p2}(I_{sp,El}^2 - I_{sp,Ch}^2)},$$

where μ_1 and μ_2 are the gravitational constants of the Earth and the destination planet respectively. Here $I_{sp,Ch}$ is the specific impulse of the chemical propulsion system, and $I_{sp,El}$ is that of the electrically propelled system. The values of ΔV_1 and ΔV_2 can be estimated from

$$\Delta V_1 = \sqrt{\left(\frac{2\mu_1}{r_{p1}} + V_{\infty,E}^2\right)} - \sqrt{\frac{2\mu_1}{r_{p1}}\left(\frac{r_{a1}}{r_{p1} + r_{a1}}\right)},$$

$$\Delta V_2 = \sqrt{\left(\frac{2\mu_2}{r_{p2}} + V_{\infty,c}^2\right)} - \sqrt{\frac{2\mu_2}{r_{p2}}\left(\frac{r_{a2}}{r_{p2} + r_{a2}}\right)},$$

and the value of ΔV_T obtained from (5.52), so that the total mass ratio (and hence fuel mass) can be acquired from (5.53).

Pioneering work on low-thrust trajectories was also done by Edelbaum [22] in the early 1960s, who developed a simplified approach to the problem of the transfer between two non-coplanar, circular orbits. With the assumption of constant acceleration, this yields the analytic solution for the required ΔV, given by

$$\Delta V = \left(V_0^2 + V^2 - 2VV_0\cos\left(\frac{\pi}{2}\Theta\right)\right)^{1/2},$$

where V_0 and V are the initial and final circular velocities of the orbits and Θ is the angle between the two orbit planes. An alternative treatment of Edelbaum's methodology can be found in [21].

5.8.5 Planetary atmospheric entry and aeromanoeuvring

Our exploration of the solar system has revealed that not only do several of the major planets have atmospheres, but also some minor bodies, such as Titan, possess a significant detectable atmosphere. Probes sent to explore such worlds by orbiting or landing on these bodies require the ability to actively or passively manoeuvre successfully in such an environment. Indeed, the opportunity is provided by planets having an atmosphere to reduce the total vehicle wet-mass in the execution of a particular mission. Table 2.6 provides data on the atmospheres of those bodies in the Solar System having a significant atmosphere.

It is evident from the discussion in Chapter 4, Section 4.4, that the principal effect of an atmosphere on a satellite's trajectory is to reduce the energy of the orbit. However, it is also possible to utilize the aerodynamic forces, in order to produce both along-track and across-track accelerations without the need for propulsive manoeuvres. The along-track effects, which are generally associated with *aerobraking*, are used and proposed for a variety of mission operations to reduce the translational energy of a spacecraft. These include direct atmospheric entry to reduce the vehicle's speed to facilitate soft landing, orbital *aerocapture* where aeroforces are used to transfer a vehicle's orbital state from hyperbolic to elliptical, and *aero-assisted orbit transfer* where an atmospheric pass is used to modify the orbit. An example of the latter is the transfer of a vehicle from high orbit to low orbit, where an aerobraking manoeuvre is performed in the periapsis region. The across-track aeroforces produce out-of-plane accelerations, and so can be used to modify the orbit plane inclination. It has also been proposed to use them to augment the beneficial effects of gravitational swing-bys (see Section 5.8.2). *Aeromanoeuvring* is a term applied generally to missions utilizing aerodynamic forces. Clearly, savings in propellant mass can be achieved, but this must be weighed in all instances against the additional mass required to protect the vehicle from the dynamic pressure and thermal effects associated with aeromanoeuvring.

A survey of aeroassisted orbit transfer is provided by Walberg [23]. He looked at three specific mission categories, that is, synergetic plane change, planetary mission applications and orbit transfer applications. All these include some degree of aeromanoeuvring which, unlike the simple swing-by missions discussed in Section 5.8.2, have a mass penalty as mentioned above. This arises from the need to incorporate in the design an aeroshield and, in some cases, propellant mass if there is a propelled trajectory between intermediate orbits.

Since Walberg's paper, there have been a number of planetary missions launched which have incorporated a degree of aeromanoeuvring. A review of the mission designs and in-flight performance of several missions: *Mars Global Surveyor, Mars Pathfinder, Galileo, Mars Microprobe, Mars Polar Lander* and *Stardust*, may be found in a special jour-nal issue [24]. More recently a review of optimization of trajectories using aeroassist specifically for orbit transfer vehicles has been provided by Bérend [25]. The engineer-ing solutions, incorporating the aeroassisted element of the mission design, were in part spurred by the Goldin initiative for NASA of providing faster, better, cheaper missions. This is as a result of aeroassist being a method of both reducing mass of a planetary satellite directly and providing a consequential cost reduction of launch vehicle services. According to Reference [24], 65% of NASA's planetary missions scheduled for the period 1995 to 2005, utilized aeroassist. The following section provides a simplified overview of the key features of the aerodynamic influences of this type of space mission.

The resurgence of interest in manned missions to Mars during the first decade of the twenty-first century has led to further studies on the use of aerocapture for large vehicles (tens of metric tons) entering the Martian atmosphere. These studies show that some significant mass savings are feasible. However the technology requirements are not met by current capabilities. The mass of the aeroshell for such a vehicle is substantial and can amount to ~15% of the vehicle entry mass. Reference [26] provides a recent review of aerocapture for manned missions to Mars, including the issue of precision landing by such methods.

Atmospheric entry

As a space vehicle approaches a planet having an atmosphere, it experiences an approxi-mate exponentially increasing atmospheric density. This provides a changing aerodynamic environment for the vehicle. Initially the Mach number (the ratio of the vehicle's speed relative to the local speed of sound in the gas) M_0 may be in the range of 20 to 50. The initial density is so low, however, that the flow field is described as a *free molecular flow*. In this regime, the molecules and atoms that constitute the atmosphere collide so infrequently, that following impact upon a vehicle surface, the molecule will not then collide with the incoming molecules. Under such conditions, shock waves are not formed about the body. However, as the vehicle progresses further into the atmosphere, a *tran-sition flow* commences. This region is difficult to describe analytically, and frequently *bridging functions* are used to describe the aerodynamic properties of the vehicle. Thick viscous shock waves are formed about the vehicle at this time. Eventually, if the vehicle penetrates sufficiently low into the atmosphere, a *continuum flow* region is encountered, which is that type of flow typified by conventional aerodynamics. The velocity, however, is still so great that the flow remains hypersonic.

Key to an understanding of the flow field and the changes that occur within it is the recognition of the amount of energy required to be dissipated. Taking the amount of energy per unit mass to be $\sim 0.5\,V^2$, if a vehicle is approaching the planet on a hyperbolic trajectory with a velocity of tens of km/s, hundreds of MJ must be dissipated. An extreme example is the entry of the Galileo probe into the Jovian atmosphere in 1995. With an approach speed of 47.5 km/s, 3.8×10^5 MJ of translational energy were dissipated in the four minutes before the drogue parachute was deployed. This generated a temperature of 15 000 K, and an estimated 90 kg of ablative material was lost from the probes forward heat shield (out of a total probe initial mass of 340 kg). At such energies, the gas that is incident upon the vehicle undergoes not only chemical reactions, but also excitation of internal energy modes such as vibration, together with dissociation and ionization. Relaxation from these excited states may arise through radiation. The time constants of these processes are large, and hence the flow field is not in equilibrium. As a result, there is great difficulty in analytically predicting the changes that arise in the gas, as the normal relationships of equilibrium thermodynamics cannot be applied. Approximations, such as assuming that the constituents (both in terms of their chemical composition and the degree to which excitation has occurred) have relative number density fractions frozen at some point in the flow, may be used to simplify the analysis. Detailed predictions are made yet more complex by the uncertain role the vehicle surface plays in the chemistry of the reacting flow. The overall net effect however, of the high velocity flow field impinging on a surface is to cause substantial heat transfer to the vehicle. It is also clear from this brief introduction that, because of the chemically reacting flow conditions, the actual heat loads a vehicle will experience will depend upon the constituents of the atmosphere itself. The dominant constituents for each of the planetary atmospheres is shown in Table 2.6. An excellent recent review of the literature concerning planetary entry gas dynamics is given by Gallais [27].

Constraints during atmospheric entry

The two principal constraints that occur in the design of an aeromanoeuvring vehicle are the peak dynamic load and the peak thermal load, together with how long these loads persist.

During entry, it is possible to write down the overall governing equations that describe the dynamics. For a ballistic entry, it is assumed that the aerodynamic forces only provide a drag force parallel to the instantaneous direction of motion, with no cross track force (equivalent to a zero lift coefficient C_L). It is possible to write down the relationship between the distance to the centre of the planet r at time t, if at that time the flight path angle of the vehicle is γ and the velocity V. Assuming the density at some reference height h is given by ρ_s, which decreases exponentially with a scale height of β, then following [28],

$$\frac{dr}{dt} = V \sin \gamma \tag{5.54}$$

where

$$\frac{dV}{dt} = -\frac{1}{2}\rho_s V^2 e^{-\beta h} \left(\frac{SC_D}{m} \right) = -\eta \beta V^2 \tag{5.55}$$

Here a dimensionless height variable η has been introduced,

$$\eta = \frac{1}{2}\rho_s \left(\frac{SC_D}{m} \right) \frac{1}{\beta} e^{-\beta h} \tag{5.56}$$

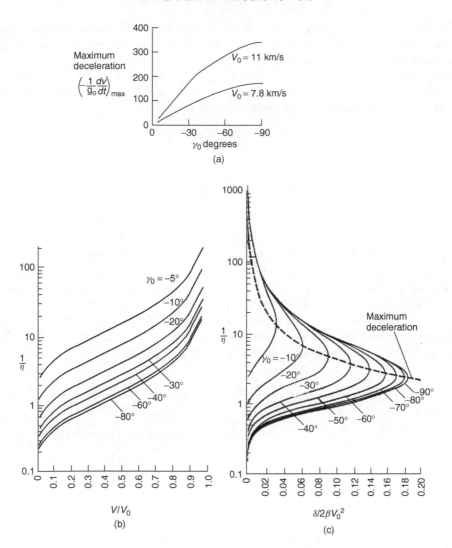

Figure 5.37 Ballistic entry characteristics: (a) variation of the maximum deceleration with entry angle γ_0 and entry speed V_0 for ballistic entry; (b) ballistic entry at large flight path angle: speed-altitude diagram for several values of the initial angle (after Vinh *et al.* [28]). (Reproduced by permission of the University of Michigan) and (c) ballistic entry at large flight path angle: acceleration-altitude diagram for several values of the initial angle (after Vinh *et al.* [28]). (Reproduced by permission of the University of Michigan)

The ballistic coefficient, (SC_D/m), where S is the wetted surface area for a vehicle of mass m having a drag coefficient C_D, is seen from equation (5.55) to provide a linear influence over the rate at which the vehicle decelerates. However, the maximum deceleration for an initial speed V_0 and entry angle γ_0 is given by δ_{max} and is found to be independent of the ballistic coefficient,

$$\delta_{max} = \frac{\beta V_0^2}{2e} \sin \gamma_0 \qquad (5.57)$$

This function is plotted in Figure 5.37.

Turning to the peak-heating load, it is clear from the discussion above that a simple analytic description is not available, if one wishes to describe the real flow situation. Approximations may be used to provide some estimate of the heating profile. These typically omit elements of the various heat transfer processes that take place in the real flow. For example, if only convective heat transfer is considered (or is indeed dominant) [28], then the peak heating rate is given by

$$\dot{q}_{max} \propto V_0^3 \sqrt{\left(\frac{m \beta \sin \gamma_0}{3 S C_D} \right)} \qquad (5.58)$$

In general, this peak heat flux will occur at a different altitude from that for the peak deceleration load.

Evidently, both the dynamics and heat loads are dependent upon the initial conditions assumed for atmospheric entry [27]. As a result, calculations that are performed must assume an overall mission profile. Thus, the preceding interplanetary manoeuvres will influence the final loads experienced by a vehicle, together with the launch date. For the Mars Pathfinder mission [29], the inertial arrival velocity could vary by 100 m/s. The worst case (highest entry velocity) trajectory is shown in Figure 5.38.

Various design solutions have been used, or proposed to accommodate both the dynamic loads and thermal heat flux. Simple passive solutions are most appropriate for vehicles that may have been in transit for several years prior to the critical minutes of atmospheric

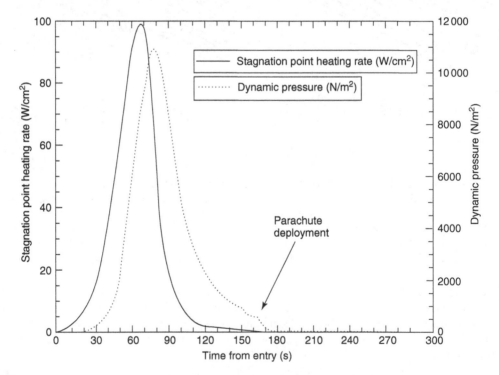

Figure 5.38 Heating rate and dynamic pressure during entry for Mars pathfinder mission (after Spenser *et al.* [29])

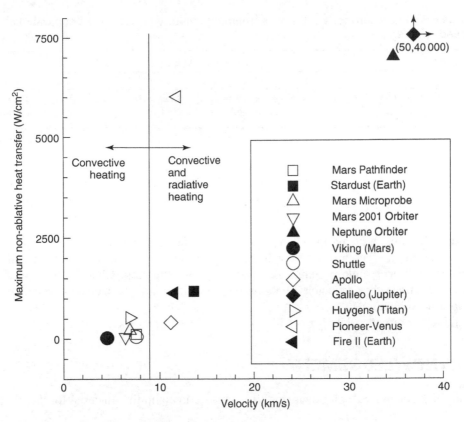

Figure 5.39 Non-ablative peak heating *versus* velocity for past and planned planetary entry vehicles

entry. Figure 5.39 shows the range of heat loads experienced for different vehicles. It should be noted that for some vehicles, such as Apollo, ablative shields are appropriate, since contamination is not an issue. However, where sensitive scientific measurements are to be taken shortly after the main phase of deceleration, the role and final fate of any ablative shield and its constituent materials, needs to be carefully examined.

Entry corridor

The foregoing has identified the major constraints arising from the mission profile, which will dictate the preferred engineering solution. There is one other major component to the final mission design that derives from the planet to be encountered. Both maximum heat and dynamic loads are a function of entry angle γ_0 as seen in equations (5.57) and (5.58) above. A design solution will specify the peak value that these loads may have. If the vehicle were to enter at an angle greater than the specified value, then it may be anticipated that the vehicle will either burn-up or break-up. However, there also needs to be consideration of the problem that will arise if the entry angle is shallower than that in the design solution. Chapman [30] defined an entry corridor in terms of the altitude of the first periapsis pass, following atmospheric entry. Undershooting the target height

Table 5.9 Entry corridor widths for entry into planetary atmospheres at parabolic speed $V_0 = (2gr_0)^{1/2}$

Planet	Corridor width (km)					
	5 g limit			10 g limit		
	$L/D = 0$	$L/D = 1$	$L/D = 1$ modulated	$L/D = 0$	$L/D = 1$	$L/D = 1$ modulated
Venus	0	43	58	13	84	113
Earth	0	43	55	11	82	105
Mars	338	482	595	644	885	1159
Jupiter	0	55	68	0	84	113

leads to burn-up or break-up, whereas overshooting leads to the vehicle re-emerging from the atmosphere. The entry corridor is then defined as the height difference of periapsis between the acceptable extremes of under- and overshooting. Clearly, this corridor is dependent upon the acceptable limits on both heat load and mechanical deceleration for a specific vehicle.

Table 5.9 provides data for corridor width for several planets.

ACKNOWLEDGEMENT

Thanks is due to Dr Hugh Lewis, for his assistance in compiling material for this chapter.

REFERENCES

[1] Klinkrad, H., Beltrami, P., Hauptmann, S., Martin, C., Sdunnus, H., Stokes, H., Walker, R. and Wilkinson, J. (2004) The ESA space debris mitigation handbook, *Advances in Space Research*, **34**(5), 1251–1259.

[2] ESA Bulletin, Ulysses Launch Issue, No. 63, August 1990.

[3] Lawden. D. F. (1963) *Optimal Trajectories for Space Navigation*, Butterworth, London.

[4] Betts, J. T. (2001) *Practical Methods for optimal control using non-linear programming*, SIAM, Philadelphia, USA.

[5] King-Hele, D. G. (1987) *Satellite orbits in an atmosphere: theory and applications*, Blackie, London.

[6] Lüders, R. D. (1961) Satellite networks for continuous zonal coverage, *Am. Rocket Soc. J.*, **31**, 179–184.

[7] Adams, W. S. and Rider, L. (1987) Circular polar constellations providing continuous single or multiple coverage above a specified latitude, *J. Astron. Sci.*, **35**, 155–192.

[8] Walker, J. G. (1971) Some circular orbit patterns providing continuous whole Earth coverage. *J. Br. Interplanet. Soc.*, **24**, 369–384.

[9] Walker, J. G. (1984) Satellite constellations, *J. Br. Interplanet. Soc.*, **37**, 559–571.

[10] Benedicto, J., Fortuny, J. and Rastrilla, P. (1992) MAGSS-14: A medium-altitude global mobile satellite system for personal communications at L-band. *ESA J.*, **16**, 117–133.

[11] Clarke, A. C. (1949) Extraterrestrial relays, *Wireless World*, October, 305.

[12] Maral, G., Bousquet, M. and Sun, Z. (2009) *Satellite Communications Systems* (5th Edition), John Wiley & Sons, Ltd, Chichester.

[13] Soop. E. M. (1983) *Introduction to Geostationary Orbits*, ESA SP-1053.

[14] Cornelisse, J. W., Schoyer, H. F. R. and Wakker, K. F. (1979) *Rocket Propulsion and Spacecraft Dynamics*, Pitman, London.

[15] Beatty, J. K., Collins Peterson, C. and Chaikin, A. (eds) (1999) *The New Solar System*, (4th Edition), Cambridge University Press, New York.

[16] Garmier, R. and Barriot. J.-P. (2001) Ellipsoidal harmonic expansions of the gravitational potential: theory and application, *Celestial Mech. Dynamical Astron.*, **79**, 235–275.

[17] Betts, J. T. and Erb, S. O. (2003) Optimal low thrust trajectories to the moon. *J. Applied Dynamical Systems (Society for Industrial and Applied Mathematics)*, **2**(2), 144–170.

[18] Falck, R. and Gefert, L. (2002) A method of efficient inclination changes for low-thrust spacecraft, NASA/TM-2002-211871, AIAA-2002-4895.

[19] Kemble, S. (2006) *Interplanetary Mission Analysis and Design*. Springer-Praxis, Chichester, UK.

[20] Betts, J. T. (1998) A survey of numerical methods for trajectory optimisation. *J. Guidance, Control and Dynamics*, **21**(2), 193–207.

[21] Chobotov, V. A. (2002) *Orbital Mechanics* (Third Edition), AIAA Education Series, Reston, USA.

[22] Edelbaum, T. N. (1961) Propulsion requirements for controllable satellites, *ARS Journal*, **31**, 1079–1089.

[23] Walberg, G. D. (1985) A survey of aeroassisted orbit transfer, *J. Spacecraft Rockets*, **22**, 3–18.

[24] Braun, R. D. (1999) Aeroassist Systems: An Important Element in NASA's New Era of Planetary Exploration, *J. Spacecraft Rockets*, **36**(3).

[25] Bérend, N., Bertrand, S. and Jolly, C. (2007) Optimization method for mission analysis of aeroassisted orbital transfer vehicles, *Aerospace Science and Technology*, **11**, 432–441.

[26] Rapp, D. (2008) *Human missions to Mars, enabling technologies for exploring the Red Planet*, Section 4.6, 127–149, Springer-Praxis, Chichester, UK.

[27] Gallais, P. (2007) *Atmospheric re-entry vehicle mechanics*, Springer-Verlag, Berlin Heidelburg.

[28] Vinh, N. X., Busemann, A. and Culp, R. D. (1980) *Hypersonic and Planetary Entry Flight Mechanics*. University of Michigan Press, Michigan.

[29] Spenser, D. A. and Braun. R. D. (1996) Mars pathfinder atmospheric entry: trajectory design and dispersion analysis, *J. Spacecraft Rockets*, **33**, 670–676.

[30] Chapman, D. R. (1959) An Approximate Analytical Method for Studying Entry into Planetary Atmospheres, NASA Technical Report R-11.

6 PROPULSION SYSTEMS

J. Barrie Moss[1] and John P. W. Stark[2]

[1] School of Engineering, Cranfield University
[2] School of Engineering and Materials Science, Queen Mary, University of London

6.1 SYSTEMS CLASSIFICATION

The broad classes of propulsion systems for space vehicles are distinguished in Figure 6.1. The opportunities for air-breathing stages in the early phases of an Earth surface launch are briefly reviewed in Section 6.2.4, but we shall focus here on systems that are not reliant on external oxidizer provision. Of the several alternatives identified, exploitation in terms of practical devices has concentrated largely on thermal and electric rockets. Primary propulsion for launch vehicles is further restricted—currently to solid- or liquid-propelled chemical rockets. The principal options are reviewed in detail in later sections and we refer only briefly to more speculative concepts.

The solar radiation pressure at 1 AU from the Sun is roughly 5×10^{-6} N m^{-2} and therefore the surface area that must be deployed in order to produce significant thrust for primary solar sailing is extremely large, even in circumstances where the gravitational force on the sail is small. Whilst missions have long been identified for which solar sailing is an attractive option—for example, the Halley's comet interception mission proposed in the mid-1970s (see Friedman et al. [1])—there has been little practical demonstration. Space vehicles of more modest dimensions may, however, be subject to significant perturbing torques, resulting, for example, from asymmetric surface deployments or simply from the cyclic variation of solar radiation pressure experienced by Earth-oriented spacecraft (see Chapter 9).

Nuclear propulsion has been the subject of very detailed studies over many years (see Gunn [2]), although these have not been pursued to significant flight demonstration. The NERVA programme of the 1960s (Nuclear Engine for Rocket Vehicle Applications) resulted in a ground-tested solid core (graphite) U235 fission powered engine delivering approximately 300 kN of thrust with a specific impulse of 825 s. With large-scale interplanetary missions still a fairly distant prospect and the commercial emphasis on near-Earth operations, direct thrust nuclear rockets do not appear to be cost effective

Spacecraft Systems Engineering, Fourth Edition.
Edited by Peter W. Fortescue, Graham G. Swinerd and John P. W. Stark.
© 2011 John Wiley & Sons, Ltd. Published 2011 by John Wiley & Sons, Ltd.

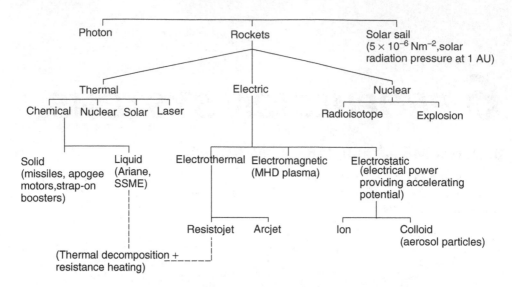

Figure 6.1 Propulsion systems classification

and raise environmental concerns that are not readily assuaged. More recent studies have focused on lower thrust orbit raising and manoeuvring propulsion as aspects of wider programmes for nuclear space power raising (see Buden and Sullivan [3]).

The development of high power lasers and intermittent interest in the establishment of complex orbiting platforms for power raising and communications in space have also encouraged a number of conceptual studies of laser sustained propulsion (see Caveney, [4]).

Crucial parameters that distinguish between the developed systems and introduce important characteristics of systems more generally can be readily identified. We show later that the beam (or kinetic) power of the rocket exhaust, P, is given by

$$P = 1/2 \, \dot{m} V_e^2 \tag{6.1}$$

where $\dot{m}$ is the exhaust mass flow rate and V_e is the exhaust velocity.

Since the thrust delivered by the rocket may be written

$$F = \dot{m} V_e \tag{6.2}$$

then

$$P = 1/2 \, F V_e \tag{6.3}$$

If we introduce the vehicle acceleration (or vehicle thrust-to-weight ratio), αg_0, where g_0 is the Earth surface acceleration due to gravity, then we may write

$$F \simeq M \alpha g_0$$

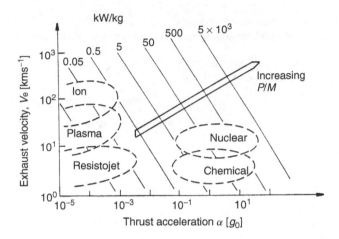

Figure 6.2   Comparative rocket performance

(M = vehicle mass), and, from equations (6.2) and (6.3),

$$P/M = 1/2\,\alpha V_e g_0 \tag{6.4}$$

Propulsion systems are readily distinguished by comparisons between their specific power, P/M, exhaust velocity, V_e, and thrust acceleration, α as illustrated in Figure 6.2. The separately powered electric rocket is characterized by high exhaust velocity and low specific power or thrust acceleration. Nuclear or chemical rockets, on the other hand, offer high powers but with relatively poor propellant utilization through their limited exhaust velocities. Booster operation from planetary surfaces is evidently restricted to these latter systems which are said to be *energy limited* - by the chemical energy stored in the propellants - but which admit high thrust for comparatively modest engine weight. Although the specific impulses of electrically powered systems are high, they are *power limited*. Whilst the energy available from the Sun is unlimited, or that from a radioactive source introduces a negligible fuel mass, the effective energy density is low and the necessary weight of the accompanying systems for electrical conversion is unacceptably large in high power applications. Electrically propelled space vehicles are thus restricted to very small accelerations. As we shall describe in Section 6.3 the more efficient expellant utilization possible with low thrust systems thus appears most naturally suited to orbit-raising manoeuvres, interplanetary transfer and spacecraft attitude and orbit control.

Whilst the role of propulsion is most clearly identified in relation to launch vehicles, spacecraft propulsion makes a number of equally important contributions to overall mission success. The range of tasks and performance requirements are illustrated below.

- Launch vehicles - main engines and 'strap-on' boosters - developing continuous high thrust for periods of minutes (approximately 10^6 N for 9 min in the case of the Vulcain 2 cryogenic main engine on Ariane 5, for example).
- Apogee motors for spacecraft orbit circularization and inclination removal (typically, 75 kN for approximately 60 seconds, developing a velocity increment of 2 km s^{-1}); perigee motors for orbit raising from low-altitude parking orbit.

- Spacecraft station-keeping, attitude and orbit control (thrust levels ranging from 10^{-3} to $10\,\text{N}$, intermittent and pulsed operation over the complete duration of the mission).

Since propulsive requirements are frequently specified in terms of ΔV (see Chapter 5) it is instructive to compare some typical values:

- ΔV into low Earth orbit (including drag and gravity losses) $\geq 9.5\,\text{km s}^{-1}$
- additional ΔV to equatorial geostationary orbit from a $30°$ inclined $\sim 4.2\,\text{km s}^{1}$ parking orbit
- additional ΔV for a Mars fly-by $\sim 3.4\,\text{km s}^{-1}$
- additional ΔV for Solar System escape (without gravitational assist) $\sim 8.5\,\text{km s}^{-1}$
- geostationary orbit station-keeping (communications satellite)
  North-South (inclination $< 0.1°$ throughout a 7–year life) $\sim 0.35\,\text{km s}^{-1}$
  East-West ($\pm 0.1°$ of nominal longitude) $\sim 0.03\,\text{km s}^{-1}$

6.2 CHEMICAL ROCKETS

The rapid growth in rocket propulsion technology following World War II is largely based on chemical rockets. Simple reaction systems, in which the propulsive force exerted on a vehicle arises from changes in system momentum through the discharge of hot products of combustion at high velocity, have proved uniquely successful in high thrust atmospheric and space applications. The burning of chemical propellants, solid or liquid, at high pressure liberates large quantities of energy in a compact volume. The subsequent expansion of these high temperature products of combustion through a convergent-divergent nozzle, converts thermal energy to directed kinetic energy for rocket propulsion. Since the rocket carries both fuel and oxidizer, the specific fuel consumption is substantially higher than that of an air-breathing reaction system such as the turbojet. It is, however, mechanically less complex than an aircraft power plant since moving parts are confined to auxiliary systems such as the propellant feed.

We show later in this section that chemical rockets are conveniently characterized by the *Tsiolkovsky equation* (in field-free space):

$$\Delta V = V_e \ln R \qquad (6.5)$$

where ΔV denotes the rocket velocity increment and R is the mass ratio, initial mass to mass at burn-out. ΔV is typically prescribed by the mission whilst V_e is essentially fixed by the choice of propellant. Only by increasing the mass ratio in equation (6.5) can the shortfall in propellant energetics be accommodated and mission objectives attained.

6.2.1 Basic principles

We first review briefly those aspects of rocket motor performance, gas dynamics and thermochemistry that most directly influence design and operation. For further details of the analysis, the reader is referred to the excellent texts by Barrere *et al.* [5] and Sutton and Biblarz [6].

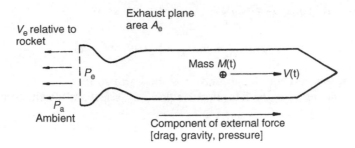

Figure 6.3 Schematic rocket

Performance parameters

Consider the rocket illustrated schematically in Figure 6.3. Applying equation (3.18) from Chapter 3 leads to

$$M\frac{dV}{dt} = \dot{m}V_e + A_e(p_e - p_a) + F_{ext} \qquad (6.6)$$

The rocket thrust F comprises two contributions, from the exhaust momentum flux and the exhaust plane pressure difference:

$$F = \dot{m}V_e + A_e(p_e - p_a) \qquad (6.7)$$

where $\dot{m}$ is the propellant mass flow rate, V_e is the exhaust velocity, $A_e(p_e - p_a)$ is the resultant force on the rocket due to the pressure difference between the nozzle exit and the ambient, and F_{ext} denotes the extra force in the direction of motion due to external forces (for example, aerodynamic drag or gravity).

We show in the following section that nozzle performance, and in particular the exhaust velocity V_e, is maximized by complete exhaust expansion to ambient pressure, whence $p_e - p_a = 0$. In space applications, the ambient pressure is either continuously varying with altitude, or is zero beyond the Earth's atmosphere. The effects of any under-expansion, $p_e > p_a$, are thus partially offset by the thrust increase accompanying this pressure difference.

From equation (6.7) the thrust at sea level is

$$F_{SL} = \dot{m}V_e + A_e(p_e - p_{SL})$$

and at altitude h it may be expressed as

$$F_h = F_{SL} + A_e(p_{SL} - p_h) \qquad (6.8)$$

In vacuo, when $p_h = 0$, it becomes

$$F_0 = F_{SL} + A_e\, p_{SL}$$

Typically $A_e p_{SL}/F_{SL}$ approaches 20%.

It is convenient to define an *effective exhaust velocity*

$$V_e^* = V_e + A_e(p_e - p_a)/\dot{m} \equiv I_{SP}\, g_0 \tag{6.9}$$

where g_0 is the acceleration due to gravity at the Earth surface.
I_{SP} is the *specific impulse*, the total impulse per unit propellant weight consumed, and is given by

$$I_{SP} = \frac{I}{M_p g_0} = \frac{\int_0^t F(t)\, dt}{g_0 \int_0^t \dot{m}(t)\, dt} \tag{6.10}$$

whence $I_{SP} = F/\dot{m}g_0$ for constant thrust and exhaust mass flow rate.
Equation (6.6) may be expressed in terms of V_e^* and, when this is constant, it may be integrated over the duration of rocket motor firing giving

$$\Delta V = V_b - V_0 = V_e^* \ln\left\{\frac{M_0}{M_b}\right\} + \int_0^{t_b} \frac{F_{ext}}{M}\, dt \tag{6.11}$$

and hence the Tsiolkovsky equation (6.5), where the *mass ratio*

$$R = M_0/M_b$$

is the ratio of the initial to burn-out mass—note that $\dot{M} = -\dot{m}$.
    The maximization of V_e^* for a specified velocity increment ΔV is essential for efficient design. We now describe the nozzle flow characteristics necessary to realize high exhaust velocity given a particular propellant selection.

Nozzle flows

We analyse the flow through a convergent-divergent nozzle, downstream from the combustion chamber and as illustrated in Figure 6.4, with the aid of the following simplifying assumptions:

- The combustion products are homogeneous and of constant composition.
- The products at temperature T and molecular weight W obey the perfect gas law relating pressure p and density ρ:

$$p = \rho T (R_0/W) \tag{6.12}$$

    where R_0 is the universal gas constant.
- The specific heat of the mixture is invariant with temperature and pressure.
- The flow is one-dimensional, steady and isentropic.

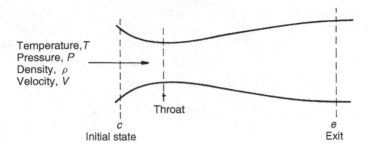

Figure 6.4 Schematic nozzle flow

The conservation equations for mass and energy may then be written:

$$\dot{m} = \rho V A \qquad (6.13)$$

$$\frac{1}{2}V^2 + C_p T = \text{constant} = \frac{1}{2}V_c^2 + C_p T_c \qquad (6.14)$$

where the subscript c denotes the initial state. For an adiabatic flow process, we have

$$p\rho^{-\gamma} = \text{constant}$$

and

$$\frac{T}{T_c} = \left(\frac{\rho}{\rho_c}\right)^{\gamma-1} = \left(\frac{p}{p_c}\right)^{(\gamma-1)/\gamma} \qquad (6.15)$$

If the initial velocity V_c is negligibly small, then

$$\frac{\dot{m}}{A} = \left\{ \frac{2\gamma}{\gamma-1} p_c \rho_c \left(\frac{p}{p_c}\right)^{\frac{2}{\gamma}} \left[1 - \left(\frac{p}{p_c}\right)^{(\gamma-1)/\gamma} \right] \right\}^{\frac{1}{2}} \qquad (6.16)$$

Clearly equation (6.16) exhibits a maximum value for $(\dot{m}/A)$ corresponding to a critical throat condition, subscript t, at which

$$\frac{p_t}{p_C} = \left(\frac{2}{\gamma+1}\right)^{\gamma/(\gamma-1)}$$

and

$$\frac{T_t}{T_C} = \frac{2}{\gamma+1}$$

and

$$\frac{\dot{m}}{A} = (\gamma \rho_t p_t)^{\frac{1}{2}} = \rho_t V_t$$

where the critical throat velocity V_t is given by

$$V_t = \left\{ \gamma \frac{p_t}{\rho_t} \right\}^{\frac{1}{2}} = \left\{ \gamma \frac{R_0}{W} T_t \right\}^{\frac{1}{2}} = a_t$$

which is the speed of sound at the throat.

In convergent-divergent (de Laval) nozzles the velocity continues to increase downstream from the throat ($p_{exit}/p_c < p_t/p_c$) but the nozzle is choked—that is, the mass flow is simply determined by throat conditions, independent of the exit flow condition. The choked mass flow rate can be expressed as a function of combustion chamber conditions (p_c, T_c) and throat area A_t,

$$\dot{m} = \sqrt{\gamma} \left\{ \frac{2}{\gamma + 1} \right\}^{(\gamma+1)/[2(\gamma-1)]} \frac{p_c A_t}{\sqrt{(R_0 T_c/W)}} \qquad (6.17)$$

We may identify a characteristic velocity

$$c^* = \sqrt{(R_0 T/W)} / \left\{ \sqrt{\gamma} \left[\frac{2}{\gamma + 1} \right]^{(\gamma+1)/[2(\gamma-1)]} \right\} \qquad (6.18)$$

whence $$\dot{m} = p_c A_t / c^* \qquad (6.19)$$

Using the energy conservation equation (6.14) we may determine the exhaust velocity V_e from

$$\frac{1}{2} V_e^2 + C_p T_e = C_p T_c \quad (V_c = 0)$$

whence, after some manipulation

$$V_e = \sqrt{ \left\{ \frac{2\gamma R_0 T_c}{(\gamma - 1)W} \left[1 - \left(\frac{p_e}{p_c} \right)^{(\gamma-1)/\gamma} \right] \right\} } \qquad (6.20)$$

We note that the exhaust velocity, V_e, increases with the following:

- increasing pressure ratio p_c/p_e - though such benefits are limited by accompanying increases in motor weight;
- increasing combustion temperature T_c - to be set against the adverse effects of higher temperatures on nozzle heat transfer and increased dissociation losses (see the later section on 'Thermochemistry");
- low molecular weight; and
- to a lesser extent, by reducing the ratio of specific heats γ - this is of limited practicality given the other influences.

It is convenient to identify a *characteristic thrust coefficient* C_F^0 such that

$$V_e = c^* C_F^0$$

where $$C_F^0 = \sqrt{ \left\{ \left[\gamma \left(\frac{2}{\gamma + 1} \right)^{(\gamma+1)/(\gamma-1)} \right] \frac{2\gamma}{\gamma - 1} \left[1 - \left(\frac{p_e}{p_c} \right)^{(\gamma-1)/\gamma} \right] \right\} } \qquad (6.21)$$

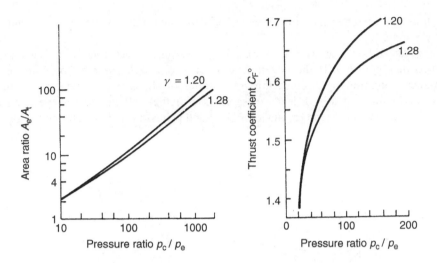

Figure 6.5   The variation of area ratio and thrust coefficient with pressure ratio

The exit-to-throat area ratio, A_e/A_t, can be determined from the continuity equation (6.13) such that

$$\frac{A_e}{A_t} = \frac{\rho_t V_t}{\rho_e V_e} = \gamma \left(\frac{2}{\gamma+1}\right)^{(\gamma+1)/(\gamma-1)} \left(\frac{p_c}{p_e}\right)^{1/\gamma} / C_F^0 \qquad (6.22)$$

The variation of area ratio and of thrust coefficient with pressure ratio is illustrated in Figure 6.5.

Nozzle performance and design

We now combine the rocket motor performance characteristics of Section 6.2 with the gas dynamics of the preceding section.

For a given combustion chamber pressure p_c and mass flow rate $\dot{m}$ (and hence throat area) the motor thrust can be optimized. From equation (6.7)

$$F = \dot{m} V_e + A_e (p_e - p_a)$$

and incremental parameter changes are related by

$$\delta F = \dot{m} \delta V_e + \delta A_e (p_e - p_a) + A_e \delta p_e$$

But conservation of momentum insists that

$$\dot{m} \delta V_e + A_e \delta p_e = 0$$

whence

$$\frac{\mathrm{d}F}{\mathrm{d}A_e} = p_e - p_a = 0 \text{ for maximum thrust}$$

The thrust is thereby maximized when $p_e = p_a$ and the nozzle flow is said to be *ideally expanded*.

The exhaust flow patterns accompanying departures from this ideal behaviour are sketched in Figure 6.6. If the nozzle flow is *overexpanded*, $p_e < p_a$, pressure recovery to ambient conditions is effected through a series of shock waves. Penetration of these shock waves into the nozzle leads to separation of the wall boundary layers, enhanced viscous losses and a reduced expansion ratio. Such behaviour is characteristic of operation

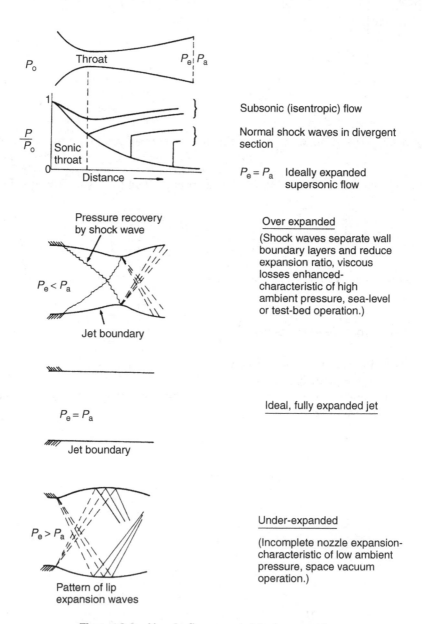

Figure 6.6   Nozzle flows: non-ideal expansion

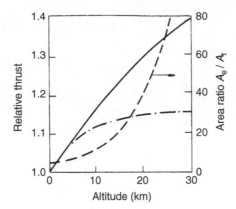

Figure 6.7 Comparative thrust levels: continuous and sea-level adaptation. Continuous adaptation _____ ; sea level adaptation _____ _ _____

at high ambient pressure, typical of sea-level or test-bed firing. *Underexpanded flows* are characterized by incomplete nozzle expansion, $p_e > p_a$, and reduced exhaust velocity. The low ambient pressures that give rise to such flows are typically realized in space vacuum operation.

Figure 6.7 illustrates the thrust improvement that would result from continuous adaptation of the nozzle geometry, leading to $p_e = p_a$, with increasing altitude, and hence reducing p_a, in comparison with the thrust of an ideally expanded nozzle at sea level. The expansion ratio A_e/A_t of the nozzle becomes very large, however, and significant weight penalties may result. Careful optimization is necessary to reconcile these competing factors.

The thrust coefficient in equation (6.21) can be modified to include the effects of non-ideal expansion in the form

$$C_F = \frac{F}{p_c A_t} = C_F^0 + \frac{A_e}{A_t}\left(\frac{p_e}{p_c} - \frac{p_a}{p_c}\right) \tag{6.23}$$

The *thrust coefficient for ideal expansion*, given p_c, p_a and A_t is, from equation (6.21),

$$\{C_F^0\}_{max} = \sqrt{\left\{\frac{2\gamma^2}{\gamma - 1}\left(\frac{2}{\gamma + 1}\right)^{(\gamma+1)/(\gamma-1)}\left[1 - \left(\frac{p_a}{p_c}\right)^{(\gamma-1)/\gamma}\right]\right\}}$$

and with equation (6.22) we write

$$\frac{C_F}{\{C_F^0\}_{max}} = \frac{C_F^0}{\{C_F^0\}_{max}} + \frac{\gamma\left(\frac{2}{(\gamma+1)}\right)^{(\gamma+1)/(\gamma-1)}\left(\frac{p_c}{p_e}\right)^{\frac{1}{\gamma}}}{C_F^0\{C_F^0\}_{max}}\left(\frac{p_e}{p_c} - \frac{p_a}{p_c}\right) \tag{6.24}$$

As Figure 6.8 indicates, the departure from ideal expansion is less severe with underexpansion than with overexpansion. However, flow separation from the nozzle, which contracts the jet in the overexpanded situation, does lead to an increase in thrust over that

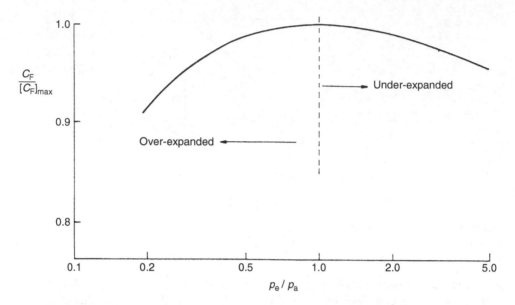

Figure 6.8 The effect on thrust coefficient of departures from ideal expansion

which would result in the absence of such separation. This observation also suggests an aerodynamic approach to varying the nozzle geometry, which is otherwise mechanically cumbersome—namely, one of controlled fluid injection through the nozzle wall which induces local flow separation and thereby modifies the nozzle contour.

As indicated earlier, nozzle adaptation has important implications for nozzle size and weight. From the manufacturing point of view, convergent-divergent nozzles that are conical represent the simplest designs. Two specific problems then arise, however:

- the exhaust is not directed in the axial direction and the flow divergence implies some loss of thrust;
- in high altitude applications the nozzle tends to be long and correspondingly heavy.

The axial thrust may be shown to be approximately

$$F = \frac{1}{2}\dot{m}V_e(1 + \cos\alpha)$$

for cone included angle 2α. For $\alpha = 15°$ the multiplicative factor $\frac{1}{2}(1 + \cos\alpha)$ is 0.986 and the effect of divergence is generally small.

Designs to reduce nozzle length, and hence weight, resulting from small divergence angles, must expand the flow from the throat more rapidly and subsequently turn the exhaust in the axial direction. Bell-shaped nozzles are designed to achieve this—see, for example, the established procedure due to Rao [7].

The benefits of continuous nozzle adaptation are particularly apparent in relation to single-stage-to-orbit (SSTO) concepts. Whilst the performance compromise introduced by

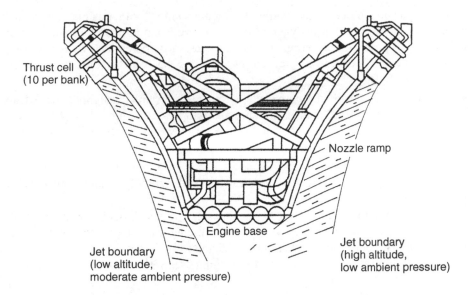

Figure 6.9 Schematic of linear aerospike nozzle

not doing so, and adopting fixed geometry nozzles, may be quite modest in a multi-stage launcher, the SSTO confronts the nozzle design with a progressively reducing ambient pressure over the complete range from sea-level static to vacuum conditions. Unlike the conventional convergent-divergent nozzle, the truncated plug nozzle or aerospike (Figure 6.9) is bounded by a solid wall (ramp) on only one side, with the combustor efflux expanding against the local ambient pressure on the other. The extent of such expansion is thereby adapted to the changing altitude and varies continuously over the vehicle trajectory. When allied to a lifting body configuration as in the Lockheed Martin VentureStar reusable launch vehicle (RLV) concept, or the X-33 demonstrator, the linear aerospike also offers opportunities for propulsion system integration and base-drag reduction [8].

Thermochemistry

We identified from equation (6.20) that the basic elements in maximizing exhaust velocity (or specific impulse) were a high combustion chamber temperature and pressure, and low molecular weight. The variation of these properties in relation to some representative propellant combinations will be illustrated later.

In developing equations for the ideal rocket motor, we assumed that, in addition to isentropic flow (no viscous or heat losses), the combustion products were of known constant composition throughout the expansion. At the high temperatures of interest, however, a significant proportion of combustion products are dissociated and the extent to which the energy of dissociation can be recovered in the expansion has a substantial impact on nozzle performance.

In the expansion process, the variation of composition with position in the nozzle depends upon both local thermodynamic state—pressure and temperature—and chemical kinetic rates, in particular, upon recombination rates. A complete description is complex and beyond the scope of this chapter, but two limiting cases are readily distinguished:

- frozen flow, in which the composition remains constant throughout the expansion irrespective of the variation in pressure and temperature; and
- equilibrium flow, in which the equilibrium composition corresponding to local conditions of pressure and temperature prevails along the nozzle.

The former implies that the chemical rates are vanishingly small, whilst the latter implies that they are infinitely fast.

As the temperature decreases along the nozzle, species dissociated in the combustion chamber recombine, releasing energy and changing the composition. Whilst specific impulse (or thrust coefficient) is therefore predicted to be larger for equilibrium flow at a particular pressure ratio (p_c/p_e), the accompanying area ratio A_e/A_t is also larger.

6.2.2 Propellants

In principle, any chemical system producing heat release in a gas flow through exothermic reaction could be used as a propellant. In practice, consideration of a range of additional factors such as the specific energy content, rate of heat release, ease of storage and handling significantly limits the choice. Heat release in a liquid-propellant rocket may be achieved through the separate injection, mixing and combustion of two liquids—fuel and oxidizer—as in a bipropellant system, or the exothermic decomposition of a monopropellant such as hydrazine or hydrogen peroxide. A bi-propellant is said to be *hypergolic* if the fuel and oxidizer react spontaneously on contact with each other. More generally, the requirements of separate propellant storage, of pumping and feed to the injector, of mixing and ignition mean that the high specific impulse is achieved at the expense of considerable complexity in design and manufacture. Significant flexibility in operation is, however, gained, including stop-start options and throttling of thrust levels. In contrast, the solid propellant rocket is of comparatively simple design. The propellant is stored within the combustion chamber in the form of shaped grains bounded by the walls of the chamber. Once ignited, combustion will generally proceed until all the propellant is consumed; the thrust-time relationship is then fixed by the grain configuration. Limited thrust regulation is possible.

Combinations of solid fuel and a liquid or gaseous oxidizer (usually identified as hybrid systems) attracted the interest of early rocketry pioneers [6] and this has continued within amateur rocket societies on grounds of comparatively low cost, modest technical complexity and reduced hazard. Hybrid motors have grown in scale in recent years, however, under the stimulus of international competitions such as the Ansari X Prize—awarded to 'Scaled Composites' SpaceShipOne in 2004 [9] which employed a nitrous oxide (N_2O)/HTPB hybrid—and the many projected developments in space tourism more generally. Research on solid fuel ramjets for missile applications has also exposed some technological challenges in common, for example, when seeking to optimize the configuration of oxidizer injector and solid propellant grain for sustained levels of high heat release.

Table 6.1   Liquid propellants

Fuel	Oxidizer	Molecular weight of products	Combustion temperature T_c (K)	Ideal specific impulse (s)	Mean density kg/m^3
H$_2$ (hydrogen)	O$_2$ (oxygen)	10	2980	390	280
	F$_2$ (fluorine)	12.8	4117	410	460
Kerosine	O$_2$	23.4	3687	301	1020
	F$_2$	23.9	3917	320	1230
	RFNA (red fuming nitric acid)	25.7	3156	268	1355
	N$_2$O$_4$ (nitrogen tetroxide)	26.2	3460	276	1260
	H$_2$O$_2$ (hydrogen peroxide)	22.2	3008	278	1362
N$_2$H$_4$ (hydrazine)	O$_2$	19.4	3410	313	1070
	HNO$_3$ (nitric acid)*	20	2967	278	1310
UDMH	O$_2$	21.5	3623	310	970
(CH$_3$)$_2$NNH$_3$ (unsymmetrical dimethyl hydrazine)	HNO$_3^*$	23.7	3222	276	1220
	* hypergolic				
Monopropellants					
N$_2$H$_4$		10.3	966	199	1011
H$_2$O$_2$		22.7	1267	165	1422

*Note: All quoted values are for $p_c = 7$ MPa with an ideal expansion to $p_e = 0.1$ MPa. Higher chamber pressures admit increases in I_{SP} — for example, at 20 MPa, LOX/LH$_2$ yields a specific impulse of ~460 s.

Liquid propellants

Table 6.1 illustrates properties of some representative propellants. We recall from equation (6.10) that in order to maximize exhaust velocity we seek propellants which give high combustion temperatures and low molecular weight products. Peak temperatures accompany the burning of approximately *stoichiometric mixtures* — that is, mixtures containing just sufficient oxidizer to convert the chemical elements in the fuel to stable combustion products. The typical variation of adiabatic flame temperature with mixture strength exhibits only a modest decline in T_c for richer-than-stoichiometric mixtures, however. It is thus possible to maintain high combustion temperatures with hydrogen as fuel, for example, and capitalize on the low fuel molecular weight by operating fuel rich.

From the table, the attraction of fluorine as an oxidizer (high combustion temperature, $W_F = 19$) is tempered by its highly corrosive properties which constrain the choice of liner materials for the combustion chamber and nozzle. Both oxygen and fluorine necessitate cryogenic storage as liquids since they have boiling points of 90 K and 85 K respectively. Long-term storage is therefore difficult and their application is restricted to launch vehicles. Nitrogen tetroxide has found increased application in space propulsion as an oxidizer, despite its high molecular weight ($W_{N_2O_4} = 92$). This is in part because it has a boiling point of 294 K at suitable storage pressures and, though highly toxic, systems for its supply and management are generally simpler and lighter.

The traditional, high thrust bi-propellant combinations of LOX/LH$_2$, employed on Saturn V, the Space Shuttle Main Engines (SSME) and the Vulcain motor of Ariane 5, for example, are complemented in apogee motors and orbital manoeuvring systems generally by monomethylhydrazine (MMH)/N$_2$O$_4$ combinations. The latter offers a specific impulse of approximately 310 s, a value that is significantly greater than that available from monopropellant hydrazine decomposition, the readily stored—but less stable—alternative (boiling point 387 K). In the context of thermal control during propellant storage, we should note that both hydrazine and nitrogen tetroxide have melting points in the neighbourhood of typical spacecraft ambient temperatures, 275 K and 262 K respectively.

Solid propellants

Solid propellants are typically of two types: either double-base, comprising homogeneous colloidal mixtures of nitrocellulose and nitroglycerine, or composite, comprising mixtures of an organic fuel and crystalline inorganic salt. Ammonium perchlorate, NH$_4$ClO$_4$, is the principal oxidizer used in composite propellants with a polymer fuel-binder, typically polyurethane or polybutadiene.

In comparison with the liquid propellants described in Table 6.1, the specific impulses for solid propellants are substantially lower—in the range 200–260 s. These values are not strongly influenced by the particular fuel binder, although high hydrogen to carbon ratios are favoured. The performance of composite propellants is improved by the addition of metals such as aluminium or beryllium in the form of finely ground particles (~10 μm).

Solid propellant charges are typically cast or extruded. The components are mixed in the form of a dough, perhaps in the presence of a plasticizer, and cast or pumped into a mould or directly into the combustion chamber. The grain geometry is fixed by a mandrel which is removed after curing and solidification.

Further information on propellant energetics and physical properties is available in Reference [6]; the complexities of solid propellant combustion are discussed extensively in Kuo and Summerfield [10].

Hybrid rockets

The oxidizers in common use in liquid bi-propellant systems (cryogenic LOX and storable NTO (nitrogen tetroxide), for example) can be used with solid hydrocarbon-based fuels such as HTPB (hydroxyl-terminated polybutadiene) to create energetic hybrids that offer much of the simplicity of design and robustness in operation of solid composite propellants together with the enhanced levels of performance and controllability typical of bi-propellants. Heat release in such heterogeneous systems is generally more sensitive to flow and mixing conditions in the turbulent boundary layer over the solid fuel, however, and they can rarely match the levels of specific impulse achieved with energetic liquid bi-propellants. Hybrid product densities are, however, higher and density-weighted specific impulses may exceed those of liquid bipropellants. Hybrid exhausts are also generally cleaner than those of solid propellants, reducing observables for tactical missiles and perhaps suggesting broader 'green' credentials. In less demanding roles that admit lower levels of thrust, nitrous oxide - which is readily gasified though storable as a liquid-when injected into simple grain geometries of PMMA (perspex) or polyethylene (PE) has found successful application with rocket enthusiasts.

6.2.3 Chemical rocket design

In this section we seek to identify some of the more important features of motor design in relation to high thrust launch vehicles or booster applications. A detailed discussion of such a major topic is inappropriate to this book which seeks more to familiarize the non-specialist with the diversity of disciplines and technologies at the heart of spacecraft engineering. More specific design issues are addressed in much greater detail in Huzel and Huang [11]. From the standpoint of spacecraft design, the involvement with the launch vehicle lies essentially in the role of procurement and the clear identification of the constraints that are imposed on the spacecraft by the launcher. These form the basis of Chapter 7, whilst the design of spacecraft propulsion systems, of more direct concern to the spacecraft engineer, is discussed in Sections 6.3 and 6.4.

If we distinguish the principal rocket components to be:

- the thrust chamber,
- the propellant feed system, and
- the propellant storage tanks,

then significant differences are evident between solid and liquid propellant devices. For the solid propellant rocket, offering high thrusts of short duration, key features of the design are the choice of propellant grain to give the appropriate thrust law and of materials selection for the nozzle and casing to combine low weight with reliable operation. The greater complexity of the liquid propellant rocket requires that consideration be given to thrust chamber design in respect of such components as the fuel injector, cooling system and propellant feed. In the case of the launch vehicle, demanding especially high propellant mass flow rates, both the gas generator and turbopump designs then become critical aspects.

Liquid propellant rockets

A typical liquid propellant rocket motor is illustrated in Figure 6.10. Liquid fuel and oxygen are pumped into the main combustion chamber by turbopumps driven by a separate hot gas generator, burning small quantities of the propellants. The starter cartridge is typically a small solid propellant gas generator.

Two important classes of engine in which fuel and oxidant are supplied by turbopumps are differentiated by their description as open or closed cycle designs. In the *open cycle*, the turbine exhaust is discharged into the nozzle downstream from the combustion chamber at modest pressure in the expanding section or simply discharged to atmosphere, making no direct thrust contribution. By contrast, in the *closed cycle* the exhaust is injected at high pressure into the combustion chamber, contributing significantly to the energy of the system. The improved performance of the latter cycle must be set against the disadvantage of more complex turbopump design necessary to operate at much greater discharge pressures.

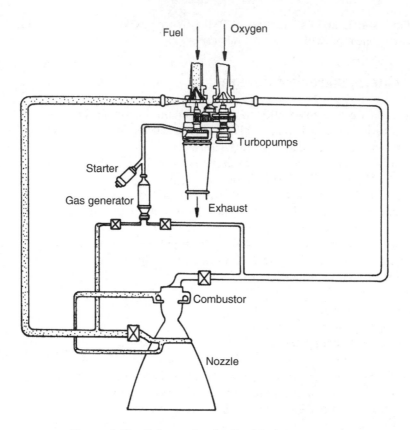

Figure 6.10 Schematic of a liquid rocket motor

Two open cycle configurations are illustrated in Figure 6.11: a separate gas generator cycle of the kind employed in the HM-7 and Vulcain engines of Ariane 5, and a coolant tap-off cycle in which vaporized hydrogen fuel from the nozzle coolant jacket drives the turbine. The turbine power is comparatively low in the latter case but the complexity of the design is much reduced. Also illustrated is the closed cycle, staged combustion cycle employed in the SSME. All the fuel and part of the oxidizer are there supplied to a pre-combustor at high pressure. The high energy fuel-rich exhaust first drives the turbopumps and is then injected into the main combustion chamber with the remaining oxidizer.

An illustrative comparison between open and closed cycle engines is presented in Table 6.2, where SSME performance characteristics are summarized together with those of the Ariane 5 Vulcain engine. The gas generator consumes approximately 3.5% of the available propellant and the closed cycle engine, also incorporating a much greater chamber pressure (approximately two-fold higher), yields a vacuum specific impulse some 5% greater. The necessarily higher turbopump discharge pressures for both fuel and oxidizer in the staged combustion cycle imply significantly more extensive design and development in turbines, pumps and ancillary equipment. A simpler closed cycle is the expander cycle in which vaporized fuel drives the turbopumps but, in contrast to the coolant bleed cycle illustrated, all of the fuel is so employed and subsequently passes into the combustion chamber. Whilst particularly suitable for LH_2, the fuel flow rate is now limited by the

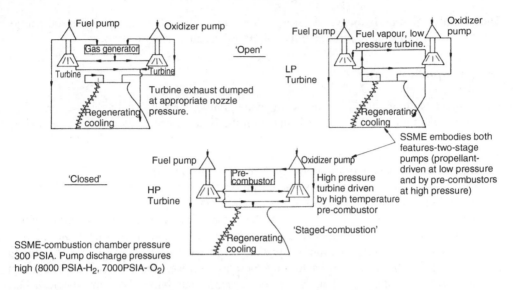

Figure 6.11 Representative engine cycles

Table 6.2 Illustrative comparison of closed- and open-cycle engines

	SSME	Vulcain (Ariane 5)
Thrust (kN):		
Vacuum	2090	1390
Sea level	1700	960
Specific impulse (s):		
Vacuum	455	432
Sea Level	363	310
Mixture ratio		
(stoichiometric 8:1	6:1	5.3:1
$2H_2 + O_2 \rightarrow 2H_2O$)		
Chamber pressure (bar)	207	108
Nozzle area ratio	77	45
Flowrates (kg/s)	468 (engine)	270
	248 (pre-combustor)	10 (gas generator)
Pump discharge pressure (bar)	309 (LOX)	125 (LOX)
	426 (LH$_2$)	150 (LH$_2$)
Burn time (s)	480	540
Mass (kg)	3022	1650

vaporization rate and hence effectively by heat transfer. Higher chamber pressures and increased thrust are thereby restricted in turn [12].

The configurations shown all employ an element of regenerative cooling. Fuel or oxidizer may be used as a coolant, flowing through a jacket surrounding the thrust chamber.

The heat absorbed in this way, necessary in prolonged firings, enhances the initial energy content of the propellants prior to injection into the combustion chamber.

Solid propellant rockets

By contrast, solid propellant rockets are comparatively inflexible in their design. The gross classification of solid rockets is made on the basis of the propellant grain geometry since, once ignited, erosive burning proceeds to propellant exhaustion. The rate at which gaseous product is discharged during combustion is characterized by the propellant surface regression rate (typically in the range 5–10 mm/s) and by the variation of total burning surface area with time. Together with the product density and chamber pressure, they determine both the thrust level and the thrust-time history.

A propellant grain is said to be neutral if the thrust remains broadly constant throughout the firing, with a burning surface area that is independent of time. Such behaviour would be characteristic of 'cigarette' burning as illustrated in Figure 6.12. In practice a convenient cylindrical geometry would insist that the burning surface area be small and hence of limited thrust. An annular grain, on the other hand, that burns from the inner surface outwards is progressive. The exposed area, and hence the thrust, increases with time. A large surface area for burning initially, combined with a period of roughly neutral thrust performance, is provided by the star-shaped cylindrical grain illustrated. The changing internal profile with time will give rise to two-stage burning, combining aspects of both progressive and regressive behaviour. The latter accompanies a decreasing exposed area, as in the case of the external burning of a rod or, with the star-shaped grain, as an initially corrugated boundary is progressively smoothed and simplified as combustion proceeds.

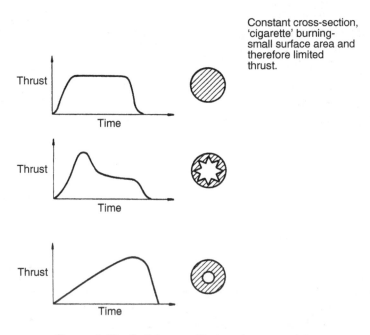

Figure 6.12 Solid propellant grain geometries

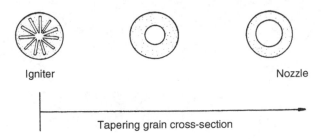

Igniter   Nozzle

Tapering grain cross-section

- Propellant mass 5.0×10^5 kg, Inert mass 8.2×10^4 kg
- Vacuum thrust 11.8×10^6 N
- Specific impulse ~ 260 seconds.

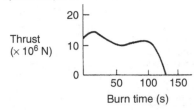

Figure 6.13 Space Shuttle solid rocket booster

Variations in grain geometry along the length of the propellant charge permit the further tailoring of the thrust-time relationship to the requirements of the mission. The configuration and performance of the solid rocket boosters used on the Space Shuttle are illustrated in Figure 6.13.

Solid propellant ignition is typically effected by a pyrotechnic or pyrogen igniter. A small quantity of heat sensitive powdered explosive is ignited electrically and the heat released in turn ignites the propellant within the igniter. Typical igniter compounds are aluminium, boron or magnesium, combined with potassium perchlorate or nitrate oxidizers. In laterally burning cylindrical grains, the igniter is placed at the end of the chamber furthest from the nozzle so that the hot products of combustion from the igniter sweep across the whole grain. Main charge ignition then occurs through convective and radiative heat transfer from these products, the balance between these processes being determined by the detailed igniter design.

6.2.4 Alternative high–speed air-breathing propulsion

A review of typical first-stage rocket performance as described in Chapter 7 reveals that approximately 55% of the launch mass is consumed in accelerating the vehicle to 20% of the orbital speed. The bulk of this propellant—consumed within the Earth atmosphere—is oxidizer. Such bare statistics provide a prima facie case for investigation of the contribution that air-breathing propulsion might make to the initial stages of a space launch.

Figure 6.14 identifies the key propulsive options for high-speed air-breathing with hydrogen as fuel. Current turbomachinery, familiar from the aircraft jet engine, becomes increasingly unattractive thermodynamically at Mach numbers (ratio of flight speed to

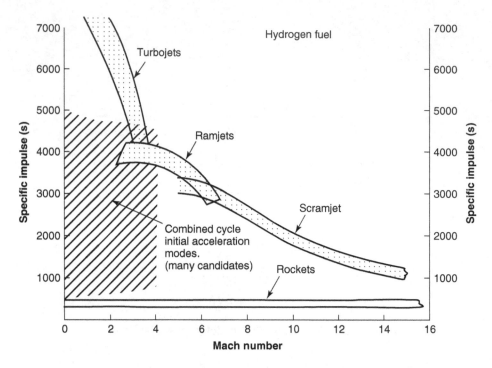

Figure 6.14   High-speed propulsion options in the Earth's atmosphere

sound speed) in excess of 3 and relatively high specific impulses can then be achieved more simply, and efficiently, by capitalizing on ram compression. In the conventional ramjet, however, the ingested air is diffused down to subsonic Mach numbers prior to mixing with fuel and combustion. Isentropic (shock-free) subsonic diffusion leads to a substantial temperature rise, from T to T_0, which increases quadratically with Mach number, M

$$\frac{T_0}{T} = 1 + \frac{1}{2}(\gamma - 1)M^2$$

For $M > 4 - 5$, heat transfer and dissociation losses in the combustor begin to erode subsonic-combustion ramjet performance significantly. However, technologies required by these engines are well established and their range of application may be further extended by the judicious use of high fuel heat capacity in a range of thermal protection and heat exchanger strategies, even embracing air liquefaction. The intimate relationship between vehicle trajectory, kinetic heating and combustion stability is illustrated in Figure 6.15. The target condition for launcher first stage burn-out identified earlier, at 30 km altitude, corresponds to a Mach number of 5.5 and, at such speeds and beyond, the *supersonic combustion ramjet* (SCRAMJET) is the more attractive prospect. The temperature rise associated with diffusion to low subsonic Mach numbers prior to combustion is avoided by burning the fuel in a supersonic air stream but the necessary technologies for efficient high-speed mixing, ignition and stable burning remain incompletely understood. Residence times in the combustor become very short and comparable with the ignition delay for the fuel.

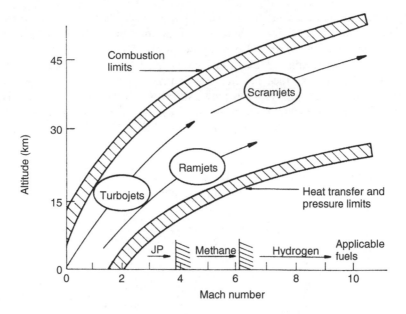

Figure 6.15   Constraints on air breathing trajectories

By comparison with rockets, air-breathing engines offer relatively modest thrust-to-weight ratios and composite engines that seek to use common components over a wide range of flight regimes and operating conditions are crucial to successful SSTO design. Cycle performance analyses and engine simulations have attracted recurrent research interest throughout the recent decades as evidenced, for example, by the two-stage SANGER [13] and National Aero-Space Plane (NASP) programmes [14]. Despite improvements in materials technology and simulation tools, advanced design remains problematic and the development costs intimidating. Current projects therefore favour more conventional rocket technology that facilitates the inclusion of reusable elements (see [15]). Hybrid propellants may prove to offer more cost effective options than those currently employed, however.

6.2.5   Propellant management

The operation of liquid propellant rockets for space vehicle applications may expose the propulsion system to a dynamical regime not usually encountered in terrestrial applications, namely that of free-fall or low residual acceleration. We shall briefly review here some aspects of the problems encountered that do not have a ready analogue in $1g_0$ environments.

The near free-fall situations typical of space manoeuvres or residual drag in Earth orbit correspond to

$$|g_{\text{local}} - a_{\text{vehicle}}| = O(10^{-6}g_0)$$

The very small weights associated with such levels of acceleration invite consideration of other forces, like surface tension, and of equilibrium conditions that might usually be considered to be insignificant except on very small physical scales.

Liquids may be conveniently characterized as wetting (surface spreading) or non-wetting. Liquid to solid surface contact angles may approach zero for perfectly wetting liquids, typical of cryogenics, but may exceed $90°$ for non-wetting liquids such as mercury (which is employed in some ion engines, as discussed in Section 6.4). The surface tension of LOX (at 90 K) is approximately 13×10^{-3} N m^{-1}, compared with 460×10^{-3} N m^{-1} for mercury and 72×10^{-3} N m^{-1} for water. The range of variation is therefore wide.

It is convenient to review the significance of the several forces prevailing in liquid propellants by introducing the following dimensionless parameters:

$$\text{Bond number, } B_0 \equiv \frac{\text{gravity}}{\text{surface tension}} \text{ forces}$$

$$= \rho L^2 g / \sigma$$

$$\text{and} \quad \text{Weber number, } W_e \equiv \frac{\text{inertia}}{\text{surface tension}} \text{ forces}$$

$$= \rho V^2 L / \sigma$$

We may then distinguish the respective inertia, capillary and gravity dominated regimes as suggested in Figure 6.16. For $B_0 > 1$, gravitational forces predominate and surface tension may be neglected, and *vice versa*. We note that for surface tension to be significant in a 1 g$_0$ (Earth surface) environment for a liquid such as water, then $B_0 < 1$ implies a characteristic length scale of less than 2 mm and effects like capillary rise are only significant at such small physical scales. For liquid oxygen in a residual micro-gravity environment, however, the same B_0 boundary corresponds to a length scale of 1 m.

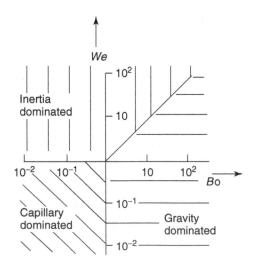

Figure 6.16   Distinctive regimes for forces acting on liquids

The equilibrium configuration of a liquid propellant in a partially filled tank under microgravity conditions is determined essentially by the minimization of droplet (or bubble) free energy which is proportional to the surface area. A wetting liquid will therefore preferentially adhere to tank walls, rather than assume a freely suspended droplet configuration. Internal surfaces and screens within tanks can be used to collect the liquid and separate it from the vapour. Active measures must clearly be adopted to ensure that liquid propellant is available at the tank outlet for rocket motor starting. The principal options for propellant storage and delivery are shown schematically in Figure 6.17 and comprise inertial (or bottoming), positive expulsion and capillary (or surface tension) systems. Supercritical storage is usually ruled out on grounds of tank weight. These options are described in more detail by Ring [16].

In addition to the problem of propellant configuration within the tank, the response to dynamic excitation in flight in the form of propellant sloshing may also be important. The viscosity of cryogenic propellants is in general low, typically by more than an order of magnitude in comparison with water (or hydrazine). The damping of free surface oscillation in the fluid which would otherwise give rise to substantial fluctuating forces and moments on the tanks may also require active provision in the form of turbulence-generating baffles.

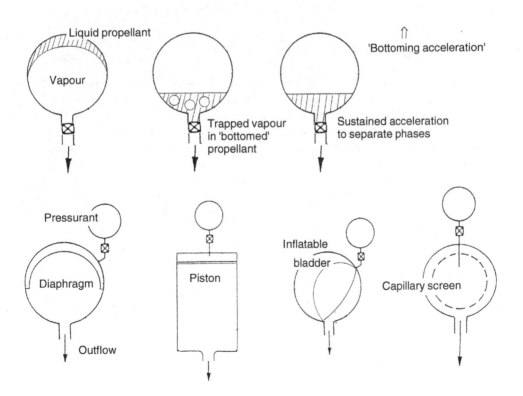

Figure 6.17 Illustrative propellant storage and delivery systems

6.3 SPACECRAFT PROPULSION

The typical functions of spacecraft propulsion, as distinct from launcher operations from the Earth surface, may be summarized, in order of reducing thrust level, as follows:

- final orbit acquisition from the initial orbit established by the launch vehicle,
- station-keeping and orbit control, and
- attitude control.

General principles in relation to both solid and liquid propellant rockets have been reviewed in earlier sections. Here we shall focus on factors influencing the choice of systems appropriate to the spacecraft propulsion roles set out above. The principal options are cold gas systems, monopropellant hydrazine, bi-propellant nitrogen tetroxide/monomethylhydrazine combinations, solid propellants and electric propulsion. We shall briefly consider each of these in turn. Further detail on preliminary design methods appropriate to such devices is provided by Brown [17]. Generally, the spacecraft's propulsion system associated with attitude and orbit control is often referred to as *secondary propulsion*, whereas the propulsion system performing functions such as orbit transfer is referred to as *primary propulsion*.

The increased interest in very small satellites such as typified by CubeSat projects, having a mass of ~ 1 kg within a cube of size ~ 0.1 m, has directed research into efficient micropropulsion functions, frequently calling for microfabrication techniques [18].

6.3.1 Cold gas systems

These systems simply comprise an inert gas, typically nitrogen, argon, freon, or a hydrocarbon like propane, which is stored at high pressure and fed to a number of small thrusters. In the absence of combustion heat release, the kinetic energy of the nozzle exhaust is then solely determined by the driving pressure in the reservoir. A schematic of a typical system is shown in Figure 6.18. Propellants are selected for the simplicity of their storage and compatibility with other facets of spacecraft operation such as the effect of exhaust plume impingement on sensitive surfaces, solar cells, sensors and detectors.

Thrust levels are small, typically of the order of 10 mN; levels chosen to provide the small impulse bits required for high pointing accuracy and stable, jitter free viewing. Minimum impulse bits of approximately 10^{-4} Ns are often necessary for better than 0.1° accuracy in attitude control on some of the larger scientific satellites.

The specific impulse from cold gas systems is comparatively small (~ 50 s)—reflecting the reduced reservoir temperature and pressure (T_c and p_c in equation 6.20). The expellant mass is relatively unimportant, however, in the fine-pointing role. Missions requiring larger ΔVs for secondary propulsion do, on the other hand, necessitate higher performance propellants in order to reduce the system mass.

6.3.2 Monopropellant hydrazine

As we described in Section 6.2, the decomposition of anhydrous hydrazine (N_2H_4), either thermally or catalytically, to the products nitrogen, ammonia and hydrogen is exothermic.

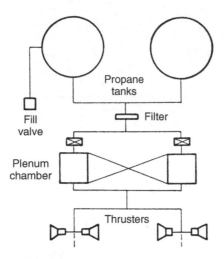

Figure 6.18 Schematic of a cold gas propulsion system

Expansion of the hot product gas through a nozzle will yield specific impulses in the range 200–250 s. The propellant is readily stored as a liquid, with a freezing point of 275 K and boiling point of 387 K, in tanks under the pressure of an inert gas such as nitrogen or helium.

A representative electrothermal hydrazine thruster configuration is sketched in Figure 6.19. The low temperature monopropellant decomposition is enhanced by a resistively-heated metal catalyst - commonly platinum/iridium dispersed on a large surface area, porous substrate of aluminium oxide. Thruster performance is enhanced by higher temperature operation but the accompanying heat transfer losses and materials compatibility problems also increase.

Thrust levels ~10 N may be required for orbit control duties and combinations of thrusters are sized accordingly. Figure 6.20 illustrates the configuration employed on a geostationary satellite for which station-keeping (N-S, using the yaw thrusters, and E-W) was a primary consideration. The propellant tanks are of a positive expulsion (elastomeric diaphragm) type, cross-linked between the paired thrusters.

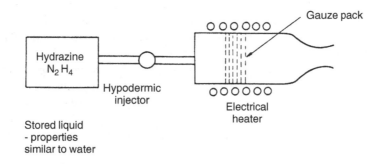

Figure 6.19 Schematic of an electrothermal hydrazine thruster

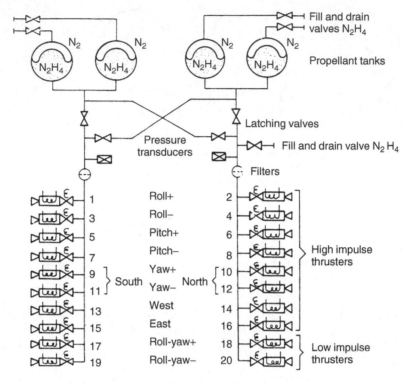

Figure 6.20   Illustrative spacecraft propulsion system using monopropellant thrusters

Improved levels of specific impulse are possible if the exothermic decomposition stage is followed by further resistance-heating in a tandem design. Temperature limitations on the nozzle materials, given the corrosive nature of the product gases, restrict the extent of additional heating but specific impulses approaching 300 s have been achieved.

6.3.3   Bi-propellant MMH/nitrogen tetroxide

The increasing size of many spacecraft, particularly into geostationary orbit, implies the allocation of substantial propellant budgets for secondary propulsion systems and greater emphasis on the level of specific impulse delivered. The combination of MMH and N_2O_4 will provide specific impulses in excess of 300 s.

A representative scheme for a geostationary spacecraft, incorporating the functions of both orbit raising and AOCS, is illustrated in Figure 6.21. The propellants are hypergolic and the layout reflects the additional complexity introduced to ensure safe handling in the propellant storage and feed to the thrusters. The propellants are stored in roughly equal volumes and are both compatible with readily available materials—typically, stainless steel, aluminium or titanium. The accompanying positive expulsion systems employ similar metals in the design of internal bellows since the oxidizer is not compatible with most elastomers.

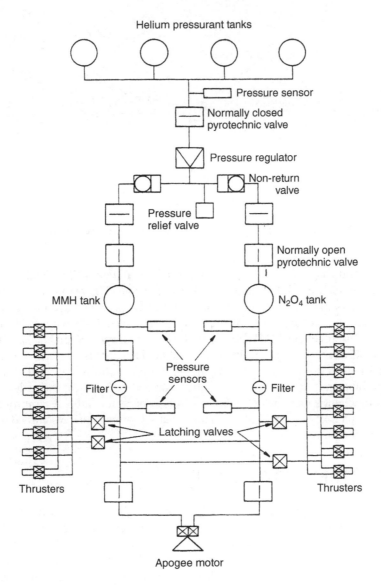

Figure 6.21 Schematic of a typical bi-propellant propulsion system

6.3.4 Solid propellant apogee motors

In the emplacement sequence for a communications spacecraft into geostationary orbit, using an expendable launch vehicle such as Ariane, the final stage injects the satellite into an elliptical transfer orbit with apogee at geostationary height. Historically, but less frequently now, the circularization manoeuvre can be achieved through a high thrust, short duration burn from a solid propellant apogee boost motor (ABM). For a satellite with an on-station mass of approximately $1000\,kg$, the necessary apogee motor fuel approaches $900\,kg$ and the propulsive ΔV is roughly $2\,km\,s^{-1}$.

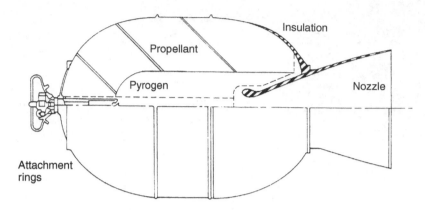

Figure 6.22   Schematic of a solid propellant apogee motor

A typical solid rocket motor for primary propulsion duties is illustrated in Figure 6.22. An example of such a solid propellant ABM is the Star 37E manufactured by Thiokol, which has a mass of 1122 kg, burns for 42 s and delivers an average thrust of 69 kN.

Clearly the emplacement roles of the solid propellant ABM, with its near-impulsive ΔV, and the bi-propellant MMH/N_2O_4, with the advantages of higher specific impulse and more controlled burning, are interchangeable. The impulsive burn requires that the spacecraft should also spin for reasons of gyroscopic stability and thrust alignment. It is therefore inherently less accurate than the extended burn, lower thrust level operation of the bi-propellant motor, which admits precise spacecraft attitude control throughout the thrusting phase. The trade-off is again that of propulsion system complexity for improved performance.

6.4   ELECTRIC PROPULSION

The opening section of this chapter revealed some crucial differences in the performance and operation of separately powered rockets. Unlike the chemical systems which we have described thus far, the energy required for expellant acceleration in an electrically propelled rocket derives from a quite separate source. Whether this source is solar radiation or nuclear fuel, it may be effectively unlimited and the constraints on performance relate to attainable thrust levels and efficient energy conversion.

Whilst electric propulsion has been an active area of development since the earliest space flights, it is only in relatively recent years that electric propulsion has regularly been adopted for commercial, scientific and military missions. Each of the technologies described below have, however, now been operated in space; some as primary systems, although some are still in the technology demonstration stage. In excess of 100 commercial spacecraft, in orbit have adopted electric propulsion systems for either attitude, orbit control or orbit raising. Missions like the Deep Space 1 probe, launched in October 1998, and ESA's SMART-1 in 2003 have demonstrated the use of electric propulsion systems in fly-by manoeuvres beyond Earth orbit. The transfer of technology from the former Soviet Union under the joint National Aeronautics and Space Administration (NASA)/Ballistic Missile Defense Organization (BMDO) RHETT programme in the early 1990s has significantly contributed to the increased use of electric propulsion technologies generally. In the following sections we review the key features and the opportunities offered by adopting electric propulsion on a vehicle, together with a brief description and performance available from individual systems.

6.4.1 Electric propulsion fundamentals

In view of the importance of the power-plant to the rocket configuration, it is convenient to analyse the performance in terms of the component masses: M_w, the power-plant mass; M_e, the expellant mass; and M_p, the payload mass. The expellant storage and feed system may be assumed to be part of the power-plant, whilst the power-plant fuel mass will be considered to be negligibly small (nuclear fuel) or inappropriate (solar powered). The configuration envisaged is shown schematically in Figure 6.23.

The power-plant supplies the exhaust kinetic energy whence the jet power is related to the exhaust velocity by an expression of the form,

$$W = 1/2 \, \dot{m} V_e^2 \tag{6.25}$$

where W denotes power-plant output, $\dot{m}$ the expellant mass flow rate and V_e the exhaust velocity (if thruster process losses are introduced, then $W_{\text{jet}} = \eta W$, but for present purposes we suppose $\eta = 1$).

We relate the power output W to the power-plant mass M_w linearly and introduce the inverse specific power α such that

$$M_w = \alpha W \tag{6.26}$$

If we suppose the exhaust mass flow rate to be constant throughout the burn time, t_b, such that

$$\dot{m} = M_e / t_b \tag{6.27}$$

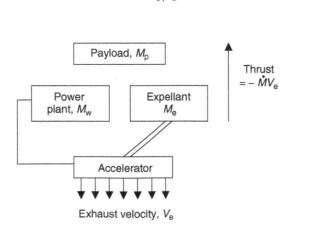

Figure 6.23 Schematic for a separately powered rocket

then

$$M_e = \frac{M_0 - M_p}{1 + \left(V_e^2 / \frac{2t_b}{\alpha}\right)} \tag{6.28}$$

and

$$M_w = \frac{M_0 - M_p}{1 + \left(\frac{2t_b}{\alpha} / V_e^2\right)}, \tag{6.29}$$

where $M_0 = M_p + M_w + M_e$ is the total rocket mass.

By identifying $\sqrt{(2t_b/\alpha)}$ as the characteristic velocity V_c and using the above relationships in the Tsiolkovsky rocket equation (6.5) it can be shown that

$$\frac{\Delta V}{V_e} = \ln\left[\frac{1 + (V_e/V_c)^2}{\frac{M_p}{M_0} + (V_e/V_c)^2}\right] \tag{6.30}$$

The rocket performance described by equation (6.30) is illustrated in Figure 6.24. For $V_e/V_c \ll 1$, increases in exhaust velocity at fixed characteristic velocity result in larger available propulsive ΔV. Such behaviour is essentially that observed with chemical

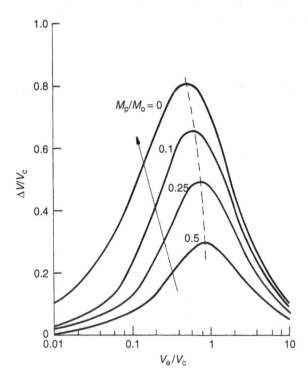

Figure 6.24   Separately powered electric rocket performance

propellants. Performance improvements accompany the better propellant utilization reflected in higher exhaust velocities. Unlike chemical propulsion, however, such benefits do not extend without limit and for $V_e/V_c \gg 1$, the ΔV passes through a maximum and then decreases. Further increases in exhaust velocity require higher powers (see equation 6.25) and therefore increased power-plant mass (see equation 6.26). Beyond a certain stage these increases in power-plant mass outweigh the further reductions in expellant mass accompanying the higher exhaust velocities.

The presence of maxima in Figure 6.24 introduces a range of optimization studies. For a particular payload ratio, M_p/M_0, it is desirable to operate in the vicinity of the maximum, broadly corresponding to $V_e/V_c \approx 1$, so that the high exhaust velocity then requires $2t_b/\alpha \gg 1$. Optimized electric propulsion usage will thus tend to imply lengthy burn times and small inverse specific power, α.

Consider the following illustration of a payload ratio $M_p/M_0 = 0.5$, a mission ΔV of, say 5 km s^{-1}, representative of an orbit-raising manoeuvre, and

$$V_c \approx V_e \approx 16 \, \text{km s}^{-1}$$

For $\alpha = 20$ kg/kW, representative of modern lightweight solar arrays, the burn time t_b would be 30 days. The mean acceleration over the thrust duration, $\Delta V/t_b$, is then $2 \times 10^{-4} g_0$. Clearly quite spectacular improvements in α would be necessary for such propulsion systems to operate from planetary surfaces. However it is quite clear that if long burn times are acceptable from a mission perspective then, according to Figure 6.24, substantial mass savings may be achievable. Generally, operational limitations for station keeping of GEO spacecraft do not include a constraint upon long duration thrust arcs. While there may be radiation induced system problems for substantial orbit raising manoeuvres for commercial satellites through the Van Allen radiation belts, these may be overcome using transfers adopting a super-synchronous orbit having a high perigee (for example ~20 000 km has been used by Boeing). It is therefore clear that many Earth orbit applications may benefit from electric propulsion. Originally however it was the opportunity presented by the high specific impulse from some types of devices, particularly the ion engine variants, which encouraged the earliest reviews (for example [19]), to focus upon the adoption of electric propulsion for interplanetary missions. We explore these aspects in the following section.

6.4.2 Propulsive roles for electric rockets

If we recall that thrust acceleration levels of $2 \times 10^{-4} g_0$ are equivalent to approximately $0.35 \times$ solar gravitational acceleration at 1 AU, then continuous thrust at these levels has a pronounced influence on interplanetary trajectories. Substantial mass savings have been demonstrated to be available for many interplanetary missions, should they adopt electric propulsion. It is perhaps therefore not surprising that papers still regularly appear in journals such as *Journal of Spacecraft and Rockets*, which contribute to the analysis of missions to Solar System bodies. The use of a gridded ion engine on the BepiColombo mission to Mercury, one of ESA's cornerstone missions due for launch in 2014, is a clear statement of the maturity of electric propulsion in such high profile missions. Many mission scenarios have been defined, not only to Mars, but also to the other major and minor planets [20, 21]. The performance of electric propulsion systems during interplanetary missions has been demonstrated following the success of the Deep Space 1 mission [22] and the ongoing mission to the Itokawa asteroid, with the Hayabusa spacecraft now on an Earth-return trajectory [23] (at the time of writing—2010). To date, however, the majority of satellites have employed electric propulsion for either station-keeping, orbit control, Earth orbit raising or attitude control.

Substantial benefits in terms of payload into final orbit may be achieved if the potential of high exhaust velocities to reduce propellant mass can be realized. Recalling Figure 6.24, the maxima are distinguished by $V_e/V_c \approx 1$ and therefore high exhaust velocity must accompany extensive burn times, $V_c^2 = 2t_b/\alpha$. Figure 6.25 presents a simple comparison between a two-impulse Hohmann transfer and low-thrust orbit raising from low Earth orbit, radius r_0, to a higher orbit, radius r. The electric rocket is typically fired continuously and near-circumferential low-thrust leads to a gradual spiralling expansion of the orbit radius. Projected burn times of months would probably be acceptable operationally for an unmanned space tug, implying a modest extension of launch procurement schedules. The implications for thruster design are more important. Extended burn times do influence the propulsive ΔV required and low-thrust orbit-raising operations will in general require a larger ΔV. Figure 6.26 compares the impulsive and circumferential low-thrust requirements; transfer to geostationary orbit requires approximately 20% more ΔV for the low-thrust mission. The requirements of accompanying manoeuvres for low-thrust orbital operations such as changes in orbit inclination were originally extensively analysed by Burt [24].

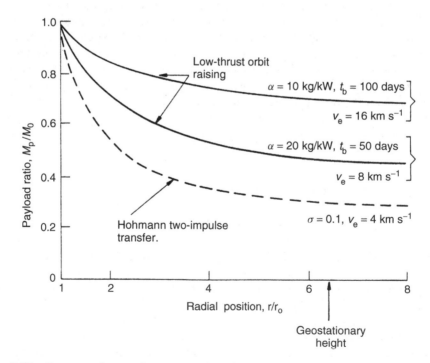

Figure 6.25 Comparative performance: chemical impulse versus low-thrust orbital transfer

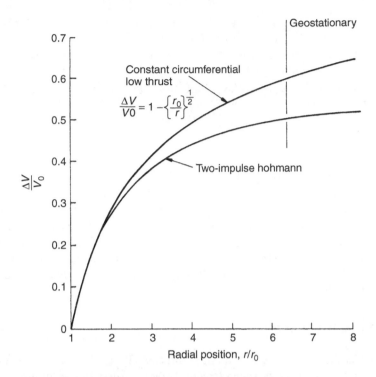

Figure 6.26 Comparative ΔV requirement for transfer between circular orbits r_0 to r

The relationship noted in Figure 6.24 between characteristic mission impulse ΔV, the characteristic velocity V_c and the exhaust gas velocity V_e identifies that the type of electric propulsion technology selected for a mission will depend upon the mission itself. In part this demonstrates why in the early years of spaceflight, research was directed to those technologies capable of achieving the highest specific impulse, especially electrostatic and electromagnetic engines.

As an example, for either typical Earth orbit raising in which the ratio of initial to final orbit is close to unity, or for N/S station keeping in GEO, *magneto-plasma dynamic* (MPD) and *thermal arc jets* demonstrate optimum performance [25]. Alternatively for interplanetary trajectories, *electrostatic ion engines* offer good performance. As noted, however, in [25] care has to be taken in the definition of system performance, specifically whether the satellite power system is sized for the propulsion module or for the operational payload or for other on-board subsystems, since this can have an important influence on the selected 'optimum' technology.

However, optimum performance is only one issue to be addressed in technology selection. More important is the issue of technology availability. In this respect, some technologies offering more modest, sub-optimum improvements in specific impulse have been developed more rapidly than those with greater potential for mass saving. Thus to date most electric propulsion systems which have been flown are of the resistojet type. Hence for Motorola's original Iridium constellation, Primex Aerospace's hydrazine resistojets having an I_{SP} of order 300 s, were used for orbit raising to the operational altitude of 780 km. Resistojets at much lower $I_{SP} \sim 50$ s, have also been developed for small satellites, where available power is very low [26].

Although practical application has thus far tended to favour resistojets having these more modest specific impulses, there are now a number of well developed and commercially available gridded ion, Hall effect and MPD engines, all principally using xenon as the propellant of choice.

6.4.3 Electric propulsion systems

The basic principles underlying electric thruster design are well established (see for example Stühlinger [27], Jahn [28], and Sutton and Biblarz [6]). Electrically powered expellant acceleration devices are of essentially three types: *electrothermal*, in which the enthalpy of the expellant is increased and converted into directed kinetic energy via a nozzle; *electrostatic*, in which charged particles, ions or colloids, are accelerated directly in an electric field; and *electrodynamic*, in which crossed electric and magnetic fields induce a Lorentz force in a plasma. We illustrate briefly designs reflecting each of these distinctive approaches.

Electrothermal thrusters

The resistojet is the simplest electrothermal thruster. Typically, the propellant is heated by passing it over a tungsten heating element. Broad features of the thruster are sketched in Figure 6.27. The more common propellants are hydrogen, nitrogen, ammonia and, in decomposition thrusters in incorporating electrical heating and chemical heat release, hydrazine (*power-augmented hydrazine thrusters* (PAEHT)). The exhaust velocity is a function of temperature, and therefore materials' integrity and life considerations limit V_e to about 10 km s^{-1}. Efficiencies in excess of 70% and thrust levels $\leq 0.5N$ have been demonstrated. Hydrogen is particularly attractive as a propellant since it is

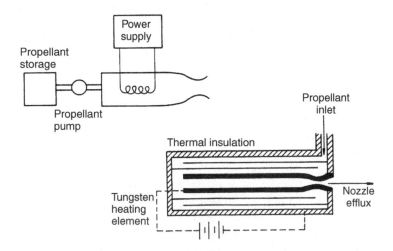

Figure 6.27   The simplest electrothermal thruster: the resistojet

non-corrosive, with both high specific heat and thermal conductivity. It is, however, difficult to store, requiring cryogenic temperatures, and dissociation to atomic hydrogen at elevated temperature leads to frozen flow losses in the nozzle.

Ammonia is readily stored without refrigeration and is dissociated to lighter species on heating, but it is corrosive in both the heater and nozzle. Higher specific impulses are available from the arc jet (~2000 s) but at reduced electrical efficiency. The expellant itself is subject to ohmic heating by passing it through an arc discharge, thereby eliminating gas-solid heat transfer. For many years the principal problem in the implementation of arc jet technology arose from the high erosion of the cathode material. This erosion is a function of both the propellant and the thruster configuration. These propulsion units are, however, intrinsically simple systems and it is a relatively easy matter to convert the on-board available voltage of most spacecraft buses, to the typical 100 V required for an arc jet. Hydrazine fuelled systems are now space proven, and are being used operationally on many satellites. Performance figures quoted by manufacturers identify an I_{SP} greater than 500 s and a thrust greater than 100 mN for an input power of order 1–2 kW.

Electrostatic thrusters

Electrostatic thrusters derive their thrust from the direct acceleration of positively charged particles in a static electric field. The stream of positively charged particles must be neutralized to avoid a charge, opposite to that carried away from the spacecraft in the beam, which would lead eventually to stalling of the thruster. Neutralization is generally achieved by a hot cathode electron source, placed in near proximity to the thruster exit plane.

A one-dimensional approximation to the underlying performance of an electrostatic thruster is simply obtained by equating the potential energy of the ion, as it leaves the ionization region in the thruster, to the kinetic energy it has achieved moving through the electrostatic field. Thus assuming a charge of q (coulombs) exists on an ion of mass m_i (kg), and the potential applied between the ion source and the exit plane of the final accelerator grid is V (volts), then:

$$\frac{1}{2}m_i v^2 = qV.$$

If it is assumed that the conversion efficiency from the electrical power plant output to kinetic energy in the ion stream is unity, then the thrust to power ratio for such a thruster is then simply:

$$\frac{F}{W} = \sqrt{\frac{2m_i/q}{V}} \tag{6.31}$$

Thus for a given electrical power and hence power plant mass (equation 6.26), the thrust is maximized for large mass to charge ratio (m_i/q). Development of these thrusters is now based exclusively upon either argon $m_i/q \sim 4 \times 10^{-7}$ or xenon $m_i/q \sim 1.4 \times 10^{-6}$ as a propellant; for each of these m_i/q values it is assumed that a single charge only is held on the ion. Each of these propellants has been used in thrusters based upon electron bombardment for the production of ions, originally called the Kaufmann engine, but now more commonly referred to as a *Gridded Ion Engine*. In an electron bombardment ion source, the electrons are emitted from an axially mounted thermionic cathode where they are attracted towards a concentric cylindrical anode. A weak, externally applied magnetic field causes the electrons to spiral within the chamber and propellant ionization results from collisions between these electrons and propellant vapour. A representative small thruster design for station-keeping is illustrated in Figure 6.28.

Significant mass savings have been predicted to accompany the application of such ion propulsion to N/S station-keeping for geostationary spacecraft—for example, in relation to the 25 mN *UK-10 xenon thruster* on Intelsat spacecraft [29].

The first use of a xenon ion engine in space was the *RITA engine* on the ESA Eureca-1 mission in 1992. Since then there has been a steadily increasing adoption of ion engines for NSSK on GEO spacecraft. The PanAmSat organization was the first to have an operational ion engine for one of its satellites PAS-5, which was launched in 1997. This ion engine, similar to one used on most subsequent flights, is the Boeing (formerly Hughes Space and Communications) company's *'XIPS' propulsion unit*. Two variants of the thruster have been flown, one having a 13 cm diameter capable of a thrust of 18 mN at an I_{SP} of 2568 s, the other with a diameter of 25 cm, with a thrust of 165 mN and an I_{SP} of 3800 s. The smaller unit has been used on several commercial communications and broadcast satellites including Astra 2A in 1998 for NSSK. In these missions the thrusters are required to fire for approximately 5 h per day in order to maintain the satellite on station. The larger unit was used on the Deep Space 1 mission.

An alternative form of ion thruster is based upon the extraction of charged fluid droplets directly from a fluid surface. Two forms of this exist: *field emission electric propulsion* (FEEP) [30], and colloid propulsion [31]. Both of these depend upon the formation of a 'Taylor cone' [32] at the surface of a fluid that is subject to a strong electric field. The fluid cone forms a capillary jet at its apex that breaks up into a fine spray of positively charged droplets. These droplets in the case of FEEP are sub-nanometre in size, whereas for colloid systems they are sub-micron in size. The colloid droplets, having significantly larger masses than those for the corresponding Kaufmann ion sources, are able to produce a higher thrust density, but at the expense of lower I_{SP}.

Applications for both nanosatellite and microsatellite applications have led to development activities towards a fully integrated micropropulsion unit with many components manufactured using MEMS technology (see also Chapter 18). One favoured solution for this is the colloid thruster. An example of this is capable of providing an I_{SP} of up to 3500 s at a thrust of 140 μN with a complete system dry mass of less than 0.14 kg, including the power control unit [33]. This system has the potential to deliver a nanosatellite into lunar orbit with a fuel load of ∼0.25 kg.

The fluid in a FEEP thruster is a metal, frequently indium or caesium, which is heated so that it becomes liquid. In a colloid system, the fluid is an electrolyte of high electrical conductivity. FEEP thrusters have been developed principally in Europe. Since it is believed that FEEP can produce a highly stable low-thrust beam having a continuous thrust level of order 10 μN, they have been identified to provide the capability of fine control for several demanding missions which require perturbation-free operation. This thruster is currently the baseline for the *Laser Interferometer Space Antenna* - LISA - a gravitational wave detector which is one of ESA's cornerstone missions, which requires a positional control of order 10 nm. In this mission the thrust must be controlled smoothly in the range 5–25 μN with a control resolution of 0.1 μN. The FEEP thruster being developed has an I_{SP} in the range 4000–6000 s. High electrical efficiency is obtained with a value greater than 95%. The power to thrust ratio is of order 60 W/mN.

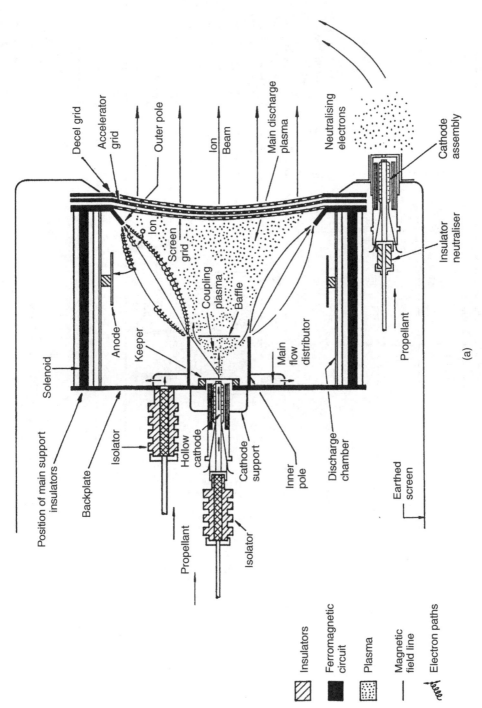

Figure 6.28 A typical ion thruster

(a)

Labels in figure:

Decel grid
Accelerator grid
Outer pole
Ion Beam
Main discharge plasma
Neutralising electrons
Cathode assembly

Position of main support insulators
Backplate
Isolator
Solenoid
Anode
Keeper
Hollow cathode
Cathode support
Inner pole
Discharge chamber

Ion
Screen grid
Coupling plasma
Baffle
Main flow distributor

Propellant
Isolator

Propellant
Insulator neutraliser
Earthed screen

Insulators
Ferromagnetic circuit
Plasma
Magnetic field line
Electron paths

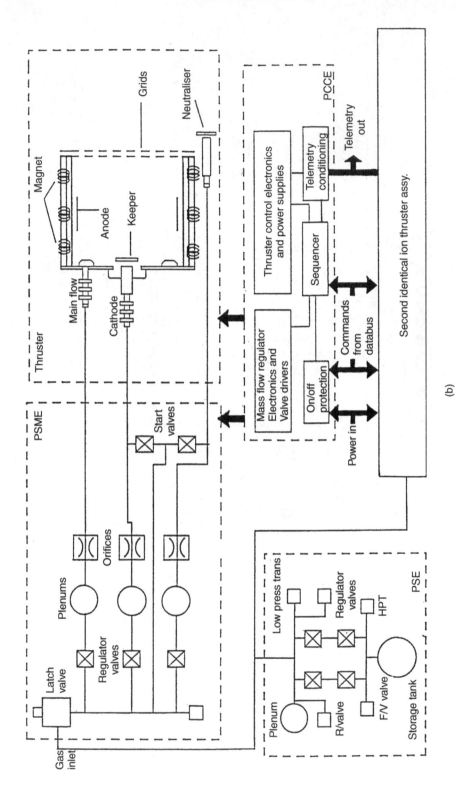

Figure 6.28 Continued

(b)

Hybrid thrusters

Several thrusters rely upon the interaction between electric and magnetic fields for their ability to produce high I_{SP} thrust. Historically in the West, considerable research emphasis was placed upon the *magnetoplasmadynamic (MPD) arc jet*. In the former Soviet Union, practical thrusters were produced which utilize the Hall Effect.

As noted above, *Hall Effect thrusters* have been the subject of a successful technology transfer programme to the USA. This has resulted in commercially available thrusters which have been adopted for several satellites. In the Hall thruster an externally provided radial magnetic field is required. Within the volume enclosed by this field a continuous axial electric discharge is maintained in the low pressure xenon propellant gas. The inter-action between the axial and radial fields generates a Hall current, perpendicular to the electric field, which is azimuthal within the volume of the thruster. It is this Hall current which interacts with the magnetic field that is responsible for the axial acceleration of the propellant. Performance is dependent upon the size of the thruster. The T-220 system developed by NASA has a thrust of 500 mN at a specific impulse of 2450 s. The input power for this performance is 10 kW, with an overall electrical efficiency of 59%. Smaller systems have also been flown, with a power input of 1.35 kW for attitude control, and on the STEX spacecraft in 1998, for orbit raising.

The *magnetoplasmadynamic (MPD) arc jet* evolved from the combination of elec-trothermal arc jet and magnetogasdynamic technologies. A neutral plasma is accelerated by means of both Joule heating and electrodynamic forces. In the full MPD engine, the self-induced magnetic field provides the dominant acceleration mechanism, with thrust being proportional to the current squared. At low power levels the propulsive efficiency is somewhat modest ($\sim$10% for a 1 kW device [34]). As the power level increases into the megawatt range, significantly improved efficiencies result ($\sim$40% at $\sim$MW [35]). For such high power levels, a pulsed operation is essential and significant developments have taken place in Japan in this area, including flight experiments. As with electrothermal arc jets, at present the major life limitation for these devices is due to cathode erosion. Typical operational characteristics are thrusts of a few newtons, with an $I_{sp} \geq 2000$ s.

Thrusters using Teflon as a solid propellant have also been developed, and generally go under the heading of *pulse plasma thrusters (PPT)*, and are a variant of the MPD thruster. These again rely upon the interaction of an electric field with a self-induced magnetic field. A capacitor is used to initiate a pulse discharge in between two electrodes separated in part by a Teflon bar. The surface of this bar ablates, because of the heating caused by the discharge. The induced magnetic field from the high current pulse discharge acts in similar manner to the MPD, to accelerate the ablated material. A very high thrust pulse lasting for some tens of milliseconds can be produced in these thrusters, with a peak thrust value in the region of several hundred newtons. These thrusters have been flown on several missions, particularly for station keeping. Specific impulse from these thrusters is similar to that available from Hall Thrusters.

REFERENCES

[1] Friedman, L. *et al.* (1978) Solar Sailing—The Concept Made Realistic, AIAA Paper 78–82.

[2] Gunn, S. (2001) Nuclear propulsion—a historical perspective, *Space Policy*, **17**(4), 291–298.

[3] Buden, D. and Sullivan, J.A. (1984) Nuclear space power systems for orbit raising and manouevring, *Prog. Astro. Aero.*, **89**, 425–459.

[4] Caveney, L.H. (ed.) (1984) Orbit Raising and Manoeuvring Propulsion: Research Status and Needs, AIAA, New York.

[5] Barrere, M., Jaumotte, A., De Veubeke, B.F. and Vandenkerckhove, J. (1960) *Rocket Propulsion*, Elsevier, Amsterdam.

[6] Sutton, G.P. and Biblarz, O. (2001) *Rocket Propulsion Elements* (7th edn), John Wiley & Sons, Ltd, Chichester.

[7] Rao, G.V.R. (1961) Recent developments in rocket nozzle configuration, *ARS J.*, **31** (11), 1488–1494.

[8] Morel, R. *et al.* (1995) The Clustered Bell Aerospike Engine: Potential, Limitations and Preparation for Experimental Validation, IAF-95-S.2.04.

[9] Scaled Composites, LLC (2009) SpaceShipOne & White Knight http://www.scaled.com/projects/tierone (accessed 1 November 2009).

[10] Kuo, K.K. and Summerfield, M. (1984) Fundamentals of solid-propellant combustion, *Prog. Astro. Aero.*, **90**, AIAA, Washington DC.

[11] Huzel, D.K. and Huang, D.H. (1992) Design of liquid propellant rocket engines, *Prog. Astro. Aero.*, **147**, AIAA, Washington DC.

[12] Brown, J.R. (1983) Expander Cycle Engines for Shuttle Cryogenic Upper Stages, AIAA-83-1311.

[13] Koelle, D. (1990) Advanced Two-Stage Vehicle Concepts (SANGER) AIAA-90-1933.

[14] Moszee, R. and Snyder, C.D. (1989) A Propulsion Development Strategy for the National Aero-space Plane, AIAA-89-2751.

[15] Tomatis, C., Bouaziz, L., Franck, T. and Kauffmann, J. (2009) RLV candidates for European Future Launchers Preparatory Programme, *Acta Astronautica*, **65** (1/2), 40–46.

[16] Ring, E. (1964) *Rocket Pressurisation and Propellant Systems*, Prentice-Hall, New York.

[17] Brown, C.D. (1996) Spacecraft Propulsion, AIAA, Washington, DC.

[18] Shea, H.R. (2009) *MEMS for Pico- to Micro-satellites, MOEMS and Miniaturized Systems VIII*, eds, D. L. Dickensheets, H. Schenk, W. Piyawattanametha. *Proc. of SPIE*, Vol. 7208, 72080M.

[19] Seifert, W.S. (1959) *Space Technology*, John Wiley & Sons, Inc., New York.

[20] Langevin, Y. (2000) Chemical and solar electric propulsion options for a cornerstone mission to Mercury, *Acta Astronautica*, **47**, 443–452.

[21] Schultz, R.A. and Benkhoffa, J. (2006) BepiColombo: Payload and Mission Updates, *Advances in Space Research*, **38**(4), 572–577.

[22] Rayman, M.D., Varghese, P., Lehman, D.H., and Livesay, L.L. (2000) Results from the Deep Space 1 Technology Validation Mission, *Acta Astronautica*, **47**, 489–502.

[23] Kuninaka, H., Nishiyama, K., Shimizu, Y., Funaki, I., Koizumi, H., Hosoda, S. and Nakata, D. (2009) *Hayabusa Asteroid Explorer Powered by Ion Engines on the way to Earth*. 31st International Electric Propulsion Conference, University of Michigan, paper IEPC-2009-267.

[24] Burt, E.G.C. (1968) The dynamics of low-thrust manoeuvres, *J. Royal Aero. Soc.*, **72**, 925.

[25] Stark, J.P.W. and Hobbs, L.W. (1989) Optimization of Electric Propulsion for GEO Missions, AIAA-89-2371.

[26] Baker, A.M., da Silva Curiel, A., Schaffner, J. and Sweeting, M. (2005) Advanced low cost propulsion concepts for small satellites, *Acta Astonautica*, **57**, 288–301.

[27] Stuhlinger, E. (1964) *Ion Propulsion for Spaceflight*, McGraw-Hill, New York.

[28] Jahn, R.G. (1968) *Physics of Electric Propulsion*, McGraw-Hill, New York.

[29] Fearn, D. and Smith, P. (1989) The Application of Ion Propulsion to Intelsat VII Class Spacecraft, AIAA-89-2275.

[30] Marcuccio, S., Genovesse, A. and Andreucci, M. (1998), Experimental performance of field emission micro-thrusters, *J. Propulsion and Power*, **14**, 774–781.

[31] Alexander, M.S., Stark, J.P.W., Smith, K.L., Stevens, B. and Kent, B. (2006) Electrospray performance of micro-fabricated colloid thruster arrays, *J. Propulsion and Power*, **22**, 620–627.

[32] Taylor, G. I. (1964) Disintegration of water drops in an electric field, *Proc. R. Soc. London*, **A280**, 383–397.

[33] Sanders, B., Van Vliet, L., Nardini, F.T., Grönland, T., Rangsten, P., Shea, H.R., Noca, M. and Stark, J.P.W. (2010) Development of MEMS based electric propulsion, Proc ESA-3F Space Propulsion 2010, ESA SP.

[34] Uematsi, K. (1984) Development of a 1 kW MPD Thruster, AIAA-87-1023.

[35] Burton, R.L., Clark, K.E. and Jahn, R.G. (1983) Measured performance of a Multi-Megawatt MPD Thruster, *J. Spacecraft and Rockets*, **20**, 299.

7 LAUNCH VEHICLES

J. Barrie Moss[1] and Graham E. Dorrington[2]

[1] *School of Engineering, Cranfield University*
[2] *School of Engineering and Materials Science, Queen Mary, University of London*

7.1 INTRODUCTION

Although transportation considerations feature prominently in infrastructure provision for a wide variety of terrestrial activities, few of these, if any, pose the difficulties encountered in transporting spacecraft (or payloads) from the Earth's surface to orbit (and back again). Whilst spacecraft design alone introduces considerable technical challenge, the technology involved in the launch will often amplify the complexity—imposing many additional mission and design constraints. The extent to which the user of launch vehicle services is able, in turn, to influence launcher development is, however, still largely unresolved.

Of particular concern to the user, in relation to spacecraft design, are the constraints that the launcher imposes on the mission. These arise especially in terms of payload mass and size, but also include the selection of launch sites and launch windows, the launch environment (mechanical and electrical—see Chapters 2 and 8) as well as issues of safety and reliability (see Chapter 19).

In this chapter, we seek first to establish the basic principles that determine launch vehicle design and performance and, in so doing, constrain payload and mission. Secondly, we outline the key features of the principal launch vehicle alternatives, as featured in Europe and the United States - notably Ariane 5 and Ares 1 - together with the growing range of smaller vehicles. We address briefly some of the key concerns of the user community in relation to launch costs, operational flexibility and reliability, ending with some speculation about future launch capability.

Spacecraft Systems Engineering, Fourth Edition.
Edited by Peter W. Fortescue, Graham G. Swinerd and John P. W. Stark.
© 2011 John Wiley & Sons, Ltd. Published 2011 by John Wiley & Sons, Ltd.

7.2 BASIC LAUNCH VEHICLE PERFORMANCE AND OPERATION

7.2.1 Vehicle dynamics

For purposes of illustration, we specialize the equations of motion to the vertical plane, parallel and normal to the flight direction, and to the motion of the centre of mass and the pitch rotation. A more comprehensive three-dimensional description of these equations is provided by Vinh [1].

The configuration envisaged and the accompanying nomenclature are shown in Figure 7.1. Parallel to the flight direction, we have

$$M \frac{\mathrm{d}V}{\mathrm{d}t} = F \cos(\alpha + \delta) - Mg \sin \gamma - D \tag{7.1}$$

and normally,

$$MV \frac{\mathrm{d}\gamma}{\mathrm{d}t} = F \sin(\alpha + \delta) - Mg \cos \gamma + L + \frac{MV^2}{r} \cos \gamma \tag{7.2}$$

where g is the local gravitational acceleration which varies with radial position of the vehicle r from the centre of the Earth, $g = g_0 r_0^2 / r^2$.

The displacements of the centre of pressure and nozzle from the centre of mass give rise to pitching moments. The accompanying angular motion may be written

$$I_P \frac{\mathrm{d}^2\theta}{\mathrm{d}t^2} = (L \cos \alpha + D \sin \alpha)l - Fl_0 \sin \delta \tag{7.3}$$

where I_P is the moment of inertia in pitch and $\theta = \alpha + \gamma$.

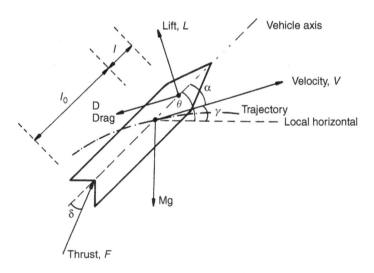

Figure 7.1 Configuration and nomenclature for rocket motion in the vertical plane

For small values of α and δ, equation (7.1) becomes

$$\frac{dV}{dt} = \frac{F}{M} - g \sin \gamma - \frac{D}{M} \tag{7.4}$$

Recall from Chapter 6 that the thrust may be written

$$F = \left(\frac{-dM}{dt}\right) I_{SP} g_0$$

whence equation (7.4) becomes

$$\frac{dV}{dt} = -I_{SP} g_0 \frac{d}{dt}(\ln M) - g \sin \gamma - \frac{D}{M} \tag{7.5}$$

On integration, this becomes

$$V = V_0 + I_{SP} g_0 \ln\left(\frac{M_0}{M}\right) - \int_0^t g \sin \gamma \, dt' - \int_0^t \frac{D}{M} \, dt'. \tag{7.6}$$

In addition to the ideal velocity increment at burn-out, say,

$$\Delta V_{\text{ideal}} = I_{SP} g_0 \ln\left(\frac{M_0}{M}\right),$$

we can now distinguish *propulsive losses* associated with gravity and aerodynamic drag

$$\Delta V_g = \int_0^{t_b} g \sin \gamma \, dt'$$

$$\Delta V_D = \int_0^{t_b} \frac{D}{M} \, dt'$$

whence

$$\Delta V = \Delta V_{\text{ideal}} - \Delta V_g - \Delta V_D \tag{7.7}$$

The relative magnitudes of these terms for an Earth surface launch are sketched in Figure 7.2. We note that the impulse requirement (minimum burn time) that might lead to reduced gravity loss is at variance with the requirement from the standpoint of drag loss, which might suggest a low velocity ascent through the denser atmosphere since $D = \frac{1}{2}\rho V^2 S C_D$. This latter requirement is broadly satisfied. The first stage propulsion of a multi-stage vehicle might typically have an initial thrust-to-weight ratio less than about 1.5, implying an initial vertical acceleration of about $0.5 \, g_0$.

Vehicles are generally launched vertically and minimization of gravity loss, together with the eventual requirement of locally horizontal payload injection, suggests that

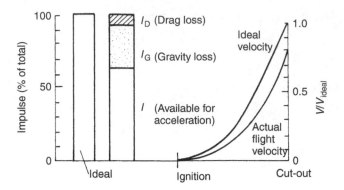

Figure 7.2   Illustration of launch losses due to drag and gravity

the flight trajectory be deflected from the vertical as rapidly as possible. The simplest manoeuvre to effect this is the gravity turn.

Neglecting all but the gravitational force in equation (7.2), we find

$$V \frac{d\gamma}{dt} = -g \cos \gamma,$$

and on integration

$$\sin \gamma(t) = \tanh \left\{ \tanh^{-1}(\sin \gamma_0) - \int_{t_0}^{t} \frac{g}{V} dt' \right\}. \tag{7.8}$$

More rapid pitching manoeuvres may be effected by thrust vectoring or the firing of attitude control thrusters. Figure 7.3 illustrates the full range of flight parameters for a typical injection into geostationary transfer orbit (GTO). The local pitch angle decreases by $60°$ in approximately the first two minutes of flight.

Exact analytical trajectory solutions are not available in general since thrust, gravity and drag are complex functions of altitude. However, approximate analytical expressions can be derived when further simplifying assumptions are introduced [2, 3]. In particular, when drag and the variation in g are ignored, and the thrust-to-weight ratio is held constant at the initial lift-off value, $a = F/Mg = F_0/M_0 g_0$, the mass ratio of a single-stage vehicle is given by [3],

$$\frac{M_b}{M_0} = \exp \left\{ -\frac{V_b}{g_0 I_{SP}} \frac{a^2}{(a^2 - 1)} \right\}$$

where V_b is the speed at burn-out. In this particular case, the burnout-to-initial mass ratio M_b/M_0 is lower than the ideal value, $\exp\{-V_b/g_0 I_{SP}\}$, and the gravity loss incurred during the ascent is

$$\Delta V_g = g_0 I_{SP} \ln\{M_0/M_b\} - V_b = V_b/(a^2 - 1)$$

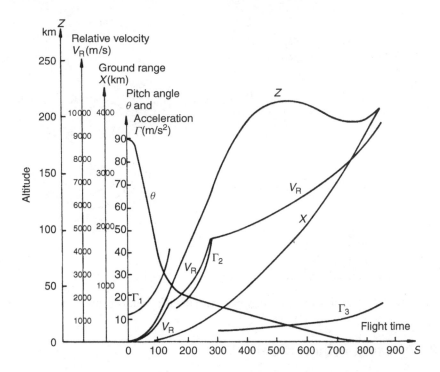

Figure 7.3   Illustrative flight profile for injection into GTO

In order to maximize the payload fraction, it is advantageous to increase the vehicle thrust-to-weight ratio—to reduce the gravity loss—provided the gains in mass ratio are not offset by increases in propulsion system mass [3]. From the user's standpoint, this helps to explain why the payload has to be designed to withstand final accelerations of about 3–4 g_0, although it should be noted many launch vehicles suffer from higher end-of-burn accelerations resulting from design limitations on engine shut-down conditions and/or the inability to throttle-back.

When the contribution of aerodynamic drag is introduced, lower accelerations are required to minimize the peak aerodynamic pressures reached in the early phases of launch. Typically the drag coefficient C_D will be highest in the transonic region and the largest drag losses will occur in the low supersonic region of flight. The actual magnitude of the drag loss will be dependent on the vehicle's aerodynamic reference area S, which is dependent on the vehicle's size—in particular the payload shroud dimensions. For a set of geometrically similar vehicles, the aerodynamic reference area will be proportional to $M^{2/3}$ and the drag loss ΔV_D will be proportional to $M^{-1/3}$ assuming C_D is roughly invariant. Consequently, for small vehicles launching small payloads, drag losses become relatively more important and the user is often forced to accept tighter constraints on payload external dimensions. In general larger vehicles are able to achieve higher payload fractions since they have lower drag losses.

At this point it is appropriate to review some representative mission requirements for spacecraft launches.

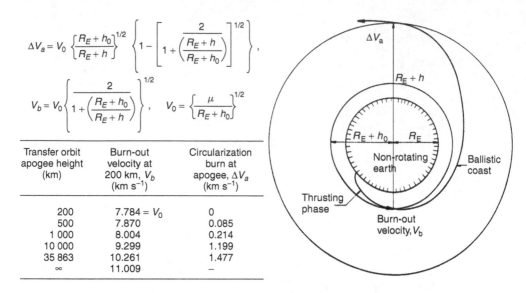

$$\Delta V_a = V_0 \left\{ \frac{R_E + h_0}{R_E + h} \right\}^{1/2} \left\{ 1 - \left[\frac{2}{1 + \left(\dfrac{R_E + h}{R_E + h_0} \right)} \right]^{1/2} \right\},$$

$$V_b = V_0 \left\{ \frac{2}{1 + \left(\dfrac{R_E + h_0}{R_E + h} \right)} \right\}^{1/2}, \quad V_0 = \left\{ \frac{\mu}{R_E + h_0} \right\}^{1/2}$$

Transfer orbit apogee height (km)	Burn-out velocity at 200 km, V_b (km s^{-1})	Circularization burn at apogee, ΔV_a (km s^{-1})
200	7.784 = V_0	0
500	7.870	0.085
1 000	8.004	0.214
10 000	9.299	1.199
35 863	10.261	1.477
∞	11.009	–

Figure 7.4   Launch vehicle burn-out velocities for spacecraft emplacement

7.2.2   Mission requirements

The standard launch vehicle must provide an efficient means of boosting a spacecraft into the planned trajectory. The payloads may vary from those of *Low Earth Orbit (LEO)* scientific satellites, through geostationary communications satellites, to Earth escape probes. Figure 7.4 distinguishes some representative launch vehicle burn-out conditions for spacecraft emplacement. As the table of values shows, the minimum velocity at a typical injection height of 200 km is approximately 7.8 km s^{-1}, whilst that for a geostationary transfer orbit must be 10.3 km s^{-1} (see also Section 5.2 of Chapter 5).

In addition to accelerating the payload to these velocities, the launch vehicle must overcome the effects of aerodynamic drag and gravity. From equation (7.7)

$$\Delta V_{\text{ideal}} = \Delta V + \Delta V_g + \Delta V_D$$

and while detailed determination of the loss terms requires computation, estimates can be made for purposes of preliminary planning using such correlations as those described by White [4]. Representative losses for the first stage of a large rocket using this approach are

$$\Delta V_g = 1.1 \, \text{km s}^{-1}$$
$$\Delta V_D = 0.2 \, \text{km s}^{-1}$$

The values quoted in Figure 7.4 embody a significant simplification to do with Earth rotation. The rotational velocity at the Equator is approximately 0.47 km s^{-1}, and an eastward launch from the Earth's surface can capitalize on this velocity. The benefit reduces with increasing latitude of the launch site and vanishes completely for launches into polar orbit.

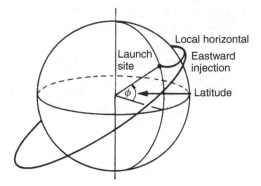

Figure 7.5 Illustration of the link between orbit inclination and the latitude of the launch site

The latitude of the launch site also has important implications for the subsequent incli-nation of the spacecraft orbit. We have identified some significant benefits in respect of launcher performance accompanying both rapid pitch angle reduction and eastward injection. Figure 7.5 illustrates a typical launch sequence. The spacecraft orbit plane is essentially fixed by the velocity vector at burn-out and, from gravitational considerations, the centre of the Earth. Without a very lengthy 'dog-leg' manoeuvre prior to insertion, accompanied by increased losses, both gravitational and aerodynamic, the inclination of the resulting orbit will then be approximately equal to the latitude of the launch site.

As we described in Chapter 5, the correct orbit inclination is a crucial orbital parameter. In the case of the 24-hour synchronous communications satellite, for example, the sub-satellite point is only stationary relative to the Earth for equatorial orbits. Inclination leads to the subsatellite point describing a closed figure of eight as illustrated in Figure 7.6 (see also Section 5.6 of Chapter 5). The amplitude of apparent drift of the satellite, north and

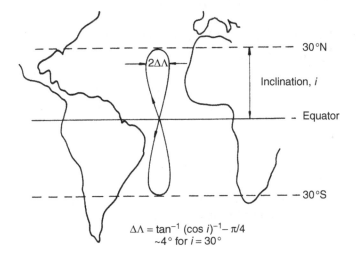

$$\Delta\Lambda = \tan^{-1}(\cos i)^{-1} - \pi/4$$
$$\sim 4° \text{ for } i = 30°$$

Figure 7.6 Drift of the subsatellite point for an inclined geosynchronous orbit

south, is $\pm i°$, where i denotes the orbit inclination. The control of orbit inclination, and specifically its reduction to zero in the case of geostationary communications satellites, is a further requirement of the launch phase.

Two guiding principles behind thrusting operations have been identified in Section 3.2.5 of Chapter 3 (see Figure 3.8). The maximum increase in vehicle speed, and hence kinetic energy, is clearly obtained by thrusting in the direction of motion, and the thrust is then used to best effect in orbit expansion. On the other hand, the maximum deviation from the initial direction is obtained if the available increment ΔV is applied at right angles to the required direction. Furthermore, the smaller the initial velocity, the larger is the deviation for a prescribed level of ΔV. Relative to the problem of orbit raising illustrated in Figure 7.4, directional changes are evidently most economically effected at apogee when orbital speeds are least; for example, 3.07 km s^{-1} at geostationary altitude of 36 000 km compared with 7.8 km s^{-1} at 200 km altitude.

We shall return to these discussions in later sections when reviewing some specific applications, but it is appropriate to note at this stage that the geographical position of the launch site will have important mission implications.

7.2.3 Multi-staged launch vehicles

It is evident from Figure 7.4 that the minimum burn-out velocity for orbital operations is approximately 7.8 km s^{-1}. Linking this requirement with equation (7.6) and neglecting drag and gravity losses, then from

$$V_b = I_{SP} g_0 \ln R \tag{7.9}$$

we may estimate the necessary rocket mass ratio R for a particular propellant combination, and hence the I_{SP}. Suppose the initial rocket mass, M_0, comprises three components, payload M_P, structure M_S and fuel M_F, whence

$$M_0 = M_P + M_S + M_F \tag{7.10}$$

If we introduce the fractional payload ratio, $p = \frac{M_P}{M_0}$ and the propellant tankage structural efficiency, $\sigma = \frac{M_S}{M_F}$, then we may write

$$R = \frac{M_P + M_S + M_F}{M_P + M_S} = \frac{1 + \sigma}{p + \sigma} \tag{7.11}$$

Substituting equation (7.11) in equation (7.9) gives

$$V_b = I_{SP} g_0 \ln\{(1 + \sigma)/(p + \sigma)\} \tag{7.12}$$

For a typical structural efficiency, $\sigma \approx 0.1$, then the maximum value of the mass ratio, R, corresponding to zero payload is approximately $R = 11$. Rearranging equation (7.12), for given V_b

$$I_{SP} = \left(\frac{V_b}{g_0}\right) \ln\{(p + \sigma)/(1 + \sigma)\}$$

and for the maximum value R, together with the minimum burn-out velocity, this implies

$$I_{SP} \geq 331 \, s.$$

Thus, even under these ideal circumstances, single-stage-to-orbit could only be attained using high energy propellant (see Chapter 6). If we incorporate a non-vanishing payload and make allowance for performance losses associated with gravity and aerodynamic drag, then low Earth orbit is only marginally attainable with the most energetic fuel/oxidizer combinations. Significant contributions, in respect of structural mass reductions, must also come from the exploitation of novel lightweight materials.

Since the mass of propellant tankage is large, significant performance benefits result from the progressive shedding of this mass by multi-staging. In these circumstances, only a small fraction of the initial tankage mass is accelerated to the final speed.

If we identify the operation of each stage by an equation of the form given in (7.12), distinguishing the ith stage by subscript, then the velocity increment produced is

$$\Delta V_i = V_{b,i} - V_{b,i-1} = I_{SP,i} \, g_0 \ln\{(1 + \sigma_i)/(p_i + \sigma_i)\} \tag{7.13}$$

We observe that the payload for the ith–stage comprises the vehicle hardware (including the satellite(s) to be launched) 'above' the ith stage. Summing over n such stages, the final burn-out velocity is then given by

$$V_b = \sum_{i=1}^{n} I_{SP,i} \, g_0 \ln\{(1 + \sigma_i)/(p_i + \sigma_i)\} \tag{7.14}$$

An elaborate optimization procedure is required if we seek, for example, to maximize this velocity subject to constraints imposed by way of payload ratio and rocket performance. We should recall that the lower stages will also be subject to drag and gravity losses. The generalized problem has been extensively analysed, for example, by White [4]. It is sufficient for purposes of illustration here to take the simplest case in which the stage specific impulses and structural efficiencies are equal ($\sigma_i = s$), whence

$$V_b = \sum_{i=1}^{n} I_{SP} \, g_0 \ln\{(1 + s)/(p_i + s)\}$$

By the use of the technique of Lagrange multipliers with the constraint imposed by specification of the overall payload ratio,

$$P = \prod_{i=1}^{n} p_i$$

it can be shown that V_b is maximized when the stage payload ratios are all equal,

$$p_i = P^{1/n} (\text{all } i) \tag{7.15}$$

and

$$V_{b\,max} = I_{SP}\ g_0\ \left\{ n \ln(1 + s) - n \ln\left(s + P^{1/n}\right) \right\}$$

This expression has the asymptotic value

$$V_{b\,max} \approx I_{SP}\ g_0\ \ln\ P^{-1} \qquad \text{as } n \to \infty. \tag{7.16}$$

The benefits introduced by multi-staging therefore do not increase without bound except for the impractical case of zero overall payload ratio, as shown in Figure 7.7. At the upper end of payload ratios of interest, P~ 0.1, there is little advantage in further partitioning beyond three stages, given the attendant increases in complexity and cost. The inclusion of the propulsive losses described earlier leads typically to a preferential loading of the first stage.

By way of illustration Figure 7.8 compares a staged vehicle with parameters based on the air-launched Pegasus vehicle, produced by Orbital Sciences Corporation (OSC), with an equivalent single-stage vehicle having the same overall mass, payload ratio and average stage properties (structure-fuel ratio and specific impulse). Table 7.1 summarizes some of the key features of the propulsive performance of the Pegasus vehicle employed in the comparison.

While the three-stage vehicle shows a substantial improvement over its single-stage equivalent, an available ΔV of only 8.1 km s^{-1} would not normally be sufficient to secure LEO once the drag and gravity losses are also incorporated. The vehicle's additional performance is derived from the air launch—which imparts an initial velocity (0.24 km s^{-1}) and the diminished losses accompanying reduced air density and angle of incidence.

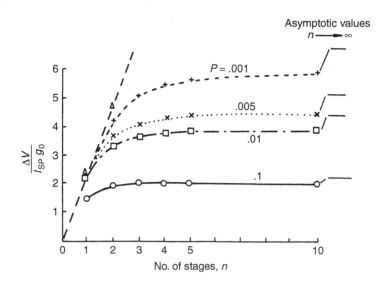

Figure 7.7 Variation of velocity increment with number of stages for fixed overall payload ratio P

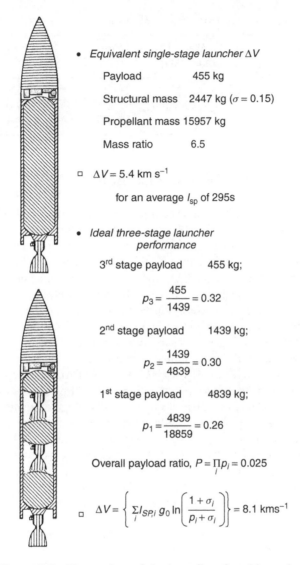

- *Equivalent single-stage launcher ΔV*

Payload	455 kg
Structural mass	2447 kg ($\sigma = 0.15$)
Propellant mass	15957 kg
Mass ratio	6.5

  ▫ $\Delta V = 5.4$ km s^{-1}

     for an average I_{sp} of 295s

- *Ideal three-stage launcher performance*

  3rd stage payload   455 kg;

  $$p_3 = \frac{455}{1439} = 0.32$$

  2nd stage payload   1439 kg;

  $$p_2 = \frac{1439}{4839} = 0.30$$

  1st stage payload   4839 kg;

  $$p_1 = \frac{4839}{18859} = 0.26$$

  Overall payload ratio, $P = \prod_i p_i = 0.025$

  ▫ $\Delta V = \left\{ \sum_i I_{SP,i} \, g_0 \ln \left(\frac{1 + \sigma_i}{p_i + \sigma_i} \right) \right\} = 8.1$ kms^{-1}

Figure 7.8   Illustration of the benefits of multi-staging

7.3   SPACECRAFT LAUNCH PHASES AND MISSION PLANNING

In the preceding sections, we have outlined some basic aspects of flight dynamics, the propulsive requirements for Earth orbit attainment and the response to the intrinsic mismatch between ΔV and chemical propellant performance as reflected in multi-staging. Spacecraft missions invariably require the emplacement of the payload in an orbit having narrowly specified parameters—altitude, eccentricity, inclination and inertial orientation. The launch phase therefore embraces each of the propulsive manoeuvres necessary to achieve the initial emplacement and these are necessarily different for each mission.

Table 7.1 Propulsive characteristics: Pegasus launcher

Stage	Specific impulse in vacuum [s]	Average thrust [kN]	Burn time [s]	Gross mass [kg]	Propellant [kg]
1	295.3	486.7	77	14 020	12 150
2	295.5	122.8	75	3400	3025
3	291.1	54.56	65	984	782

7.3.1 Geostationary orbit emplacement

A typical three-stage expendable launcher injects the satellite into GTO with a perigee at 200 km altitude and apogee at 36 000 km (see also Section 5.6 of Chapter 5). Upper-stage burn-out occurs at the first equatorial plane crossing. Following a ballistic coast, the transfer orbit apogee then coincides with the second crossing of the equatorial plane, or node. This transfer orbit is inclined, the inclination being dependent upon the latitude of the launch site and the launch azimuth. As indicated earlier, a due east launch (launch azimuth $= 90°$) will result in a transfer orbit inclination equal to the latitude of the launch site, while any other launch azimuth must increase the orbit inclination.

Positioning of the geostationary apogee over the equator admits the possibility of emplacement in final equatorial geostationary orbit by a further single motor firing. The satellite is fitted with an *apogee boost* (or *kick*) *motor* (ABM/AKM) specifically to effect this combined manoeuvre of orbit circularization and inclination removal. The transfer orbit apogee velocity is approximately 1.60 km s^{-1} while the equatorial geostationary orbital velocity is 3.07 km s^{-1}. Table 7.2 illustrates the impact of launch site latitude, and hence transfer orbit inclination, on ABM impulse requirement.

Near-equatorial launch sites such as the European Space Agency site at Kourou in French Guiana ($\sim 5°$ N) offer significant propulsive advantages over the Eastern Test Range [Kennedy Space Center (KSC)] in Florida (latitude $\sim 28°$ N) or the Japanese or Russian sites at even higher latitudes. The option to launch Soyuz rockets from Kourou from 2010 will exploit this advantage.

Injection into the final orbit by the satellite ABM is not usually effected at the first transfer orbit apogee, approximately five hours after launch. The manoeuvre is constrained by the requirement to attain the operational longitude station with a high level of accuracy and within a defined period. Precise determination of the satellite orbit and attitude by ground station tracking is necessary in order to correctly orientate the motor prior to commanding its firing. Several apogees will then pass—the elliptical transfer orbit period is approximately 10 hours—before the ABM burn. The satellite must be ground controlled

Table 7.2 ABM impulse requirements for geostationary orbit emplacement

	Transfer orbit inclination (degrees)			
	0	10	30	50
ABM ΔV (km s^{-1})	1.47	1.52	1.86	2.38

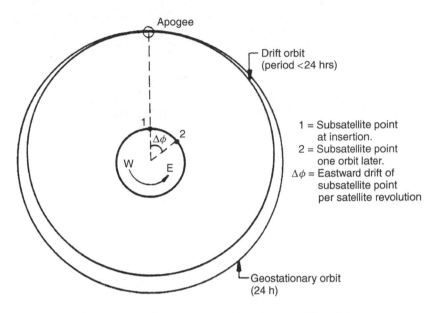

Figure 7.9   Station acquisition from a drift orbit

during this phase and have sufficient electrical power to maintain communications and some on-board systems, for example, attitude and orbit control. If this phase is solely battery powered, then such factors will determine the acceptable length of time spent in transfer orbit.

In view of the possible accumulation of injection errors and the requirement to position the satellite precisely on longitude station, final positioning can be achieved from a drift orbit. If the satellite orbit after ABM firing is arranged to be very slightly elliptical, with perigee somewhat less than geostationary height, then the orbit period will be slightly less than the Earth rotational period. Furthermore, as Figure 7.9 illustrates, the subsatellite point at geostationary height will then drift gradually eastwards relative to the Earth. Short duration firing of the satellite's station-keeping thrusters over several days then permits the precise positioning of the satellite. The use of liquid bi-propellants—rather than solid propellants—in apogee motors [Liquid Apogee Motor (LAM)] offers an alternative strategy for station acquisition involving more extended motor firings at lower thrust levels that make less demands on the satellite ACS (see Section 5.6 of Chapter 5).

7.3.2   Low Earth orbit emplacement

Similar, if less elaborate, manoeuvres apply to most other near-Earth missions. The more energetic and accurate is the launch vehicle injection, the smaller is the secondary satellite propulsion requirement and the greater is the true payload fraction. Preliminary mission planning is essentially iterative, however (see Chapter 5). The identification of broad mission objectives—Earth observation, communications, science—will all involve the specification of the satellite orbit. Launch vehicle performance summaries that describe the orbits accessible to a particular payload mass introduce the first constraint. Detailed

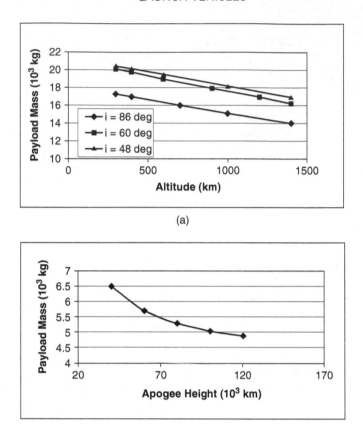

(a)

(b)

Figure 7.10 Illustrative trends for Ariane 5 ES payload capability into (a) low Earth circular and (b) highly elliptical orbits

consideration of the distinctive launch capabilities of Ariane 5 and other launch vehicles is presented later, but we illustrate briefly here some general principles.

Figure 7.10 illustrates typical users' manual information for Ariane 5 ES launches into low Earth circular and highly elliptical orbits. As might be expected, the payload capability into circular orbits of increasing altitude but fixed inclination decreases steadily. Further reductions accompany the raising of the apogee height. Changes in inclination also have a substantial impact on payload capability. If the 500 km orbit is inclined at $86°$ rather than $48°$, for example, the payload is reduced by roughly 3 tonnes, reflecting the diminishing benefits derived from Earth rotation. In other respects, the east coast of South America presents little restriction to eastward launches of any azimuth between zero and $90°$. However, this is clearly launch site specific. Figure 7.11 identifies the *launch azimuth limits* that apply at the Western Test Range (WTR) (Vandenberg) in Southern California. Direct injection is here restricted to launches towards the south-west, over the Pacific Ocean and away from continental USA, giving access only to retrograde orbits ($i > 90°$). Only along the $164°$ azimuth limit is there a possibility of an eastwards launch and a small contribution from Earth rotation.

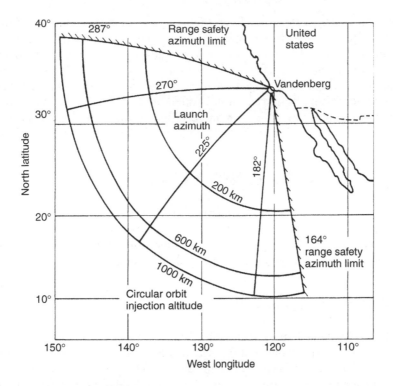

Figure 7.11   Range safety limits and launch trajectories from the WTR

7.3.3   Configuration interactions

Whilst payload mass may be considered to assume an overriding importance in determining the mission parameters, in many circumstances the spacecraft configuration can also be constrained by the size and shape of the available payload volume. Multi-stage expendable launch vehicles are essentially cylindrical, and aerodynamic considerations naturally restrict the payload fairing (or envelope) to a shape resembling a cone-cylinder combination. Spacecraft designed to be launched by such vehicles tend to be slender—at least in their stowed configuration—and solar arrays, communications antennas and scientific instruments may have to be folded, furled or telescoped to conform to the fairing and then deployed on station. The inclusion of mechanisms necessary to effect such deployment then adds substantially to the complexity and vulnerability of the payload design.

Although the launch phase duration may be measured in hours in a mission with an operational timescale of years, the launch often produces the most demanding environment faced by the mechanical design. The longitudinal acceleration is high—for example, in excess of $4.5g_0$ in the case of Ariane 5 at solid rocket burn-out. Sensitive elements of the payload and deployable equipment must therefore withstand both the mean acceleration and the structural vibration accompanying motor firing and stage separation (see also Chapter 8).

As we indicated in Section 7.3.1, geostationary orbit emplacement may also incorporate quite major satellite thrusting phases at apogee (ABM firing). Where motors employ solid propellants, they deliver relatively high thrusts in firings of short duration (see Chapter 6). In the absence of fine control of thrust, the spacecraft and motor are spun-up to an angular rate of ~ 100 rpm. This both provides a measure of gyroscopic stiffness for guidance and reduces the effects of any thrust misalignment. In consequence, however, deployment of lightweight, flexible structures like solar arrays is further delayed and substantial power-raising in transfer orbit is often prevented. The enhanced mission flexibility that accompanies lower thrust, more readily controlled and higher specific impulse bi-propellant rocket motors is a major factor in their development as spacecraft propulsion.

7.4   THE ARIANE 5 LAUNCH VEHICLE

In the preceding section, we have described some general features of launch vehicle performance and operation from the user or payload perspective. Here we describe the Ariane 5 launch vehicle in more detail but from the same standpoint. The reader will be referred to other sources for launch vehicle design information.

7.4.1   Vehicle design summary

While earlier expendable launch vehicles evolved from missile designs and were adapted to a wide range of missions, much of the Ariane programme has been specifically directed towards the task of geostationary communication spacecraft emplacement. Despite subsequent broadening of the performance envelope through the development of Ariane 1 to 5, each embracing increased payload mass and envelope, the defining mission remains that of injection into GTO. In the presentation here, we concentrate on the capabilities of Ariane 5 in its two key configurations—Ariane 5 ECA, providing direct access to GTO, and Ariane 5 ES, servicing LEO [in particular, the launch of the automated transfer vehicle (ATV) to the International Space Station (ISS)]. The latter will complement the continuing use of the Soyuz rocket—soon to be launched from French Guiana also—to transport crews to and from the ISS.

Ariane 5 comprises a central core stage, incorporating the Vulcain cryogenic engine, to which are attached two solid propellant boosters. While these components are common to both variants, the ECA and ES versions employ different upper stages—a cryogenic motor for the more powerful ECA and a re-ignitable storable propellant motor for the ES version. The stage mass breakdown for Ariane 5 ECA is summarized in Table 7.3.

Table 7.3   Ariane 5 ECA mass breakdown (tonnes)

Stage	Dry mass	Propellant mass	Cumulative mass
Solid rocket booster (each EAP)	39.8	237.7	555 (72%)
Cryogenic main core (EPC)	14.7	143 (LOX) 27 (LH_2)	739.7 (96%)
Cryogenic upper stage (ESC-A)	4.5	10.0 (LOX) 4.9 (LH_2)	769.1 (100%)

Note: Additional mass arises from items of inter-stage structure and the equipment bay.

Table 7.4   Ariane 5 ECA engine characteristics

	Solid rocket booster	Vulcain 2: main cryogenic core	HM7B: cryogenic upper stage
Vacuum thrust [kN]	7000	1390	67
Vacuum I_{SP} [s]	275	432	446
Chamber pressure [bar]	61	108	35
Burn time [s]	130	540	945
Propellant type	HTPB	LOX/LH$_2$ (5.3:1)	LOX/LH$_2$ (2 :1)

Note: HTPB = Hydroxyl-terminated Polybutadiene

The two boosters provide the bulk of thrust during the initial phase of the launch, burning in parallel with the main stage Vulcain 2 engine. The latter then enters a lengthy sustained phase following booster separation before transition to the upper stage HM7B motor. Whilst the Vulcain engine was developed specifically for the Ariane 5 programme, the upper stage motor is that employed on Ariane 4. Both motor designs are based on gas generator cycles as described in more detail in Chapter 6.

The component performance is summarized in Table 7.4. The Ariane 5 launcher represents a significant departure from the earlier configurations in the Ariane family by incorporating two large (EAP) solid boosters, each delivering a vacuum thrust of 7000 kN at a specific impulse of 275 s. The Vulcain 2 core stage is a cryogenic LH2-LOX engine, delivering a vacuum thrust of 1390 kN at a specific impulse of 432 s. To accommodate the low Earth manoeuvring requirements of the ATV launch, a smaller re-ignitable mono-methylhydrazine (MMH)/NTO upper stage replaces the HM7B in the ES variant.

7.4.2   Mission performance

Ariane 5 is launched from the near-equatorial Centre Spatiale Guyanais (CSG) in Kourou, French Guiana, at a latitude of 5.2°N. As described earlier, such a location is particularly favourable for geostationary satellite emplacement but launches can also be made from within azimuth limits −10.5° to 93.5°. Illustrative performance data are shown in the Ariane 5 users' manual [5] for purposes of preliminary mission planning, including a variety of elliptical and Sun-synchronous orbits in addition to the primary programme objective of GTO emplacement.

Table 7.5 illustrates the primary mission performance of the two variants into GTO and injection of the ATV. Ariane 5 ECA can launch two spacecraft into GTO with a combined mass of 9.6 tonnes, including the additional interface and supporting masses of the dual launch system.

The dual launch capability has been an important feature of the Ariane system throughout its history and this is maintained within the Ariane 5 programme in the form of SYLDA5 and SPELTRA. A brief description of the Systeme de Lancement Double Ariane (SYLDA) and payload envelopes is provided later in this chapter.

If we focus specifically on the dual launch into GTO, both satellites are injected into the same orbit although they may be oriented and spun-up independently as required (up to

Table 7.5   Illustrative Ariane 5 performance (tonnes)

Ariane 5 ECA	
Liftoff mass	780
Payload into GTO (dual launch)	9.6
Ariane 5 ES ATV	
Liftoff mass	760
Payload mass (260×260 km; $51.6°$ inclination)	21 (max)

$30°/s$ in longitudinal mode) after upper-stage engine cut-off. The ACS can provide either three-axis stabilization or spinning modes. A minimum relative velocity of $0.5\,\mathrm{m\,s^{-1}}$ is established between the two satellites at separation and a safe distance is determined between them before apogee motor firing. The argument of perigee at injection is $178°$ and the apogee height is $35\,786$ km. The GTO inclination is $6°$ and the longitude of the first descending node occurs over West Africa. The injection phase is inertially stabilized with high accuracy. For example, typical values cited for the standard deviations of key injection parameters are: 40 km for the semi-major axis, 4.5×10^{-4} for eccentricity and $0.02°$ for inclination.

For smaller payloads into LEO, multiple separations are also possible with the aid of user-defined payload dispensers. If required, the dispenser can be mounted inside the SPELTRA.

7.4.3   Dual-launch system

Building on the experience with earlier versions, the dual launch capability of Ariane 5 has been used extensively for GTO missions. Whilst improvements have been introduced in relation to the system weight and complexity in SYLDA-5, the key features of the SYLDA concept as employed on Ariane 4 remain the same and are illustrated in Figure 7.12. A load carrying, carbon-fibre reinforced shell, mounted within the fairing, encapsulates the lower satellite and supports the upper one.

The shell incorporates two separable parts, retained in flight by a clamp band, providing satellite adapter rings. The two satellites are quite independent, without mechanical or electrical interfaces. The sequence of events leading to the release and injection into orbit of the two satellites is shown in Figure 7.13. Following upper stage engine cut-off, the attitude control system orientates the stage correctly and spins up. Pyrotechnic cutters are fired and release the spring-loaded upper passenger. The upper portion of the shell is itself then released pyrotechnically and the lower satellite is exposed and spring released.

The payload accommodation has steadily increased in size over the course of the Ariane programme and Table 7.6 summarizes the envelope dimensions and masses of the basic SYLDA-5 and SPELTRA configurations. A range of alternative adaptors and spacers are identified in the users' manual [5], permitting some tailored use of the free space.

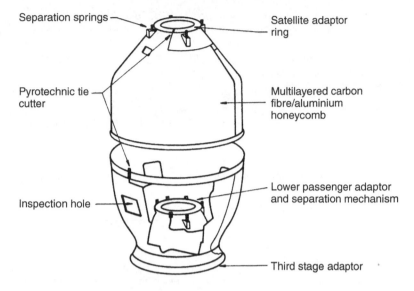

Separation springs

Satellite adaptor
ring

Pyrotechnic tie
cutter

Multilayered carbon
fibre/aluminium
honeycomb

Inspection hole

Lower passenger adaptor
and separation mechanism

Third stage adaptor

Figure 7.12   SYLDA

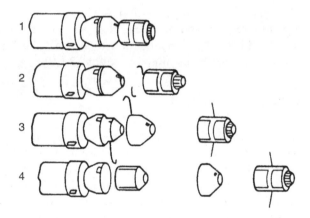

1

2

3

4

Figure 7.13   Ariane dual-launch separation phases

7.5 US CREWED LAUNCH SYSTEMS

After almost 30 years service as the USA's principal means of launching humans into
orbit, the Space Shuttle fleet is to be retired in 2010. In anticipation of this event, a
radical rethink of the US human space activity emerged, resulting in the Constellation
programme, the launch components of which comprise the Ares 1 man-rated launcher and
the Ares 5 heavy lift launch vehicle. However, the changing political winds generated by
the incoming Obama administration, has spawned uncertainty about Ares 1 as a means of

Table 7.6   Masses and envelopes for SYLDA-5 and SPELTRA

SYLDA-5	
Maximum diameter	4.5 m
Cylindrical section height	3.2 m
Empty mass	440 kg
SPELTRA	
Short version: maximum diameter	5.4 m
cylindrical height	4.1 m
empty mass	704 kg
Long version: maximum diameter	5.4 m
cylindrical height	5.6 m
empty mass	820 kg

US human access to space. Furthermore, it is proposed to provide funding to US industry to encourage innovative solutions to launching human cargo into orbit. It is against this backdrop of political and funding uncertainty that we present this section reviewing US crewed launch systems.

The Space Shuttle system, as a unique semi-reusable man-rated launcher, warrants significant mention in this review, and alongside this we have chosen to describe the attributes of the Ares 1 vehicle. However, it is fair to say that the future shape of US human access to orbit is undefined at the time of writing (2010).

7.5.1   Shuttle STS

The US Shuttle Space Transportation System (STS) was originally intended to be a fully reusable, low cost system to transport crew and a wide variety of payloads (up to 25 tonnes) to and from LEO. During the design process in the early 1970s, the STS evolved, from a piggyback configuration with two winged-reusable vehicles, to its familiar present form, as shown in Figure 7.14. The present configuration comprises the reusable winged Orbiter, coupled with an expendable External Tank (ET) containing cryogenic propellants (liquid hydrogen and liquid oxygen in a mass ratio of 1:6). The original winged-reusable first stage was replaced with two parallel-burning *Solid Rocket Boosters* (SRBs) each made from four segments attached together by clevis joints. The Orbiter uses three throttled Space Shuttle Main Engines (SSMEs) with a pre-burner cycle, each with a vacuum specific impulse of 455 s and maximum thrust of about 2.1 MN. The cryogenic propellant required for these engines is cross-fed from the ET. After ET separation, the Orbiter uses an Orbital Maneuvering System (OMS), fuelled by storable nitrogen tetroxide and monomethyl hydrazine propellants, to provide the impulse to acquire and circularize the required orbit. The same OMS is used to de-orbit. During re-entry, the Orbiter develops significant lift in order to reduce peak entry temperatures and deceleration rates. The Orbiter (the primary structure of which is aluminum alloy) is protected during re-entry by a Thermal Protection System (TPS). On the lower surfaces of the Orbiter, the TPS comprises an array of small ceramic tiles which have proven to be problematic to maintain

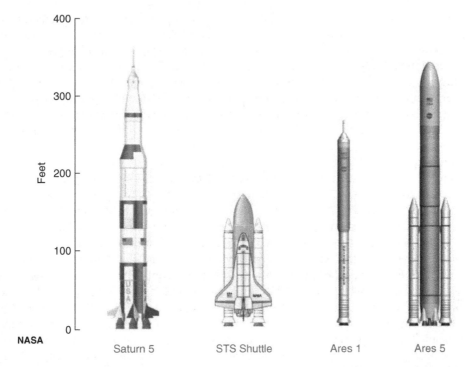

Figure 7.14   US launch vehicle configurations, in particular STS Shuttle and Ares 1
(Reproduced by permission of NASA)

and refurbish during the repeated operations of the small Orbiter fleet (each Orbiter has
been flown about 25 times).

At the time of writing, the Shuttle STS has flown 128 successful missions, and experienced two catastrophic failures, over a 29 year period (1981–2009). The first failure of
'Challenger' STS-51-L (1986) involved a burn-through of an O-ring sealed clevis joint on
one SRB early in the ascent, resulting in a catastrophic explosion of the ET. The second
failure of 'Columbia' STS-107 (2003) occurred after the TPS on the leading edge of the
port wing was penetrated by the impact of a foam wedge from the ET during ascent.
Unfortunately, since no on-orbit inspection and repair was carried-out, the Orbiter was
subsequently destroyed during re-entry.

Despite the proven flexibility of the Shuttle STS for crewed orbital operations, its flight
costs have proven to be much higher than was originally envisaged [6]. Consequently, as
mentioned above, NASA is planning to cease operations in 2010.

7.5.2   Ares 1

Continued uncertainty surrounds the provision of US launcher capability to support future
manned missions. If development continues to operations as NASA has envisaged, however, the Ares 1 launch vehicle will provide human access to orbit in around 2015, and
beyond [7, 8], and this is predicted to have considerably lower operational costs than the
Shuttle STS—see Figure 7.14.

It was originally intended that Ares 1 would comprise key Shuttle elements with minimal modification, to reduce development costs. That is, a first stage derived from the four-segment Shuttle SRB and a cryogenic second stage using a single SSME. In early feasibility study reports, it was stated that this particular configuration would offer the highest overall crew safety and mission reliability, despite the fact that it requires in-flight ignition of the second stage cryogenic engine (unlike the Shuttle STS). As the design evolved, a decision was made to replace the relatively expensive SSME with an updated version of the re-ignitable J2 engine originally used on the Saturn 5 launch vehicle. The new engine, J2X, with a gas generator cycle, is predicted to have a vacuum specific impulse of 448 s with a thrust of 1.3 MN (about 62% of the SSME thrust in vacuum at 100% throttle). As a consequence of this thrust reduction, the first stage of Ares 1 had to be increased in size. The current baseline adopts a five segment *Reusable Solid Rocket Motor* (RSRM) that is about 25% heavier than a Shuttle SRB. The burn time of the RSRM is similar to the Shuttle SRB (about 120 s), so that burn rates have been increased significantly—demanding a new ground-test firing programme.

The total height of the Ares 1 will be about 100 m, somewhat taller than Shuttle, but still within the constraints of the modified Vehicle Assembly Building. The gross lift-off mass will be about 46% of the Shuttle's, well within the capabilities of the Apollo crawler transport that was used to take the vehicle to the launch pad. The first stage RSRM will be recovered using a triple parachute system downrange in the Atlantic. The second stage is expended, and expected to burn-up during re-entry over the Indian Ocean. The first successful flight of the Ares 1-X test vehicle employed a modified four segment Shuttle SRB with a dummy upper-stage [8].

The primary payload of Ares 1 is intended to be the *Orion Crew Exploration Vehicle* (CEV)—an Apollo-like, conically-shaped, ballistic re-entry capsule attached to a service-propulsion module [9]. Orion is currently baselined to carry four crew to LEO. When stacked on top of Ares 1, the capsule will be shrouded and fitted with a launch abort rocket system to permit safe separation from the lower stages in the event of a catastrophic failure. Re-entry of the Orion capsule will essentially be ballistic with a lift-to-drag ratio less than 0.3 [10]. Hence peak decelerations will be dictated by equation (5.57) in Chapter 5. Recovery by parachute on land, as well as in the ocean, has been evaluated.

The Ares 1 launcher is also viewed as being integral to NASA's Constellation Programme, with the main motivation to renew manned lunar exploration by about 2020. An associated heavy-lift launcher, Ares 5 [11], would also be required to deliver the cis-lunar propulsive stage that will be docked with Orion after rendezvous in LEO. However, as discussed above, the complete funding of this programme remains uncertain. The comparative mass breakdown for Ares 1 and 5, as originally conceived, is summarized in Table 7.7.

7.6 SMALL LAUNCHERS AND REUSABLE SUB-ORBITAL VEHICLES

7.6.1 Small launchers

The continuing interest in launching small satellites into LEO—including constellations of such satellites, whether for scientific or Earth observation purposes—has prompted

Table 7.7   Ares 1 and Ares 5 mass breakdown (tonnes)

	Ares 1	Ares V
Gross lift-off mass	927.1	3705
RSRM 1st stage mass	732.5 (5 segments)	2 × 794 (5.5 segments)
Cryogenic 2nd stage mass	152.7	279
Cryogenic propellant mass	115.4 (LOX) 21.7 (LH$_2$)	213 (LOX) 38 (LH$_2$)
Inter-stage mass	4.1	12
Payload to LEO	27 (Orion CEV)	140-180
Payload to lunar transfer	-	57

Table 7.8   Vega stage characteristics

	1st stage: P80	2nd stage: Zefiro 23	3rd stage: Zefiro 9	4th stage: AVUM
Height [m]	10.5	7.5	3.85	1.74
Diameter [m]	3.0	1.9	1.9	1.9
Propellant mass [tonnes]	88	23	10	0.55
Thrust [kN]	3040	1200	313	2.45
Vacuum I_{SP} [s]	279	289	294	315
Burn time [s]	107	71	117	317
Propellant type	HTPB	HTPB	HTPB	UDMH/NTO

Note: HTPB = Hydroxyl-terminated Polybutadiene; UDMH = Unsymmetrical Dimethylhydrazine; NTO = Nitrogen tetroxide

renewed commercial interest in small launchers, offering flexible operation and making more limited infrastructure demands. A number of such vehicles are now available commercially or are in development. These typically comprise three- or four-stage solid propellant rockets providing a launch capability into Sun-synchronous LEO for payloads of approximately 1 tonne. This class of launchers is typified in the USA by the Lockheed Martin Athena or the OSC Taurus, a ground-launched version of the Pegasus discussed earlier [6]. The key features of the ESA-developed Vega [12], in this same category and scheduled for its first flight in late 2010, are summarized in Table 7.8.

With a lift-off mass of 137 tonnes, Vega will typically launch a 1500 kg payload into a 90° inclination circular orbit at a height of 700 km.

7.6.2   Reusable sub-orbital vehicles

A number of commercial ventures have recently aspired to offer fare-paying passengers a sub-orbital flight experience. The most promising of these ventures is the Virgin Galactic project which plans to use a subsonic aircraft, WhiteKnightTwo, to carry a winged, reusable rocket-powered vehicle, SpaceShipTwo, to 50 000 ft. At this altitude, the nitrous oxide, hybrid propulsion system will be ignited and the rocket vehicle will climb to an apogee of about 100 km, carrying six passengers. During the supersonic descent the vehicle cranks its wing to provide inherent aerodynamic stability. The wing is then folded back to permit a conventional landing on a runway. The feasibility of this unique, patented recovery system has already been proven in the previous test flights of SpaceShipOne [13].

7.7 RE-ENTRY INTO EARTH'S ATMOSPHERE

In addition to the transport of payloads and personnel to orbit, some missions will also require their secure recovery from orbit—whether simply scientific samples for analysis or crew return from the ISS, and potentially further afield. With the exception of the Shuttle, which affords the opportunity to use aerodynamic lift to control the re-entry trajectory, recovery from orbit has been principally achieved using ballistic re-entry capsules.

Ballistic recovery essentially involves using aerodynamic drag to decelerate a vehicle from an entry velocity V_0, prior to parachute deployment, and subsequent impact on the Earth's surface with an acceptable velocity. The maximum deceleration rates involved in purely ballistic re-entry are relatively high compared to the launch ascent accelerations, typically exceeding about $-8g_0$, depending on the entry angle (which must be within prescribed maximum and minimum limits). In practice, re-entry capsules are flown at incidence, relative to the flight direction, to generate a lift force similar in magnitude, though generally less than, the drag force. This permits the adoption of trajectories that reduce the peak deceleration and peak heat transfer rates (see Figure 7.15), as well as providing some control on the dimensions of the landing footprint.

The essential features of atmospheric re-entry were first considered by Allen and Eggers [15] and employed extensively in the Apollo programme, see also Section 5.8.5. In the

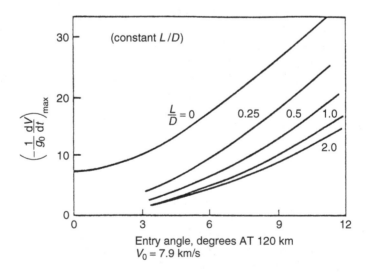

Figure 7.15 Variation of maximum deceleration with (re) entry angle for various L/D ratios. (Reproduced by permission from W.H.T. Loh, (1968) *Re-entry and Planetary Entry*, Springer-Verlag, [14].)

absence of lift forces, the deceleration rate of a vehicle re-entering the atmosphere at an entry angle γ is,

$$m\frac{dV}{dt} = -\frac{1}{2}\rho V^2 SC_D + W \sin\gamma \tag{7.17}$$

If atmospheric density is approximated by a simple exponential model,

$$\rho/\rho_s = \exp(-\beta h)$$

where h is the vehicle altitude above some reference height at which the density is ρ_s, γ is assumed to be shallow and held constant at the value γ_0, and the drag force is assumed to be much larger than the vehicle weight, then it can be shown that the vehicle velocity will vary as,

$$V = V_0 \exp\left[\frac{-\rho}{2(m/C_D S)\beta \sin\gamma_0}\right] \tag{7.18}$$

Furthermore, as seen in Section 5.8.5 of Chapter 5, the peak deceleration rate is given by

$$\left|\frac{dV}{dt}\right|_{max} = \frac{\beta V_0^2 \sin\gamma_0}{2e} \tag{7.19}$$

where e is the natural base of logarithms. Then for purely ballistic re-entry from a LEO, the peak deceleration rate is essentially dependent only on the entry angle, γ_0.

Lees [16] showed that blunt-shaped bodies are needed in order to reduce re-entry heating, since stagnation point heat transfer rates are inversely proportional to the square root of the nose radius of curvature r_{nose} in the laminar hypersonic regime,

$$\dot{q} = k\rho^{1/2}V^3/r_{nose}^{1/2} \tag{7.20}$$

Even adopting such features, thermal protection systems are also needed to prevent excessive heating of the payload. In particular, the most prevalent protection schemes employ ablative heat shields in which significant heat is absorbed during vaporization of the surface material (for example, of an epoxy resin), as adopted on the Apollo command module entry capsule. In addition to absorbing energy through phase change, the vehicle surface is further shielded by the efflux of ablating material entering the boundary layer (see Figures 7.16 and 7.17).

Whereas the peak deceleration rate is independent of the so-called 'ballistic coefficient', m/SC_D, Eggers and Allen [15] showed that the peak heating rates are proportional to its square root (see equation (5.58) of Chapter 5) and quadrupling the frontal area of the vehicle will halve the peak heat transfer rate. The successful return to Earth of dust samples collected during encounter with a cometary tail, as in the sample return capsule of the Stardust spacecraft in 2002, have benefitted from improved thermal protection systems (TPS) employing a *'phenolic-impregnated carbon ablator'* (PICA) [17]. Such materials are now to be incorporated, for example, in the Orion crew exploration vehicle design [18].

Peak heat transfer rates may also be reduced by modifying the flight trajectory using lift during the re-entry phase (see Figure 7.15). The Shuttle Orbiter re-enters the atmosphere at high angle of attack (about $40°$) and at hypersonic speeds it attains a lift/drag ratio

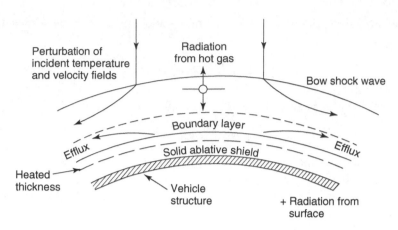

Figure 7.16 Schematic of an ablating surface thermal protection system

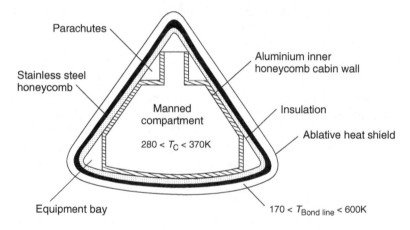

Figure 7.17 Schematic of the Apollo command module thermal protection system

of about 2. By also rolling the vehicle about its longitudinal axis, while maintaining the high angle of attack the lift vector can be modulated. In this way the rate at which the vehicle descends into the atmosphere, and hence the heating rate, can be controlled. This also serves to determine the across-track range that can be achieved.

The Orbiter employs a reusable thermal protection system that is zoned according to the local heating levels—see Figure 7.18. The nose and wing leading edges receive the highest heating rates and are consequently made from *reinforced carbon-carbon* (RCC). Most of the body and wing surfaces are covered by pure silica-fibre tiles coated with a borosilicate glass. On the upper surface of the wing and the payload bay doors, where heating rates are much reduced, flexible coated Nomex felt is sufficient.

About 8% of the Orbiter dry mass is taken-up by the thermal protection system—which results in a significant payload penalty from the standpoint of a user who does not require recovery. Furthermore, the Orbiter often requires significant post-flight maintenance to check, remove and replace the ceramic tiles.

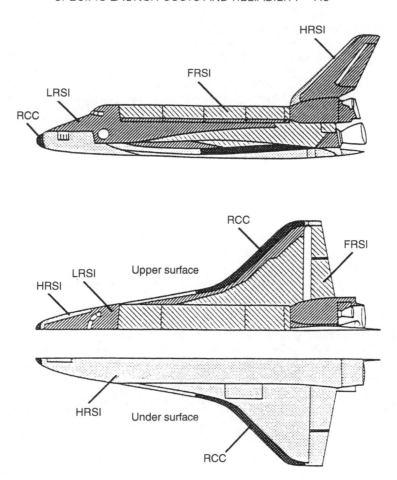

Figure 7.18   The Shuttle Orbiter thermal protection system. RCC: reinforced carbon-carbon; HRSI: high-temperature reusable surface insulation (silica tiles); LRSI: low-temperature reusable surface insulation; FRSI: flexible reusable surface insulation (coated Nomex)

Whilst programmes have been in place to evaluate advanced lifting re-entry vehicles based on wingless lifting body designs (such as the X-38 Crew Return Vehicle at NASA's Dryden Flight Research Center, cancelled in 2002), these have not progressed beyond early flight testing at reduced physical scale [19].

7.8   SPECIFIC LAUNCH COSTS AND RELIABILITY

The commercial exploitation of both low-Earth and geostationary orbits has focused attention on specific launch costs (cost per kg of payload delivered) as a figure of merit. The launcher and related services, such as insurance, continue to introduce costs which may be comparable with those of the spacecraft payload itself. For example, specific launch costs into GTO of ~25 k$/kg (see Table 7.9) remain high and so continue to constrain the commercial and scientific exploitation of space.

Table 7.9 Illustrative launch costs into LEO and GTO (FY2000 US$) [6, 20]

Launch vehicle	LEO payload [kg]	Launch cost [M$]	Specific cost [k$/kg]
Athena 2	2060	24	11.7
Pegasus XL	440	14	31.8
Rockot	1850	14	7.6
Taurus	1380	19	13.8
	GTO payload [kg]		
Atlas 2AS	3720	98	26.3
Delta 2 (7920)	1800	55	30.6
Long March 2C	1000	23	23
Soyuz	1350	39	28.9
Ariane 5ES	6800	165	24.3
Long March 3B	5200	60	11.5
Proton	4630	90	19.4
Sea Launch (Zenit3)	5250	85	16.2

Table 7.10 Launch Success Rate Data (to December 2003) [6]

Launch vehicle	Number of launches	Successful launches	Launch success rate
Athena 1/2	7	5	0.71
Pegasus XL	25	21	0.84
Rockot	2	2	1.00
Taurus	6	4	0.67
Atlas 2/3	30	30	1.00
Delta 2	110	108	0.98
Soyuz	713	692	0.97
Ariane 5	17	13	0.76
Long March 3	26	21	0.81
Proton	295	262	0.89
Sea Launch (Zenit3)	11	10	0.91

The insurance charges accompanying launch essentially reflect the reliability of the particular vehicle. Mission success levels of >90% are typical of established systems (see Table 7.10) although small rockets do tend to remain generally less reliable. Whilst these levels can evidently be improved further, progress is incremental and substantial reductions in overall costs and reliability will only be realized if there are major changes in the design of future launch vehicles.

Given the widespread application of high-energy propellants currently, notably LH_2/LOX delivering a vacuum specific impulse of 432 s in the case of the Vulcain 2 engine on Ariane 5, and payload fractions of ≤3%, there is arguably limited opportunity for very substantial reductions in the launch costs of expendable vehicles to be derived from improved technology.

The present high levels of cost continue to provide much of the stimulus for feasibility studies and technology demonstration programmes directed to the incorporation of reusability into launch vehicle design [21]. The operational economies that might accompany vehicle-centred logistical support have encouraged numerous studies of single- and two-stage-to-orbit concepts in recent decades—some incorporating air-breathing elements. However, few have progressed beyond conceptual design. The short-lived NASA X-33 demonstrator programme [22–24] represented one of the more determined attempts in the 1990s to exploit the best contemporary lightweight materials and novel propulsion technologies, such as the linear aerospike nozzle [25] (see Chapter 6).

In most cases, the principal drawback to reusability is that it adds a significant mass penalty to an already performance-stretched system. Furthermore, although the operations of the Space Shuttle have shown that reusability permits some improvement—for example, in permitting post-flight subsystem checks and continuous upgrades—there is little indication that fully reusable launch systems would deliver much greater reliability or introduce major costs reductions. From a user standpoint, other factors, such as a reduction in the time between commissioning and launch date or improved orbit injection accuracy, may therefore be more important launch service considerations.

In this chapter we have sought to identify the complex tasks involved in transporting a payload from the Earth's surface to a designated orbit. We have outlined many of the basic principles that determine launch vehicle design and performance and, in so doing, illustrated how they constrain both payload and mission. On present evidence, rocket technology will continue to underpin this process and the constraints identified seem likely to persist for at least the next decade.

REFERENCES

[1] Vinh, N.X. (1981) *Optimal Trajectories in Atmospheric Flight*, Elsevier, New York.

[2] Ruppe, H.O. (1966) *Introduction to Astronautics*, Vol.1, Academic Press, New York.

[3] Dorrington, G.E. (2000) Optimum Thrust-to-Weight for Gravity-Turn Trajectories, *J. Spacecraft and Rockets*, **37**(4), 543–544.

[4] White, J.F. (1962) *Flight Performance Handbook for Powered Flight Operations*, John Wiley & Sons, Inc., New York.

[5] ARIANE-5 User Manual (Issue 5) (2005) Arianespace.

[6] Isakowitz, S.J., Hopkins, J.P. and Hopkins, J.B. (2004) *International Reference Guide to Space Launch Systems*, (4th edn), AIAA, Reston, VA.

[7] Davis, J.D. and McArthur, J.C. (2008) NASA Ares 1 Crew Launch Vehicle Upper Stage Overview, AIAA 2008-4897, 44th *Joint Propulsion Conf.*, Hartford, CT.

[8] Huebner, L.D. *et al.* (2009) Ares 1-X Flight Test Vehicle Similitude to the Ares 1 Crew Launch Vehicle, *Acta Aston.*, **65**, 1706–1716.

[9] Broome, J.M. and Johnson, W. (2007) Orion Entry, Descent and Landing Performance and Mission Design, AIAA 2007-6430, *Guidance, Navigation and Control Conf.*, Hilton Head, SC.

[10] Moss, J.N., Boyles, K.A. and Greene, F.A. (2006) Orion Aerodynamics for Hypersonic Free Molecular to Continuum Conditions, 14th *AIAA Int. Space Planes and Hypersonic Systems and Tech Conf.*, Canberra.

[11] Stahl P.H, Sumrall, P. and Hopkins R., (2009) Ares V launch vehicle: An Enabling capability for future space science missions, *Acta Astron.*, **64**, 1032–1040.

[12] ESA (2005) *Vega: the European small launcher*, BR-257.

[13] Rutan, E.L. (2006) Winged Spacecraft, US Patent, 0108479, 25 May 2006.

[14] Loh, W.H.T. (1968) *Re-entry and Planetary Entry*, Springer-Verlag, Berlin.

[15] Allen, H.J. and Eggers, A.J. (1958) *A Study of the Motion and Aerodynamic Heating of Missiles Entering the Earth's Atmosphere at High Supersonic Speeds*, NACA TR 1381.

[16] Lees, L. (1956) Laminar heat transfer over blunt bodies at hypersonic flight speeds, *Jet Propulsion*, **26**(4) 259–269.

[17] NASA Stardust homepage. http://www.nasa.gov/stardust (accessed 1 March 2010).

[18] Orion Crew Vehicle homepage. http://www.nasa.gov/orion (accessed 1 March 2010).

[19] Dryden Flight Research Center (2001) http://www.nasa.gov/centers/dryden/news/NewsReleases/2001/01-75.html (accessed 1 March 2010).

[20] Space Transportation Costs: Trends in Price per Pound to Orbit 1990-2000, (2002) Futron Corporation.

[21] Tomatis, C., Bouaziz, L., Franck, T. and Kauffmann, J. (2009) RLV candidates for European future launchers preparatory programme, *Acta Astron.*, **65** (1/2), 40–46.

[22] Eldred, C.H. *et al.* (1997) Future space transportation systems and launch vehicles. In *Future Aeronautical and Space Systems* (ed. Noor, A.K. and Venneri, S.L.), Progress in Astronautics and Aeronautics, 172, Chapter 8, 373–408, AIAA, Reston, VA.

[23] Sumrall, J. *et al.* (1998) *VentureStar: reaping the benefits of the X-33 program*, IAF-98-V.3.03, *49th Int. Astro. Congress*, Melbourne.

[24] Powell, R.W. and Cook, S.A. (1998) *The road from the NASA access-to-space study to a reusable launch vehicle*, IAF-98-V.4.02, *49th Int. Astro. Congress*, Melbourne.

[25] Harmon, T. (1999) *X-33 Linear Aerospike on the Fast Track in Systems Engineering*, AIAA-99-2181, *35th Joint Propulsion Conference*, Los Angeles.

8 SPACECRAFT STRUCTURES

John M. Houghton

EADS Astrium, Stevenage

8.1 INTRODUCTION

The discipline and methods of spacecraft structural design are based upon aircraft design with a stronger emphasis upon minimum weight, vibration interaction and material selection considerations for space use. The major goals of minimum mass and maximum reliability must be met with minimum cost and schedule. The structure must achieve its goals for the static and dynamic loading of the testing and launch phases, and ultimately in the zero gravity operational environment. What makes spacecraft structural engineering perhaps unique is that its goals are strongly dependent on other subsystems such as thermal design, attitude control, communications and power.

Structural design starts with configuration definition, material selection and manufacturing method selection, moving on to analysis and verification with an increasing reliance being placed upon analysis as experience grows.

8.2 DESIGN REQUIREMENTS

Previously successful structure design concepts (such as those shown in Section 8.7) give a good starting point for future development. The relative importance of the factors discussed in the following subsections need to be assessed before development of a structure design concept.

8.2.1 Launch vehicle interface

Launch vehicle selection has a major influence on geometric and mass limits. Some launchers have the capability of launching more than one spacecraft at a time. Figure 8.1 shows the Ariane 5 dual launch configuration. Note that there is no connection between the fairing and the spacecraft. All accelerations and consequent inertia loads (except acoustic)

Spacecraft Systems Engineering, Fourth Edition.
Edited by Peter W. Fortescue, Graham G. Swinerd and John P. W. Stark.
© 2011 John Wiley & Sons, Ltd. Published 2011 by John Wiley & Sons, Ltd.

Horschol

Sylda

Planck

Figure 8.1   Ariane 5 Dual Launch System. (Reproduced by permission of
ESA – Guarniero)

are imparted to the spacecraft via the launch vehicle interface. Figure 8.2 shows that up
to eight smaller spacecraft may be launched by the ASAP5 adaptor for Ariane 5.

Competition between launch vehicle suppliers encourages the use of standard interfaces,
giving the customer the option of using alternative launch vehicles with the same interface
design. A commonly used adaptor is shown in Figure 8.3. Figure 8.4 shows a sectional
view of a manacle clamp used to attach the spacecraft (see also Chapter 15).

Up to 12 accurately machined clamp blocks are placed to form a segmented ring over
correspondingly accurately machined wedge shaped lips on both the launch vehicle and
spacecraft side of the interface. The clamp band is then tightened over these blocks by
two pyrotechnic bolts. This gives an even distribution of load around the circular launch
vehicle interface and a reliable release mechanism.

One route to an efficient structural design is to provide continuity to a clamp band
circular load path with a circular central thrust structure in the spacecraft. Additional
structure must then be designed to support all spacecraft equipment from the central thrust
structure. This may be seen in many of the configuration examples given in Section 8.7.

Many launch vehicles also offer discrete pyrotechnic bolt interfaces using between
three and eight bolts. This type of interface concentrates the load at these discrete bolt

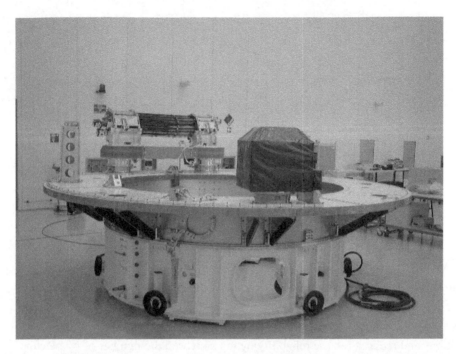

Figure 8.2   Ariane Structure for Auxiliary Payloads (ASAP5) for Ariane 5. (Reproduced by permission of EADS Astrium Ltd.)

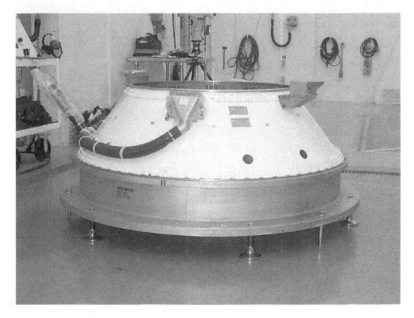

Figure 8.3   Launch vehicle to spacecraft adaptor. (Reproduced by permission of EADS Astrium Ltd.)

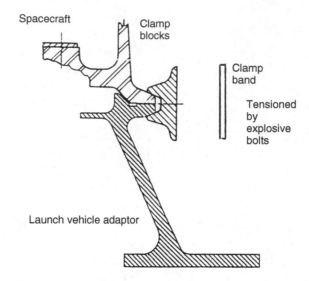

Figure 8.4 Spacecraft attachment clamp band. (Reproduced by permission of Arianespace)

positions and reduces release reliability by the increased the number of pyrotechnics used. Spacecraft interface ring strengthening, to resist buckling due to clamp band tension, is avoided and a non-circular interface may be used if desired. The influence of this is illustrated by the difference between the Eurostar Structure configuration of Figure 8.19, which is designed for a clamp band attachment, and the Swarm Configuration of Figure 8.20 which is designed for a three discrete bolt interface.

8.2.2 Equipment mounting

A flat, bolted interface is used for most items of equipment, dictating the need for large flat areas on the structure where a significant number of equipment units are carried. The flat panels seen in many of the examples in Section 8.7 are designed primarily for this purpose. Large areas of panel may need to be supported by perpendicular panels or struts to strengthen and stiffen the panel and limit the magnitude of vibration for equipment.

The primary structure may have to be specifically designed to efficiently carry large or heavy equipment, such as larger propellant tanks, which require strong and stiff attachments.

8.2.3 Environmental protection

The structure design trade-off may be biased towards a loaded skin structure with a composite section as described in Section 8.6 (as opposed to a framework structure) to meet the requirements for micrometeorite, debris or radiation protection.

8.2.4 Alignment

The required accuracy of alignment can vary widely, from a broad tolerance for a communication spacecraft providing a global service to a very tight tolerance for accurate location pointing. Distortion has three main sources. Firstly, all alignment and testing of the spacecraft will be done terrestrially under a 1 g loading. The structural elastic distortion between critical equipments generated by the 1 g environment must be calculated and possibly verified by mathematical model correlation with static load testing. Secondly, there will be a small movement or hysteresis in the structure as a result of launch vibration. This can usually only be judged from previous test experience. Thirdly, when on-station, there will be temperature variations throughout the structure, and differences in temperature from the time of ground alignment will generate thermo-elastic distortions. The distortions these generate through the material expansion coefficients may be calculated using the structure finite element model. A requirement for moderately high pointing accuracy can bias material choice to carbon fibre reinforced plastic structures for antenna reflectors, for example, where the operating temperature range can be large ($-160°C$ to $+160°C$). The highest thermal stability requires the use of ceramic structures for items such as optical benches, but use of these materials must be approached with caution due to their low strength and fracture toughness.

8.2.5 Thermal and electrical conductivity

Thermal control may require heat pipes to be embedded in structural honeycomb panels or thick conductive doublers to be added to dissipate heat from high-powered equipment. Thermal conductivity or isolation may be required through a structure and at structural joints.

The structure may be required to provide a ground return path for electrical circuits. Electromagnetic compatibility (see Chapter 16) may be required to prevent electrostatic discharge. These requirements give a preference for metallic structures, but composites may still be a better option with electrical or thermally conductive strapping.

8.2.6 Accessibility

All stages of spacecraft assembly, integration and verification (see also Chapter 17) must be carefully considered, with any likely equipment refurbishment requirements, to establish the level of accessibility required inside the structure and whether any parts must be interchangeable, testable or transportable with equipment installed.

8.2.7 Mass

The cost of engineering and manufacture to achieve minimum mass must be compared with launch cost to give a guide to the effort it is worth spending to achieve a reduction. For the example of a geosynchronous communications spacecraft, the most commonly used 'Western World' expendable launch vehicles cost approximately \$25 000 to launch 1 kg of spacecraft mass to GTO (see also Chapter 7). This can be overridden by absolute mass limits. Launch costs are much lower for LEO missions giving relatively stronger emphasis on minimum cost.

8.3 MATERIAL SELECTION

8.3.1 Material properties

The selection of an appropriate material for an application requires knowledge of the advantage to be gained from each property of the material and where each limitation must be recognized.

Selection criteria can encompass the following:

- specific strength,
- specific stiffness,
- stress corrosion resistance,
- fracture and fatigue resistance,
- thermal expansion coefficient and conductivity,
- ease of manufacture.

Specific strength and stiffness

An elementary comparison of materials can be made by examining their proof strengths (the stress at which a material will sustain a permanent deformation of 0.1 or 0.2% when the load is removed). Specific proof strength is this stress divided by the material density. Values for a selection of widely used lightweight materials are given in Table 8.1. For metals, titanium shows up well but it is only used where strength is the dominant criteria—at joints and areas of local load concentration. For all other applications of lightweight structures, overall strength is determined by buckling. The material property that determines the buckling strength of slender structures is stiffness.

The material efficiency criteria derived for the three design cases in Figure 8.5 are included in columns D, E and F of Table 8.1 to aid material selection [1]. The efficiencies of titanium and aluminium alloys are similar despite the greater strength of the former. Ashby and Jones have given an explicit treatment of this subject together with illuminating case studies [2].

Stress corrosion resistance

Stress corrosion cracking (SCC) can develop in a terrestrial environment containing a corrosive medium and moisture when materials are subject to sustained tensile loading, particularly in the short transverse grain direction. Tensile loading conditions can exist even when in storage, due to weight, residual stress or joint assembly preload. The potential effect is a failure occurring at stress levels below values normally considered safe for the material. An ESA design guideline [3] points out that the corrosive environment need not be severe, and failed parts may not show visible evidence of corrosion. It also

Table 8.1   Sample materials properties

C. R. corrosion resistance
S. C. C. stress corrosion cracking

Caution: There is considerable variation of the properties of materials according to conditions (ageing, temper, form and structure orientation). Consult manufacturers' data

	Density (kg/m³)	Young's modulus E (GPa)	Yield strength f (MPa)	Selection criteria				Thermal expansion (μm/m K^{-1})	Fracture toughness (MPa m)	Fatigue strength (MPa)	Comment
				E/ρ	$E^{1/2}/\rho$	$E^{1/3}/\rho$	f_y/ρ				
Aluminium alloy											
6061.T6	2700	68	276	24	2.9	1.5	98.6	23.6	186	97	Good C.R
7075.T6	2800	71	503	26	3.1	1.5	186.3	23.4	24	159	Prone to SSC in T6 Form
Magnesium alloy											
A2 31B	1700	45	220	26	3.9	2.1	129.4	26			Prone to SCC
ZK 60 A.T5 extrm	1700	45	234	26	3.9	2.1	137.6	26		124	
Titanium alloys											
T1-6A1-4V (annealed)	4400	110	825	25	2.4	1.1	187.5	9	75	500	
(solution treated and aged)			1035						42	690	
Beryllium alloys											
S 65 A	2000	304	207	151	8.7	3.4	103.5	11.5		500	Hot pressed sheet ⎫ Low fracture toughness
SR 200 E			345							690	⎭
Ferrous alloys											
INVAR		150	275/415					1.66			Low expansion Ferromagnetic
Stainless steel AM 350 (SCT850)	7700	200	1034	26	1.84	0.8	134.3	11.9	40/60	550	
304L Ann	7800	193	170	25	1.8	0.7	21.8	17.2			Austenitic
Composites											
KEVLAR 49 0° (Aramid fibre) 90°	1380 1380	76* 5.5	1379† 29.6	55 4	6.3 1.7	3.1 1.3	999.3 21.4	−4 57			Structure members Pressure vessels Rocket casings
Graphite epoxy sheets (undirectional) GY70/934	1620	282	586	174	10.4	4.0	361.7	−11.7 (Longitudinal) 29.7 (Transverse)			Sheet
Column ref: (see text)	A	B	C	D	E	F	G	H	I	J	

*Tensile modulus
†Tensile strength

Case	Deflection of a beam	Buckling of a strut	Buckling of a panel
	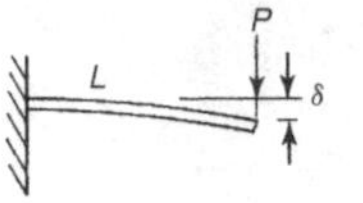	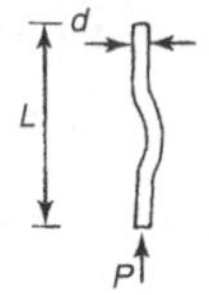	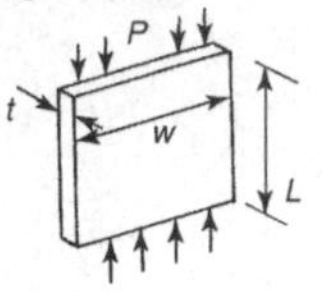 (w fixed) (t and L variable)
Characteristic equations	$\delta = \dfrac{PL^3}{3EI}$ $I = \dfrac{wt^3}{12}$ Weight = $L.W.t.\rho$.	$P_e = \dfrac{\pi^2 EI}{L^2}$ $I = \dfrac{\pi d^4}{64}$ Weight = $\dfrac{\pi d^2}{4} L\rho$	$P = \text{const.} \dfrac{E}{1-\sigma^2}\left(\dfrac{t}{W}\right)^2 . tw$ Weight = $w.t.L.\rho$
Structural efficiency $= \dfrac{\text{load}}{\text{weight}}$	$\dfrac{P}{WT} = \text{const.}\left(\dfrac{\delta t^2}{L^4}\right).\dfrac{E}{\rho}$	$\dfrac{P}{WT} = \text{const.}\left(\dfrac{P_e}{L^4}\right)^{1/2}\dfrac{E^{1/2}}{\rho}$	$\dfrac{P}{WT} = \text{const.}\left(\dfrac{P_e}{L^2}\right)^{2/3}\dfrac{E^{1/3}}{\rho}$
Structure loading coefficient	$\left(\dfrac{\delta t^2}{L^4}\right)$	$\left(\dfrac{P_e}{L^4}\right)^{1/2}$	$\left(\dfrac{P}{L^2}\right)^{2/3}$
Material efficiency criterion	$\dfrac{E}{\rho}$	$\dfrac{E^{1/2}}{\rho}$	$\dfrac{E^{1/3}}{\rho}$
Nomenclature	P = Load δ = Deflection σ = Poisson's ratio (0.3)	P_e = Critical buckling load (Euler theory) L = Length	d = Diameter ρ = density w = Width t = thickness E = Young's WT = Weight modulus

Figure 8.5   Material selection criteria — stiffness

gives a qualitative rating of commonly used metallic materials as having good, moderate or poor resistance.

Weld zones require particular consideration due to microstructure discontinuities in the heat affected areas, particularly if a filler material has been used.

Precautions in design and material selection include:

- choosing alloys less susceptible to SCC,
- minimizing sustained tensile stress, particularly in the short transverse grain direction,
- specifying the need for close inspection of areas of stress concentration and welds,
- heat-treating components to remove residual stresses due to manufacturing processes,
- avoiding material combinations, which promote galvanic corrosion,
- avoiding exposure to atmospheric conditions, which can induce corrosion.

Sublimation/erosion

Normally the sublimation of metals does not pose any major problems in the space environment, although for thin films the rate at which their thickness decreases may be significant. Erosion by atomic oxygen, particularly in respect of polymeric materials in LEO, can be an important factor in material selection [4, 5]. This is a subject of investigation and research, in view of the possible long-term problems which could ensue in the 25-year life of a space station, for example.

Ease of manufacture/modification

Some materials present individual problems during component manufacture. For example, some forms of beryllium and the cutting fluids used present toxic material control problems, which preclude late modification in unsuitable facilities. It is not unusual for additional mounting points or cut outs for cable run clearances to be needed at a very late stage in integration.

Advanced materials

The aerospace industry has driven the development of numerous advanced materials in the search for more mass efficient structures. In most cases cost is a significant factor since, particularly in the space sector, designs are very low volume/high quality applications. Many materials developments now fail to reach maturity simply on the grounds of economics. There are exceptions where the need for a technical solution is paramount. A good example is high stability ceramic optical bench material.

A wide range of materials with developed characteristics is discussed in useful detail with a commentary on fabrication, jointing and applications in [6, 7]. These materials can generally be grouped as follows:

- **Polymer composites**
  Continuous fibre reinforcements:
    Carbon
    Aramid
    Glass
  Matrix:
    Epoxy
    Polyimide
    Bismaleimide
  Thermoplastic:
    Carbon in polyether ether ketone (PEEK)
    Aramide-based (PEI)
- **Advanced metal matrix and ceramic materials**
  Magnesium alloys and their composites
  Aluminium alloys and their composites
  Titanium alloys and their composites
  Super alloys and their composites
  Intermetallic materials
  Refractory materials
  Beryllium
  Ceramic matrix composites
  Glass and glass-ceramic matrix composites
  Carbon-carbon matrix composites

These materials offer substantial advantages for some specific requirements, provided their characteristics are well understood. Care must be taken with the non-linear stress strain response and brittleness of some of them. The long term effects of on-orbit temperature and radiation have been evaluated [8] and must be considered when sizing components.

8.3.2 Materials applications

Metals

Ferrous alloys have numerous applications in which their properties of high strength, corrosion resistance and toughness are required.

Austenitic stainless steels are used for propulsion and cryogenic systems due to their excellent low temperature toughness. Other alloys are used in optical and precision structures where properties can be selected to match expansion criteria in a dynamic thermal environment.

Susceptibility to hydrogen embrittlement is a potential hazard for ferrous alloys, particularly where they have been treated in plating solutions. The result is similar to SCC. The corrective treatment is a severe bake-out within a limited time period—observing of course that the materials are not affected.

Some types of stainless steel and invar are magnetic. This can be a problem when a spacecraft carries electromagnetic sensors.

Fibre reinforced composites

Advantage can be taken of the high strength offered along the fibre. In a single ply of unidirectional fibres, the ply strength and stiffness in the fibre direction is about 60% of the fibre strength and stiffness due to the presence of the relatively low strength and stiffness resin. In the other two orthogonal directions, strength and stiffness is limited to the properties of the matrix resin. A multi-ply laminate can be built by adding unidirectional plies aligned at an angle to the first. A laminate can be created with structural properties tailored to the application. The high unidirectional strength and stiffness, such as is quoted in Table 8.1 will be reduced in any one direction by approximately the ratio of the number of plies in that direction to the number of plies in the complete laminate. Methods for the analysis of individual ply stresses, and failure prediction of a laminate have been developed [9, 10]. Computer programs are commercially available from agencies such as ESA, which enable much quicker analysis, particularly where optimization is required.

Carbon epoxy materials are used quite extensively in fabrication by hand lay up of uni-directional or woven plies to produce flat or curved panels often of honeycomb sandwich construction. Large deployable antenna reflectors use the combination of high stiffness, lightweight and low coefficient of expansion (to hold shape under temperature extremes) to particular advantage.

Filament winding and tape placement of strut tubes and spacecraft central thrust tubes are used where the direction of loading is well defined.

A note of caution, particularly with carbon fibre-based composites, is that hygroscopic absorption can add up to 2% water by weight in a normal atmosphere which can reduce strength. Once exposed to the space environment they lose the water and exhibit small dimensional changes. Methods of managing this include total control of the component's environment, baking-out prior to alignment, and barrier coatings to seal the moisture in or out. None of these methods are totally convincing. Careful choice of the resin system can also limit this effect.

Carbon composite materials are brittle, requiring careful consideration of stress concentrations produced by features such as holes, sudden changes of section, grooves or fillets which will reduce the static strength by the magnitude of the corresponding stress concentration factor. This contrasts with the more generally used high strength aluminium or titanium alloys, which have sufficient ductility to make them insensitive to most stress concentrations under static loading. For ductile materials, stress concentrations are of more concern under cyclic fatigue loading.

For this reason, and because complex shapes are more difficult to manufacture in composites, they tend to find application in relatively simple components—strut tubes (but not end fittings) and central thrust structures—but not attachment fittings and flanges. Complex components such as strut end fittings are sometimes made in composites, but this is usually when stiffness, and not joint strength, is the design driver.

Unlike metals, the tension and compression bulk modulus of composite materials may differ significantly, complicating the vibration response.

Composite materials (metal matrix)

The limiting factors of an epoxy matrix can be substantially overcome by employing high strength fibres in a diffusion bonded metal matrix. Although this technology may represent the ultimate direction for 'designed' materials, current costs are high and care is required with non-destructive testing (NDT). Some materials suffer from low yield strength or elongation to failure, which is an indication of brittleness, notch sensitivity and poor fatigue strength. However, applications are emerging, one particular example being lightweight mirrors and optical benches [11].

In general, a clear rule to be borne in mind is to choose materials (or their close equivalents) that appear in the ESA handbooks [6, 7] or American standards [12, 13]. This gives all of the design and approval data necessary, and the confidence that no non-compliances will be discovered later in a programme when any required material qualification testing is completed. When this is not possible, national defence specifications should be used for guidance. In all cases, material tractability, including treatment history from billet/raw material to finished product, is one of the quality assurance records to be maintained.

8.3.3 Section properties

Hollow or reduced section members such as tubes and 'I' beams give much better mass efficiency than solid bars. Panel deformations such as corrugations can give greatly increased stiffness and resistance to buckling compared with flat sections. It is the art of the design engineer to use materials most efficiently in this way when considering all of the duties required of a section.

Honeycomb panels have relatively low weight and high bending stiffness. Figure 8.6 shows the general configuration of a honeycomb sandwich.

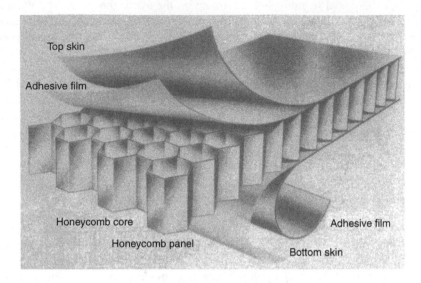

Figure 8.6   Honeycomb panel. (Reproduced by permission of Hexcel Corporation)

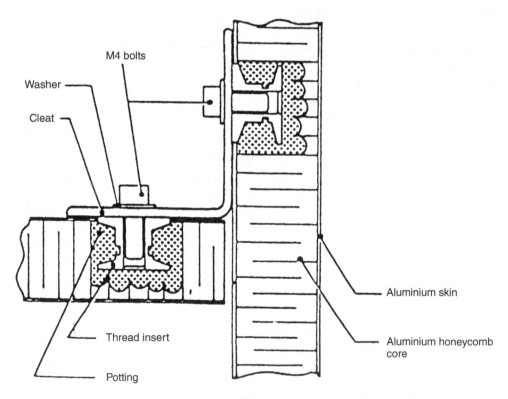

Figure 8.7   Inserts and panel cleat joints. (Reproduced by permission of EADS Astrium Ltd.)

A variety of materials and material combinations may be employed with the mix of skin thickness, core depth and density necessary to give the desired overall section properties.

It is important to take care when designing the load attachment points for honeycomb panels. Figure 8.7 shows a widely used blind potted insert. The potting compound transmits bolt tension load to the core while bearing transfers of shear loads from the insert to the hole in the panel skin.

8.3.4 Quality Assurance

The final proof of a structure's integrity involves a combination of analysis, inspection and test. Non-destructive testing using X-ray techniques can be employed to find voids in materials and joints, and ultrasonic techniques can be used to find lack of adhesion in bonded joints or laps in wrought materials. In the case of low volume production of spacecraft structures, considerable care must be taken to ensure that integrity can be assured in the face of variability due to individual ply angle errors hidden within a laminate, bonding or complex section. Added to this is the problem of ensuring that the structure has not degraded during test.

8.4 ANALYSIS

8.4.1 Building a finite element model

A *finite element model* for analysis of the structure is an essential part of the design development process. It serves the following purposes:

- to determine load distributions internal to a structure with multiple load paths;
- to predict spacecraft overall natural frequencies;
- to predict the level of spacecraft response to low frequency vibration;
- to be used by the launch vehicle contractor to determine the mission-specific maximum loads, accelerations and deflections and thereby determine sine test input level reductions (notching); and
- to apportion stiffness and corresponding strength requirements to sub-structured appendages.

Figure 8.8 shows a notional structure divided into a large number of discrete quadrilateral elements. Material properties, shape, degrees of freedom and connection to the next element are defined [14, 15].

Within each element the rate of change of stress is limited by the element formulation. Every corner point that may be connected to several elements is called a node. Between any two nodes on the edge of an element, there is a constant state of stress and strain. The mesh may not be fine enough to capture rapidly varying peaks of stress, and an overly coarse mesh will overestimate the stiffness of a structure by not representing small radius of curvature deflections. The temptation is, therefore, to use a very fine mesh to be sure that peaks are not missed and stiffness is not misrepresented. However this approach must be tempered by the mathematical size of the complete model, defined by the number of degrees of freedom it contains. Every node has six degrees of freedom for

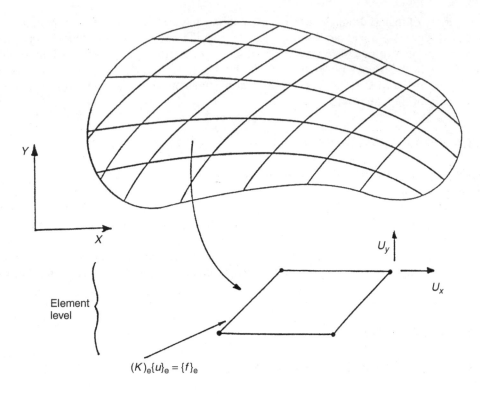

Figure 8.8 Finite element modelled area

movement—three orthogonal translations and three rotations. The total number of degrees of freedom in a complete model is six times the number of nodes (this may be reduced by introducing constraints which eliminate some unimportant degrees of freedom). It is the total number of degrees of freedom in a model that determines the mathematical size of the computation required. This, in turn, is limited by the power of the computer being used.

Models may be composed of a number of element types. A selection of these is shown in Figure 8.9.

The simplest of these are axial elements (to represent struts) which can be made mathematically more complex to carry torsion and moments. The two main types of planar elements are membrane and plate elements. The former cannot sustain out of plane bending and shear, while the latter can. The third major type is solid or three-dimensional elements, which are required to predict complex three-dimensional stresses and deflections. They are not widely used in lightweight structures - only at relatively heavy fittings such as propellant tank attachments and launch vehicle interface brackets.

A complete spacecraft finite element model is constructed from sub-models, which are ideally provided by the designers of the respective hardware subassembly who have the best knowledge of its structural behaviour. The spacecraft finite element model construction therefore emulates the construction of the hardware. Also, the maturity and accuracy of the model reflects the maturity of the design. At the concept design stage, some gross assumptions may be made. The accuracy of the model should be kept current with the design, such that during spacecraft testing the model accuracy can be verified by test.

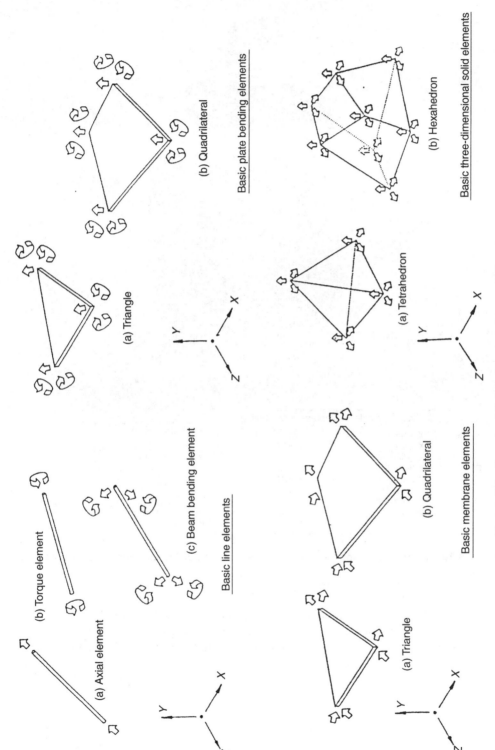

Figure 8.9 Standard finite elements for modelling

8 Super elements	Degrees of freedom
–X reflector	7145
–X feed	13867
+X reflector	7782
+X feed	13819
–Y solar array	21121
+Y solar array	21121
Top floor antenna	58387
Platform	32498
Total	175740

Element types	
24671	quad 4
7114	tria 3
9371	bar
6936	elas
348	hexa
33302	grids

Figure 8.10 Complete spacecraft finite element model. (Reproduced by permission of EADS Astrium Ltd.)

Figure 8.10 shows the model of a telecommunication spacecraft. In this case *super elements* were used to reduce the computing power required. The complete spacecraft is analysed in an automatic analysis sequence as a series of subassemblies, which are then automatically assembled as though they were large elements—hence the name super elements.

8.4.2 Launch loads

Launch loads are divided into the following categories and are generated by different flight events (see also Chapter 2). All are treated slightly differently and are critical for different parts of the structure.

Table 8.2 Launch quasi-static loads for Ariane 5

Critical flight event	Acceleration (g)			
	Longitudinal		Lateral	Additional line load (N/mm)
	Static	Dynamic	Static + Dynamic	
Lift off	−1.8	+/− 1.5	+/− 2	10
Maximum dynamic pressure	−2.7	+/− 0.5	+/− 2	14
SRB end of flight	−4.55	+/− 1.45	+/− 1	20
Main core thrust tail off	−0.2	+/− 1.4	+/− 0.25	0
Max. tension case: SRB jettisoning	+2.5		+/− 0.9	0

Note: The minus sign with longitude axis values indicates compression.

Quasi-static loads

These are generated by a uniform level of acceleration throughout the whole spacecraft that may exist for just an instant. They include a vibration, or transient, component but the frequency content of this will be separated from the natural frequencies of the space-craft, and therefore will not cause significant magnification of acceleration throughout the spacecraft. Maxima occur at transonic buffeting and at main engine shut down, rather than at lift off. Just before engine shut down, the thrust level is maintained at a maximum, while the launcher fuel mass is at a minimum—consequently, the vehicle acceleration reaches a maximum. Quasi-static loads are critical for the major load bearing parts of the structure spacecraft such as the central thrust tube shown in the configuration examples. Table 8.2 give the quasi-static loads for Ariane 5.

Lateral loads may act in any direction simultaneously with longitudinal loads.

Sine vibration

The sine vibration environment for Ariane 5 is given in Table 8.3. It is a relatively simply defined envelope of all of the many complex low frequency transient vibrations that may occur during launch. For a qualification sine vibration test, a spacecraft in launch configuration is attached to a large 'shaker' which starts vibrating at 5 Hz. The frequency of the vibration is increased at a rate of 2 octaves (doublings) per minute. In one minute the frequency increases from 5 Hz, through 10 to 20 Hz. This continues until 100 Hz is reached when the test stops.

Any lightweight spacecraft will have resonant frequencies in this range which can cause responses in the spacecraft 20 or more times the specified input. If these responses are in excess of the predicted launch transient responses, the defined test input may be reduced, or *notched* at critical response frequencies by agreement with the launcher agency. Such an agreement requires demonstration by the spacecraft-and-launcher coupled loads transient analysis that the reduced response within the spacecraft in the notched sine test is still greater than that to be expected from launch transients. A strong emphasis is therefore placed upon the accuracy of the mathematical model of the spacecraft supplied by the spacecraft designer, which may also be verified by the sine vibration test.

Table 8.3 Sine vibration environment for Ariane 5. (Reproduced by permission of Arianespace)

Sine	Frequency range (Hz)	Qualification levels (0-peak)	Protoflight levels (0-peak)	Acceptance levels (0-peak)
Longituinal	2–5*	12.4 mm	12.4 mm	9.9 mm
	5–50	1.25 g	1.25 g	1 g
	50–100	1 g	1 g	0.8 g
Lateral	2–5	9.9 mm	9.9 mm	8.0 mm
	5–25	1.g	1 g	0.8 g
	25–100	0.8 g	0.8 g	0.6 g
Sweep rate		2 oct./min	4 oct./min	4 oct./min

Figures 8.11 and 8.12 show two results from a spacecraft coupled loads analysis. They both show axial force at the spacecraft interface varying with time over 1.5 s from engine thrust cut-off.

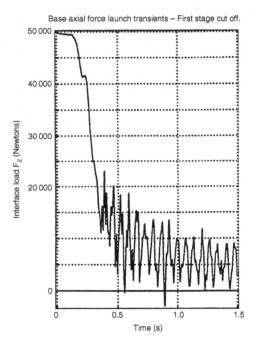

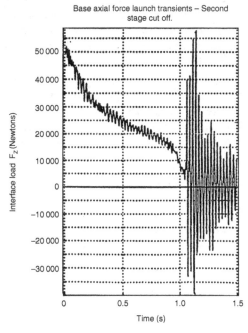

Figure 8.11 Base axial force launch transient for first stage cut-off. (Reproduced by permission of Arianespace)

Figure 8.12 Base axial force launch transient for second stage cut-off. (Reproduced by permission of Arianespace)

In comparison, the decay transient vibration content for first stage engine cut-off is relatively small at a lower frequency. This is because the launcher mass is much higher than it is at second stage cut-off. The transient magnitude is so great in second stage cut-off that the force becomes negative (in this case tensile). This is the reason for the specification of a tensile case in the quasi-static loads (Table 8.2). These are just two examples of coupled loads analysis recovery. Hundreds of accelerations, displacements and forces may be recovered for many different flight events.

Acoustic noise and random vibration

The largest acoustic noise excitation occurs at the point of lift-off when the reflected noise from the launch pad and ambient air pressure are greatest. An Ariane 5 acoustic noise spectrum is shown in Table 8.4.

Octave (doubling) bands are used to partition the spectrum. Acoustic noise can be critical for the design of lightweight structures with large area and low mass, such as dish-type antenna reflectors, solar arrays and spacecraft side panels with low mass attachments. The magnitude of the random vibration response, in terms power spectral density, can be calculated or measured in the frequency domain. This is the mean square of acceleration in each one Hertz bandwidth of the spectrum. Acoustic test results from similar, previous designs are often used to estimate the responses of complex structures.

Under 100 Hz, the launch vehicle random vibration environment for a complete spacecraft is usually covered by the sine vibration environment testing. Above 100 Hz acoustic environment testing can be critical for large projected area spacecraft while spacecraft random vibration may be critical for smaller spacecraft (less than 1000 kg) but this is dependent on stiffness and mass of large area components.

Random vibration testing is widely used during development and qualification of spacecraft equipment to be sure that no problems will be found at the relatively late stage of integrated spacecraft acoustic noise or random vibration testing. The required test level is judged from previous test experience. An example is given in Table 8.5.

Figure 8.13 superimposes this equipment test level on the response during the eventual spacecraft acoustic test, showing how close the acoustic test response is to the equipment

Table 8.4 Spacecraft acoustic environment for Ariane 5. (Reproduced by permission of Arianespace)

Octave band centre frequency	Qualification Level (dB)	Protoflight Level (dB)	Acceptance level (flight) (dB)
(Hz)	ref: 0 dB $= 2 \times 10^{-5}$ Pascal		
31.5	131	131	128
63	134	134	131
125	139	139	136
250	136	136	133
500	132	132	129
1000	126	126	123
2000	119	119	116
Overall level	142.5	142.5	139.5
Test duration	2 minutes	1 minute	1 minute

Table 8.5 Equipment random vibration spectrum from spacecraft acoustic testing (1 min duration for protoflight testing, 2 min duration for qualification testing)

Frequency (Hz)	Level
20–150	+6 dB/oct.
150–1000	0.25 g²/Hz (18.7 g RMS)
1000–2000	−6 dB/oct.

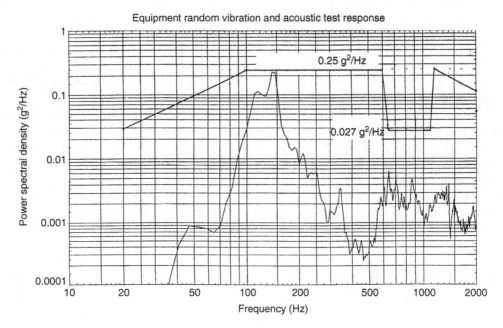

Figure 8.13 Equipment random vibration test and acoustic response. (Reproduced by permission of Arianespace)

random vibration test at 140 Hz, and that an equipment test notch from 600 to 1000 Hz was justified.

Shock loads

Figure 8.14 shows the shock envelope at the spacecraft interface for the Ariane 5 upper stage separation and fairing jettison.

The accelerations are alarmingly high, but they would only generate a damaging load to a very stiff item with a strong resonance at the given frequencies close to the shock source. High frequency shock energy is attenuated very rapidly with distance from the source and by structural joints between the shock source and the responding item. High frequency shock is usually of more concern to the function of equipment such as relays or glass oscillators, rather than structural strength.

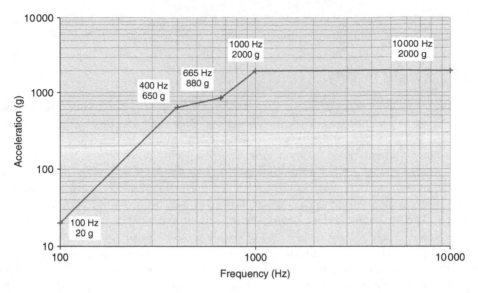

Figure 8.14 Shock response spectrum at the separation plane. (Reproduced by permission of Arianespace)

8.4.3 On-station requirements

On-station loads are of a much smaller magnitude than launch cases but they can be equally demanding on structure design. Structural transmission of *micro-vibration* from sources, such as momentum wheel bearing rumble or thruster firing, to sensitive equipment such as lasers or telescopes may be of critical concern.

Large appendages, such as antenna reflectors or solar array panels, may have a very low natural frequency when deployed. A minimum deployed natural frequency between 0.5–2 Hz is often required to avoid attitude control instability. Although a very low frequency, this requirement may be difficult to meet. Consequently it may be critical in the design of the deployed appendage interface and local backing structure.

Items such as strut tubes and honeycomb core cells, when manufactured in atmospheric pressure, could become inadvertent pressure vessels in the vacuum of space. They must be vented, or if venting is not practicable, designed as a pressure vessel.

All non-metallic materials must be space-qualified, primarily with respect to out-gassing under sun light in vacuum. The release of volatiles is doubly undesirable, since they may degrade the performance of the residual material and may redeposit on adjacent sensitive equipment.

8.4.4 Test loads

Testing discussed in Section 8.5 will apply loads of greater magnitude than those encountered in flight to give confidence that minor variations in load, material properties and build differences will not cause failure. Also, the test method may not be able to simulate the conditions of flight exactly, and over-test in some areas may be necessary to ensure adequate testing in others.

It is essential that the structure designer has a clear test plan before the structure design is completed, to be sure that the structure is adequate for test as well as flight.

8.4.5 Safety factors

The maximum load that can occur in service with worst weather conditions, roughest launch vehicle vibration and so on is termed the *limit load*. To be sure that structural strength is not a concern at the time of launch the structure is designed and qualification-tested to levels in excess of the limit load. The first safety level above limit is the *proof level*. This is equal to limit load multiplied by a *proof factor*. The most frequently used proof factor is 1.1, but factors between 1.0 and 1.5 are also used. The structure must not sustain any significant permanent deformation once a proof load has been applied and removed. The second safety level above limit is the *ultimate level*. A minimum ultimate safety factor is 1.25 for Ariane 5. Factors up to 4.0 are used for some pressure vessels. The structure must not rupture, collapse or undergo any gross permanent deformation under ultimate load. A commonly used parameter to express the adequacy of a structure under these failure criteria is the *reserve factor*. A reserve factor at any critical location is equal to the allowable load or stress divided by the applied load or stress. If the reserve factor is greater than 1.0 then the structure is adequate for purpose, judged by this particular criterion. Another widely used term is *margin of safety*. A margin of safety is equal to the reserve factor minus 1.0. A positive margin of safety indicates adequacy, and a negative one, inadequacy. In addition to the above safety factors, some safety agencies require minimum margins on particular failure modes. For example, a minimum margin of 0.25 may be required on first ply failure for composite materials or slender structure buckling. The level of the safety factors may be seen as a measure of the mass criticality of the structure *versus* the schedule and cost risk that a customer is prepared to take.

8.4.6 Stiffness requirements

To be sure that the quasi-static and transient vibrations responses, discussed in the 'Sine vibration' subsection, do not exceed the levels indicated in the launch vehicle preliminary design manual, the spacecraft minimum natural frequency requirements must be well separated from those of the launch vehicle forcing functions. Table 8.6 gives the minimum frequency requirements for spacecraft launched by Ariane 5. If these requirements are not met, the spacecraft dynamic coupling with the launch vehicle will be stronger, causing the quasi-static loads and dynamic transients to increase. To satisfy these requirements, possible apportioned requirements are 30 = 80 Hz for fixed base major spacecraft appendages and 100 Hz for equipment. The spacecraft system designer will change these figures in conjunction with the corresponding quasi-static loads from the results of the spacecraft sine response analysis.

8.4.7 Fracture control

Fracture control is required for ESA spacecraft and for pressure vessels in commercial spacecraft to satisfy launch range safety of expendable launch vehicles.

Table 8.6 Minimum natural frequency of vibration requirements

The fundamental frequency in the lateral axis of a spacecraft hard-mounted at the interface must be as follows with an off-the-shelf adapter:

S/C mass (kg)	Launcher interface diameter (mm)	1st fundamental lateral frequency (Hz)	Transverse inertia wrt separation plane (kg.m^2)
< 4500	< Ø2624	≥ 10	≤ 50,000
	Ø2624	≥ 9	
4500 ≤ M M ≤ 6500	≤ Ø2624	≤ 8	≤ 90,000
M > 6500	Ø2624	≥ 7.5	≤ 535,000
	< Ø2624	TBD	TBD

No local mode should be lower than the first fundamental frequenices.

Longitudinal frequencies

The fundamental frequency in the longitudinal axis of a spacecraft hard-mounted at the interface most be as follows:

≥ 31 Hz for S/C mas <4500 kg
≥ 27 Hz for S/C mass ≥ 4500 kg

No local mode should be lower than the first fundamental frequency.

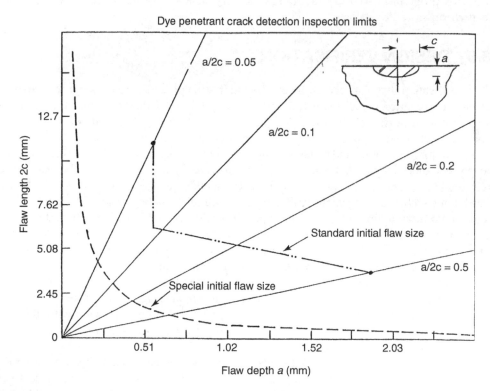

Figure 8.15 Dye penetrant crack detection inspection limits. (Reproduced by permission of NASA)

The method requires a careful crack detection inspection. It is then assumed that cracks may exist, in the most critical locations, at the crack detection limit of the method used. Figure 8.15 gives the detectable limit defined by NASA standards for dye penetrant inspection of a semi-elliptical surface flaw. A *standard initial flaw* is one that can be detected by a competent trained inspector. A *special initial flaw* is one that can be detected by an experienced inspector who has passes an examination by NASA or ESA representatives using samples that contain known flaws.

Fracture control analysis uses a concept called *stress intensity*, which is equal to the stress, times a geometric factor for location, multiplied by the square root of the characteristic crack length. The whole essence of fracture mechanics is coupled to strain energy, stress concentration and material state [16].

Crack propagation analysis is performed to predict the crack growth using empirical data, which shows that a crack will grow a tiny amount every time a load or stress is applied and removed. No safety margin factor is applied to the load magnitude, but a safe life factor of 4 is applied to the actual number of load cycles in the service spectrum to be sure that a safe life margin exists. The structure is considered to be *damage tolerant* if the crack does not grow to critical size after application of this load spectrum. A critical crack is one that will give unstable growth when the limit load is applied once. Unstable growth will result if the applied stress intensity is greater than the material fracture toughness. ESA has published a computer program with manual [17]. This based upon a NASA program FLAGRO that embeds industry standard methods, data and techniques to perform safe life analyses.

8.5 DESIGN VERIFICATION

Test verification that a spacecraft meets its major strength and stiffness requirements will usually be required unless the safety margins are all greater than one or if the design is very similar to a previously tested design. In the latter case, a *qualification by similarity* analysis will demonstrate that the previous test covers the new assembly.

The major requirements to be test-verified are static strength, sine and acoustic vibration responses and strength. A vibration test will also verify the accuracy of the mathematical model. Figure 8.16 shows the static test assembly for the *Eurostar 3000* structure. Producing such an arrangement with adequate hydraulic loading jacks, their control systems, recording strain gauges, displacement transducers and data logging systems is a complicated and expensive job. A dedicated *test model* is often used for this test because of the cut outs and load application points required specifically for the test. Load spreader plate attachments are used to diffuse load from the limited number of jacks into lightweight structures, which in flight will carry a more distributed load.

Sine and acoustic tests may be performed on *engineering models* fitted with mass dummies. They are often performed as protoflight tests on flight hardware with all flight equipment fitted as part of the final *Assembly, Integration and Verification* process (see also Chapter 17).

Verification that the mathematical model is accurate may be obtained by comparing the sine response analysis with the spacecraft sine test results. There are several aspects to be considered here. Firstly, comparison of natural frequencies and mode shapes will ensure that the model has the correct overall mass and stiffness. Comparing mode shape or the variations of accelerations in any one particular mode of vibration will ensure that the

Figure 8.16 Eurostar 3000 structure in the static test rig. (Reproduced by permission of EADS Astrium Ltd.)

model represents the correct distribution of stiffness throughout the spacecraft. Comparing the magnitude of response will ensure that the correct level of damping has been used.

There may be instances when errors may be present in the model, and modes of similar frequency cannot be separated by base excitation. Figure 8.17 illustrates a possible problem.

The left-hand Figure shows a second lateral mode of a spacecraft predicted at 60 Hz. The right-hand Figure shows a payload panel mode predicted at 70 Hz. On a base excitation sine test these could combine into a single mode at 65 Hz, making it impossible to isolate and correct the model error. It is in such instances that a multipoint modal survey test would be helpful.

For a modal survey test, the spacecraft is attached to a seismic block. This is a large concrete block mounted on springs or air bags in a pit to isolate the spacecraft from any extraneous ambient ground vibration generated by heavy factory equipment or passing

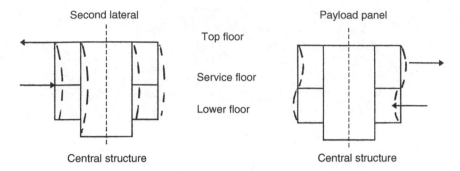

Figure 8.17 Vibration mode isolation by multipoint modal survey testing

road traffic. The spacecraft modes are excited individually by hand- or frame-held exciters at the positions indicated by the arrows. The frequency, phase, position and relative magnitude of excitation are varied until a mode is isolated by trial and error. Having isolated the second lateral, the exercise can be repeated for the panel mode, so isolating and correcting the model error [18].

8.6 IMPACT PROTECTION

The predicted growth of the millimetre and sub-millimetre size space debris environment, and the greater understanding of the meteoroid population, have led to a rise in interest in shielding for unmanned spacecraft in LEO. The impacts, typically in the range 5–20 km/s are capable of damaging and perforating spacecraft external structures [honeycomb panel, often covered by multilayered insulation (MLI)].

A notable early strategy for shielding unmanned spacecraft was that developed for Radarsat [19]. The design changes effected—re-routing of critical harnesses, enhancement of MLI using Nextel—are included in the design recommendations made in the Orbital Debris Handbook [20].

More recently, similar enhanced MLI configurations have been tested for use on the Iridium satellites and on the International Space Station (ISS) [21].

Typically a space debris and meteoroid shield is based on a Whipple bumper. This consists of a thin bumper shield of aluminium, metallic laminates or composite materials and a thicker aluminium back-up wall separated by a gap (s). The outer bumper layer disrupts the projectile by either shattering, melting or vaporizing it. The spacing allows the debris cloud to be distributed over a relatively large area, lowering the impact loading on the back-up wall. Placing a range of materials between the bumper and back-up wall (Kevlar, MLI, Spectra, aluminium meshes and Nextel) can augment the shield. Multiple thin bumpers ensure that the projectile is repeatedly shocked (and thus fragmented, melted or vaporized) before impacting on the back-up wall. Figure 8.18 shows the three main types of shield.

Different types of bumper material have been tested for their ability to disrupt the projectile into a debris cloud. Metal and ceramic matrix composites were found to give better projectile disruption than the equivalent areal density aluminium bumper, resulting in less damage to the back-up wall. In addition to Kevlar and Nextel, honeycomb and

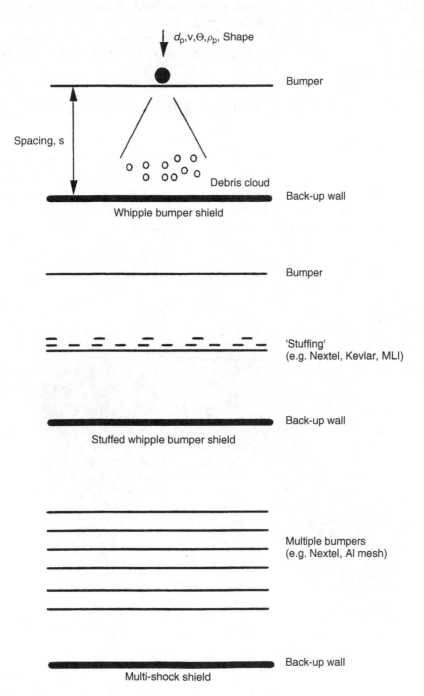

Figure 8.18 Different types of Whipple bumper shields. (Reproduced by permission of EADS Astrium Ltd.)

aluminium/titanium laminates have been used as intermediate layers in Whipple bumpers. The ability of laminates to provide good projectile disruption due to shock impedance mismatch has been noted [22, 23]. Advanced shield designs, Multi-shock shields (MSS), Multiple-mesh bumper shield (MMBS) and Mesh double bumper shields (MDBS) have been investigated [24, 25].

8.7 CONFIGURATION EXAMPLES

8.7.1 Eurostar

This GEO communications spacecraft is shown in Figure 8.19. The apogee engine and pressurant tank are fitted inside the primary load bearing CFRP filament wound central cone and cylinder assembly while propellant tanks are supported symmetrically around it. Main thrust axis inertia loads from the propellant tanks are carried to the launch

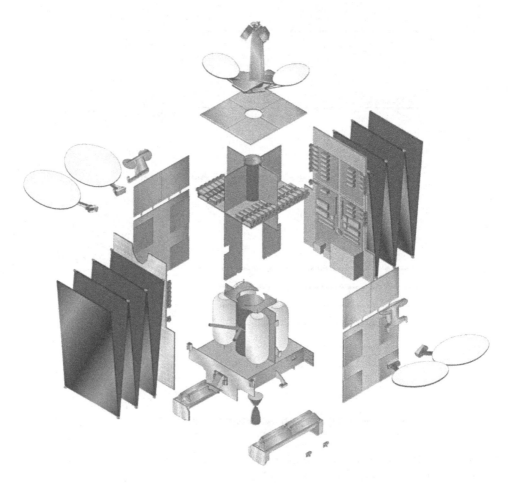

Figure 8.19 Eurostar 3000 structure configuration. (Reproduced by permission of EADS Astrium Ltd.)

vehicle interface ring by carbon fibre tubular struts with titanium end fittings. This is a classical use of carbon fibre—the load direction is well defined and the construction lends itself to a well-controlled production process. Titanium alloy end fittings are used in this application for their high strength. Steel fittings are used in the highest load missions to achieve the maximum strength in a confined geometric space. The flat panel configuration allows good access and logistics for separate panel integration at different sites by the removal of sidewalls carrying whole systems. The solar cells are fitted to external arrays, leaving the side members of the spacecraft free for heat radiation from the high power payload equipment mounted on the inside of the exterior panels. A matrix of embedded and surface heat pipes are used to spread heat from local hot spots on these panels. The solar arrays are rotated in GEO to ensure that they are always facing the Sun.

8.7.2 Swarm

The three-point, bolted launch vehicle interface and the requirement to launch three space-craft inside a launch vehicle fairing have led to the long narrow shape illustrated in Figure 8.20. An all carbon fibre composite structure has helped to minimize structure mass and magnetic interference for the Swarm mission, which is to measure fluctuations in the Earth's Magnetic field. Panel skins 0.48 mm thick, with plies biased towards an axial layup, are used at the base of the spacecraft with double thickness doublers at the launch vehicle attachment brackets. Skins 0.24 mm thick, with quasi-isotropic or nearly equal strength and stiffness properties in all directions, are used at the top of the space-craft. All panels are of sandwich construction for compressive stability, and out-of-plane stiffness to accommodate attached equipment load-bearing. The exception is the spine longeron which varies in thickness from 8 mm at the launch vehicle interface bracket attachment to 4 mm at the top boom interface.

8.7.3 Lisa Pathfinder Propulsion Module

The Lisa Pathfinder Propulsion Module (PRM), shown in Figure 8.21, is based on the Eurostar core structure, with the dual function of supporting propulsion equipment and a science module. The relatively simple central thrust cone uses a filament wound carbon fibre composite cone of similar design but different geometry to the Eurostar structure previously described. The outer box structure used by Eurostar to support, protect and transmit heat from payload equipment is discarded and replaced by an exterior thermal blanket support structure. The all-CFRP science module (SCM), which gives good thermo-elastic stability to the instrument, separates from the PRM in the final approach to the Lagrange point.

8.7.4 BepiColombo

Figure 8.22 shows the BepiColombo spacecraft configuration, which is an ESA mission to Mercury. The structure comprises three modules with a total launch mass of 3000 kg. The influence of the purpose of each of these different modules can be seen in the design. The Mercury Transfer Module (MTM) has a central cone to support the upper modules and

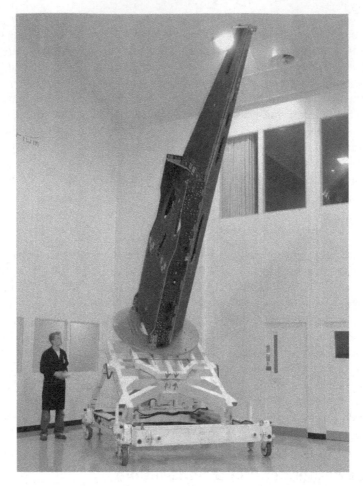

Figure 8.20 Swarm structure configuration. (Reproduced by permission of EADS Astrium Ltd.)

carries five propellant tanks. The Mercury Planetary Orbiter (MPO), with a mass of 367 kg, has a highly asymmetric outer structure to give a large space-facing radiator. Note that the shear walls have been oriented to complete the shear flows to the outer box structure that support the higher mass equipment. The upper Mercury Magnetosphere Orbiter (MMO) is a flat cylinder with a mass of 165 kg. A sun shield protects it from sun light and radiation while attached to the MPO. When orbiting independently it will measure the magnetic field around Mercury. The MMO is spin stabilized, carrying louver protected radiators on its top and bottom surfaces while solar cells are mounted on its sides.

8.7.5 The International Space Station (ISS)

On-orbit assembly of the ISS began in 1998 with the launch of Zarya. It has been continuously inhabited since November 2000. The centre straight section to be seen in Figure 8.23

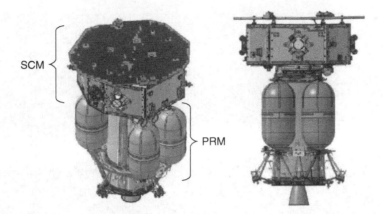

Figure 8.21   Lisa Pathfinder Propulsion Module. (Reproduced by permission of EADS Astrium Ltd.)

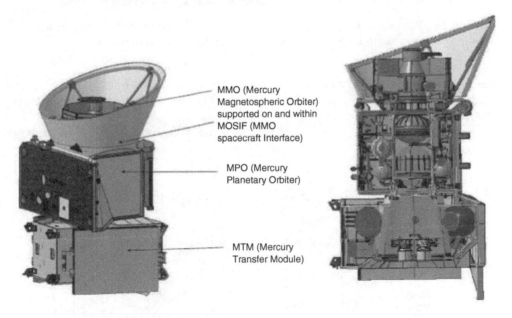

Figure 8.22   The BepiColombo spacecraft configuration. (Reproduced by permission of EADS Astrium Ltd.)

is the integrated truss structure which is 108.5 m long and assembled in space from 10 separately launched segments. The segments were designed to survive their individual launch loads and to give the completed space station the necessary strength and stiffness for stable attitude control during orbit boosting maneuvers to counteract atmospheric drag in low earth orbit. The truss structure supports solar arrays, thermal radiator, cable trays and EVA equipment. There are 13 pressurized modules, with a total volume of 820 m². 850 hours of extra vehicular activity have been required to assemble and maintain the

Figure 8.23 The International Space Station, as imaged from the departing Space Shuttle Discovery during STS 131. (Reproduced by permission of NASA)

space station so far. All structural components of the space station were designed to man-rated fracture control requirements. Pressurized modules have Whipple style micrometeor protection.

One of the typical building blocks of the ISS is the 'Unity' connecting module shown in Figure 8.24.

Figure 8.24 The Unity connecting module — International Space Station. (Reproduced by permission of NASA)

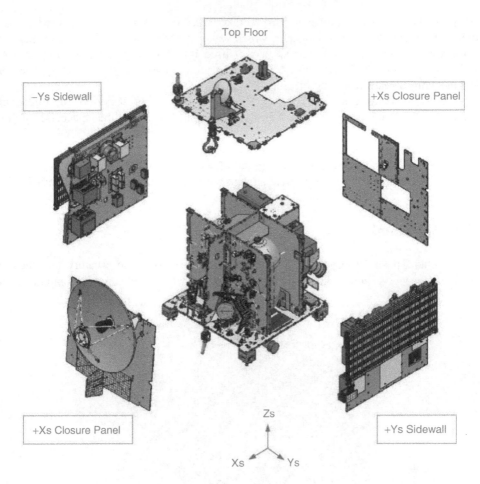

Top Floor

−Ys Sidewall

+Xs Closure Panel

Zs

Xs   Ys

+Xs Closure Panel

+Ys Sidewall

Figure 8.25 Mars Express structure configuration. (Reproduced by permission of EADS Astrium Ltd.)

8.7.6   Mars express

This spacecraft configuration, illustrated in Figure 8.25, was designed to carry seven instruments and one lander into Martian orbit.

The mission was a test case for the development of new working methods to speed up spacecraft production and to minimize mission costs. 116 kg was allowed for the instruments and 60 kg for the lander. Off the shelf technology, or technology developed for the Rosetta comet mission (see Chapter 5) was used wherever possible.

The instruments were accommodated within the spacecraft bus, an aluminium honeycomb box 1.5 m long by 1.8 m wide by 1.4 m high. The lander, Beagle 2, was attached to the outside of the bus. Total mass of payload, lander, spacecraft and on-board fuel was 1070 kg at launch.

Most of the propulsive power needed to take Mars Express from Earth to Mars was provided by the Soyuz Fregat launcher, which separated from the spacecraft after placing it on a Mars-bound trajectory. The spacecraft used its onboard propulsion to inject into Martian orbit and for orbit corrections. The spacecraft had two Eurostar 2000 bi-propellant tanks carrying a total of 427 kg of propellant at launch.

For up to six hours during the spacecraft's 7.5 h Martian orbit period, the high gain antenna is pointed towards Earth. During the remaining 1.5 h, the spacecraft points towards Mars.

The spacecraft structure uses an arrangement of internal shear panels instead of a central thrust structure. This generated critical joints at the points where the shear walls cross the circumference of the launch vehicle attachment ring, requiring substantial brackets at these positions.

8.7.7 Herschel

The Herschel Space Observatory is the largest infrared space observatory launched to date. It is equipped with a 3.5 m diameter mirror, and instruments on an optical bench

Figure 8.26 The Herschel spacecraft service module. (Reproduced by permission of ESA)

(just below the mirror) which are housed in the central cryostat section that cools the bench close to absolute zero. The spacecraft is approximately 7.5 m high and 4 × 4 m in overall cross section, with a launch mass of around 3.4 tonnes. The bottom section of the spacecraft shown in Figure 8.26 is the service module, which houses 'warm' service and payload equipment. Structural attachment to top and bottom of the cryostat is by a truss of carbon fibre reinforced plastic struts which transmit all structural loads, while providing thermal isolation and allowing thermo-elastic radial distortion relative to the cryostat. The payload module is fitted with a sunshield, which protects the telescope and cryostat from solar visible and infrared radiation and also prevents stray light from the Earth from entering the telescope. The sunshield also provides a mounting for the solar array for electrical power generation.

In summary, these examples illustrate the ambitious range and scale of projects currently challenging the spacecraft designer. Most spacecraft structures are designed using flat panels, often honeycomb sandwich, or cylindrical and conical outer or inner shells. Mouldings are used when complex shapes are advantageous. When selecting the spacecraft structure configuration to meet the requirements of any mission, the golden rule is 'keep it simple'. Development of design and realization of detailed requirements as the project progresses invariably increases complexity.

8.8   THE FUTURE OF SPACE STRUCTURES

An examination of the history of structural design shows a progression to the more efficient use of materials. This trend will continue as load path assessments become more refined and tailored application-specific, high strength materials are developed.

8.8.1   Composite structures

Material and manufacturing process improvements, for composites in particular, generally increase specific strength and stiffness, while improved thermal stability for items such as optical benches is being achieved with carbon-carbon and ceramic materials.

Carbon fibres derived from mesophase pitch give improved conductivity and modulus over standard carbon fibres derived from polyacrylonitrile (PAN), but they have lower compressive strength. This is good for optical benches but can be a problem at joints in highly loaded primary structures. Introducing carbon nanotubes into composite resins and adhesives or growing them onto the sides of carbon fibres improves thermal and electrical conductivity, strength and stiffness.

Draping, weaving and stitching of fibres are being used to generate complex shapes which can be complemented by resin transfer moulding or a resin injection process.

By integrating systems into composite structures, we may approach the concept of the intelligent structure. Already optical fibres can be embedded into a carbon fibre matrix allowing signals to pass along skins. References [26, 27] are sources of information on the developing field of electro-active and shape memory materials. Currently crude actuators are developed to sense and correct the onset of buckling.

8.8.2 Deployable structures

Inflatable structures that cure in orbit provide an interesting option for deployable structures. Further development is needed to achieve satisfactory rigidity. Various techniques are being developed:

- mechanical including kapton and aluminium laminates,
- physical techniques including shape memory materials and solvent evaporation which suffers from out-gassing problems,
- chemical, thermal and/or UV curing,
- cas catalysed polymers.

Potentially the task of designing for launch can be made easier, since the structure may be stowed compactly within the launch vehicle, and an extremely low mass may be achieved, designed to withstand the zero g environment only when deployed. Careful consideration needs to be given to loads created in hinges or members during automatic or remote deployment, where sections can build up momentum as they are swung into place and are then suddenly stopped when positioned.

A similar argument prevails for attitude control or orbit maintenance manoeuvres by the space station. Not only are these likely to excite very low frequency cross-coupled oscillations within the structure, but the designer is faced with a continuously changing situation as mass is added, moved or subtracted from the configuration. This is potentially a fertile area for the application of distributed thrusts and an 'expert system' or 'intelligent' attitude control system to phase the individual firings to counter an undesirable dynamic response.

It can be expected that new families of wire tensioned structures will emerge that can only be used in a zero g environment.

8.8.3 Multiple spacecraft structures

A greater diversity of structural design concepts is emerging. With the interest in small spacecraft for low Earth orbit constellations or small-scale, single experiments, the designer is required not only to produce an efficient structure to maximize payload but also to respond to the ever-increasing commercial pressure of schedule and cost. Moulded, single component structures that can be stacked for multiple launches are an attractive proposition. The possibility of large numbers of spacecraft for the proposed global constellations (see Chapter 5) requires the design to encompass the needs of mass production, and ease of integration and test. The concept of *six sigma* is being explored to give more robust manufacturing methods. A six sigma process reduces the probability of defects occurring to less than 3.4 per million or six standard deviations above a 50% chance in a statistical normal distribution. As an example, bolted joints can be six sigma, while bonded joints will not. This is because there is significant scope for defects in the bond due to a range of possible causes.

REFERENCES

[1] Crane, F. A. and Charles, J. A. (1984) *Selection and Use of Engineering Materials*, Butterworths, London.

[2] Ashby, M. F. and Jones, D. R. H. (1980) *Engineering Materials—An Introduction to Their Properties and Applications*, Pergamon, Oxford.

[3] ESA *Space Product Assurance* Material selection for controlling stress corrosion cracking ECSS-Q-70-36, January 1998.

[4] Kinnersley, M. A., Stark, J. and Swingard, B. (1989) Development of, and initial results from a high fluence, high velocity atomic oxygen source. 4th*European Symposium on Spacecraft Materials in the Space Environment*, CERT/CNES/ESA, 701.

[5] *LDEF-69 Months in Space*. Second Post Review Symposium, NASA CP-3914 (1992).

[6] *Structural Materials Handbook, Vol 1 Polymer Composites* (1994) ESA-PSS-03-203.

[7] *Structural Materials Handbook, Vol 2 New Advanced Materials* (1994) ESA-PSS-03-203.

[8] Edwards D. L. and Chipara M, (2005) Polymeric Materials for Solar Sail: The combined Effects of Polymer Thickness, Radiation and Temperature, *Materials Research Society Symposium Proceedings*, Volume **851**.

[9] Vasiliev, V. V. and Morozov, E, (2005) *Mechanics and Analysis of Composite Materials*, Elsevier.

[10] Reddy, J. N. and Miravete, A. (1995) *Practical Analysis of Composite Laminates*, CRC Press, Boca Raton, Florida.

[11] Mattais Krodel, C. *Engineering Material for Optics and Structures* in W. A. Goodman (ed.), *Optical Materials and Structure Technologies*, 11Proc SPIE, 5868, 58680 - 2005.

[12] *Composite Materials Handbook*, MIL-HDBK-17-3F.

[13] *Metallic Materials Properties Development and Standardisation*, MMPDS -04 US Department of Transport.

[14] Zienkiewicz, O. C, Taylor, R.L, Zhu, J. Z (2005) *The Finite Element Method: Its Basis and Fundamentals*. Elsevier Publishing, Oxford.

[15] MacDonald, B. J, (2007) *Practical Stress Analysis with Finite Elements*, Glasenvin Publishing, Dublin.

[16] Anderson, T. L. (1995) *Fracture Mechanics: Fundamentals and Applications*, CRC Press, Boca Raton, Florida.

[17] ESA *ESACRACK User's Manual* ESA PSS-03-209.

[18] Fletcher, K., *Hexapod Mechanical Assembly Modal Survey Test*, Proceedings of the 5th International Symposium on Environmental Testing for Space Programmes, Noordwijk, The Netherlands (ESA SP-588, June 2004).

[19] Terrillion, F., Warren, H. R. and Yelle, M. J. (1991) Orbital Debris Shielding Design of the Radarsat Satellite. IAF-91-283, *42nd Congress of the International Astronautical Federation*, 5-11 October, Montreal, Canada.

[20] Klinkras, H. (2002) *Update of the ESA Space Debris Mitigation Handbook* CR021539.

[21] Christiansen, E. L. (1998) *Design Practice for Spacecraft Meteoroid/Debris Protection*, Presented at the IAT/NASA Hypervelocity Shielding Workshop, March 9–11, Galveston Texas.

[22] McMillan, A. R. (1963) *An Investigation of the Penetration of Hypervelocity Impact Projectiles into Composite Laminates*. Proceedings of the 6th Symposium on Hypervelocity Impact, Cleveland Ohio, Vol **III**, 309–356.

[23] Stilp, A. J. and Weber, K. (1997) Debris Clouds Behind Double Layer Targets, *International Journal of Impact Engineering*, **20**, 765–778.

[24] Cour-Palais, B. G. and Crews, J. L. (1990) A Multi-shock Concept for Spacecraft Shielding, *International Journal of Impact Engineering*, **10**, 135–146.

[25] Christiansen, E. L. (2009) NASA/TM-2009021478 *Handbook for Designing MMOD Protection*, NASA Johnson Space Center, Huston, TX.

[26] Bandyopadhyay, B., Manjunath, T. C., Umapathy, M., (2007) *Modelling Control and Implementation of Smart Structures*, Springer, Heidelberg.

[27] Srinivasan, A. V. and McFarland, D. M. (2000) *Smart Structures Analysis and Design*, Cambridge University Press, Cambridge.

9 ATTITUDE CONTROL

Peter W. Fortescue and Graham G. Swinerd

Aeronautics and Astronautics, Faculty of Engineering and the Environment, University of Southampton

9.1 INTRODUCTION

Although the prime purpose of the attitude control system (ACS) is to orientate the main structure of the spacecraft correctly and to the required accuracy, it is worth considering it also as a momentum management system. Angular momentum is a commodity that can be acquired and disposed of, or stored.

In terms of angular momentum management, the ACS designer must decide what his or her best policy is, and provide the hardware to achieve it. He or she will need to assess the momentum implied by the pointing requirements set by the mission objectives, and to specify torquers and storage to provide it.

When one looks at the types of ACS in existing spacecraft, it is found that a profusion of different methods are employed to meet this seemingly simple objective. The structures may be spinning or not, or may be doing so only during certain phases of the mission. The ACS may or may not use momentum bias and/or momentum storage; it may use a variety of torquing methods, in combination or on their own. The number of options open to the designer is large. What then influences the designer to select one solution as opposed to another?

The decisions that he or she makes are not based solely upon the considerations presented here. The designer will be influenced by the experience and the history of the company in which he or she works. A feature of attitude control is that different technical solutions may give very similar performance, and the designer will choose well-trodden paths, the type of solution of which his or her company has experience. And rightly so.

There are complete books on the subject of attitude dynamics and control [1–3], and this solitary chapter cannot present a comprehensive account of the subject. The approach adopted is to look at the fundamentals, to progress from the system level downwards and to identify the design options. Detailed design is not covered.

Spacecraft Systems Engineering, Fourth Edition.
Edited by Peter W. Fortescue, Graham G. Swinerd and John P. W. Stark.
© 2011 John Wiley & Sons, Ltd. Published 2011 by John Wiley & Sons, Ltd.

9.2 ACS OVERVIEW

9.2.1 The design objective

The orientation required of the spacecraft's structure will be determined by the mission. The structure will be seen as the mounting base for the payload(s), and for several 'housekeeping' subsystems that have objects that must be pointed in specific directions. Among the latter will be solar arrays to be pointed at the Sun, thermal radiators to be pointed at deep space, antennas to be pointed at ground stations and, prior to firing, thrusters to be pointed in the correct direction. Add to this the pointing requirement of the payload and it can be seen that there is an essentially three-dimensional problem to solve, at the end of which will emerge a proposed layout for the spacecraft with locations for the objects to be pointed and a specification for the orientation of their mounting base, the main structure.

The required orientation will frequently be related to an Earth-based frame of reference, such as, 'One face of the spacecraft must point down the local vertical'. The ACS designer must then analyse the orbit and the mission in order to assess the motion that is required of the structure and the disturbance torques to which it will be subjected. For this objective, the structure requires an angular rate equal to $\dot{\theta}$, where θ is the true anomaly. (See Section 4.2 and Figure 4.3 of Chapter 4.)

There will often be several alternative configurations that will meet the overall objective for a given payload, and these may pose very different problems. The Intelsat series of satellites illustrates this point. Their payloads are all for communications purposes and have antennas to be pointed at locations on the Earth. But their configurations include spinners, dual-spin and three-axis-stabilized types.

In general, pointing mechanisms are to be avoided. However, they are often needed in order to enable an object to remain pointing in one direction whilst the main structure changes its orientation, or *vice versa*. For example, a solar array needs to remain pointing at the Sun whilst the main structure turns to align with the local vertical.

The required accuracy of orientation will be set by the payload. The accuracy with which its direction can be controlled will be less than that to which it can be measured, and this is in line with the requirement. For example, astronomers will need to know where a space telescope's axis is pointing to great accuracy (typically arc seconds), but the control of its direction may be less accurate, related to its field of view.

A full accuracy specification for both measurement and control of the main structure's attitude may then be determined from the various pointing requirements.

9.2.2 Mission-related system considerations

The ACS designer will need to know the required time-history of the rotational motion, the angular rate ω which is required of the spacecraft, and of any parts of it that can move independently on bearings. The angular momentum $\mathbf{H}_c$ and the torque $\mathbf{T}$ needed to produce it may then be calculated from the Newtonian law (from equation 3.28 of Chapter 3):

$$\mathrm{d}\mathbf{H}_c/\mathrm{d}t = \mathbf{T} \qquad\qquad (9.1)$$

where $\mathbf{H}_c$ is the angular momentum referred to the centre-of-mass C, detailed in equations (3.33) to (3.36) of Chapter 3, and $\mathbf{T}$ is the torque.

A circular orbit provides an interesting example. The local vertical will rotate at a constant rate equal to 1 revolution per orbit about the normal to the orbit plane. If one face of a structure with no momentum bias is required to point down the vertical, then no torque is needed to maintain this condition provided that a principal axis is aligned with the normal to the orbit plane. If the normal is not a principal axis, then equation (9.1) will indicate that a cyclic torque is needed.

For an elliptical orbit, on the other hand, the local vertical will not have a constant angular velocity. An assessment over one orbit indicates that the angular rate will vary as shown in Figure 9.1. It has a cyclic variation about a mean level of one revolution per orbit.

If one face of the structure is required to point down the vertical and a principal axis is aligned with the normal to the orbit plane, then no torque is needed to maintain the mean angular rate (constant angular momentum). But torques *will* be needed to produce the fluctuations in momentum about the mean value.

If the structure is not rotating, frequently the case for astronomical missions, then again no torque is needed to maintain this condition once it has been achieved. This will be approximately true of solar arrays too, since their angular velocity is likely to be virtually zero as they point towards the Sun, even if the main structure is rotating.

The torque requirements for each part of the spacecraft may be assessed as above, on the assumption that it is rigid, and these will be the *minimum* torques needed. Extra torques will be required in order to combat the uncontrolled (disturbance) torques such as that due to solar radiation pressure (SRP), movement of fuel, mechanisms, friction in bearings and so on, and so these must be evaluated too (see Section 9.4). The assessment must cover *all* phases of the mission, and all three axes of the spacecraft.

It may be expected that the largest torques will be needed during the early stages, between final separation and being on station. The *maximum* torque capability will

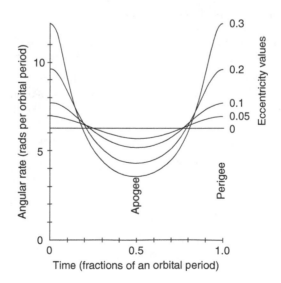

Figure 9.1   Angular rate of the local vertical for satellites in elliptical orbits

govern the time taken to execute manoeuvres such as the repointing manoeuvres shown in Figure 9.7. If manoeuvre times are important, then these may determine the specification for this feature of the torquers.

Momentum storage will normally be provided on spacecraft of medium to large size, especially when there is a requirement for fluctuating torque or momentum or when there are tight tolerances on pointing accuracy (see Table 9.2). A momentum storage device is basically a wheel fitted with a torque motor, each enabling one component of momentum to be stored (but see Section 9.4.7). On small satellites they are less likely to be used because of their mass and cost, attitude control then being achieved by using external torquers.

In operation, the wheel torque motor will be the prime means of achieving the pointing requirement of the structure, correcting *any* errors, however caused, in a feedback sense. In so doing, it will insert or extract momentum from the wheel. When the wheel reaches an extreme of its permitted range of speeds—its store of momentum has been filled—then its speed must be restored to its normal value, using external torquers to counter the torque on the wheel so as to maintain attitude control. This process is known as *momentum dumping*.

Assessment of the required amount of storage will be based upon the mission, and it will be chosen to achieve infrequent use of the dumping process. Orbit considerations such as its eccentricity, imply rotation as shown in Figure 9.1, for example, and a momentum component ($I\omega$) along the normal to the orbit plane. Dumping will be required during every orbit unless the store can accommodate at least half of the difference between the maximum and the minimum values of $I\omega$. In addition, it must accommodate consequential momentum arising from disturbance torques. Table 9.1 shows a list of the potential sources of these.

Manoeuvres such as the repointing one shown in Figure 9.7 must also be assessed in a similar way.

Table 9.1 Disturbance torques

External torques source	Height range over which it is potentially dominant
Aerodynamic	<about 500 km*
Magnetic	500–35 000 km
Gravity gradient	500–35 000 km
Solar radiation	>700 km*
Thrust misalignment	all heights
Internal torques source	
Mechanisms	
Fuel movement	
Astronaut movement	
Flexible appendages	
General mass movement	

*Values depend upon the level of solar activity.

9.2.3 Momentum bias

The level of momentum that is involved in a storage system will be quite small, and much smaller than will bring any significant benefit from gyroscopic rigidity. This will come from *momentum bias*, which makes the direction of one axis of the spacecraft highly resistant to change. (See Section 3.4 of Chapter 3.)

For example, if the mission requires that one axis of the spacecraft shall always lie in the direction of the normal to the orbit plane, or at right angles to the Sun vector, then the designer might include momentum bias in that direction. The magnitude of the bias, H_b, whilst not critical, is likely to be an order of magnitude greater than the storage system will handle. A torque T at right angles to it will cause the axis to precess at a rate T/H_b (see equation 3.29 of Chapter 3). The attitude response *about* the bias direction will not be altered.

Momentum bias devices may be used for momentum storage too. If more than one bias device is used, and this may well be done in order to enhance the reliability, then it must be remembered that their momenta add vectorially to produce only *one* gyroscopically rigid axis.

9.2.4 The ACS block diagram

The block diagram in Figure 9.2 shows the major components of a general ACS system. The links between components identify major interactions, with arrows indicating that there is a cause–effect relationship; it is convenient to think of them as channels along

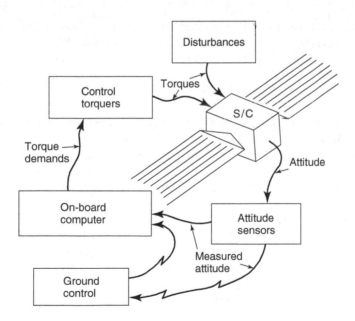

Figure 9.2 Block diagram for an attitude-control system

which information flows. For example, the main structure of the spacecraft is subjected to time-varying torques from torquers, and will respond with attitude motion that will be detected by the sensors. Outputs from these will be sent to computers, on-board and at the ground control station, and the information will be used to determine the torques that should be applied to the structure.

9.3   THE SPACECRAFT ATTITUDE RESPONSE

In the context of the ACS block diagram, the spacecraft may be seen as an object that changes its orientation in response to a torque, which is the simple message conveyed in Figure 9.2. The way in which it responds will depend upon whether it has momentum bias, and the response equations are contained in Sections 3.3 and 3.4 of Chapter 3.

Whilst the decision to include bias and the method of doing so will be decided at system level (see Section 9.2), it is as well to recall the main types of structure and their characteristics with which ACS designer will have to deal. For each type, he or she will be concerned with the following:

- the equilibrium state,
- the response to a steady torque,
- the stability, and the existence of any oscillatory modes.

9.3.1   The three-axis-stabilized spacecraft (with no momentum bias)

The dynamic equations for this spacecraft, treated as a rigid body, are covered in Section 3.4.1 of Chapter 3, using principal axes for the analysis. With small angular velocities the responses about these axes are largely uncoupled and may be approximated by

$$I_{xx}\dot{\omega}_x = T_x, \quad I_{yy}\dot{\omega}_y = T_y, \quad I_{zz}\dot{\omega}_z = T_z \tag{9.2}$$

Each of these equations corresponds to a root location diagram* having a single root at the origin, as shown in Figure 9.6(a).

Any of the torque components on its own will produce an acceleration about its own axis. Combinations, however, will not do so; they will produce a cross-coupled response except when the moments of inertia are equal. Cross-couplings will increase as the angular velocity increases.

This type of spacecraft usually has flexible solar arrays attached to the main structure as shown in Figure 9.3, with lightly damped flexure modes with a low fundamental frequency. When selecting the algorithms for the ACS computer, up to about 20 of these modes may be mathematically modelled, in order to ensure ideally that they are stabilized, but at least that they are not destabilized.

Another typical configuration for this type is that of the Hubble Space Telescope.

---

* A root location diagram is an Argand diagram on which is marked the root(s) of the Characteristic Equation of a linear differential equation.

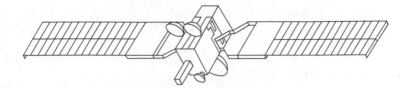

Figure 9.3   Typical configuration of a three-axis stabilized spacecraft

9.3.2   Pure-spin spacecraft

The dynamics of a spinner are covered in Section 3.4.2 of Chapter 3. It is assumed here that the whole vehicle is spinning about its z-axis with an angular rate S, and that its mass distribution is axially symmetric so that $I_{xx} = I_{yy}$. A typical 'spinner' is shown in Figure 9.4. The response to a torque will be of interest in one or both of the two sets of axes.

The response in axes fixed in the structure will be of interest when the torque components are in these axes or when the attitude sensors measure components in these axes. The equations, developed from equation (3.40) of Chapter 3, are

$$I_{xx}\dot{\omega}_x + S\omega_y(I_{zz} - I_{xx}) = T_x$$
$$I_{xx}\dot{\omega}_y - S\omega_x(I_{zz} - I_{xx}) = T_y \qquad (9.3)$$
$$I_{zz}\dot{S} = T_z$$

Separating ω_x from ω_y leads to

$$I_{xx}(\ddot{\omega}_x + \omega_{\text{nut}}^2\omega_x) = \dot{T}_x - \omega_{\text{nut}}T_y \quad \text{and} \qquad I_{xx}(\ddot{\omega}_y + \omega_{\text{nut}}^2\omega_y) = \dot{T}_y + \omega_{\text{nut}}T_x \qquad (9.4)$$

where $\omega_{\text{nut}} = S[(I_{zz}/I_{xx}) - 1]$ is the frequency (rads/s) of the nutation mode referred to in Section 3.4.2 of Chapter 3, when observed in the spacecraft's axes. This mode will need to be damped, whereupon the eventual constant components ω_x, ω_y in response to constant torques T_x, T_y will become

$$\omega_x = -T_y/S(I_{zz} - I_{xx}) \quad \text{and} \quad \omega_y = T_x/S(I_{zz} - I_{xx}) \qquad (9.5)$$

When these are combined with the spin motion, the result is a coning rotation of the z-axis at the spin frequency S when viewed from outside the spacecraft.

One of the major reasons for spinning a spacecraft is to counter the effect upon the trajectory that is caused by a thrust offset when a high-thrust motor is being used, such as during orbit-changing manoeuvres. Ideally, it is assumed that the thrust vector will pass

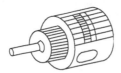

Figure 9.4   Typical configuration of a pure-spin spacecraft

through the centre-of-mass along the z-axis. Equations (9.4) indicate that the torque due to an offset will then cause coning and spinning, which is acceptable, with an average thrust in the correct direction.

The z-component of torque T_z produces an acceleration about that axis, which is uncoupled from the other two axes.

The response in non-spinning axes will be of interest when the spin axis has to be redirected and when the reason for spinning the structure is to provide gyroscopic rigidity *via* momentum bias. One of the axes is aligned with the spin axis (z-axis); the x- and y-axes do not rotate with the structure. The response equations become (see equations 3.47 of Chapter 3)

$$I_{xx}\dot{\Omega}_x + I_{zz}S\Omega_y = T_x$$

$$I_{xx}\dot{\Omega}_y - I_{zz}S\Omega_x = T_y \qquad (9.6)$$

$$I_{zz}\dot{S} = T_z$$

Constant torque components T_x, T_y produce precessional angular rate components

$$\Omega_x = -T_y/I_{zz}S, \quad \Omega_y = T_x/I_{zz}S \qquad (9.7)$$

These are the rotations that will be observed from outside the spacecraft, and they demonstrate gyroscopic rigidity in that as the bias ($I_{zz}S$) increases, so the response to a given torque decreases. The nutation mode, observed in these non-spinning axes, has frequency

$$\omega_{\text{nut}} = S(I_{zz}/I_{xx}) \qquad (9.8)$$

and will need to be damped (see Section 9.3.4).

The root location diagrams corresponding to equations (9.3) and (9.6) are shown in Figure 9.6. The response of the spin-rate S corresponds to a single root at the origin, exactly the same as that for all the principal axes in the case when there is no momentum bias (equation 9.2). On the other hand the equations for the components ω_x, ω_y generate a pair of (oscillatory) roots at $\pm jS((I_{zz}/I_{xx}) - 1)$; the corresponding components Ω_x, Ω_y in *non*-spinning axes generate roots at $\pm jS(I_{zz}/I_{xx})$. These represent the nutation mode (Section 3.5.1 of Chapter 3).

9.3.3 Hybrid and dual-spin spacecraft

The dynamics of these craft are covered in Section 3.4.3 of Chapter 3. In the dual-spin version, momentum bias is provided by rotating a piece of structure on a bearing attached to the non-spinning part; it will be assumed that the spinning part has an axially symmetric mass distribution (see Figure 9.5).

In the hybrid spacecraft, the bias is provided by a momentum wheel (MW) (see Section 9.4.7).

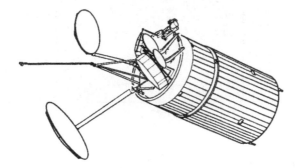

Figure 9.5   Typical configuration of a dual-spin spacecraft

The response equations for these types of spacecraft, with their bias H_z along the z-axis, are equations (3.57) of Chapter 3. For low body rates, they approximate to

$$
\begin{aligned}
I_{xx}\dot{\omega}_x + H_z\omega_y &= T_x \\
I_{yy}\dot{\omega}_y - H_z\omega_x &= T_y \\
I_{zz}\dot{\omega}_z + \dot{H}_z &= T_z
\end{aligned}
\tag{9.9}
$$

Constant torque components T_x, T_y produce precessional angular rates:

$$
\omega_x = -T_y/H_z, \quad \omega_y = T_x/H_z
\tag{9.10}
$$

The torque component T_z about the bias direction produces an acceleration response about that direction, which does not cross-couple into the other axes. In the case of MWs, the total bias may be the vector sum of the biases from three or more wheels in different directions; nevertheless it is the component of torque in the total bias direction that produces the acceleration response described above.

The *nutation mode* has a frequency

$$
\omega_{\text{nut}} = H_z/\sqrt{(I_{xx}I_{yy})}
\tag{9.11}
$$

and needs to be damped (see Section 9.3.4).

The root location diagrams for spacecraft with momentum bias, referred to non-spinning axes (equations 9.9) are shown in Figures 9.6. Motion about the z-axis—the bias direction—in response to T_z, has a single root at the origin as in Figure 9.6(a).

The coupled motion about the x- and y-axes in response to T_x and/or T_y has (in Figure 9.6b) two roots representing the nutation roots. As the amount of bias decreases to zero, the roots move to the origin, as in Figure 9.6(a).

Reorientation manoeuvres in response to torques will differ depending upon whether the manoeuvre changes the direction of the momentum bias, and will depend upon whether the torquer is an On/Off device or whether its magnitude is controllable.

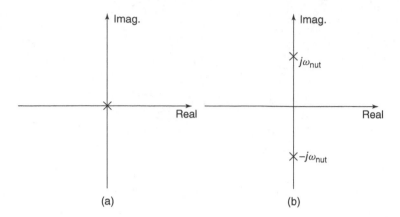

Figure 9.6 Root location diagrams: (a) ω_z in response to T_z; (b) ω_x, ω_y in response to T_x, T_y

Torque pulses to produce a manoeuvre about a bias direction—or about any principle axis if there is no bias—may be achieved by torque pulses as shown in Figure 9.7(b).

Torque pulses used to change a bias direction—the z-axis of the pure-spinner, dual-spinner and hybrid spacecraft—will only produce a change of direction whilst the torque acts. At its start and at its termination it will excite the nutation mode as shown in Figure 9.7(a). A torque pulse lasting for an integer number of periods of the nutation oscillation will cancel the oscillation when it ceases.

Torquers whose torque magnitude is controllable admit the possibility of avoiding excitation of the nutation mode. The cross-coupling between pitch and yaw that is characteristic of the mode can be cancelled by introducing cross-coupling in the computer algorithms that control the torquers. Reference [4] shows how the technique of complex summation of the T_x and T_y equations can lead to a suitable algorithm when the spacecraft is axisymmetric.

9.3.4 Oscillatory modes

A characteristic of the space environment is that oscillatory modes have very little damping. The ACS has to avoid undue excitation of these and must include means of damping them. Damping may be enhanced by means of energy dissipation or by active control techniques.

The energy dissipation method is based upon the fact that the energy present in an oscillatory mode is exchanged between kinetic and potential types during each cycle. A constant total amount corresponds to a constant amplitude of oscillation that is proportional to the square root of the energy. By using the oscillation to excite an energy dissipator, the amplitude will decrease and the mode becomes damped. Active damping entails sensing the oscillation and applying a suppressing torque in the correct phase.

Nutation damping may be implemented either way. A passive damper may consist simply of a metallic tube containing a viscous fluid. The fluid is chosen to have a nearly constant viscosity over a wide range of temperatures keeping it from solidifying at low temperatures. A variety of fluids have been used, including alcohol and freon.

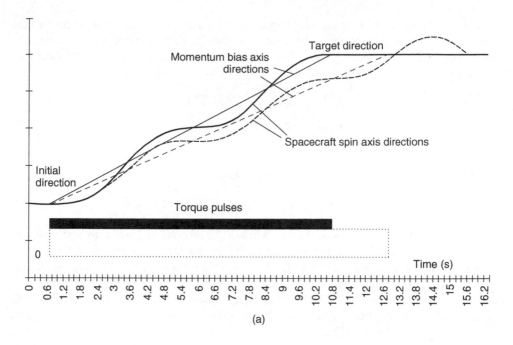

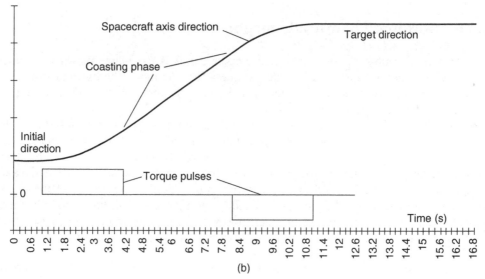

Figure 9.7 Torque responses: (a) repointing a momentum bias axis and (b) repointing about a principal axis, with no momentum bias

The geometry of the tube is often a ring, and it is mounted at a location to ensure a good coupling with the nutational motion. When nutation occurs, the frictional force between the fluid and the tube introduces energy dissipation that dampens the nutational mode. For the long-term stability of the dual-spin spacecraft, it should be mounted in the non-spinning part. (see Section 3.4.3 of Chapter 3.)

Active damping may be achieved by applying a torque component proportional to the angular velocity component along the same axis, that is, T_x proportional to ω_x. The axes may be the spinning (body) axes or non-spinning ones. For the hybrid and dual-spin spacecraft, equations (9.9) then become

$$I_{xx}\dot{\omega}_x + H_z\omega_y = T_x = -K_x\omega_x$$
$$I_{yy}\dot{\omega}_y - H_z\omega_x = T_y = -K_y\omega_y \qquad (9.12)$$

where K_x, K_y are the feedback constants expressing the torque per unit angular rate.

The characteristic equation, in terms of the Laplace operator s, is

$$I_{xx}I_{yy}s^2 + (I_{xx}K_y + I_{yy}K_x)s + (K_xK_y + H_z^2) = 0 \qquad (9.13)$$

This represents damped harmonic motion when K_x and/or K_y are non-zero and positive.

In practice, there will be other oscillatory modes present, and excited by the same torquers. These may be separated into the so-called rigid-body ones, and those associated with mass movement and flexure, as described in Section 3.5.2 of Chapter 3. The mathematical modelling covered in this chapter is restricted to the rigid body and thus will cover nutation and libration but not the other modes. A more comprehensive coverage is given in References [1] and [2].

9.3.5 Summary of attitude response

The ACS is a feedback system in which the spacecraft is the plant, shown as a block in a block diagram (Figure 9.2) in which torque affects attitude. The dynamic equations linking these are presented in Chapter 3, and above they are shown when the torque components are about the principal axes.

Whilst the torques will excite oscillatory modes, and these must ultimately be stable, their prime purpose is to control the orientation of the axes of the spacecraft. Each component of torque will have an effect, and by resolving into principal axes, their effects are largely uncoupled from each other. These simple responses to each separate component of torque are useful to be preserved as the dominant responses of the block. There are just two types—the precessional and the non-precessional. Figure 9.7 illustrates typical responses to torque pulses in order to achieve repointing in these two cases.

The *non-precessional response* shown in Figure 9.7(b) obeys, for example,

$$I_{xx}\dot{\omega}_x = I_{xx}\ddot{\phi} = T_x \qquad (9.14)$$

The rotational response is an angular acceleration about the same principal axis as the torque. This applies to each axis when there is no momentum bias, but applies only to motion about the bias axis when it is present.

The *precessional response*, approximated by omitting the nutation mode, is the steady-state response, which is given by:

$$\Omega_y = \dot{\theta} = T_x/H_z \quad \text{and} \quad \Omega_x = \dot{\phi} = -T_y/H_z \qquad (9.15)$$

The response equations with nutation included, in state-space form, are

$$
\mathrm{d}/\mathrm{d}t
\begin{bmatrix} \Omega_x \\ \Omega_y \\ \phi \\ \theta \end{bmatrix}
=
\begin{bmatrix}
0 & -H_z/I_{xx} & 0 & 0 \\
H_z/I_{yy} & 0 & 0 & 0 \\
1 & 0 & 0 & 0 \\
0 & 1 & 0 & 0
\end{bmatrix}
\begin{bmatrix} \Omega_x \\ \Omega_y \\ \phi \\ \theta \end{bmatrix}
+
\begin{bmatrix}
1/I_{xx} & 0 \\
0 & 1/I_{yy} \\
0 & 0 \\
0 & 0
\end{bmatrix}
\begin{bmatrix} T_x \\ T_y \end{bmatrix}
\qquad (9.16)
$$

where $\dot{\theta} \approx \Omega_y$ and $\dot{\phi} \approx \Omega_x$.

9.4 TORQUES AND TORQUERS

The torques, arising from moments of forces about the centre-of-mass, must be identified as being *external* or *internal* to the spacecraft. The former will affect its total angular momentum, whereas the latter will affect only the distribution of momentum between its moving parts. The case has already been made that it is *necessary* to include controllable external torquers whereas internal ones, with their momentum storage facility, are *optional* (see Section 3.3.2 of Chapter 3).

The main sources of torques, occurring either naturally or as disturbances, are introduced below and summarized in Table 9.1. The magnitude of torques in space is small when compared with terrestrial standards. Even very small ones become significant when there is no friction to oppose them and when the orientation has to be very accurate.

Some of the phenomena listed as disturbance torques in Table 9.1 may be used as a means of achieving the required orientation of the spacecraft. For this, they will normally need to be controllable; a possible exception is the gravity-gradient torque, which will establish an Earth-facing equilibrium orientation passively, with the axis of least inertia along the local vertical (see Section 9.4.3).

Table 9.2 summarizes the main advantages and disadvantages of various types of torquer.

9.4.1 Thrusters (external type)

Orbit-changing thrusters provide potentially the largest source of force on spacecraft, and potentially the largest source of torque. Being external, the torque will affect the total momentum. Ideally, the thrust vector passes through the centre-of-mass, but inevitably there is a tolerance on this and consequently a disturbance torque arises.

The main means of countering the effects of this torque when large thrust levels are present are either to spin the vehicle about the intended thrust direction (cf. Section 9.3.2) or to provide means of controlling the achieved thrust direction. At lift-off, for example, the latter method must clearly be used. This involves mounting some of the thrusters in gimbals, or using secondary fuel injection into the rocket nozzle, and controlling the thrust direction so as to achieve the required trajectory. Later stages of the flight of booster rockets may adopt the alternative method of spinning the vehicle in order to average out the effect of thrust offset upon the trajectory. Thruster firings used for changing the orbit, such as from Low Earth Orbit (LEO) to Transfer Orbit, and again from Transfer Orbit

Table 9.2 Types of torquer

Type	Advantages	Disadvantages
External types	Can control momentum build-up	
Gas jets	Insensitive to altitude Suit any orbit Can torque about any axis	Requires fuel On-off operation only Has minimum impulse Exhaust plume contaminants

Magnetic	No fuel required
	Torque magnitude is
	   controllable
		No torque about the local field
		   direction
		Torque is altitude and latitude
		   sensitive
		Can cause magnetic
		   interference

Gravity gradient	No fuel or energy needed	No torque about the local vertical Low accuracy Low torque, altitude sensitive Libration mode needs damping
Solar radiation	No fuel required	Needs controllable panels Very low torque
Internal types	No fuel required Can store momentum Torque magnitude is controllable	Cannot control momentum build-up
Reaction wheels (RW) Momentum wheels (MWs)	Continuous, fine-pointing capability Provide momentum bias	Non-linearity at zero speed
Control moment gyroscope (CMG)	Suitable for three-axis control Provides momentum bias	Complicated Potential reliability problem

to Geostationary Earth orbit (GEO), are sometimes preceded by a spin-up manoeuvre, followed by de-spin after the orbit changes are complete.

Thrusters with very much lower levels of thrust are in common use in attitude-control systems for providing controllable external torquing, and hence controlling the total momentum of the spacecraft. For this purpose, they will be mounted in clusters on the surface of the vehicle, pointing in different directions in order to provide three components of torque. They have a number of advantages and disadvantages compared with their main rival, the magnetic torquer.

Their main advantage is that their torque level is independent of altitude and there is potentially no limit to its magnitude. However, the magnitude is not controllable when installed; only the switch-on duration. This torquing system integrates well with the station-keeping requirement for thrusters, since a common fuel and control system can be used.

Thrusters have a number of disadvantages. If they are used as the prime means of orientating a spacecraft, their restriction to an on−off type of control leads to a limit cycle occurring, a similar process to that occurring in station-keeping manoeuvres (see

Sections 5.6.3 and 5.6.4 of Chapter 5) but with a much shorter period. If they are used to achieve accurate pointing, then the torquers will need to provide small and consistent impulses, and their minimum switch-on time of several milliseconds leads to a low maximum torque being needed. Typically a minimum thrust impulse of order 10^{-4} Ns may be required, with a thrust level as low as 10^{-2} N. When the prime means of attitude control is a reaction wheel or momentum wheel (Section 9.4.7), then thrusters may be used for momentum dumping. In this case, their thrust level will be higher than that stated above, and they may be used perhaps only once every several orbits.

A variety of thruster systems may be used, ranging from cold gas to electric propulsion, as indicated in Sections 6.3 and 6.4 of Chapter 6.

The fact that thrusters use fuel is another disadvantage. Although a large amount of fuel is not normally needed for attitude control it will eventually be exhausted, and a number of spacecraft have reached the end of their useful life because of this.

9.4.2 Magnetic torque (external type)

The magnetic field generated by a spacecraft interacts with the local field from the Earth and thereby exerts an external couple on the vehicle. This is a similar effect to that of a compass needle that attempts to align itself with the local direction of the field. If the spacecraft's magnetism is represented as a dipole whose magnetic moment is **m**, then it reacts with the local flux density **B** to produce a torque **T** given by

$$\mathbf{T} = \mathbf{m} \times \mathbf{B} \tag{9.17}$$

Care must be taken that electric currents and spurious magnetic effects do not cause a significant disturbance torque (see Chapter 16).

Electromagnets may be used to provide a controllable external torque. Their strength can be controlled by means of the current I. Their reaction with a local field **B** leads to the couple:

$$\mathbf{T} = nIA(\hat{\mathbf{c}} \times \mathbf{B}) \tag{9.18}$$

where

n is the number of turns
A is the cross-sectional area of the coil,
$\hat{\mathbf{c}}$ is the unit vector in the direction of the coil's axis.

Rod-like electromagnets will normally be used, such as those shown in Figure 9.8. A range of strengths is available to suit the mission requirements, and they may be used in an on–off or a proportional control manner, for attitude control or momentum dumping.

Three orthogonal magnets enable the direction and magnitude of the dipole to be controlled. Their mounting locations should be away from instruments that are sensitive to magnetic fields and separate from each other in order to avoid cross-coupling.

Magnetic torquers are in common use in satellites orbiting at altitudes up to geostationary altitude, but their utility decreases at the higher altitudes since the strength of the Earth's field reduces with height. The field's strength and direction also vary with the

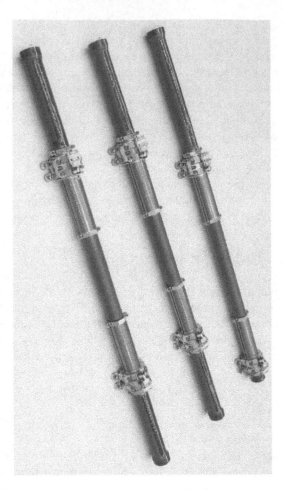

Figure 9.8 Magnetic torquers. The three torque rods shown are those used on the
EURECA spacecraft, and are each in excess of a metre in length. (Reproduced by
permission of Dutch Space BV)

position of the spacecraft in its orbit in general, and when using magnetic torquers it is
common practice to carry a magnetometer to measure the local field.

A feature of magnetic torquers is that they cannot produce a torque component about
the local field direction. In a polar orbit, any required torque direction can always be
achieved at some point in the orbit since the field direction changes round the orbit.
In the equatorial plane, however, the field lines always lie horizontally, north–south.
Consequently a spacecraft whose orbit lies in this plane cannot use magnetic torquers
to counteract the north–south component of their disturbance torque, or to dump this
component of momentum. A reorientation about the north–south direction is achievable
nevertheless, by a rotation of the torque vector about this direction. In the resulting coning
motion, the reorientation will slowly take place.

An advantage of magnetic torquers is that they require no fuel and so have virtually
unlimited life. They do of course require electrical power. But there is no exhaust pollutant
and by providing a couple they are not sensitive to movement of the centre-of-mass.

9.4.3 Gravity-gradient torque (external type)

This source of torque occurs because in a gravitational field that gets weaker with increase in height, a body will only be in stable equilibrium if its axis of minimum inertia is aligned with the local vertical.

The gravitational force dF on an increment of mass dm is

$$dF = \frac{\mu \, dm}{r^2} \tag{9.19}$$

where

μ is the Earth's gravitational constant $= 0.3986 \times 10^{15} \mathrm{m}^3/\mathrm{s}^2$,

r is the distance from the Earth's centre.

By summing the moments about the centre-of-mass C, the torque components may be shown to be

$$
\begin{aligned}
T_x &= (3\mu/2r^3)(I_{zz} - I_{yy}) \sin 2\phi \cos^2 \theta \\
T_y &= (3\mu/2r^3)(I_{zz} - I_{xx}) \sin 2\theta \cos \phi \\
T_z &= (3\mu/2r^3)(I_{xx} - I_{yy}) \sin 2\theta \sin \phi
\end{aligned}
\tag{9.20}
$$

where ϕ and θ refer to the roll (about the x-axis) and pitch (about the y-axis) angles, respectively, using the aircraft axis convention (see Figure 3.A2 in the Appendix of Chapter 3).

These torques contribute to the total disturbance torque in general, but an oscillatory 'libration' mode will occur if they govern the motion about the equilibrium state (see Section 3.5.1 of Chapter 3). For small oscillations of an axisymmetric spacecraft ($I_{yy} = I_{xx}$), the motion is like a conical pendulum, whose frequency is

$$\omega_{\mathrm{lib}} = \sqrt{[(3\mu/r^3)(1 - I_{zz}/I_{xx})]} \text{ rad/s} \tag{9.21}$$

Gravity-gradient torque provides a passive self-aligning torque that has been used (e.g. LDEF and Seasat), but the libration does need damping to be incorporated. The torque levels will be low unless a long thin configuration is used, or in the case of tethered satellites.

9.4.4 Aerodynamic torques (external type)

Aerodynamic torques are dominated by the drag force, which is dependent on frontal area A. Their total moment about the centre-of-mass C may be assessed by considering the projection in the direction of travel. If the spacecraft surface comprises a collection of small incremental areas dA, each with unit normal vectors $\hat{\mathbf{n}}$, such that the position vector of the centre-of-area of each such small area with respect to C is $\mathbf{r}$, then the aerodynamic torque is given by

$$\mathbf{T}_{\mathrm{aero}} = \int_A \mathbf{r} \times d\mathbf{F}_{\mathrm{aero}}, \quad d\mathbf{F}_{\mathrm{aero}} = \frac{1}{2}\rho V_a^2 C_D \, (\hat{\mathbf{n}} . \hat{\mathbf{V}}_{\mathbf{a}})(-\hat{\mathbf{V}}_{\mathbf{a}}) dA$$

where ρ is the atmospheric density, $\mathbf{V_a}$ is the air-relative velocity of the vehicle (and $\hat{\mathbf{V}}_\mathbf{a}$ denotes its unit dimensionless vector), and C_D is the drag coefficient (normally taken to be about 2.2). The area A over which the integral is performed is the spacecraft surface for which $\hat{\mathbf{n}}.\hat{\mathbf{V}}_\mathbf{a} \geq 0$—in other words, the surfaces exposed to the incoming flow of atmospheric particles. Usually this integral is not amenable to simple solution for a surface A associated with a complex spacecraft configuration. A commonly used alternative to assess the aero-torques is to represent the spacecraft as a collection of simple geometrical elements (e.g. flat plates over which the integral can be performed easily), and construct a simple summation over all such surfaces exposed to the incoming particles. For a configuration model comprising n flat plates, each with unit normal $\hat{\mathbf{n}}_i$ and area A_i, then

$$\mathbf{T}_{\mathrm{aero}} = \sum_{i=1}^{n} \mathbf{r}_i \times \mathbf{F}_{\mathrm{aero},i}, \quad \mathbf{F}_{\mathrm{aero},i} = \frac{1}{2}\rho V_a^2 C_D (\hat{\mathbf{n}}_i.\hat{\mathbf{V}}_\mathbf{a}) A_i (-\hat{\mathbf{V}}_\mathbf{a}) \tag{9.22}$$

for values of i for which $\hat{\mathbf{n}}_i.\hat{\mathbf{V}}_\mathbf{a} \geq 0$. Here $\mathbf{r}_i$ is the position vector of the centre-of-area of the ith plate with respect to C. This technique gives an adequate assessment of the aero-torques, but does not take account of issues such as shadowing of one element by another, or of multiple collisions of particles with the spacecraft.

For zero torque, spacecraft designers will aim to balance the terms in the summation (9.22), but engineering tolerances, shifts of the centre-of-mass and thermal distortion will usually give rise to a residual torque.

The torque is height-dependent, and is not an important effect above about 600 to 700 km, depending upon the spacecraft configuration and the level of solar activity.

9.4.5 Solar radiation pressure (external type)

Solar radiation produces a force on a surface, which depends upon its distance from the Sun; it is independent to first order of the height above the Earth. Large flat surfaces with a significant moment arm about the centre-of-mass, such as solar arrays, may produce a significant torque. Since light carries momentum, when it is reflected at a surface this represents an exchange of momentum with the surface, which gives rise to the SRP. This is not to be confused with the pressure exerted on the spacecraft by the solar wind (the stream of high-energy ionized particles emanating from the Sun), which at Earth orbit can be orders of magnitude less than light pressure. The force exerted on the surface can be calculated by a vector difference between the incoming and outgoing momentum fluxes. If $\hat{\mathbf{s}}$ is the unit vector from the spacecraft to the Sun, and $\hat{\mathbf{n}}$ is the outward unit normal of an incremental area dA, then the force due to solar radiation is [1]

$$d\mathbf{F}_{\mathrm{SRP}} = -P\cos\theta dA \left[(1 - f_s)\hat{\mathbf{s}} + 2\left(f_s\cos\theta + \tfrac{1}{3}f_d\right)\hat{\mathbf{n}}\right], \quad \hat{\mathbf{s}}.\hat{\mathbf{n}} \geq 0$$

where P is the mean momentum flux $\sim 4.67 \times 10^{-6}\,\mathrm{Nm}^{-2}$ at the Earth, $\theta = \cos^{-1}(\hat{\mathbf{s}}.\hat{\mathbf{n}})$ is the angle of incidence of the radiation and f_s, f_d are the coefficients of specular and diffuse reflection, respectively.

The torque may be obtained by integrating the moment of the above force over the spacecraft's illuminated surface area. In practicality, this integration can be problematic, so a similar method to that described for aero-torques (Section 9.4.4) may be adopted.

For example, if the configuration is approximated by a collection of $i = 1$ to n flat plates with area A_i, normal $\hat{\mathbf{n}}_i$ and incidence θ_i, then the SRP torque can be expressed as

$$\mathbf{T}_{\text{SRP}} = \sum_{i=1}^{n} \mathbf{r}_i \times \mathbf{F}_{\text{SRP},i}, \quad \mathbf{F}_{\text{SRP},i} = a_i \hat{\mathbf{s}} + b_i \hat{\mathbf{n}}_i \tag{9.23}$$

where the summation is performed over the plates that are illuminated. Here $\mathbf{r}_i$ is defined as in equation (9.22), and

$$a_i = -PA_i \cos\theta_i (1 - f_{\text{s},i}), \quad b_i = -2PA_i \cos\theta_i (f_{\text{s},i} \cos\theta_i + \tfrac{1}{3}f_{\text{d},i})$$

Both aero and solar radiation pressure torques are in principle usable for counteracting momentum build-up, and for trimming attitude (e.g. ASTRA 2B), but usually the spacecraft is designed such that the forces balance to give zero torque.

9.4.6 Mass movement (internal type)

The movements of masses within a spacecraft may directly exert torques upon the main structure. These are classified as internal torques and do not affect the total momentum. The movements may also alter the location of the centre-of-mass within the spacecraft, and change the inertia matrix.

The centre-of-mass C has been identified as a key reference point for establishing the dynamic behaviour (Chapter 3). Moving the point affects the balance of the vehicle in dynamic ways. It also affects the torques due to forces on the vehicle, but not the couple of the magnetic torquer. In principle, the centre-of-mass location could be controlled in order to balance out the disturbance torques.

A major source of mass movement is that of the fuel. The tanks are normally located in such a way that as their contents are used up the centre-of-mass does not shift. Fuel movement within the tanks causes a different sort of problem in that it moves in a dynamic way in response to the motion of the spacecraft—fuel slosh—affecting its modal characteristics.

Mass movements from one position to another, such as the erection of solar arrays and other appendages and movement of astronauts and so on, have an effect upon attitude, which is best assessed by using the fact that angular momentum is conserved.

9.4.7 Momentum storage torquers (internal type)

Torquers associated with momentum storage such as RWs and MWs are essentially internal torquers, suitable for attitude control but not for controlling the total momentum.

These devices are purpose-built precision-engineered wheels that rotate about a fixed axis, with a built-in torque motor. Figure 9.9 shows a photograph of such a wheel with its protective cover removed. The basic design comprises a reliable bearing unit (with a life expectancy of typically 15 years), a spoked flywheel mass and a DC motor in a vacuum-tight, evacuated housing. The stator is controlled by the drive electronics, which may be seen under the wheel. The mechanism aspects of RWs and MWs are discussed in Chapter 15.

Figure 9.9   TELDIX Space Wheel RSI 68. (Reproduced by permission of Rockwell Collins Deutschland GmbH)

It is difficult to generalize about the range of wheel characteristics available for use in 'large' spacecraft (mass of the order of a tonne or more), but we may attempt to summarize it as—wheel diameter ~20 to 40 cm; wheel mass (including torque motor) ~3 to 10 kg; momentum storage capacity ~5 to 70 kg m^2/s. With the relatively recent trend to use wheels (see Chapter 18) to control 'small' satellites (mass of the order of 50 kg), much smaller wheels are required, and these are often designed and manufactured as part of the smallsat programme.

Reaction wheels have a nominally zero speed, and may be rotated in either direction in response to the control torques called for by the spacecraft's ACS. However, it should be noted that at low or zero angular rate, the wheel displays a non-linear response due to 'sticking friction', which can impose an irregular motion on the spacecraft in this region. This problem is often circumvented by setting the nominal operating speed of the wheels above zero rate, at a few rpm.

Momentum wheels on the other hand have a high mean speed (in the range typically 5000 to 10 000 r.p.m.) in order to provide momentum bias. The control torques will then slow down or increase the wheel speed, the permissible amount being about 10% of the mean value.

Both types of wheel provide momentum storage, and need to be used in conjunction with external torquers, as described in Section 9.2.2. (See also Chapter 15.)

For three-axis control, three orthogonal reaction wheels will be the minimum requirement. A redundant fourth is normally added at an equal angle to the other three, in order to avoid a single-point failure. When more than one MW is used, the total bias is the *vector* sum of contributions from the separate wheels.

The principle of MWs has been extended by the development of more advanced forms, such as control moment gyroscopes (CMGs). By mounting the wheel in gimbals fitted with torque motors, all three components of torque may be developed from a single wheel. This can be done to a limited extent with sophisticated wheels mounted on five-degree-of-freedom magnetic bearings (see Chapter 15). There is potential for incorporating attitude sensing with momentum storage and momentum bias in sophisticated devices of this type.

9.5   ATTITUDE MEASUREMENT

9.5.1   Attitude: its meaning and measurement

The meaning of 'attitude' or 'orientation' usually presents no conceptual difficulties. There must be some datum frame of reference, and once this has been chosen then the attitude of a spacecraft refers to its angular departure from this datum. A right-handed set of axes is normally used in order to define a frame of reference, and if both a datum set and a set of spacecraft axes are chosen, then the attitude may be defined in a way that may be quantified.

Specifying attitude may be done in a number of ways such as Euler angles, direction cosines, quaternions and so on [3]. Three pieces of information are needed. A common way is to use the three Euler angles that are defined in the same way as is standard practice for aircraft. These are the angles of yaw ψ, pitch θ and roll ϕ, as measures of the rotations about the z-, y- and x-axes, respectively, in that sequence, which are needed to bring the datum axes into alignment with those of the spacecraft. Figure 3.A2 in the Appendix of Chapter 3 illustrates these rotations.

For a spacecraft in circular orbit, whose z-axis is nominally down the local vertical and whose x-axis is nominally in the direction of travel, the aircraft's standard is frequently used. For other applications, a star-fixed (inertial) set would be better. There is, however, no universally accepted standard for specifying a spacecraft's attitude.

A potential problem when using Euler angles as defined above is that there is a singularity when the pitch angle θ is 90°. Whilst a set of angles $(\psi, 90°, \phi)$ may be chosen to specify any such attitude, it is not a unique set. For example, if the aircraft's set is used for Shuttle, namely, x is horizontal, z is vertically down and y completes the right-handed set, then when it is on the launch-pad its pitch attitude is 90° and its yaw and roll angles cannot be uniquely specified. Such a problem may be overcome by choosing a more suitable datum set.

It is worth noting that angles, and consequently attitude, are not vector quantities. The combination (ψ, θ, ϕ) should not be thought of as three components of a vector. On the other hand, the rates of change of $\dot{\psi}$, $\dot{\theta}$, $\dot{\phi}$ can be interpreted as vector quantities whose directions are along the (non-orthogonal) axes about which the rotations take place (see Figure 3.A2 in the Appendix of Chapter 3). Resolving $\dot{\psi}$, $\dot{\theta}$, $\dot{\phi}$ along spacecraft axes enables the components of the spacecraft's angular velocity ω relative to the datum axes to be expressed as

$$\omega_x = \dot{\phi} - \dot{\psi} \sin \theta$$
$$\omega_y = \dot{\theta} \cos \phi + \dot{\psi} \cos \theta \sin \phi \qquad (9.24)$$
$$\omega_z = \dot{\psi} \cos \theta \cos \phi - \dot{\theta} \sin \phi$$

The inverse relationship is

$$\dot{\psi} = (\omega_y \sin\phi + \omega_z \cos\phi)/\cos\theta$$

$$\dot{\theta} = \omega_y \cos\phi - \omega_z \sin\phi \qquad\qquad (9.25)$$

$$\dot{\phi} = \omega_x + (\omega_y \sin\phi + \omega_z \cos\phi)\tan\theta$$

When the angles are small, then $\dot{\psi} \approx \omega_z$, $\dot{\theta} \approx \omega_y$ and $\dot{\phi} \approx \omega_x$.

Equations (9.25) indicate how, by integration, the attitude in the form of the Euler angles (ψ, θ, ϕ) may be obtained from measured components of angular velocity. The singularity at $\theta = 90°$ shows up in the form of $\tan\theta$ and will lead to problems with the integration as θ approaches this value.

9.5.2   Measurement system fundamentals

Fundamentally, the measurement of attitude requires the determination of *three* pieces of information that relate the spacecraft axes to some datum set, whether they are in the form of Euler angles or in other forms. The measurement subsystem must include sufficient sensors to enable the information to be extracted with the necessary accuracy and with reasonable simplicity. This must be done in all phases of the mission.

There are *two categories of sensor*, and they are commonly used to complement each other in a measurement system:

- The *reference sensor* gives a definite 'fix' by measuring the direction of an object such as the Sun or a star, but there are normally periods of eclipse during which its information is not available.
- *Inertial sensors* measure continuously, but they measure only *changes* in attitude, effectively relative to a gyroscope. They therefore need a fix—a calibration from reference sensors. In between fixes, their errors progressively increase because of random drifts.

A *measurement system* may be formed by using reference and inertial sensors to complement each other. In a simple combination, the reference sensors will calibrate the inertial sensor at discrete times and the latter will then effectively 'remember' the reference object's direction until the next calibration. This allows a period in eclipse to be covered. The accuracy of the system will fluctuate, being that of the reference sensor at the calibration instant, and steadily degrading until the next calibration, as shown in Figure 9.10.

It is clear that the achievement of good system accuracy calls for good accuracy from the reference sensors, and a low degradation (drift) rate from the inertial sensors. In practice, it is likely that the mixing will take place in a computational Kalman filter to minimize errors, the design of which is a specialist topic.

Complete attitude information requires *three* pieces of information as explained above. Reference sensors that are based upon detecting the direction of a single vector are incapable of providing all three pieces. A sun sensor cannot detect any rotation of spacecraft about the Sun vector, for example. Two vector directions, ideally orthogonal, are needed for complete attitude information to be obtained from simultaneous measurements.

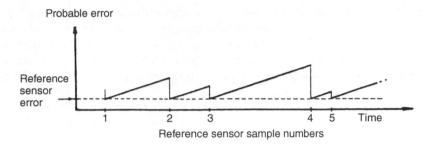

Figure 9.10   Error history for a mixed reference/inertial sensor system

A star sensor may have the ability to track more than one star within its field of view (FOV), thereby monitoring the two or more vector directions that are needed for full attitude information. The angle between these directions will be far from the ideal of 90°, however.

In some cases, the two different vector directions can be monitored by just one sensor, but with the measurements separated by a time interval. Then the change in attitude during the interval must also be measured, by using inertial sensors, and the time interval must be short enough for the build-up of errors from these sensors to be acceptable.

A star scanner is one example. It scans the heavens and so one sensor uses different star directions for complete attitude determination. The time lapse between viewing two stars in orthogonal directions will be short enough to avoid significant build-up of errors.

An Earth sensor (detecting the local vertical) can be used in a similar fashion. The local vertical rotates in space through 90° in a quarter of a circular orbit, a lapse of about 23 minutes in LEO and 6 hours in geostationary orbit. The Sun vector's '90° time' is three months—too long for use in this way! In practice, the Earth vector direction will be monitored continuously; the '90° time' merely gives an indication of the degradation in accuracy due to the sensor system's reliance upon inertial sensors.

The accuracy required will normally be set by the payload, its pointing direction and the required measurement accuracy. There will be an ultimate accuracy of measurement that is determined by the object used by the sensor; stars provide the most accurate sources, with the Sun and Earth being progressively less accurate by virtue of the angle that they subtend at the sensor and the fuzziness of the Earth horizon. A rough guide to accuracies is shown in Table 9.3.

Table 9.3   Potential accuracies of reference sensors

Reference object	Potential accuracy
Stars	1 arc second
Sun	1 arc minute
Earth (horizon)	6 arc minutes
RF beacon	1 arc minute
Magnetometer	30 arc minutes
Navstar Global Positioning System (GPS)	6 arc minutes

Note: This table gives only a guideline. The GPS estimate depends upon the 'baseline' used (see text).

The extent to which the ultimate accuracy is realized in a given instrument depends upon its design, and this will be related to its intended use. Each sensor will have a limited FOV, beyond which it gives no information. In getting a spacecraft to its intended attitude it will normally be necessary to include very wide-angle low-accuracy sensors for use when steering the craft towards the state in which the accurate sensor has its objective in its field of view.

Each phase of the mission must be addressed when the list of sensors is being compiled. In the early stages, the visibility conditions will be quite different from those when it is on station, partly because the orbit is different, but also because of stowed arrays and so on. The sensor list must also cover the possibility that the spacecraft attitude may need to be recaptured following a failure of some sort.

The *datum axes* that are used when defining the attitude of the spacecraft are normally related to its payload. A telescope is likely to need star-based datum axes for specifying its attitude, for example. For the control system, it is the error measured from the intended attitude that is important and in many systems, sensors will be chosen to measure this directly. For example, the pointing of the solar arrays towards the Sun will use a Sun sensor. Earth-facing hardware will possibly use the local vertical as a datum, or maybe RF beacons on the ground that have been set up for the purpose. The error in measuring the pointing direction of the payload will be reduced if the sensor uses a related reference object.

Errors in measuring the payload's pointing direction will also depend upon the physical separation between it and the sensor. If each is mounted on the same base, then the mounting tolerances of both instruments will be sources of error, and so will distortions of the base due to thermal or other effects. Error paths of this type should be kept to a minimum and must be carefully assessed when the location of sensors measuring payload-pointing directions is being considered. For extreme accuracy, the payload pointing will be calibrated in space.

9.5.3 Types of reference sensor

There are numerous different engineered forms of sensor. Only a brief review of the selection on offer is given here.

Sun sensors

The Sun subtends an angle of about 30 arc minutes at Earth, and provides a well-defined vector, which is unambiguous because of the intensity of the radiation. Sensors range from mere presence detectors that determine whether the Sun is in a specified FOV, to instruments that measure its direction to an accuracy of better than one arc minute. An example of the latter is the *reticle slits mask detector*, shown schematically in Figure 9.11. The sensor comprises a transparent block of known refractive index, coated in a thin opaque film. An entry slit is etched into the upper surface, and reticle slits in the base, and the block mounted on a sequence of photo-cell detectors. When exposed to the Sun, a plane of sunlight intersects the base so that some of the photo-cells receive photons giving a '1' output, whereas others do not and so are assigned a '0'. The digital output can be mapped onto a unique entry angle, using the Gray code, allowing the plane in which the Sun lies to be determined. Two such detectors, mounted on a three-axis stabilized

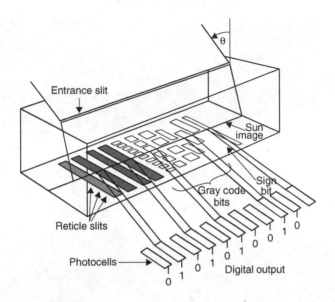

Figure 9.11 Sun sensor with digital output

spacecraft with their entry slits orthogonal to each other, give sufficient information to determine the spacecraft-Sun vector in the body axes. One detector mounted on a spinning spacecraft is sufficient to determine the solar aspect angle—between the Sun vector and the spin axis.

Another type of digital sun sensor again has orthogonal entry slits, which permits photons to be incident upon two orthogonally-mounted linear charge coupled device (CCD) arrays—again allowing the spacecraft-Sun vector to be determined.

An alternative method to determine the Sun vector on a spinning vehicle is to use a '*V slit*' *sensor*, an example of which is illustrated in Figure 9.12. This shows the sensor array on the *STRV*-1 (Space Technology Research Vehicle), which is a spinning microsatellite (~50 kg mass) built by QinetiQ at Farnborough, UK. The V slit Sun sensor is indicated by the arrow. The Sun is viewed through the two slits, set at an angle as illustrated. Detection of the Sun through the first, 'vertical' slit gives a measure of the rotational 'azimuth' of the spacecraft relative to the Sun. With knowledge of the spin rate of the vehicle, the time taken for the Sun to appear in the second slit, gives a measure of the 'elevation' of the Sun above or below the plane in which the instrument rotates. Other types of optical/detector combinations are available.

Earth sensors

The Earth, radius R_e, subtends an angle $2\sin^{-1}[R_e/(R_e + h)]$ at a spacecraft at height h. At 500 km altitude this is about 135° (so that an Earth-presence detector is not very useful!), and this falls to about 17.5° at geostationary height. The determination of the *nadir vector* (the spacecraft-to-geocentre direction) can be achieved, however, by sensing the position of the Earth's horizon, and Earth-horizon sensors provide the means of doing this. For example, the nadir vector can be determined by bisecting the directions to the horizon at the ends of a diameter of the Earth's disc.

Figure 9.12   The V slit Sun sensor on the STRV-1 microsatellite. (Reproduced by permission of QinetiQ Ltd)

Earth sensors are usually designed to operate in the infra-red, often in the $\sim 15\,\mu m$ CO_2 absorption band due to the following reasons:

- There is less variation between maximum and minimum radiance compared to the visible band.
- The *terminator* (the dividing line between night and day on the Earth's disc) disappears in the infra-red. Although the night-time is generally cooler than the day, the variation is small compared to absolute zero. The modelling of the complex and variable geometry of the visible terminator is also avoided.
- The 'infra-red Earth' is always present as a reference object, even when the spacecraft is in eclipse.
- However, *bolometers* (heat detectors) are required, which generally have a slow response.

The *static Earth-horizon sensor*, which is illustrated schematically in Figure 9.13, is a conceptually simple device used on three-axis stabilized spacecraft in high circular orbits, in particular, GEO. The infra-red image of the Earth is focused on the focal plane of the instrument. With nominal attitude, the image falls equally upon heat-detecting arrays, as shown. However, if the spacecraft attitude has errors in pitch or roll, the 'signal' will decrease in some arrays and increase in others. The instrument cannot detect errors in yaw (rotations about the local vertical).

A *scanning Earth-horizon sensor* is generally used in LEO to determine the nadir vector. Typically, a narrow FOV ('pencil beam') is swept across the Earth horizon to detect the abrupt change in infra-red signal between viewing 'deep space' and Earth. The

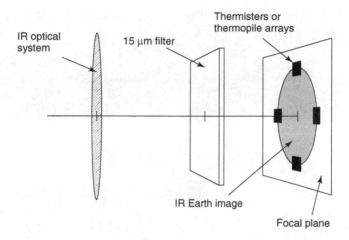

Figure 9.13 Schematic of a static Earth-horizon sensor

beam scans the horizon by means of an internal mechanism or by using the rotation of a spinning spacecraft. Figure 9.14 shows schematically a typical scan geometry. The pencil beam traces out a cone, of known semi-angle γ, that intersects the Earth horizon at two locations, one corresponding to acquisition of the infra-red signal and the other to the loss of signal. With knowledge of the scan rate, the measurement made is an estimate of the fraction of the scan period between acquisition and loss of signal. This, combined with

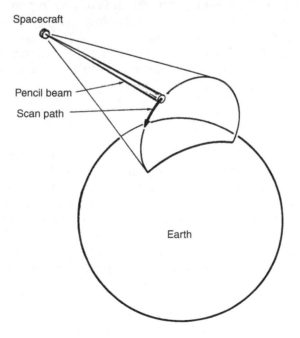

Figure 9.14 Scanning technique for an Earth horizon sensor

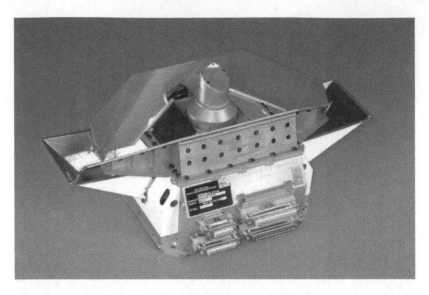

Figure 9.15 A scanning infra-red Earth-horizon sensor for LEO. The unit illustrated is the STD16 sensor. (Reproduced by permission of EADS SODERN)

knowledge of γ and the Earth's angular size, can be used to estimate the angle between the spin axis of the scan mechanism and the nadir vector. Figure 9.15 shows a scanning infra-red Earth-horizon sensor, which has two such scan cones on opposite sides of the spacecraft. If their pulses are of equal duration, then the local vertical lies in the plane bisecting the cone angles, and this can form the basis of a nulling control system for roll motion. Pitch sensing uses the fact that the local vertical lies in the plane defined by the bisector of the pulse from each conical scan.

For this system, one cone angle will be suitable for only a limited range of altitudes. In some sensors, the angle may be adjusted to accommodate the changes in altitude as the spacecraft moves from LEO to geostationary height.

Star sensors

Star sensors are the most accurate reference sensors in common use for measuring attitude. Accuracies of 1 arc second or better may be obtained. But the large number of stars means that sophisticated techniques are needed in the instrument and its associated computer in order to identify any particular star in its FOV. This problem is usually overcome by relying on a degree of *a priori* knowledge of the attitude of the spacecraft, or by adjusting the detection threshold of the sensor to reduce the number of 'targets'. Star sensors once had the reputation of being massive, power-hungry and expensive, but the advent of CCD technology has led to considerable improvements in their system characteristics.

Star sensors may be classified as

- *Star scanners* for mounting on a spinning spacecraft. The spin provides a means for the instrument to scan the sky. The characteristics of stars passing through the FOV can be compared with a star directory in order to determine attitude. A 'V slit screen' can be used (see section on Sun sensors) to determine the spacecraft-star vector.

- *Star trackers* for mounting on a three-axis stabilized spacecraft. Their detectors/ controllers enable them to select, locate and track one stellar image with precision. Clearly one such fix is insufficient to determine the spacecraft attitude uniquely. Two such trackers 'staring' in orthogonal directions, as used on the US Space Shuttle, will provide an optimal, unique attitude estimate.
- *Star mappers* for mounting on a three-axis stabilized spacecraft. Their FOV is suffi- cient to include several stars. They locate and record the position of each, so allowing the spacecraft's orientation about the sensor's axis to be determined.

Considerable improvements in star sensors, referred to above, have been achieved with the use of solid-state detectors. CCD and APS (active pixel sensor) detectors have led to lower mass, volume and power characteristics for the same performance in terms of accuracy. Figure 9.16 shows the Hydra star-sensing system, with three optical heads, developed by EADS SODERN. Each head operates using a 1024×1024 APS array, and can track 15 stars simultaneously at 10 Hz. Generally, star trackers/mappers provide information to refine knowledge of spacecraft attitude to high accuracy. This unit can also determine angular velocity, up to 10°/second, paving the way to gyro-less spacecraft operation. Its mass is 1.35 kg per optical head, plus 1.75 kg for the electronics unit, and a typical power consumption with three heads is 11 W at 20°C.

Figure 9.16 The Hydra APS star sensor, which integrates three optical heads into one system. (Reproduced by permission of EADS SODERN)

Radio frequency beacons

Direction-finding techniques may be used to detect the direction of an RF source, with an accuracy of order 1 arc minute. There are several techniques by which this can be done.

For example, Ulysses, rotating at a nominal 5 rpm, carried an antenna whose axis is offset from the spin axis. The intensity of the signal that it received from a ground station was thereby modulated at the spin frequency. The actual spin rate and its phase, and the angle between the spin axis and the ground station direction, was then derived respectively from the frequency, the phase and the depth of modulation.

Magnetometers

The magnetometer is a robust instrument but with an accuracy that is limited to about $0.5°$. It measures the direction and possibly the strength of the local magnetic field. But the field is not well mapped and has abnormalities that make the sensor of limited use for attitude sensing. To gain attitude information, the measured field is compared to a magnetic field model held in the on-board processor. The magnetometer is also used in conjunction with magnetic torquers as described in Section 9.4.2.

Attitude determination using global navigation satellite systems (GNSS)

GNSS, such as the Navstar GPS system, is commonly used for the determination of orbital position, but it can also be used to determine spacecraft attitude. Figure 9.17 illustrates the principle. A set of GPS patch antennae (at least three), slaved to a GPS receiver, are located on an 'upward looking' face of the spacecraft. If **L** is the baseline vector between two such antennae, and **e** is the unit vector along the line of sight to a GPS satellite P, then the range from P to each of the antennae differs by an amount

$$\delta\rho = \mathbf{L}.\mathbf{e} = L\cos\theta$$

The measurement made is the phase difference between the two antennae, which allows $\delta\rho$ and therefore θ to be estimated. A similar procedure using each antenna baseline and

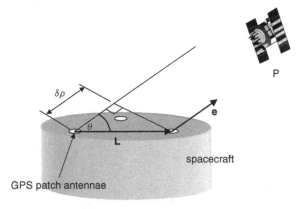

Figure 9.17 The principle of attitude determination using GNSS

multiple GPS satellites, allows the attitude of the spacecraft to be constructed. Accuracies of the order of ~0.1° are possible, although this will depend on the length of the baseline. Increasing this is not without its problems, however, as there is more flexure in long baselines that will compromise the accuracy. The phase difference measurement is also inherently ambiguous because of the unknown number of GPS signal wavelengths received between antennae. Finally *multi-path* — unwanted local reflections of the signal — can also pose problems. Despite these difficulties, however, attitude determination using GNSS is an established technique, which has been demonstrated in orbit [5].

9.5.4 Inertial sensors

Gyroscopes form the basis of the inertial sensing system for attitude. The conventional, mechanical gyro has a rotor mounted in a single gimbal in an environment that is very carefully controlled. In the rate- and rate-integrating types, the gimbal is torqued so that it follows the motion of the spacecraft. The torque is then a measure of the angular rate about the instrument's sensitive axis (see equation 3.29).

A set of three orthogonal rate-gyros will measure the components $(\omega_x, \omega_y, \omega_z)$ of the spacecraft's inertial angular velocity; a fourth at a skew angle is normally carried to avoid a single-point failure. The output of a rate-integrating gyro (RIG) is the integral of the angular velocity component, such as $\int \omega_x \, dt$ and so on. Only when the direction of a RIG axis remains fixed in space does its output represent the angular displacement about the axis. In more general motion equations (9.25) apply.

In high-quality sensors, drift rates of less than 0.001 degree/h are obtainable. Recent years have seen the development and introduction of gyroscopic sensors without moving mechanisms. The best known of these is perhaps the *Ring Laser Gyroscope (RLG)*. This comprises a small triangular prism of ceramic glass, the vertices of which are truncated to form optically flat surfaces. Laser light is then introduced and split at one vertex, so that two beams are internally reflected around the prism in opposing directions. If the prism is rotated around the axis normal to the triangle, then the path length of light traversing the prism in the same sense as the rotation is longer than that for the beam in the opposite direction. This difference, detected by examining the interference between the two beams, gives a measure of the inertial angular rate about the sensitive axis. Figure 9.18 shows a RLG test rig, in which the housing of the gyro pack has been cut away to reveal the triangular prisms within. Such a RLG pack usually comprises three such prisms, with mutually perpendicular sensitive axes, to give angular rates about each of the three spacecraft axes. The drift rate for such a device is typically of the order of 10^{-3} degrees per hour.

A device that uses a similar principle is the *Fibre Optic Gyroscope (FOG)*, which is generally less massive than the RLG, the need for the prismatic block having been eliminated. In this instrument, laser light is fed simultaneously into both ends of a long ($\sim$ km) fibre optic coil. When the coil rotates about its axis, the counter-rotating beams again travel different distances before they reach the detector, and the phase difference gives a measure of the rotation rate.

Another class of gyroscopic device is based upon the response of a mechanical resonator to the Coriolis force generated by rotation, first analysed by Bryan in 1890 [6]. The principle of operation of one such device, the *Hemispherical Resonator Gyroscope (HRG)*, can be thought of in terms of a resonating wine glass. When the glass is made to 'sing' (resonate), a standing wave is produced on its surface. When the glass is rotated

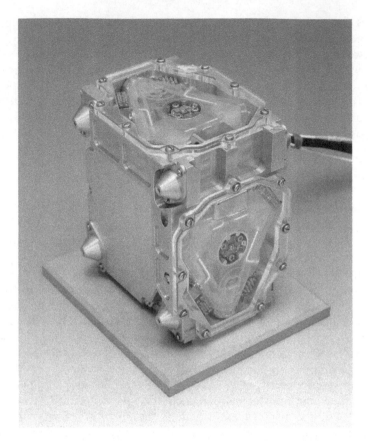

Figure 9.18 A ring laser gyro pack, showing two orthogonal triangular prisms.
(Reproduced by permission of SELEX Galileo, UK)

about the stem (the sensitive gyro axis), the nodes and anti-nodes of this wave precess
at a rate dependent upon the inertial angular velocity. The HRG therefore comprises a
hemispherical fused silica or quartz shell, which is electrostatically actuated and sensed to
determine the shell's response and measure the wave's precession. Such devices have been
used in flight (e.g. the *Cassini* mission to Saturn), and have a performance that matches
advanced RLGs. The principle of the HRG has been extended to other resonator geometries, to simplify manufacture. Gyroscopic devices, sometimes referred to as *Piezoelectric
Vibratory Gyroscopes (PVGs)*, are available. These units are relatively inexpensive, with
an overall size similar to, or smaller than a matchbox. This type of device makes use
of vibrational modes of a piezoelectric body, which can have a variety of shapes such
as a beam, a tuning fork or a circular ring. Generally two modes of vibration are used
which have material particles moving in perpendicular directions. One of the two modes is
excited by applying an alternating electric voltage. If the device is then rotated in inertial
space, the Coriolis effects will excite the second mode, through which the angular rate
can be estimated from the electrical signals resulting from the second mode. A review
of this type of device is given by Yang [7], to which the reader is referred for more
detail. Furthermore, very small devices based upon this principle can be produced using

advanced microfabrication techniques. At the time of writing, however, the performance of these microgyros is insufficient for many space applications.

9.6 ACS COMPUTATION

9.6.1 The computer

The development of digital computers for use in spacecraft has proceeded rapidly. They must perform reliably in the radiation environment of space, and many space-qualified processors exist. Further development is providing more power and speed, and the capability of being programmed in higher-level languages. These on-board computers (OBCs) link with ground control computers, which will normally host their software development tools. The availability of powerful computers means that spacecraft are given greater autonomy, and many of the sophisticated control techniques that find applications in ground-based systems may be used on spacecraft.

Robustness is a requirement for ACS and other on-board systems. For example, the ACS must potentially operate with large flexible structures such as solar arrays, whose natural frequencies cannot be established accurately before launch. Fixed algorithms will tolerate only limited variation from their expected value. The ability to reprogram the OBC from Ground Control permits any necessary adjustment of the control algorithms to be made following calibration of the spacecraft's parameters after launch. For full autonomy or immediate response to any changes that occur such as hardware failures, adaptive control techniques may be used. Indeed, the uploading of mission or control software is routinely performed to adapt to changing requirements or to enhance performance.

Computer power will also benefit the attitude measurement subsystem. The fusion of sensor data to achieve maximum accuracy *via* the Kalman type of filter requires computer modelling. In addition, they can provide the substantial data backup, which is needed when star mappers and scanners are used.

9.6.2 Active attitude control using a PID algorithm

The sophisticated techniques mentioned above are beyond the scope of this text, and for these the reader is referred to more specialized texts [3, 8]. However, there are many control situations that are satisfactorily achieved by using a simple *Proportional, integral and differential (PID)* algorithm, and the following discussion addresses the question of controlling a spacecraft's attitude using this.

The principle of active control involves the measurement of the spacecraft's attitude using installed sensors, and comparing this with the desired attitude as explained in Section 9.2.4. The *error signals* — the difference between the measured and the desired attitude — are then used in appropriate algorithms within the OBC to determine corrective torques (see Figure 9.2). Euler angles ϕ, θ, ψ are commonly used as measures of the errors when they are small and may also be used when only one of them is large, but alternative measures are direction cosines or quaternions.

A torque acting upon a spacecraft may cause an angular acceleration or, when momentum bias is used, an angular velocity. A PID algorithm can be used in both cases, but it is discussed here in the context of zero-momentum-bias when the torque causes an

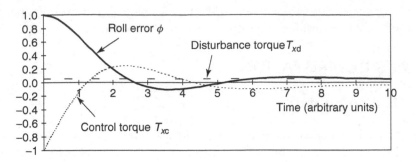

Figure 9.19 The response of the roll error ϕ and the control torque T_{xc} due to a PD algorithm (no integral term) in the presence of a disturbance torque T_{xd}

acceleration of the existing error towards zero—its desired value. Equation (9.14) shows that $I_{xx}\ddot{\phi} \approx T_x$, and similarly $I_{yy}\ddot{\theta} \approx T_y$ and $I_{zz}\ddot{\psi} \approx T_z$ under the constraints mentioned in the previous paragraph.

The algorithm attempts to make the pointing error—ε say—obey a differential equation of form

$$\ddot{\varepsilon} + 2\zeta\omega_n\dot{\varepsilon} + \omega_n^2\varepsilon = 0 \qquad (9.26)$$

In this, the undamped natural frequency ω_n (rads/sec) governs the speed of the response to a disturbance, and the damping ratio ζ, ideally of order 0.5, affects how rapidly the (typically) oscillatory response dies away. For example, the overshoot in the response is about 16% when $\zeta = 0.5$ (see Figure 9.19).

The implementation of a PID control algorithm for the roll error ϕ

Although we treat the roll angle ϕ in the remainder of this section, the equations for the pitch θ and yaw ψ take the same form.

The total roll torque T_x will in general include a disturbance torque T_{xd} as well as the control torque T_{xc} from the actuator, leading to

$$I_{xx}\ddot{\phi} = T_x = T_{xd} + T_{xc} \qquad (9.27)$$

Using a control torque T_{xc} that involves only the proportional and differential parts of the PID algorithm, namely,

$$T_{xc} = -K_{xp}\phi - K_{xd}\dot{\phi} \qquad (9.28)$$

and inserting this into equation (9.27) leads to

$$\ddot{\phi} + (K_{xd}/I_{xx})\dot{\phi} + (K_{xp}/I_{xx})\phi = T_{xd}/I_{xx} \qquad (9.29)$$

This has the same form as equation (9.26) except that its right-hand side is not zero.

The response of equation (9.29) to an initial error, caused by a command for a new orientation say, is shown in Figure 9.19. When the transient has died away, then the roll attitude will have an error $\phi = T_{xd}/K_{xp}$. This will reduce to zero when the integral term in the PID controller is included. The full algorithm, in general form, is

$$T_{xc} = -K_{xp}\phi - K_{xd}\dot{\phi} - K_{xi} \int \phi \, \mathrm{d}t \qquad (9.30)$$

Some practical issues are involved:

- The torquers above were assumed to be ones whose torque magnitude is controllable. RWs answer this description (Table 9.2). There will, however, be a limit to the amount of torque that these devices can provide. Therefore, high values of the PID constants will lead to a call for large torques to achieve a rapid response.
- The spacecraft will not be entirely rigid as was assumed in the theory above. Its structure will have oscillatory flexure modes, the natural frequencies of which will be very low if there are large flexible appendages such as solar arrays. Their damping ratios may be only of order 0.015—definitely stable, but only just. The output of all attitude sensors will include a measure of modal excitation, the amount depending upon their location within the modal shapes. It is important that the control algorithm does not lead to a destabilizing feedback to these modes. In large spacecraft, it will be necessary to include many modes in the mathematical model when designing the final form of the algorithms, but the modal data in this model is likely to be imperfect prior to launch. Some calibration and adjustment of the controller parameters may be necessary in orbit.
- The angular rate term $\dot{\phi}$ may be provided by a rate gyroscope. It may alternatively be calculated by differentiating the signal from a roll *angle* sensor. Any sudden repointing demand, however, will create an immediate change in the roll *error*, and differentiating this step function should be avoided if possible. In general, differentiation will amplify any high-frequency 'noise' and also the content of any high-frequency flexure modes in the signal. Some attenuation of this must be included in the differentiation process.
- When large repointing manoeuvres are required, cross-coupling between the roll, pitch and yaw motions may become detrimental. Lower values of the PID constants will alleviate this, but an alternative is to carry out these large manoeuvres as a sequence of roll followed by pitch and then yaw, say, so that only one of these angles is large at any one time.

In summary, although the preceding theory indicated no limit to the magnitude of the PID parameters K_{xp}, K_{xi}, K_{xd}, they will nevertheless be limited by the practical considerations, and possibly severely so.

On/Off control

Thrusters used for applying a torque for attitude control can only be used in an on/off manner, with different jets being used for positive and negative torques about each axis.

The pulses required for repointing a momentum bias direction as opposed to a zero-momentum bias axis are very different, as illustrated in Figure 9.7. The zero-momentum bias case used above will also be discussed here.

A PID algorithm controlling the level of torque is theoretically capable of producing zero attitude error, even when there is a constant disturbance torque, as has been shown above. With on/off control the limitation of having only three torque levels—clockwise, zero or anticlockwise—means that the control objective becomes 'the maintenance of an attitude error within acceptable bounds'. Furthermore, this must be done with a minimum of fuel usage.

With on/off control the error ϕ is expected to settle into a *limit cycle* under steady conditions as shown in Figure 9.20. A simple 'P' control law has been used, the appropriate torquer being switched on when the roll error exceeds a specified threshold value, $|\phi| \geq \phi_0$. With no momentum bias there is an angular acceleration while $|\phi| \geq \phi_0$, and constant angular rate otherwise. Under these conditions, the speed at which the error crosses between the boundaries where $\phi = \pm\phi_0$ is the same in each direction, and so the period and hence the fuel consumption will depend on its speed when the thrusters are first switched off. Figure 9.20 illustrates this. The control strategy shown in Figure 9.20(a) has twice the initial roll-rate as that of Figure 9.20(b), and consequently twice the fuel consumption per cycle, and a shorter period.

This shows that a switching signal that is proportional to the error alone is not satisfactory. However, it will be used to demonstrate some other aspects of on/off control.

The effect of a disturbance torque can be beneficial, as shown in Figure 9.21. Its optimum value is that which returns the error to the same switching level, $\phi = +\phi_0$ in

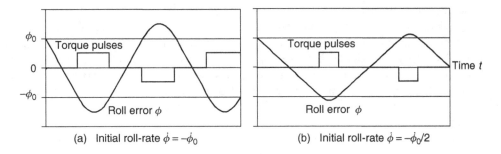

(a)   Initial roll-rate $\dot{\phi} = -\dot{\phi}_0$ (b)   Initial roll-rate $\dot{\phi} = -\dot{\phi}_0/2$

Figure 9.20   Roll error limit cycle for a simple P controller, in the absence of a disturbance torque

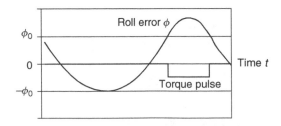

Figure 9.21   Roll error limit cycle with optimum disturbance torque

this case, without needing to switch on the reverse thruster. Greater disturbance torques will increase fuel expenditure.

The PID controller for an On/Off system

Figures 9.20 and 9.21 involve the torquer switching ON when $|\phi| \geq \phi_0$. Implementing this will be achieved by a computer-generated signal ε that switches on the appropriate torquer when $|\varepsilon| \geq \phi_0$. The equation for the switching signal above was $\varepsilon = \phi$, a PID control signal with the proportional 'P' term only.

The inclusion of a differential term to form a 'PD' controller means that as ϕ approaches the switching level $\phi = \phi_0$ say, the torquer will switch ON *before* reaching it, and as it returns it will again switch OFF early since the sign of $\dot{\phi}$ will have changed. This progressively reduces the speed of entry into the zero torque 'corridor' as shown in Figure 9.22. In theory, this could result in an indefinite zero angular rate within the desired corridor, but a disturbance torque will not allow this to happen.

The inclusion of an integral term, to create a full PID controller, can cause a response such as is shown in Figure 9.23 under a constant disturbance. It has the effect of lowering the switching boundary progressively under steady disturbance torque conditions, until the average error over a limit cycle is zero.

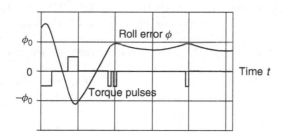

Figure 9.22   Roll error response with a PD controller, in the presence of a disturbance torque

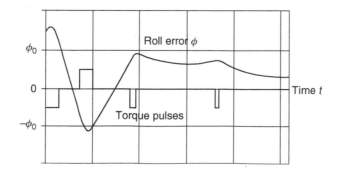

Figure 9.23   Roll error with full PID control, with a disturbance torque

Some practical considerations

- Torquer pulses will not be the instantaneous sharp-edged ones that have been assumed above. Furthermore, there will be a minimum switch-on time.
- Fuel consumption is an important issue, and there may be a compromise between achieving a short period limit cycle with small average error—'bouncing along close to a switching value' as in Figures 9.22 and 9.23—and making relatively large excursions across the switching 'corridor' as in Figure 9.21 in order to achieve as long a period as possible. Furthermore, the parameters of the PID controller can be modified if a change of mission emphasis occurs, say from 'high pointing accuracy' to 'fuel conservation'.

As we have indicated, there are a number of limitations to the brief treatment presented here. For a text in which these are addressed in the context of spacecraft control, the reader is referred to Reference [3]. More generally, the topic of automatic control has a vast literature. Reference texts include [9] and [10], the latter being a brief and approachable text for those wishing to acquire the basics rapidly.

REFERENCES

[1] Wertz, J. R. (1978) *Spacecraft Attitude Determination and Control*, Kluwer, Dordrecht, The Netherlands.
[2] Hughes, P. C. (1985) *Spacecraft Attitude Dynamics*, John Wiley & Sons, Inc., New York.
[3] Sidi, M. J. (1997) *Spacecraft Dynamics and Control—A Practical Engineering Approach*, Cambridge University Press, New York.
[4] Fortescue, P. W. and Belo, E. M. (1989) Control decoupling analysis for gyroscopic effects in rolling missiles, *J. Guidance, Control Dynam.*, **12** (6), 798–805.
[5] Chu, Q. P. and Van Woerkom, P. Th. L. M. (1997) GPS for low-cost attitude determination: A review of concepts, in-flight experience and current developments, *Acta Astronautica*, **41**, pp. 421–433.
[6] Bryan, G. H. (1890) On the beats in the vibrations of a revolving cylinder or bell, *Proceedings of the Cambridge Philosophical Society*, Vol. 7, pp. 101–111.
[7] Yang, J. (2005) A review of analyses related to vibrations of rotating piezoelectric bodies and gyroscopes, *IEEE Transactions on ultrasonics, ferroelectrics, and frequency control*, **52**, pp. 698–706.
[8] Gelb, A. (1974) *Applied Optimal Estimation*, MIT Press, Cambridge, MA.
[9] Nise, N.S. (2008) *Control Systems Engineering* (5th edn), John Wiley & Sons, Inc., New York.
[10] Schwarzenbach, J. (1996) *Essentials of Control*, Pearson Education Ltd, UK.

10 ELECTRICAL POWER SYSTEMS

John P. W. Stark

School of Engineering and Material Science, Queen Mary, University of London

10.1 INTRODUCTION

Provision of electrical power for space vehicles is, perhaps, the most fundamental requirement for the satellite payload. Power-system failure necessarily results in the loss of a space mission, and it is interesting to note that many of the early satellite systems failed due to such a loss. The demand for power has increased and is characterized by enhanced spacecraft operational complexity and sophistication. The earliest spacecraft, such as Vanguard 1, typically required a power raising capability of only ~1 W, whereas current communications satellites typically require three orders of magnitude greater than this. Evolving trends suggest that a further two orders of magnitude may still be needed. However, prediction of future power demand must always be treated with a great deal of caution. In the early 1980s there was great focus upon large systems, particularly with the infrastructure elements associated with space stations, containing both manned and unmanned elements. The potential for Solar Power Satellite systems (SPS), large orbiting power stations generating GW of electrical power for terrestrial use, was investigated in many studies, initially supported in the United States of America by both NASA and DoE, and in Europe by the European Space Agency (ESA) [1]. Political and technical issues have slowed down such developments, and very little research continues on such concepts [2], even within the context of concern over global warming. Manned systems do require higher power levels, perhaps of the order of 300 kW. Commercial communication services and broadcast systems are still seeing a slowly evolving trend to higher power levels, but enhanced system performance is reducing power demands. However, much focus is now also upon smaller satellite systems (see Chapter 18) where cost and system efficiency are critical system drivers.

The best methods of raising power can be broadly related to power level and mission duration as shown in Figure 10.1. It is apparent that photovoltaic (solar cells) or radioisotope thermoelectric generators (RTGs) are appropriate for the power requirements typical of present generation spacecraft, namely, tens of kilowatts for missions of several years. For shorter periods, fuel cells are advantageous, and for periods of less than a few days

Spacecraft Systems Engineering, Fourth Edition.
Edited by Peter W. Fortescue, Graham G. Swinerd and John P. W. Stark.
© 2011 John Wiley & Sons, Ltd. Published 2011 by John Wiley & Sons, Ltd.

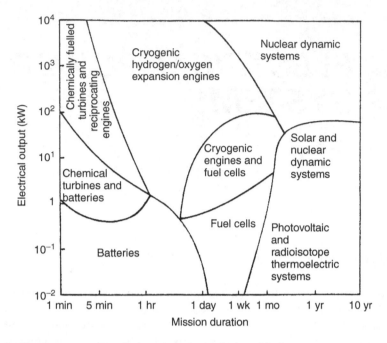

Figure 10.1 Power outputs: mission duration relationship between energy source and appropriate operational scenario [3] (From Angrist, S. W. (1982) *Direct Energy Conversion*, 4th edn, Copyright Allyn and Bacon, New York)

batteries come to the fore. It is not surprising, therefore, to discover that batteries are used in launch vehicles to provide the primary energy source, fuel cells are used in short duration missions of the order of a few days (e.g. the Space Shuttle) and both photovoltaic devices and RTG are used for general spacecraft operation, dependent upon the mission. It should be noted that nuclear sources of power such as RTGs and nuclear dynamic systems are used for military applications, but are not generally acceptable for civilian vehicles in Earth orbit. Indeed, during the 1990s, an increasing public awareness of environmental issues resulted in protests concerning the launch of nuclear-based sources for scientific purposes—for example, the launch of the Cassini–Huygens mission. Before the individual elements of a spacecraft power system are considered, the overall power system configuration will be described briefly.

10.2 POWER SYSTEM ELEMENTS

In general a spacecraft power system consists of three main elements: primary and secondary energy sources, and a power control/distribution network. These are shown schematically in Figure 10.2.

The *primary energy source* converts a fuel into electrical power. On early space flights and on launch vehicles, batteries have provided this. Strictly these systems do not have a fuel element, in that a battery is a device that stores energy rather than performing a direct energy conversion process.

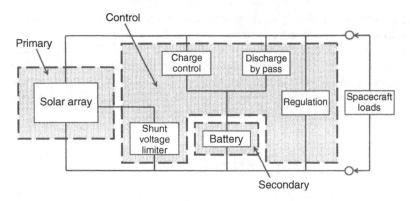

Figure 10.2 Schematic of typical spacecraft power system block elements

The majority of present-day spacecraft use a solar array as the primary energy source. The 'fuel' in this case is solar radiant energy, which is converted *via* the photovoltaic effect (see Section 10.3.1) into electrical energy. On manned missions of short duration, fuel cells have been used most frequently as the primary source; for longer duration flights associated with space stations such as Mir and ISS, the combination of solar arrays and fuel cells has been adopted. These electrochemical devices perform a controlled chemical reaction, in such a way that electrical energy may be derived rather than heat energy. The fuels used for space operation are hydrogen and oxygen yielding water as the reaction product: this may then be drunk by astronauts. Nuclear systems utilize either a radioactive decay process (RTG uses this) or a nuclear fission process as the energy source. RTG makes use of the thermoelectric effect, whereas fission reactors operate in a manner similar to terrestrial nuclear power plants [4].

The *secondary energy source* is required to store energy and subsequently deliver electrical power to the satellite system and its payload, when the primary system's energy is not available. The most usual situation when this condition arises is during an eclipse period when the primary system is a solar array. The eclipse's duration depends on the spacecraft orbit (see Section 5.3.2 of Chapter 5). Typically for Low Earth Orbit (LEO), a 35-minute eclipse occurs in each orbit for low-inclination satellites: in Geostationary Earth orbit (GEO), eclipses occur only during equinoctial periods, with a maximum duration of 1.2 h in a 24-hour period. For such short times, batteries demonstrate the highest efficiency. However, for systems that require high-power levels, typically 100 kW, a solar array/regenerative fuel–cell combination has improved characteristics over a solar array/battery combination. Regenerative fuel cells operate in a closed fuel cycle: H_2/O_2 fuel is consumed to form water on the 'discharge' cycle and electrolysis of water is performed during the 'charge' cycle, with power for this being derived from the solar array. The use of regenerative fuel-cells with a solar array has a particular application to lunar base scenarios, resulting in significant mass savings over battery/solar array options [5].

The *power control and distribution network* is required to deliver appropriate voltage–current levels to all spacecraft loads when required. Several salient features should be noted.

The primary power source always degrades during the mission. Thus, at its start an excess of power will be generated, and it is necessary to provide an ancillary load to dissipate it. The simplest way to do this is to use a resistive load, generally external to the main spacecraft structure in order to simplify the thermal design.

Both primary and secondary power system characteristics will change during the mission, leading to a requirement for voltage and/or current regulation. The changes arise due to both degradation effects, such as cell failures, and also illumination variations caused by changing solar array aspect angles with respect to the Sun. The customary approach is to use a voltage shunt regulator across the array.

Charge control of a battery system is particularly important to maintain the lifetime and reliability of battery units. It generally necessitates both current and voltage control. A variety of techniques may be used to sample the state of a battery and these will be discussed in Section 10.5. Discharge control is also required in order to limit current output.

This chapter provides an overview of each of these systems. For further detailed information, the reader is referred to two texts dedicated to spacecraft power systems [6, 7].

10.3 PRIMARY POWER SYSTEMS

10.3.1 Solar arrays

A solar array is an assembly of many thousand individual solar cells, connected in a suitable way to provide dc power levels from a few watts to tens of kilowatts. For a detailed description of both terrestrial and space solar arrays, Rauschenbach [8] is recommended.

Each *solar cell* assembly has a semiconductor $p - n$ junction as shown schematically in Figure 10.3. For spacecraft applications, the base material typically has a resistivity of between 10^{-3} and $10^2\,\Omega$ cm. Using silicon, for example, it might be doped with boron to form the p-type material (electron deficient), and with phosphorous for the n-type material (electron excess).

With no illumination, the junction achieves an equilibrium state in which no current flows. But when it is illuminated with suitable radiation, photons with sufficient energy will create electron–hole pairs, and the radiation is converted to a potential across the cell with usable electrical power. The incident photon energy needed for this must exceed a *band gap* that depends upon the material, as shown in Table 10.1. Photons with excess energy dissipate it as heat within the cell, leading to reduced efficiency.

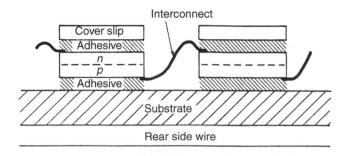

Figure 10.3   Schematic of a typical solar cell assembly

Table 10.1 Properties of semiconductor materials

Material	Band gap (eV)	Maximum wavelength(μm)
Si	1.12	1.12
CdS	1.2	1.03
GaAs	1.35	0.92
GaP	2.24	0.554
CdTe	2.1	0.59

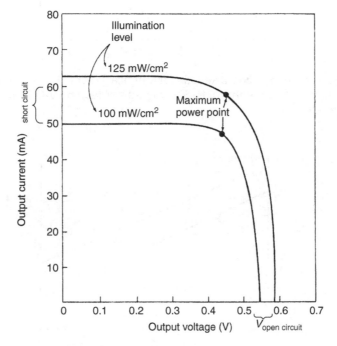

Figure 10.4 Current–voltage characteristic for a typical solar cell. The short-circuit current is dependent upon both the illumination level and the size (area) of the cell (From Angrist, S. W. (1982) *Direct Energy Conversion*, 4th edn, Copyright Allyn and Bacon, New York)

Characteristic voltage-current curves for cells are shown in Figure 10.4. Typically open-circuit voltages for silicon cells lie between 0.5 and 0.6 V under solar illumination. The plot of power against voltage (Figure 10.5) has a clear maximum, with a particularly rapid fall once the optimum voltage is exceeded. On some spacecraft maximum-power-point tracking is used to operate the array most efficiently. Increase in cell temperature results in decreasing open-circuit voltage with only a modest increase in short-circuit current. The theoretical maximum efficiency of both silicon (Si) and gallium arsenide (GaAs) cells is shown as a function of temperature in Figure 10.6. It shows the particular sensitivity of Si to temperature and also the improved performance of GaAs at high temperatures. GaAs is increasingly replacing Si as the material selected for satellite solar arrays, as for example

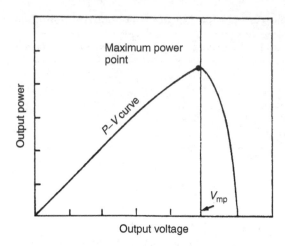

Figure 10.5   Power–voltage characteristic for a typical solar cell (From *Solar Cell Array Design Handbook* by Rauschenbach, H. S. Copyright © 1980 by Van Nostrand Reinhold. All rights reserved)

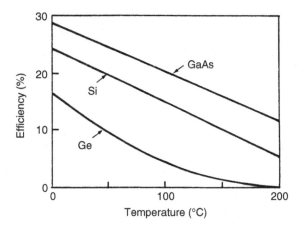

Figure 10.6   Theoretical cell efficiency as a function of temperature for three semiconductor materials (From *Solar Cell Array Design Handbook* by Rauschenbach, H. S. Copyright © 1980 by Van Nostrand Reinhold. All rights reserved)

on the Iridium system and more recently the Spacebus 4000 geostationary communications satellite bus. Triple junction GaAs technology can currently achieve an efficiency of between 28% and 30% in production level cells. The high temperature potential for efficient energy conversion with GaAs identifies the potential use of a focusing optical arrangement for GaAs cells such that these cells may be illuminated by an intensity greater than the nominal radiation intensity of $1.4\,kW/m^2$ in Earth orbit. Fewer cells would then be required to provide a given power level, which could result in array cost reductions since the cell cost is a large proportion of the total cost. Concentration ratios as high as 100 : 1 for GaAs cells have been investigated in such studies. Systems have been flown, for example the Deep Space 1 mission (1998), that had a concentration ratio of 8. Improvements in design since then have identified the capability for greatly improved specific power (300 W/kg), benefiting additionally from increased screening from damaging radiation (see below) making such arrays suitable for operation in the Jovian environment [9].

The *n-type contact* on the upper surface of the cell is in the form of a multiple-finger arrangement. These fingers are required for efficient current collection, whilst maintaining good optical transparency (typically ~60%). They are connected at a bar, along one edge of the cell. Frequently titanium/silver (Ti/Ag) is used for this.

Radiation damage is a problem with solar cells. In general, Si cells having higher base resistivity (~10 Ω cm) are the most tolerant of radiation. However, cells with a lower base resistance (~2 Ω cm) have a higher short circuit voltage, leading to fewer cells in series to achieve the operational voltage. Selection of material is therefore mission dependent. Furthermore, cells have the *n*-type material uppermost, since on early space flights it was discovered that cells having the *p*-type material as the upper region rapidly suffered from radiation damage. Thin cells suffer less than thicker ones, but at present they have a lower conversion efficiency. GaAs cells are more radiation tolerant than Si and for this reason there is considerable interest and effort in their development.

The *cover glass* provides environmental and radiation protection. For design purposes, the particle fluence of a spacecraft's radiation environment may be expressed as an equivalent fluence of monoenergetic 1 MeV electrons (see Figure 10.7). Degradation of cell output to this irradiation is generally available from manufacturers' data; Figure 10.8 shows typical degradation curves for cells with a variety of thicknesses.

The effectiveness of the cover glass depends on its density and thickness. Suitable glass microsheet is commercially available in several thicknesses from 50 μm to 500 μm [10]. Their absorption of radiation follows approximately an exponential law, so that the intensity of radiation after traversing a depth x into the glass is

$$I \sim I_0 e^{-k\rho x} \tag{10.1}$$

where I_0 is the initial radiation fluence at $x = 0$, ρ is the density of material and k is an energy-dependent absorption coefficient. For fused silica, Figure 10.9 shows the effect of changing glass thickness. It should be noted that in order to evaluate the total radiation fluence absorbed by the solar cell, both front-side and rear-side fluences must be calculated. On rigid panel arrays mounted on honeycomb structure, and on solar arrays bonded to the spacecraft wall (e.g. spinning spacecraft) the rear side is effectively screened from radiation, whilst on lightweight arrays this is not the case. Using the data supplied in Figures 10.7–10.9, it is possible to derive the area of active solar cells required to meet a specific mission requirement of end of life (EOL) performance. Suppose that an output of 1 kW is required at EOL for a satellite in a circular, equatorial orbit at 1000 km altitude. Assume that the cell to be used is made of silicon, 150 μm thick, having the properties shown in Figure 10.8. The mission duration is to be 5 years. Note that in the discussion that follows, the numerical values are approximate.

The starting point is to use Figure 10.7 to find the total damage equivalent 1 MeV electron fluence for a cell protected by a 150 μm cover slip. At 1000 km, this Figure shows that the damage equivalent due to protons is 1.7×10^{14} electrons/cm/year and that due to electrons is 2×10^{12} electrons/cm/year. The total flux in 5 years is therefore

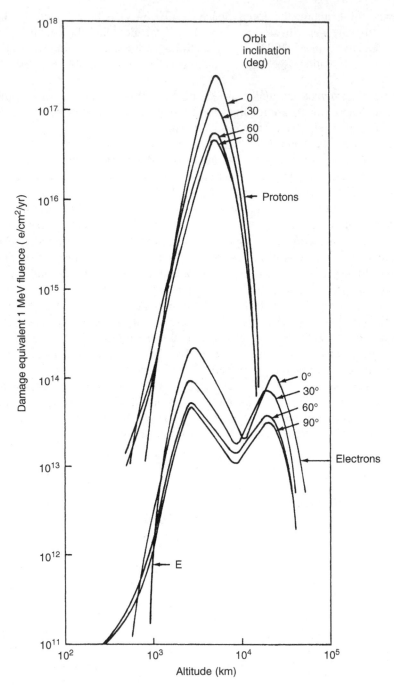

Figure 10.7 Damage equivalent 1 MeV fluence caused by electrons and protons due to trapped particles, to silicon cells protected by 150 μm fused silica covers and infinitely thick rear shielding [8] (From *Solar Cell Array Design Handbook* by Rauschenbach, H. S. Copyright © 1980 by Van Nostrand Reinhold. All rights reserved)

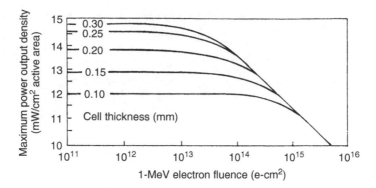

Figure 10.8 Effects of thickness and fluence on conventional non-p$^+$ silicon solar cell performance [8] (From *Solar Cell Array Design Handbook* by Rauschenbach, H. S. Copyright © 1980 Van Nostrand Reinhold. All rights reserved)

$5(1.72 \times 10^{14})$ electrons/cm/year, or 8.6×10^{14} electrons/cm/year. It is evident that the damage due to protons is much greater than that from electrons, as noted in Section 2.3 of Chapter 2. From Figure 10.8, the power per unit area is 11.5 mW /cm^2. It is however, noticeable from this Figure that a significant deterioration in the performance of the cell is evident at such a high radiation dose. A reduction in dose may be achieved by increasing the cover slip thickness. For example, if a cover slip of 500 µm were to be used, then the flux will be reduced according to the data presented in Figure 10.9 by a factor of 0.6. Thus the expected radiation dose absorbed in the cell after 5 years would then only be 5.2×10^{14} electrons/cm, yielding an EOL performance of 12 mW/cm^2, an improvement of approximately 4%. The mass increase associated with the use of this cover slip, assuming as a first approximation that the density of the cover slip is the same as that of the cell, will be (mass of 500 µm cover slip plus 150 µm cell)/(mass of 150 µm cover slip plus 150 µm cell), equivalent to a factor of nearly 2.2. This increase in mass needs to be considered however alongside the cost increase associated with the alternative of a 4% increase in the number of cells. The thinner cover slips result in an active area of $(1000/0.0115)$ cm^2, or 8.7 m^2, whilst the thicker protection requires an area of only 8.3 m^2.

Additional features required of the cover glass are that it provides good optical coupling between free space and glass and also between glass and adhesive, and that it provides suitable wavelength selection, limiting the UV flux to the adhesive layer and the cell. These features are achieved using an anti-reflection coating, such as magnesium fluoride on the upper surface, perhaps with an additional indium oxide conductive coating as described in Chapter 2. A UV-filter coating may be applied to the underside of the cell, to reflect UV radiation. For a cover glass with cerium doping, additional UV filtering is unnecessary.

For *efficient cell operation* and insensitivity to radiation, a shallow junction depth (typically less than 10 µm) is required. Various Si cell configurations have been investigated to improve conversion efficiency. These include the following:

- Back-surface reflectors (BSRs), used, for example, on the ISS and the Spot series of spacecraft, to reflect unabsorbed radiation from the rear side of the (p)-region back through the cell. This reduces cell heating.

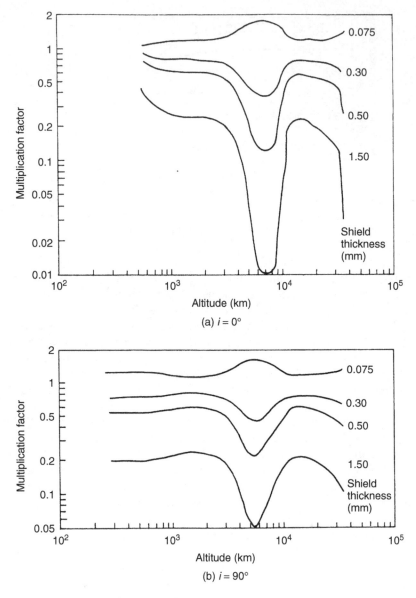

Figure 10.9 Multiplicative factors to be applied to damage fluence on a solar cell as a function of cover slip (shield) thickness, and operational orbit height (From *Solar Cell Array Design Handbook* by Rauschenbach, H. S. Copyright © 1980 Van Nostrand Reinhold. All rights reserved)

- The introduction of a p$^+$-region or back-surface field (BSF) at the rear of the p-regions. This exhibits higher output due to enhanced carrier collection efficiency, but the improvement is lost under high fluence damage. This technology has been used on the array for the Hubble Space Telescope (HST) and Envisat.
- The use of a textured front surface of the cell reduces reflection from the cell surface.

Solar arrays using Si cells are made from individual cells that historically have been rectangular, $2\,cm \times 4\,cm$, having a conversion efficiency of $\sim$12 to 14%. To improve packing efficiency, much work on high-efficiency large area Si cells has been completed, with beginning of life efficiency in excess of 20%. The cells used on ISS are 8 cm square with an efficiency of $\sim$13% [11]. The cells have a thickness of between 50 and 250 μm, $\sim$200 μm being used for the majority of arrays. Silicon cells having a thickness of 50 μm are now able to convert with a power density of $120\,W/m^2$; thinner GaAs cells $\sim$5 μm have a beginning of life efficiency of $\sim$28%, with 25% at the end of life for a 15 year GEO-type mission. There is much development now on highly efficient multi-junction thin film cells. Whilst the intrinsic efficiency of such cells is lower (of order $\sim$10−15%), the specific power is greatly improved with figures quoted as high as $1000\,W/kg$ [12]. Further emphasis on thin film technology for solar arrays is continuing as reported through the Space Photovoltaics Research and Technology Conference series [13].

Interconnections between cells represent a major array failure hazard. This arises because of the thermal cycling inherent upon entry/departure from sunlight to eclipse. Since the materials used for cell and substrate are different, differential expansion takes place during the rapid temperature change ($\sim$100°C in a few minutes). Thermal stress-relieving loops are required to reduce such failure mechanisms as interconnect lift-off and fracture.

Atomic oxygen effects on exposed interconnects have been mentioned earlier, in Chapter 2. A particular problem of the solar cell interconnection is that historically these have always been made of thin silver foil. Silver has a high capture efficiency for atomic oxygen, resulting in the formation of a variety of silver oxides. The process results in thinning due to flake-off of the oxides and hence an increase in interconnection resistivity. This leads to a loss of power. In the case of the HST array, the interconnects were originally designed to be pure silver. The delay in the original launch date (because of the Challenger accident) permitted a redesign in which silver was used only as a surface layer on a molybdenum interconnect. During flight, the silver eroded revealing the molybdenum, which is oxidation resistant. The power output from the array was thus maintained at the predicted value for the life of the mission.

A variety of *substrate materials* have been used and proven in space. Frequently Kapton with glass- or carbon- fibre reinforcement, $\sim$100 μm thick, forms the immediate interface with the cell, which may be mounted on a honeycomb panel for rigidity (e.g. Tracking and Data Relay Satellite (TDRS) solar array). Flexible cell blankets have also been used to reduce mass. The original HST solar array was a typical example, with glass fibre reinforcing the Kapton. In general, for the largest arrays, flexible substrate materials offer mass savings. But it is interesting to note that early work [14] has shown that for power levels up to 6 kW, mass savings occur when using advanced rigid arrays, wherein Kapton is reinforced with carbon fibre. Dutch Space has continued the development of advanced rigid solar arrays using both Si and GaAs, with in-orbit power (beginning of life (BOL)) in excess of 8 kW.

As noted earlier, individual cells produce power at a voltage of $\sim$0.5 V, and it is necessary to connect many cells in series. Reliability is then achieved by additional parallel coupling at each cell; typically three or four cells form a parallel combination. This series−parallel arrangement is called a *solar cell string*. Further protection is afforded using shunt diodes that provide current bypass paths should individual cells become shadowed. Shadowing can cause cell failures since if a cell is unable to generate power because of loss of illumination, then the entire string voltage may appear as a reverse bias voltage across the cell.

System level interactions

Now consider the *system level interactions* between the solar array design and the vehicle itself. The relatively low conversion efficiency of an array results in the need for large areas of solar cells to intercept sufficient solar radiation for the power demand. Examples of typical array configurations, associated with generic stabilization types are illustrated in Chapter 9.

For a spinning satellite, using either dual spin or simple spin ACS, the 'drum' size evidently limits the power that can be generated. The drum itself is limited by the launch volume. A possible solution is to introduce an additional mechanism such as the 'drop skirt' as used on Intelsat VI, where a larger-diameter hollow cylinder is deployed to expose additional cell area. Thruster plume impingement in such a configuration can, however, cause both disturbance torques and contamination.

Other aspects to note for the spinning satellite solar array are coupled with the thermal environment. Assuming a typical spin rate of ~50 rpm, the average temperature of the array can be maintained at a lower value than for the three-axis configuration. This leads to an increased efficiency of the individual cells, and thus a decrease in the required active cell area. However, since not all the array is instantaneously illuminated, a factor of ~π times the number of cells is required for the same collected power. Since also on the spinner, the array is mounted essentially on the body of the spacecraft, the temperature excursions noted on the array (between sunlit and eclipse phases) are reduced, having a significant impact on reducing the thermal shock characteristic of a three-axis deployed solar array.

Considering the three-axis-stabilized satellite, the solar array requires a mechanism to deploy the stowed array following launch and then orientate it appropriately to track the Sun. These deployment mechanisms may be of a simple extending telescopic construction, or of the 'Coilable' variety (See Chapter 15). In the Astromast, the carbon-fibre members of the mast are deformed by coiling them into a stowage container. They may then be driven out by a screw mechanism to the un-deformed (extended) configuration.

Tensioning wires are then required to achieve an acceptable minimum fundamental frequency of the array largely because of AOCS requirements. On three-axis-stabilized vehicles, power take-off from the array generally, but not always, requires a rotary degree of freedom between the satellite and the array, in order that pointing requirements of the array and the payload may be met. This requires two elements—the mechanical rotation device to allow the body to move relative to the array (with appropriate sensing systems) and an electrical power take-off device (See Chapter 15). The provision of power using a solar array clearly has many design interactions with the rest of the spacecraft system.

Table 10.2 summarizes performance characteristics for several solar arrays. A method for array sizing is given in Section 10.6.

10.3.2   Fuel cells

Fuel cells provided the primary power source for the Shuttle orbiter. Originally they were designed as part of the Mercury, Gemini and Apollo US manned missions. Table 10.3

Table 10.2 Solar array performance figures

Array	Type	BOL power (kW)	Specific power (W/kg)	Power density (W/m^2)
XMM	Rigid	2.5	32	215
Astra 2B	Rigid	9.2	52	409
Comets	Flexible	6.3	34	146
ISS (Wing)	Flexible	32	29	135

Note: XMM: X-ray multi-mirror mission.

Table 10.3 Performance summary of fuel cells for space use

System	Specific power (W/kg)	Operation
Gemini	33	
Apollo	25	
Shuttle	275	2500 h at P_{ave}
SPE technology	110–146	>40 000 h
Alkaline technology	367	>3000 h
Alkaline technology	110	>40 000 h

Note: SPE solid polymer electrolyte.

shows how their performance has evolved since the earliest days of manned space flight. A fuel cell converts the chemical energy of an oxidation reaction directly into electrical energy, with minimal thermal changes. From a system viewpoint, a major advantage is its flexibility. For example, it provides power during both sunlit and eclipse periods, and the fuel has a high-energy density and thus provides a compact solution compared with a solar array. The evident disadvantage is the need to carry fuel.

The *hydrogen/oxygen fuel cell* has been used for space applications, a product of the reaction being water. This is clearly useful for manned missions. A schematic diagram of such a cell is shown in Figure 10.10. This technology has also been proposed for lunar base requirements in association with solar arrays [5] and lunar rovers [15].

The voltage that appears at the terminals of an ideal cell is given by

$$E_r = \frac{-\Delta G}{nF} \tag{10.2}$$

where ΔG is the change of Gibbs free energy occurring in the reaction, n is the number of electrons transferred and F is the Faraday constant (product of Avogadro number and elementary charge) equal to 9.65×10^4 C/mol. For the reaction of the H_2/O_2 cell, two electrons are transferred per mole of water formed and ΔG has the value of -237.2 kJ/mole at 25°C. The reaction takes place spontaneously.

Thus the reversible voltage of the ideal cell is $237.2 \times 10^3/(2 \times 9.65 \times 10^4) = 1.229$ V.

In practice this is not realized because there are various irreversibilities, termed polarization losses. Figure 10.11 shows a typical current–voltage curve for a hydrogen/oxygen

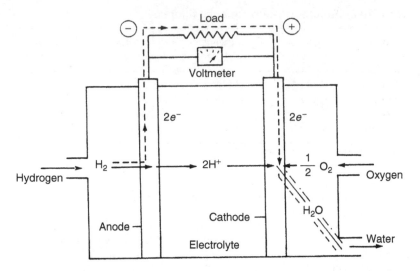

Figure 10.10 Schematic of a hydrogen/oxygen fuel cell. At the anode–electrolyte interface, hydrogen dissociates into hydrogen ions and electrons. The hydrogen ions migrate through the electrolyte to the cathode interface where they combine with the electrons that have traversed the load [3] (From Angrist, S. W. (1982) *Direct Energy Conversion*, 4th edn, Copyright Allyn and Bacon, New York)

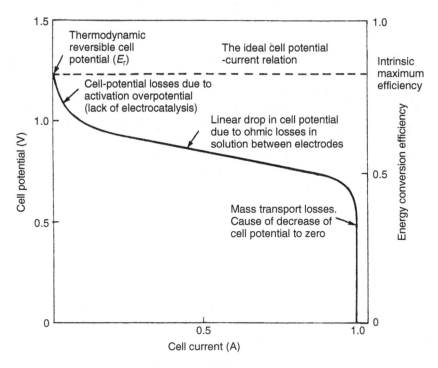

Figure 10.11 Typical cell potential and efficiency–current relation of an electrochemical electricity producer showing regions of major influence of various types of overpotential losses (Source [16])

fuel cell. Initially, as soon as a current is drawn from the cell, a rapid drop in voltage occurs. This is associated with the energy required to activate the electrode reactions. For the H_2/O_2 fuel cell these are three-phase: gas (fuel), solid (electrode) and liquid (electrolyte—this can also be a solid). It is necessary for the reactants to be chemisorbed onto the electrode, a process that requires breaking and forming new chemical bonds, and energy; hence the voltage drops. The process is called activation polarization. The magnitude of voltage drop is given by the Tafel equation:

$$\Delta V_{ACT} = a + b \ln J \tag{10.3}$$

where J is the current density at the electrodes, and a and b are temperature-dependent constants for the reaction/surface description.

As the current drawn increases, a linear voltage–current relation is noted. This is simply because of the resistive nature of the electrolyte.

At high current values problems arise because of the transport of reactants to the reaction sites, a feature that is particularly important at porous electrodes since pressure gradients are set up that limit flow rates. Additionally, species concentrations are not uniform and thus ionic species can create a back emf. This concentration polarization provides the ultimate limit on current density that may be achieved by a fuel cell. All of these electrochemical polarization processes are common to both batteries and fuel cells, and their voltage–current characteristics are very similar.

Early fuel-cell systems were primarily based upon the technology of solid polymer electrolyte (SPE). For the Gemini series, 1 kW was produced at a specific power of 33 W/kg, within a volume of $0.05\,m^3$, and the objective to extend missions to greater than four days was achieved. However, the water produced was not of drinking quality because of degradation of the fuel-cell membrane.

The Apollo system, also used for Skylab, was based upon matrix aqueous alkaline technology and achieved a power level of 1.5 kW at a specific power of 25 W/kg. It had to operate whilst the vehicle was on the lunar surface, at a temperature greater than 394 K. The selected system, a Bacon fuel cell, operated at 505 K.

Shuttle developments, also based upon the alkaline technology, improved the specific power by an order of magnitude, $\sim$12 kW, 275 W/kg. Further, the start-up time for this cell was 15 min with shutdown being instantaneous, whereas for Apollo, 24-hour start-up periods were required with 17-hour shutdown. Table 10.3 summarizes past and present fuel-cell status. Regenerative fuel cells wherein water is also electrolysed are not yet space-proven.

10.3.3   Radioisotope thermoelectric generators (RTG)

For deep-space missions, the use of fuel cells is precluded by their long duration. Solar arrays produce less power as they move away from the Sun, by a factor of approximately $(r_E/r_{S/C})^{1.5}$, where $r_{S/C}$ and r_E are the distances from the Sun to the spacecraft and to the Earth respectively. This factor comes about from the combined effect of the reduction in the intensity of the illumination from the Sun as the spacecraft moves away from it, partly offset by the beneficial effect of the reduction in the temperature of the solar cells.

For spacecraft travelling further than Jupiter solar arrays show disadvantages from a system viewpoint, compared with radioisotope generators.

The operation of a RTG is based on the thermoelectric effect noted by Seebeck, that it is possible to generate a voltage between two materials, A and B (either conductors or semiconductors) if a temperature difference is maintained (see Figure 10.12). This is analogous to a thermocouple. Practical RTG space systems utilize two semiconductor materials—one p-type, the other n-type—in order to exploit the effect.

The power output from such a device is a function of the absolute temperature of the hot junction, the temperature difference that may be maintained between the junctions and also the properties of the materials. Because such devices are relatively inefficient (less than 10%), one major problem in their design is removing waste heat.

The heat source used in space systems is derived from the spontaneous decay of a radioactive material. As this decays, it emits high-energy particles that can lose part of their energy in heating absorbing materials. Suitable fuels are listed in Table 10.4, which shows the half-life ($\tau_{1/2}$) for each of the fuels, namely, the time required for the amount of a given radioactive isotope in a sample to halve. Thus over a period of time t, the power available from such a fuel decreases by an amount given by

$$P_t = P_0 \exp\left(\frac{-0.693}{\tau_{1/2}}t\right) \tag{10.4}$$

where P_t is the power at time t after some initial time t_0.

Table 10.4 indicates that high specific power levels are available from sources with shorter half-lives (and hence shorter duration missions). For deep-space missions a long

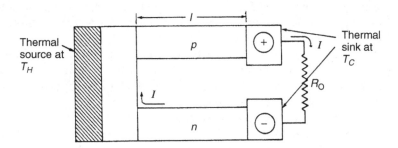

Figure 10.12 Schematic diagram of a semiconductor radioisotope generator (From Angrist, S. W. (1982) *Direct Energy Conversion*, 4th edn, Copyright Allyn and Bacon, New York)

Table 10.4 Possible fuels and their performance for radioisotope generators (From Angrist, S. W. (1982) *Direct energy conversion*, 4th edn, Copyright Allyn and Bacon, New York)

Isotope	Fuel form	Decay	Compound power density (W/g)	$\tau_{1/2}$ (yr)
Polonium 210	GdPo	α	82	0.38
Plutonium 238	PuO$_2$	α	0.41	86.4
Curium 242	Cm$_2$O$_3$	α	98	0.4
Strontium 90	SrO	β	0.24	28.0

life isotope is essential; for example, the design life for the Cassini–Saturn orbiter is 11 years, after which time the electrical power source is required to be 628 W. For these missions Plutonium is used exclusively.

The advantages of RTGs over other systems include the following:

1. They provide independence of power production from spacecraft orientation and shadowing.
2. They provide independence of distance from the Sun (deep-space missions are possible).
3. They can provide low power levels for long periods of time.
4. They are not susceptible to radiation damage in the Van Allen belts.
5. They are suitable for missions with long eclipse periods, for example, lunar landers.

The disadvantages of RTG systems need also be considered, and include

1. They adversely affect the radiation environment of the satellite whilst in orbit. This will influence the spacecraft configuration significantly as may be seen from Figure 10.13, which shows the Galileo spacecraft. In this instance, the RTG needs to be deployed on a lengthy boom away from the main satellite bus.
2. Careful handling procedures are required during satellite integration owing to the radiation hazard posed by the radioactive source.
3. High temperature operation is required for efficient energy conversion. This impacts upon the thermal environment of the vehicle, and again on vehicle configuration.
4. RTGs are a source of interference for plasma diagnostic equipment that may be carried as part of the scientific objectives of the mission.

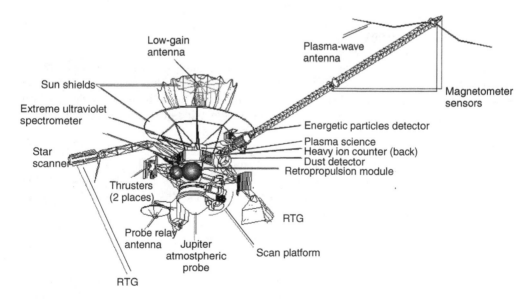

Figure 10.13   The Galileo spacecraft configuration, showing the position of the RTG sources (Courtesy of NASA/JPL/Caltech)

Table 10.5 RTG system performance

Name	Power (W)	kg/kW
Cassini (1997 launch)	628	195
Galileo probe/Ulysses (GPHS RTG, late 1980s)	285	195
Nimbus/Viking/Pioneer (SNAP 19, mid 1970s)	35	457
Apollo lunar surface experiment:		
SNAP-27, early 1970s	25	490
SNAP 9A, 1960s	73	261

5. At the political level there has been increasing concern expressed at the inclusion of radioactive material on board a satellite. This is principally of concern because of the potential for such a source to be dispersed in the atmosphere, should there be a launch failure.

System for Nuclear Auxiliary Power (SNAP-19), which powered the Viking lander vehicle to Mars, had a specific power of 2.2 W/kg, with a thermal/electric efficiency of ~5%. The output electrical power was 35 W. Table 10.5 summarizes data on RTG systems.

10.3.4 Other primary power systems

Two other primary power systems have been developed for operation—nuclear fission and solar heat. The former has been extensively used in the former Soviet space programme for military purposes. The latter has not flown in space yet, but a substantial effort has taken place in recent years in developing such systems for potential use on the International Space Station.

Nuclear fission systems

These systems operate in a similar way to conventional ground-based nuclear power stations, in that fissile material such as uranium-235 is used as a heat source. In space systems, this is used to drive a thermoelectric converter as noted in the preceding section. Specific features of space-based systems relate to the fail-safe requirement, particularly during launch.

Whilst the USA has invested substantially in the SNAP systems (even number SNAP designates a reactor system), these have not been used regularly. The main focus of their activity in recent years has been on the SP100 system [17]. It is notable, however, that the purchase of Russian technology by the US suggests that this US programme may not be pursued vigorously in the future.

Solar heat systems

The use of solar energy directly in the form of heat can provide system advantages. The heat energy can be used to drive a heat engine and then a rotary converter to electricity (solar dynamic), or directly be used as a heat source for a thermoelectric converter (solar thermoelectric).

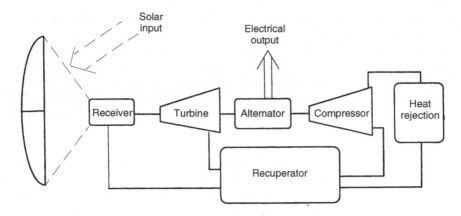

Figure 10.14   Solar dynamic Brayton cycle

Solar dynamic systems have had the greatest concentration of effort for the ISS. Design studies show that their end-to-end conversion efficiency is approximately 25% greater than for photovoltaics. This results in a reduced need for deployed collection area by about 25%, and consequently in reduced aerodynamic drag for LEO satellites. In the original concept for the ISS, primary power for the initial in-orbit capability was to be 75 kW, derived from photovoltaics. Power expansion was then assumed to be provided by solar dynamics in two units of 25 kW.

Solar dynamic systems, resulting in less drag, lead to lower fuel usage for orbit maintenance. This reduces the cost of station operation, principally by reducing the demand for refuelling flights. An additional cost benefit arises from the lower maintenance costs envisaged for solar dynamic systems compared with photovoltaic systems. Over the lifetime of the Space Station, cost savings of several billions of dollars have been identified [18].

The concept studied by NASA, and which was the original baseline for the ISS solar dynamic system, is based upon the Brayton Cycle engine [18]. A block diagram of this is shown in Figure 10.14. The working fluid for this all-gas phase cycle is helium and xenon in such a proportion that the equivalent molecular weight is 40. This all-gas phase cycle minimizes problems of handling wet vapours (leading to erosion) and gravitational effects in transporting fluids. Storage of power within the concept was to be performed thermally, using the latent heat of fusion for a lithium fluoride/calcium fluoride mixture. This phase change occurs at 1042 K. Storing power in this way provides mass savings compared to battery technology, because of the high quality (temperature) heat energy.

Funding problems for the ISS have led to a cancellation of the power extension requirements. There has, however, been recent ongoing research through NASA, associated with the Constellation programme. This has led to revaluation of advanced radioisotope power conversion technologies wherein a radioisotope output is harnessed to thermoelectric, Brayton and Stirling cycle options.

10.4   SECONDARY POWER SYSTEMS: BATTERIES

Batteries have been used extensively for the secondary power system, providing power during periods when the primary one is not available. As a back-up for a solar array this means that the batteries must provide power during eclipses, and that the array must recharge the batteries in sunlight.

In GEO operations, eclipses only occur during the two equinoctial periods producing eclipse seasons for the spacecraft. These last about 45 days at each equinox. Initially they are short ($\sim$ minutes), but lengthen to a maximum of 1.2 h before decreasing again. The total of $\sim$90 eclipses, thus, occur irregularly with significant periods of time when no battery operation is required. With more than 22 h of sunlight available in each orbit, a trickle charge solution is possible.

In LEO, on the other hand, the spacecraft may be in eclipse and thus require battery power for 40% of each orbit. Although the precise duration will depend on orbit inclination, it is fairly regular, and the eclipse cycle results in typically 5000 to 6000 charge/discharge cycles of the battery per year. This results in the array-power sizing needing to be nearly twice the nominal load requirement (see Section 10.6).

In summary, LEO operations require a large number of low-depth discharges, whereas in GEO a few deep discharges suffice. This inevitably influences battery type, resulting in the past trend of using nickel-cadmium (Ni–Cd) or silver–zinc (Ag–Zn) cells for LEO operation and nickel–hydrogen (Ni–H$_2$) cells for GEO operations. Cell cycle life, specific mass (kW h/kg) and volume (kW h/m^3) all influence the acceptability of a particular battery technology. However, work on the other materials, particularly Li-ion and Li–SO$_2$, is continuing [19] and alternative technologies continue to be implemented on spacecraft (see, for example, the annual Goddard Space Flight Centre (GSFC) battery workshop proceedings [20]); see Table 10.6 for present battery technology performance characteristics. The use of Li-Ion battery technology has come to the fore in recent years, and its use in scientific and application satellites alike has become increasingly common and is indeed the baseline for the Eurostar 3000 platform.

A rather specialist battery requirement specification led to the Li–SO$_2$ battery system that was used to power the Huygens probe, for the period of the mission following separation from the Cassini Saturn orbiter. This separation occurred in January 2005, and followed a period of seven years when the probe had been in hibernation. The battery was then required to provide power during a low-power coast of 22 h, followed by a high-power load for 2.5 h during the descent through Titan's atmosphere. The design requirements were for the probe to remain operational for 30 min on the surface. However, the battery continued to power the probe for 70 m. For this, the overall battery system contained five individual battery units. Each of these consisted of two modules of 13 Li–SO$_2$ cells in series. Each cell had a capacity of 7.5 A-h.

The detailed electrochemistry of batteries is covered in References [21] and [22]. The main function of battery operation, which is of importance to spacecraft design, is the way in which the reliability and charge efficiency are related to charge control. Parameters of critical importance are the charge/discharge rate, the depth of discharge (DOD), the extent of overcharging and the thermal sensitivity to each of these parameters. The associated relationship between performance and control environment is a feature common to all battery technologies, although some are more sensitive than others; this is particularly so for Ni–Cd technology which historically has been used on many spacecraft. The technology that is now becoming the baseline for many missions however is the Li-ion based battery. This was first flown as a primary battery in 2001 on the ESA Proba-1 mission that operated in LEO [24], and far exceeded the cycle life performance. This technology, in addition to offering high specific power, is also capable of being operated

Table 10.6 Performance of battery technologies for space use [23]

Type	Specific energy (W h/kg)	Mission examples
Ni-Cd	28–34	Sampex
Ni-H$_2$	30–54	Odyssey
Ag-Zn	100	Pathfinder
Li-Ion	90	MER Rover
Li-SO$_2$	90–150	Galileo
Li-SOCL$_2$	200–250	Sojourner

Table 10.7 Hubble space telescope (HST), Intelsat VII and Eurostar 3000 battery summary

Parameter	HST	Intelsat VII	Eurostar 3000
Technology	Ni-H$_2$	Ni-H$_2$	Li-ion
Specific energy (W h/kg)	57.14	61.26	175
Capacity (A-h)	96	91.5	50
Cell dimensions:			
Diameter (cm)	9.03	8.89	5.3
Length (cm)	23.62	23.67	25.0
Cell mass (kg)	2.1	1.867	1.1

at high depth of discharge with a large number of cycles. It has been adopted for GEO spacecraft and is the adopted battery technology for the Eurostar 3000 platform, in which it is designed to operate with a depth of discharge of 80%. This far exceeds that of Ni–H$_2$ technology which typically operates with a DOD of ∼50%. The resulting battery has a specific power greater than 110 W h/kg, and due to the deeper discharge provides additional mass saving. The performance of the cells used on Eurostar 3000 is shown in Table 10.7, which also includes data for the Ni–H$_2$ batteries for HST and Intelsat VII.

10.5 POWER MANAGEMENT, DISTRIBUTION AND CONTROL

The basic features of power control were outlined in Section 10.2. A key aspect of the power management system is that it must be designed to operate with both a primary and a secondary power system whose characteristics are changing with time as outlined in the previous sections. Several philosophies for power management are outlined in Figure 10.15.

The *electrical 'bus'* may be required to provide a variety of voltages to meet the needs of the various equipment. Generally within Europe, the trend has been to have a regulated dc power bus, typically at 28, 50 or 100 V (ECSS-E-ST-20C). For example, the bus of the second generation Meteosat system has a bus voltage of 28.2 V ± 1%, and for Artemis it is 42.5 ± 0.5 V. Both of these satellites were designed in Europe. In contrast, US spacecraft generally use unregulated buses; for example, the NASA standard for unmanned spacecraft provides a voltage in the range 21 to 35 V dc. Present spacecraft are being designed with

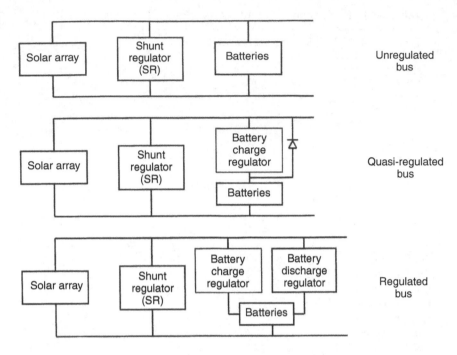

Figure 10.15 Schematic bus concepts

higher bus voltages (~150 V) to reduce resistive losses and harness mass; for example, the nominal bus voltage for the ISS is 160 V. For both regulated and unregulated systems, dc–dc converters are required to provide the variety of voltages needed; this conversion frequently takes place at equipment level rather than centrally.

An ac bus is sometimes used to augment the dc one. The hybrid system can provide mass savings due to both the simplicity of conversion from ac to a variety of dc levels, and also the fact that it is possible to run the power distribution harness at higher voltages if an ac supply is used. Indeed it has been noted [25], that the equivalent wiring cross-section is six times smaller for a three-phase ac network than for a dc one. An ac distribution is mainly applicable to high-power spacecraft and where a large number of dc voltages are required at equipment level. The voltages in ac buses generally have a square waveform, an example being Hipparcos. A notable exception was on the Shuttle Space-lab, where the ac bus was more sophisticated, providing a three-phase sinusoidal voltage at 400 Hz.

As noted above, the bus can be either regulated or unregulated; in some circumstances it can be quasi-regulated. The fully regulated bus provides voltage regulation during both normal sunlight operation (battery charge cycle) and during an eclipse (discharge cycle). With quasi-regulation, the regulation occurs only during sunlight operation.

The primary units

The main units used in a power system are described below for a typical configuration used on European communications spacecraft. The terminology may differ slightly for US spacecraft. Figure 10.16 shows the overall power system layout.

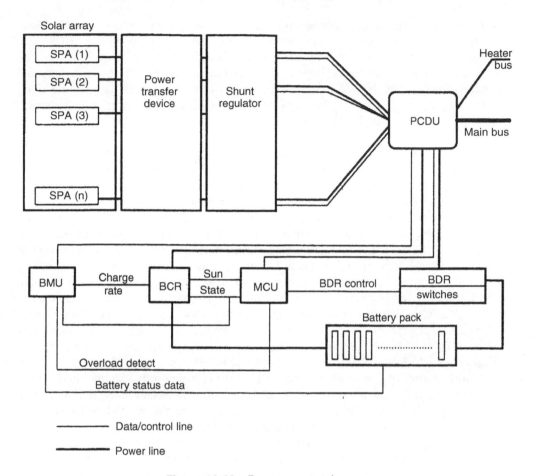

Figure 10.16 Power system layout

- *Array regulator*. As the power available from an array varies during a mission, or the power demanded by the payload varies, it is possible to switch in or out segments of the solar array. Switching out is achieved by grounding the individual segments in the shunt regulator module. The array itself can be structured into various modules, sometimes called *solar power assemblies* (SPA). The regulation of power output from the array bus is typically achieved by using either a pulse-width modulation scheme, or a sequential switching-shunt regulation, known as S^3R [26]. A considerable number of satellites have used this latter approach. The voltage sensing that is used to control the shunt regulator module is termed the *mode control unit* (MCU). An additional strategy used on some spacecraft is maximum power-point tracking. As can be seen from Figure 10.5, the power output from an individual cell shows a distinct maximum. Maximum power-point trackers control the operating point of a string of cells by varying the voltage at which the string operates. As an example [27] of a solar array regulator, the unit on the Artemis satellite, which has a sunlight load of 3.3 kW and an eclipse load of 2 kW, provides shunt regulation. This regulator has a mass of 6.5 kg and dissipates 117 W at a nominal load of 2.65 kW. During eclipse operations the regulator requires 18 W.

- *Battery control.* Three units are typically associated with battery control. These are the *battery management unit* (BMU), the *battery charge regulator* (BCR) and the *battery discharge regulator* (BDR). The BMU's functions are to monitor the battery's temperature and voltage as well as individual cell voltages, pressures and temperatures. It is the interface between the power subsystem and the data-handling subsystem, and also provides control inputs to the charge regulation of the batteries, carried out by the BCR. The principal function of the BCR is to provide a constant current charge of the battery during sunlight operation, whilst that of the BDR is to supply a constant current to the spacecraft bus during eclipse operation. Control of this current is derived from the MCU, typically with further protection from the BMU. Whilst the charge/discharge rate of the battery may be controlled in a fairly simple manner through current regulation, the monitoring of the state of charge in a battery is more complex. The principal methods that may be used to sense charge-state are cell voltage, cell temperature, or cell pressure. It should be noted that the level of full charge noted by each of these methods results in a different level of overcharging. Pressure and temperature sensing results in overcharging by 20 to 30%, whereas voltage sensing may indicate 10 to 20% overcharge. A principal problem with voltage sensing arises because of the voltage–charge–temperature characteristics that may cause significant errors in determining the state of charge of the battery. Again taking the Artemis spacecraft as an example, the overall battery control unit has a mass of 25.5 kg. In sunlight, the power required to operate this is 63.5 W and in eclipse, 282 W. The efficiency of the BDR is 89%, and the BCR is 91%.
- *Power control and distribution unit (PCDU).* This unit provides monitoring and protection for the bus current. Protection is normally achieved either by current limiting or by fusing, the latter generally requiring a redundant path to be switched into operation, normally by command from ground-control.
- *Power conversion unit (PCU).* This unit supplies the individual voltage/current characteristics required for loads. The typical low voltage outputs (e.g. ±15 V and 5 V) will be regulated using solid-state switches that are pulse-width modulated. This unit must also be able to cope with transient protection for over- and under-voltage and in-rush current limiting when units are switched on or off.

10.6 POWER BUDGET

The foregoing sections have outlined the principles of the technology behind a power system for use on a spacecraft system. In this section the methodology used to provide the size of a power system is outlined.

10.6.1 Mission specific design issues

The starting point for any power system is in the definition of spacecraft electrical loads. In general these will not be constant throughout the mission, or even throughout a single orbit. The initial analysis must therefore take into account the mission profile and hence the power demand. The three critical issues that need be considered are the orbit parameters, the nature of the mission (communications, science, or other) and the mission duration.

The *orbit selection* has a major influence upon the radiation environment experienced, and hence the degradation anticipated in any solar array-based solution. Further, the orbit will define the duration of eclipse periods (see Section 5.3.2 of Chapter 5), which together with the number of eclipses anticipated, will define battery requirements and the degradation that might be expected to occur during the mission. Clearly in deep-space missions an investigation of the most appropriate technology for the primary power source will be influenced by the orbit.

The *nature of the mission* will have significant impact on the type of loads expected. Thus for a communications satellite, independently of the orbit specification, it can be anticipated that the primary payload will be required at all times. Further, the power demand in eclipse may well exceed that in sunlight, owing to the need both to operate the payload and to meet the additional burden from active, or power augmented elements, in the Thermal Control System (TCS). Navigation and broadcast satellites will have similar requirements to these. In contrast, a remote sensing spacecraft during eclipse may well not require the whole payload to be operational, particularly if there are passive optical instruments. For such satellites, there may be a very great range of power demands owing to the mission requirements, especially if there are active microwave instruments. These instruments may operate for only limited portions of the orbit, perhaps when communication of data to a ground station is available; this can provide a very high peak load for the power system. Scientific satellites may have very specific payload operational requirements, with additional complexity being provided by the need to download scientific data only when the vehicle is within sight of a ground station. Deep-space missions will inevitably face the temperature extremes that may require additional power for environmental control if the spacecraft either goes very close to or very far from the Sun.

The *mission duration* will provide a major influence on the degradation of the power system. The two most significant influences are

1. The total radiation dose expected which may determine, for a satellite carrying a solar array, the amount of shielding required, and hence influence the specific mass of the power system.
2. The number of eclipse cycles, which will influence the system reliability and its degradation. Solar cell failure through open-circuit losses will clearly increase with increased number of thermal cycles, driven by the entry to and from eclipse. Battery degradation will progress with number of eclipse cycles; indeed for any given technology there is only a maximum number of charge/discharge cycles that a battery can sustain before failure.

10.6.2   Power budget evaluation

The format of a typical power budget is shown in Table 10.8. In this each of the subsystems are identified, with, in this example, power shown simply in terms of eclipse and sunlight loads and peak loads. It is notable that the power subsystem itself also places a load on the spacecraft.

The method adopted for deriving suitable values to insert in this table is as follows: Suppose data is available for each subsystem. Initially such data will exclude the power subsystem. For each subsystem, data must be provided for both the sunlit orbit phase,

Table 10.8 Typical structure of a power budget

Subsystem	Peak power	Sunlight power P_{sun}	Eclipse power $P_{eclipse}$	Intelsat VIIa (%)	Average GEO comms.satellite (%)
AOCS				5.0	3.6
Power				10.4	11.2
Thermal control				4.9	6.4
Comms.				n/a	n/a
Data handling				0.6	1.6
Payload				79.1	77.2
Average total power				100	100

having, say an average value of P_{sun} and the eclipse phase having an average value of $P_{eclipse}$. Since the satellite must be provided with power throughout the mission, this specifies the EOL requirement.

In reality this approach is highly simplified, since there will be specific events that may place a high transient load on the power system. One such event is the firing of a pyrotechnic device followed by the actuation of some form of deployment mechanism. As an example, the firing of a pyrotechnic release, followed by the deployment of a boom on the Ulysses spacecraft, provided a transient load of nearly two thirds of the total available power from the RTG at the BOL.

In the principal operational phase of a mission, generally not all the spacecraft equipment will be operating at one time. As a result the potential power demand, identified by summing all the spacecraft loads, will never be a realistic value for the peak demand load. Indeed in many cases it will be found that such a summation exceeds the total power availability from the power bus. One subsystem that has widely varying requirements during a mission is the thermal system. This subsystem must meet both a hot and cold case, which may require very different levels of heater input. Again taking the Ulysses mission as an example, the 'hot' case heater power (8.7 W) was only one third that of the 'cold' case heater load (24.7 W).

For telecommunications spacecraft operating in GEO there is a fairly well-defined power profile between subsystems. Increasingly, as noted in Chapter 6, electric propulsion is being used on such missions for station keeping control, which results in an increase in the power required for the propulsion subsystem. Power for propulsion in Table 10.8 is included in the AOCS subsystem in the power profile. This is shown as a percentage for each of the subsystems for recently launched GEO communications satellites. The specific profile for Intelsat VIIa is also shown.

10.6.3 Approximate power system sizing

A simplified block diagram for the power system is shown in Figure 10.17, in which the efficiencies of various components are also identified. Representative values for the efficiencies may be found in Sections 10.3, 10.4 and 10.5.

Consider a general case for sizing a power system. Assuming that the orbit period is τ, with the time spent in sunlight τ_{sun}, and the time spent in eclipse $\tau_{eclipse}$, then the power

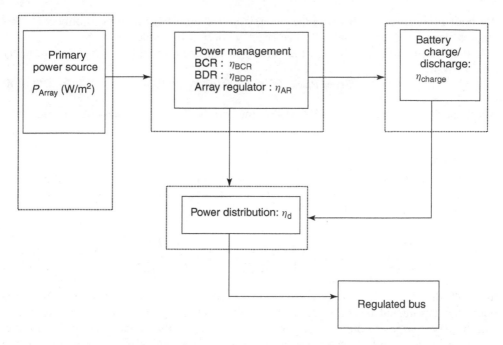

Figure 10.17 Definition of system efficiencies

required from the array to meet the eclipse load is clearly given by P_{charge}, where

$$P_{charge}\tau_{sun} = \frac{1}{\eta}P_{eclipse}\tau_{eclipse} \qquad (10.5)$$

and η is the product of the efficiency terms shown in Figure 10.17,

$$\eta = \eta_{BDR}\eta_{BCR}\eta_{AR}$$

The total power required to be available from the array is thus approximately given by

$$P_{array} = P_{sun} + P_{charge} \qquad (10.6)$$

In the limit, if all the efficiencies are equal to unity, it is apparent from equations (10.5) and (10.6) that if the eclipse power demand is the same as that in sunlight then the array power is simply given by

$$P_{array} = P_{sun}\left(\frac{\tau}{\tau_{sun}}\right) \qquad (10.7)$$

Thus for LEO orbits where the fraction of the orbit in eclipse is large, typically of the order of 30 min out of 90 min, the array needs to be oversized relative to the sunlight provision by a factor τ/τ_{sun} of approximately 1.5. Hence, in this approximate scenario, the array power in LEO is required to be 50% in excess of the bus load. This may be contrasted with the GEO case where a maximum eclipse duration of ∼70 min in the

24- hour orbit arises. In this case the 'oversizing' of the array amounts to only 5%. This feature clearly has significant influence over the design of a solar array.

If the orbit period is τ (hours), the battery-stored energy will be given approximately by E_B (W-hrs)

$$E_B = P_{\text{eclipse}}(\tau - \tau_{\text{sun}})/(\eta_{\text{charge}}DOD) \tag{10.8}$$

where the *DOD* is the depth of discharge of the battery. The battery mass can be estimated by dividing the stored energy (W-hrs) by the energy density (W-hrs/kg) for the chosen battery technology. For example, for Li-ion the energy density is approximately 120 to 175 W-hrs/kg.

In any particular case, the individual loads that will be switched in and out need to be considered. We can then modify equation (10.5) to obtain the total energy required from the array. This may be written in the form $\varepsilon_{\text{array}}$ where

$$\varepsilon_{\text{array}} = P_{\text{array}}\tau_{\text{sun}} = \frac{1}{\eta_{\text{sun}}}\left(\sum_{i=1}^{k} P_i t_i\right) + \frac{1}{\eta_{\text{ecl}}}\left(\sum_{i=k+1}^{n} P_i t_i\right) \tag{10.9}$$

in which the $P_i, i = 1,\ldots,n$, gives the typical power profile for payload and subsystem operation throughout the orbit. A typical profile may be as illustrated in Figure 10.18. The battery charge requirement during sunlight is explicitly excluded from this profile, since the eclipse profile (second term in equation 10.9) is equivalent to the battery charge energy (see equation 10.5). For the purposes of a first estimate of the array size, the efficiency factors have typical values of $\eta_{\text{sun}} \sim 0.8$, from array to loads, and $\eta_{\text{ecl}} \sim 0.6$, from batteries to loads.

In terms of calculating the array size, allowance needs to be made for any pointing-angle offset of the array relative to the sun line. This results in the array area being given by A_{array} where

$$A_{\text{array}} = P_{\text{array}}/(S \cos \delta\theta \eta_{\text{cell}}\eta_{\text{packing}}(1 - D)) \tag{10.10}$$

Here S is the solar flux ($\sim$1400 W/m^2 in a near-Earth orbit); $\delta\theta$ is the array pointing error with respect to the Sun, which will typically be of order 1°, but is highly dependent on

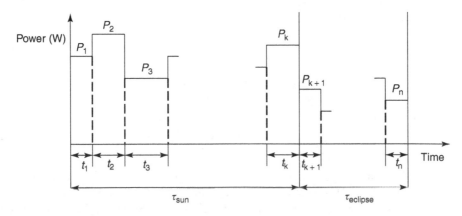

Figure 10.18 Approximate power profile for payload and subsystem operation throughout the orbit

the mission; η_{cell} is the solar cell efficiency; $\eta_{packing}$ is the cell packing efficiency that is typically 0.90; D is the array degradation factor over the spacecraft lifetime, which may be calculated in the manner described in Section 10.3.1.

In a specific design, the sizing of the array given by the equivalent of equation (10.7) will be modified to include specific mission profiles for 'hot' and 'cold' cases. These differing cases arise during the year, as a result of variation in solar insolation, which will influence the array temperature (see Chapter 11); this variation occurs due to the influence of the angle between the orbit plane and the ecliptic. As a result, the final EOL array and battery capacity can be specified. To define the BOL characteristics, loss factors due to radiation damage, micrometeorite damage and battery hysteresis losses must included. Thus this process evidently requires the array and battery capacity to be oversized initially to meet the EOL power demands. Typically these loss factors may rise to 25% of the initial provision, although this figure is highly dependent upon both mission duration and the type of orbit, with Highly Elliptical Orbits (HEOs) being particularly affected by radiation damage.

REFERENCES

[1] *SPS Concept Development and Evaluation Programme Reference System Report* (1978) US DOE and NASA DOE/ER 0023.

[2] Landis, G. A. (2006) Reevaluating Solar Power Systems for Earth, *IEEE 4th World Conference on Photovoltaic Energy Conversion 2006*, NTRS-2007-0005136.

[3] Angrist, S. W. (1982) *Direct Energy Conversion*, Allyn and Bacon, New York.

[4] Bennett, G. (1995) Summary of the US Use of Space Nuclear Power, ESA SP 369, 163–169.

[5] Freeh, J. E. (2009) Analysis of Stationary Photovoltaic Based Surface Power Systems Designs at the Lunar South Pole, NASA/TM-2009-215506.

[6] Hyder, A. K., Wiley, R. L., Halpert, G., Flood D. J. and Sabripour, S. (2003) *Spacecraft Power Technologies*, Imperial College Press, London.

[7] Patel, M. P. (2004) *Spacecraft Power Systems*, CRC Press, New York.

[8] Rauschenbach, H. S. (1980) *Solar Cell Array Design Handbook*, Van Nostrand Reinhold, New York.

[9] Branhorst, H. W., Rodick, J., Ferguson, D. C. and O'Neill, M. J. (2008) A SEP Mission to Jupiter Using the Stretched Lens Array, Space Propulsion 2008, NTRS-2008-0031048.

[10] Taylor, H., Simpson, A. F. and Dollery, A. A. (1984) CMX-50: A New Ultra-thin Solar Cell Cover for Lightweight Arrays, ESA SP-173, 211–214.

[11] NASA - CR 189215 (1990).

[12] Merrill, J., Haugsen, P., Senft, D. and Tlomak, P. (2005) NASA/CP 2005-213431.

[13] For example, *19th Space Photovoltaic Research and Technology Conference NASA/CP 2007-214494*.

[14] Mawira, D. (1982) Advanced Rigid Array, ESA SP-173, 9–14.

[15] Knorr, W., Theurer, G. and Schwartz, M. (1995) A Regenerative Fuel Cell System for a Lunar Rover, ESA SP-369, 21–26.

[16] Bockris, J. O. M. and Srinivason, S. (1969) *Fuel Cells and Their Electrochemistry*, McGraw-Hill, New York.

[17] Terrill, W. and Haley, V. (1986) Thermoelectric Converter for SP-100, *21st IECEC*, 1950–1955.

[18] *Solar Dynamic Power System Development for Space Station Freedom*, NASA RP1310 (1993).

[19] Wong, W. A. (2004) Advanced Radioisotope Power Conversion Technology Research and Development, AIAA-2004-5515, also NASA TM - 2004-213352.

[20] NASA - CP - 2006-214599 (2006).

[21] NASA RP 1052 (1979).

[22] NASA SP 172 (1968).

[23] Surampudi, R. (2006) Overview of Energy Storage Technologies for Space Applications, *42nd Power Sources Conference 2006*, NTRS-2009-0019718.

[24] Spurett, R. and Thwaite, C. (2003) Project for on-board autonomy (PROBA) lithium-ion battery, *Proceedings of the 2003 Space Power Workshop*, Huntsville, Alabama.

[25] Eggers, G. (1985) AC Buses for LEO - A Viable Alternative, ESA SP-230, 17.

[26] Knorr, W. (1998) Power System for 2nd Generation Meteosat, ESA SP-416, 11–16.

[27] ESA SP-369 4th European Space Power Conference (1995).

11 THERMAL CONTROL OF SPACECRAFT

Chris J. Savage

European Space Research and Technology Centre (ESTEC), European Space Agency

11.1 INTRODUCTION

Spacecraft thermal control—that is the control of spacecraft equipment and structural temperatures—is required for two main reasons: (1) electronic and mechanical equipment usually operate efficiently and reliably only within relatively narrow temperature ranges and (2) most materials have non-zero coefficients of thermal expansion and hence temperature changes imply thermal distortion.

Spacecraft equipment is designed to operate most effectively at or around room temperature. The main reason for this is that most of the components used in spacecraft equipment, whether electronic or mechanical, were originally designed for terrestrial use. It is also much easier and cheaper to perform equipment development and, eventually, qualification and flight acceptance testing at room temperature. Typically, operating electronic equipment requires to be maintained in a temperature range between about $-15°C$ and $+50°C$, rechargeable batteries between about $0°C$ and $+20°C$ and mechanisms (solar array drives, momentum wheels, gyroscopes etc.) between about $0°C$ and $+50°C$. There are, of course, exceptions to this—for example, some detectors within astronomical telescopes that need to be cooled to very low temperatures.

Many spacecraft payloads require very high structural stability, and therefore thermally induced distortion must be minimized or strictly controlled. For example, the search for ever-higher resolution from space-based telescopes means that temperatures stable to within a fraction of a Kelvin are often required within telescope systems several metres in size.

Heat is generated both within the spacecraft and by the environment. Components producing heat include rocket motors, electronic devices and batteries. Initial ascent heating effects are minimized by the launch vehicle's nose fairing. Heat from the space environment is largely the result of solar radiation. Heat is lost from the spacecraft by radiation. The balance between heat gained and lost will determine the spacecraft temperatures.

Spacecraft Systems Engineering, Fourth Edition.
Edited by Peter W. Fortescue, Graham G. Swinerd and John P. W. Stark.
© 2011 John Wiley & Sons, Ltd. Published 2011 by John Wiley & Sons, Ltd.

The configuration of a spacecraft is dictated by many factors and 'thermal control' is only one of them. The task for the thermal control engineer consists, in fact, of three main parts. Firstly, *analysis*—they must be able to analyse a given spacecraft configuration and predict equipment and structural temperatures for all phases of the mission. Secondly, *design*—in the rather likely circumstance that the results of the analysis show temperatures falling outside allowed limits, the engineer must devise suitable solutions, for example, by modifying heat-flow paths or implementing heaters, radiators and so on. Finally, *testing*—the engineer must perform sufficient and appropriate testing to confirm the accuracy of the analysis and of the thermal predictions for the mission.

Modern spacecraft, particularly those exploring the Solar System or involving human crews, will often be composed of hardware supplied by different organizations, often from different parts of the world. Customers and those responsible for the various stages of spacecraft integration will usually require such hardware to be developed according to recognized standard practices and procedures. For Europe, the controlling authority for such standardization is the *European Cooperation for Space Standardization* (*ECSS*). The requirements for the specific discipline of spacecraft thermal engineering are covered in ECSS-E-ST-31C [1].

11.2   THE THERMAL ENVIRONMENT

An important characteristic of the space environment is its high vacuum. Spacecraft are generally launched into orbits where the residual atmospheric pressure, and hence drag, is very small (although often not negligible—the International Space Station (ISS) requires regular re-boosting to compensate for air drag). Fortunately for the thermal control engineer, the very low level of drag implies also the absence of any significant *aerodynamic heating*. For an orbiting spacecraft, aerodynamic heating and indeed any *convective interaction* between spacecraft and environment can be ignored.

The Earth's atmospheric pressure falls approximately exponentially with altitude, dropping more than six orders of magnitude in the first 100 km above sea level [2]. Spacecraft in orbit around the Earth usually orbit at altitudes higher than 300 km where the residual atmospheric pressure is typically less than 10^{-7} mb. During the launch phase, the transition from being fully protected within the launch vehicle to autonomous operation in space is the result of a compromise. The sooner the nose fairings can be jettisoned, the more payload a given vehicle can launch. However, if they are jettisoned too early, dynamic pressure and aerodynamic heating will damage the spacecraft. Fairings are normally jettisoned at the point where residual aerodynamic heating is the same as or less than the incident solar heating. This occurs at around 100 km altitude, depending on the characteristics of the launch vehicle and trajectory. Hence, once again, aerodynamic heating effects can be ignored by the spacecraft thermal designer. During re-entry or aerobraking manoeuvres (see Chapter 5), specific protection is provided, which ensures that these phases do not drive the spacecraft thermal design.

A spacecraft in space can interact with its environment only by radiation and this interaction is characterized by the exchange of energy by means of the following (see Figure 11.1):

- direct solar radiation;
- solar radiation reflected from nearby planets (albedo radiation);

- thermal energy radiated from nearby planets (planetary radiation);
- radiation from the spacecraft to deep space.

The spacecraft will experience thermal equilibrium when the sum of the radiant energy received from the first three sources listed above, together with any thermal dissipation within the spacecraft, is equal to the energy radiated to deep space. It is this balance that will determine the physical temperature of the spacecraft.

11.2.1 Solar radiation

The solar radiation parameters of interest to the thermal design engineer are (1) spectral distribution, (2) intensity and (3) degree of collimation. The spectral distribution can be considered constant throughout the solar system and the solar irradiance, or spectral energy distribution, resembles a Plank curve with an effective temperature of 5800 K (see Chapter 2). This means that the bulk of the solar energy (99%) lies between 150 nm and 10 μm wavelength, with a maximum near 450 nm (in the yellow part of the visible spectrum).

The solar radiation intensity outside the Earth's atmosphere and at the Earth's average distance from the Sun (1 AU) is called the solar constant and is about 1371 ± 5 W/m^2. The solar radiation intensity J_s at any other distance d from the Sun can be found from the simple relationship

$$J_s = \frac{P}{4\pi d^2} \tag{11.1}$$

where P is the total power output from the Sun, 3.856×10^{26} W. Table 11.1 shows the resulting variation in solar intensity that can be expected at the average distance from the Sun of each of the planets in the solar system.

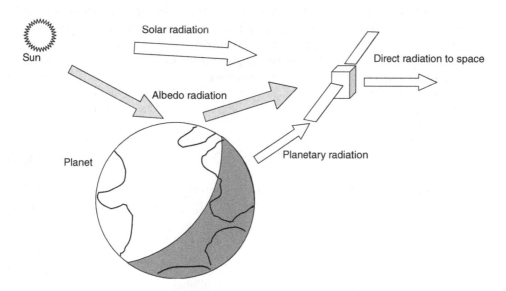

Figure 11.1 Typical spacecraft thermal environment

Table 11.1   Planetary solar constants and albedo values [3, 4]

Planet	Solar radiation intensity, J_s (percentage of solar intensity at 1 AU)	Planetary albedo, a
Mercury	667	0.06–0.10
Venus	191	0.60–0.76
Earth	100	0.31–0.39
Moon	100	0.07
Mars	43.1	0.15
Jupiter	3.69	0.41–0.52
Saturn	1.10	0.42–0.76
Uranus	0.27	0.45–0.66
Neptune	0.11	0.35–0.62
Pluto	0.064	0.16–0.40

The angle subtended by the Sun in the vicinity of the Earth (at 1 AU from the Sun) is about 0.5°. This means that the sunlight incident on a spacecraft can, for thermal control purposes, be regarded as a parallel beam emanating from a point source. This is not true, however, for spacecraft whose mission takes them very close to the Sun.

The fraction of the solar radiation that is reflected from the surface and/or atmosphere of a planet is known as the *planetary albedo*. Its value is highly dependent on local surface and atmospheric properties. For example, for the Earth, it varies from as high as 0.8 from clouds to as low as 0.05 over surface features such as water and forest [3, 4]. Fortunately for the thermal engineer, such changes occur rapidly in relation to the thermal inertia of most spacecraft, and an orbital average value, a, can be used for thermal design purposes. For the Earth, this is in the range 0.31–0.39. Table 11.1 lists the albedo values for the planets of the solar system. The reader should be aware that measuring the albedo of the more distant planets is not an easy task and that the quoted figures should be treated with caution. Although the spectral distribution of albedo radiation is not identical to that of the Sun, as is evidenced by the diverse colours of planetary surface features, the differences are insignificant for thermal engineering purposes and can be ignored.

The intensity of the albedo radiation, J_a, incident on a spacecraft is a complex function of planet size and reflective characteristics, spacecraft altitude and the angle β between the local vertical and the Sun's rays. This can be expressed in terms of a *visibility factor* F as follows

$$J_a = J_s a F \qquad (11.2)$$

For the purpose of calculating albedo radiation inputs, the Earth can be regarded as a diffuse reflecting sphere, in which case the visibility factor varies approximately as shown in Figure 11.2.

It is emphasized that the above treatment is approximate. For complex spacecraft, particularly in low orbits, accurate calculation of albedo inputs may need to be performed as a function of orbital position for each external surface element. These are complicated calculations for which specific software tools are available.

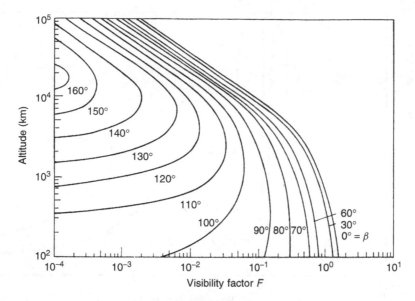

Figure 11.2   Spacecraft albedo irradiation. β is the angle between the local vertical and the Sun's rays

11.2.2   Planetary radiation

Since the planets of the Solar System all have non-zero temperatures, they all radiate heat. Because of its relatively low temperature, the Earth radiates all of its heat at infrared wavelengths, effectively between about 2 and 50 µm with peak intensity around 10 µm. For this reason, the radiation is often referred to as *thermal radiation*. The spectral distribution of the Earth's thermal radiation is shown in Figure 11.3 [3]. The atmosphere is essentially opaque over much of the infrared spectrum, with important transparent windows at around 8–13 µm. The radiation that a spacecraft sees is hence composed of radiation from the upper atmosphere, radiating with an effective black-body temperature of 218 K, with, superimposed upon it, radiation passing through the infrared windows from the Earth's surface. Since terrestrial temperatures vary with time and geographical location, the intensity J_p of the thermal radiation incident on orbiting spacecraft can also be expected to vary with time and position around the orbit. In fact, due to the Earth's large thermal inertia with respect to diurnal and seasonal changes and the spacecraft's large thermal inertia with respect to its orbital period, only very small errors occur if averaged values are used. For most practical purposes, the thermal engineer can assume that the Earth radiates with an intensity of 237 W/m² and that the thermal radiation emanates uniformly from the whole cross-sectional area of the Earth.

Since the intensity falls with altitude according to the inverse-square law, the approximate value of J_p in W/m² at a given altitude can be found from

$$J_p = 237 \left(\frac{R_{\text{rad}}}{R_{\text{orbit}}} \right)^2$$

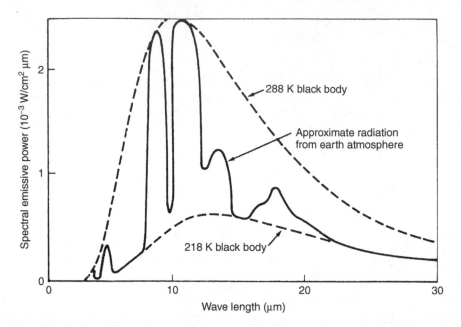

Figure 11.3 Typical spectral emissive power for the thermal radiation from Earth. (*Note*: The 288 K black-body curve approximates the radiation from the Earth's surface, and the 218 K black-body curve approximates the radiation from the atmosphere in those spectral regions where the atmosphere is opaque)

where R_{rad} is the radius of the Earth's effective radiating surface and R_{orbit} is the orbit radius. The precise value of R_{rad} is not easy to determine and for most practical purposes it can be assumed equal to the radius of the Earth's surface, R_E. For other planets, care needs to be taken to verify the validity of these assumptions on a case-by-case basis. For example, Mercury, with a sidereal day of the same order of magnitude as its year (59 and 88 Earth-days, respectively), sustains temperature differences of hundreds of degrees between sunlit and shadowed sides, with a slow-moving terminator. Its orbit is also sufficiently eccentric for its solar constant to more than double between aphelion and perihelion, giving rise to large seasonal variations.

11.2.3 Spacecraft heat emission

The spacecraft itself has a finite temperature, so it will also radiate heat to space. Since the spacecraft temperature will be similar to that of the Earth (if the thermal engineers have done their job properly), it too will radiate all its heat in the infrared region of the spectrum.

11.3 THERMAL BALANCE

As already noted, the temperature of a spacecraft depends on the balance between the heat received from external and internal sources, and the heat radiated to space. In order to control spacecraft temperatures, it is necessary to control the heat absorbed, the heat radiated or (usually) both.

If spacecraft were *black bodies*, that is, radiated as black bodies and absorbed all the radiation that fell on them, they would acquire a certain temperature and that would be the end of the story. Thermal control would be impossible, except perhaps by varying the internal heat dissipation. However, spacecraft are not black bodies but absorb only a fraction α of incident energy. They also emit as a *grey body*, radiating a fraction ε of the radiation of a black body at the same temperature. This may be expressed as

$$J_{\text{absorbed}} = \alpha J_{\text{incident}} \tag{11.3}$$

$$J_{\text{radiated}} = \varepsilon\sigma T^4 \tag{11.4}$$

where α and ε are known as the *absorptance* and the *emittance*, respectively, and σ is the *Stefan-Boltzmann constant* equal to $5.67 \times 10^{-8}\,\text{W}\,\text{m}^{-2}\,\text{K}^{-4}$.

For a spacecraft with no internal heat dissipation, an effective absorbing area (projected area facing the Sun) A_α, and emitting area A_ε, thermal equilibrium is achieved when

$$A_\alpha J_{\text{absorbed}} = A_\varepsilon J_{\text{radiated}} \tag{11.5}$$

which from equations (11.3) and (11.4) gives

$$A_\alpha \alpha J_{\text{incident}} = A_\varepsilon \varepsilon\sigma T^4$$

so that the equilibrium temperature T is given by

$$T^4 = \frac{A_\alpha}{A_\varepsilon} \frac{J_{\text{incident}}}{\sigma} \left(\frac{\alpha}{\varepsilon}\right) \tag{11.6}$$

Since A_α, A_ε and σ are constants, and for a given value of J_{incident}, the value of T can be controlled by varying the value of α/ε.

In fact, things are not quite as simple as that, as α and ε are not independent variables. Over any given wavelength range, the laws of thermodynamics require that $\alpha = \varepsilon$ (Kirchoff's law). Thus for any surface, its *absorbtivity* at a given wavelength is equal to its *emissivity* at the same wavelength.

Fortunately for the thermal control engineer, absorptivity and emissivity generally vary with wavelength and we have already learnt that the radiation environment of a spacecraft is basically composed of radiation either at 'visible' wavelengths or in the infrared. It is this feature that makes spacecraft thermal control possible. For the spacecraft thermal control engineer:

α means the absorptance of a surface illuminated by solar radiation (peak intensity at about 0.45 µm)—it is therefore often referred to as the '*solar absorptance*'.

ε means the emittance of a surface radiating in the infrared region (peak intensity at about 10 µm)—it is therefore often referred to as the '*infrared emittance*'.

Note that, according to Kirchoff's law, the absorptance of a spacecraft to planetary radiation is equal to its infrared emittance, ε.

By way of an example, let us consider a simple spacecraft in Low Earth Orbit (LEO). For the sake of convenience, let us assume a polar orbit that does not suffer from eclipses (a so-called dawn-dusk orbit—see Chapter 5) and let us furthermore assume that our spacecraft has a high thermal inertia and is isothermal. We have

$$
\begin{aligned}
\text{heat received directly from the Sun} &= J_s \alpha A_{\text{solar}} \\
\text{albedo contribution} &= J_a \alpha A_{\text{albedo}} \\
\text{planetary radiation contribution} &= J_p \varepsilon A_{\text{planetary}} \\
\text{heat radiated to space} &= \sigma T^4 \varepsilon A_{\text{surface}} \\
\text{internally dissipated power} &= Q
\end{aligned}
$$

where $A_{\text{solar}}, A_{\text{albedo}}$ and $A_{\text{planetary}}$ are the projected areas receiving, respectively, solar, albedo and planetary radiation, and A_{surface} is the spacecraft total surface area. If we assume that J_s, J_a, J_p and Q remain constant, our spacecraft will acquire an equilibrium temperature T given by

$$(A_{\text{solar}} J_s + A_{\text{albedo}} J_a)\alpha + A_{\text{planetary}} J_p \varepsilon + Q = A_{\text{surface}} \sigma T^4 \varepsilon.$$

Hence

$$T^4 = \frac{A_{\text{planetary}} J_p}{A_{\text{surface}} \sigma} + \frac{Q}{A_{\text{surface}} \sigma \varepsilon} + \frac{(A_{\text{solar}} J_s + A_{\text{albedo}} J_a)}{A_{\text{surface}} \sigma}\left(\frac{\alpha}{\varepsilon}\right) \qquad (11.7)$$

So once again we see that the spacecraft temperature is dependent on the ratio α/ε, particularly for spacecraft for which Q is small. For simplicity, let us assume a spherical spacecraft, radius r, for which

$$A_{\text{surface}} = 4\pi r^2, A_{\text{solar}} = A_{\text{albedo}} = A_{\text{planetary}} = \pi r^2, Q = 0$$

$$J_a = 0.33 F J_s \quad \text{(Earth albedo } a = 0.33\text{)},$$

$$J_p = 220\,\text{W/m}^2 \text{(corresponding to an orbit altitude of around 240 km)}.$$

Then equation (11.7) reduces to

$$T^4 = 9.70 \times 10^8 + 4.41 \times 10^6 (1 + 0.33F) J_s \left(\frac{\alpha}{\varepsilon}\right)$$

For $J_s = 1371$ W/m^2, F $= 0.15$ (from Figure 11.2) and a black paint finish for which $\alpha/\varepsilon = 1$, our spacecraft equilibrium temperature is about 293 K or 20°C.

If we now turn the orbit plane until the spacecraft passes through the Earth's shadow, the heat absorbed from the Sun (directly and as albedo) will be reduced. Assuming a sufficiently high thermal inertia, a new equilibrium temperature will be obtained. Let us consider the case in which the Earth-Sun vector lies in the plane of the orbit. This will evidently give the minimum time in sunlight, which, for a LEO spacecraft at an altitude of 240 km, is about 59% of its orbit period. Under these conditions, an average albedo visibility factor can be estimated from Figure 11.2 for the illuminated part of the orbit, F ~ 0.7. Note that the albedo radiation is zero during eclipse. The new equilibrium temperature will then be obtained from

$$T^4 = 9.70 \times 10^8 + 5.43 \times 10^6 J_s \left(\frac{\alpha}{\varepsilon}\right) f$$

Table 11.2   Equilibrium temperatures for a simple spacecraft in LEO

Surface finish	White paint $\alpha = 0.15$ $\varepsilon = 0.9$	Black paint $\alpha = 0.9$ $\varepsilon = 0.9$	Electroplated gold $\alpha = 0.25$ $\varepsilon = 0.04$
No eclipse	$-61°C$	$+20°C$	$+176°C$
Maximum eclipse	$-70°C$	$-2°C$	$+138°C$

where f is equal to the fraction of the orbit that is illuminated by the Sun, 0.59 in this case. It will be seen that our black-painted spacecraft has fallen in temperature to about $-2°C$. Table 11.2 shows the results of similar calculations for a white paint finish (low α/ε) and electroplated gold (high α/ε). It is evident that, by using different surface finishes in different ratios, spacecraft temperatures can be controlled over quite large ranges. Table 11.3 lists α and ε values for a number of common spacecraft surface finishes [5, 6]. It should be realized that, particularly for metal surfaces, the values of α and ε may be very dependent on preparation and surface treatment.

Real spacecraft are, of course, far more complicated than the one discussed above. They are certainly not isothermal and often contain components (e.g. the solar arrays) with a relatively low thermal inertia, which will change temperature significantly around an orbit (particularly when entering or leaving an eclipse). Whilst the overall spacecraft thermal

Table 11.3 α and ε values for several surfaces and finishes [5, 6]

Surface	Absorptance (α)	Emittance (ε)	α/ε
Polished beryllium	0.44	0.01	44.00
Goldized kapton (gold outside)	0.25	0.02	12.5
Gold	0.25	0.04	6.25
Aluminium tape	0.21	0.04	5.25
Polished aluminium	0.24	0.08	3.00
Aluminized kapton (aluminium outside)	0.14	0.05	2.80
Polished titanium	0.60	0.60	1.00
Black paint (epoxy)	0.95	0.85	1.12
Black paint (polyurethane)	0.95	0.90	1.06
—electrically conducting	0.95	0.80–0.85	1.12–1.19
Silver paint (electrically conducting)	0.37	0.44	0.84
White paint (silicone)	0.26	0.83	0.31
—after 1000 hours UV radiation	0.29	0.83	0.35
White paint (silicate)	0.12	0.90	0.13
—after 1000 hours UV radiation	0.14	0.90	0.16
Solar cells, GaAs (typical values)	0.88	0.80	1.10
Solar cells, silicon (typical values)	0.75	0.82	0.91
Aluminized kapton (kapton outside)	0.40	0.63	0.63
Aluminized FEP	0.16	0.47	0.34
Silver coated FEP (SSM)	0.08	0.78	0.10
OSR	0.07	0.74	0.09

Notes: SSM, Second Surface Mirror. OSR, Optical Solar Reflector.

balance is determined by its external surface characteristics and the radiative environment, the internal thermal balance determines equipment temperatures and is hence of crucial importance to the thermal engineer. Calculation of the internal thermal balance, involving radiative and conductive exchanges between all the spacecraft components, is complex and is covered in some detail in the next section.

11.4 THERMAL ANALYSIS

11.4.1 Thermal mathematical model (TMM)

Spacecraft are generally very complex structures within which temperatures are varying continuously as a function of location and time. Calculating these temperature fields in rigorous detail is, for all practical purposes, impossible. In order to progress further, it is first necessary to simplify the problem. This is done by generating an approximate representation of the spacecraft that is amenable to mathematical treatment. Such a representation is known as a *thermal mathematical model* (TMM).

In order to construct a TMM, the spacecraft is considered as being composed of a number of discrete regions within which temperature gradients can be neglected. These regions are known as *isothermal nodes*. Each node is characterized by a temperature, thermal capacity, heat dissipation (if any) and radiative and conductive interfaces with the surrounding nodes. Nodes that can 'see' space directly will also have radiative interfaces with the external environment.

11.4.2 Conductive heat exchange

The conductive heat flow rate, Q_c, is given by

$$Q_c = \frac{\lambda A}{l} \Delta T \tag{11.8}$$

where λ is the thermal conductivity, A the cross-sectional area, l the conductive path length and ΔT the temperature difference. The term $\lambda A/l$ is known as the *thermal conductance*, h_c, and hence the temperature difference can be written as

$$\Delta T = Q_c \frac{1}{h_c} \tag{11.9}$$

In most engineering applications, A and possibly λ may vary significantly along the path length. If the conductive path is considered as a number of discrete conductive paths connected in series, the temperature difference can be rewritten as

$$\Delta T = Q_c \left(\frac{1}{h_1} + \frac{1}{h_2} + \frac{1}{h_3} + \ldots \right) = Q_c \frac{1}{h_c}$$

and hence the effective thermal conductance, h_c, for the path can be found from

$$\frac{1}{h_c} = \frac{1}{h_1} + \frac{1}{h_2} + \frac{1}{h_3} + \ldots \tag{11.10}$$

For a spacecraft composed of n isothermal nodes, the heat conducted from the ith to the jth node is given by

$$Q_{c_{ij}} = h_{ij}(T_i - T_j) \tag{11.11}$$

where h_{ij} is the effective conductance between nodes i and j and T_i and T_j are the temperatures of the ith and jth nodes, respectively.

11.4.3 Radiative heat exchange

Radiative heat exchange between two surfaces is determined by three important parameters—the surface temperatures, the radiative view factors and the surface properties. For diffuse surfaces, the amount of radiation leaving a surface i and absorbed by a surface j can be shown [7] to be of the form

$$Q_{r_{ij}} = A_i F_{ij} \varepsilon_{ij} \sigma (T_i^4 - T_j^4) \tag{11.12}$$

where A_i is the area of the surface i, F_{ij} is the view factor of surface j as seen from surface i and ε_{ij} is a parameter known as the *effective emittance*. Note that it is assumed in the above that the value of the view factor F_{ij} remains constant over the surface i.

View factors

The *radiative view factor* F_{ij} is defined as the fraction of the radiation leaving one surface that is intercepted by another. It follows that from any node i inside a spacecraft, the sum of the view factors to surrounding equipment must be unity,

$$\sum_{j=1}^{k} F_{ij} = 1 \tag{11.13}$$

where k is the number of surrounding surfaces.

Consider two surfaces A_1 and A_2 as indicated in Figure 11.4 and let us assume that they are both diffuse surfaces. The radiation emitted from A_1 in the direction of A_2, per unit solid angle and per unit area of A_1, is given by

$$I_1 = I_0 \cos \phi_1 \tag{11.14}$$

where I_0 is the radiation intensity normal to A_1.

For the elementary surfaces δA_1 and δA_2, the total radiation leaving δA_1 and falling on δA_2 is given by

$$\delta Q_{r_{12}} = I_0 \delta A_1 \cos \phi_1 \frac{\delta A_2 \cos \phi_2}{s^2}$$

Hence, the total radiation reaching A_2 from A_1 is given by

$$Q_{r_{12}} = I_0 \int_{A_1} \int_{A_2} \frac{\cos \phi_1 \cos \phi_2}{s^2} dA_1 dA_2$$

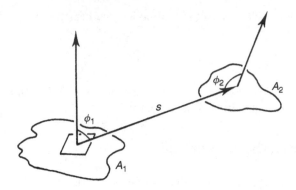

Figure 11.4   View factor geometry between surfaces A_1 and A_2

However, from equation (11.14) it can be easily shown that the total radiation leaving A_1 is given by

$$Q_{r_{\text{tot}}} = A_1 I_0 \int_0^{\pi/2} 2\pi \sin\phi \cos\phi \, d\phi = A_1 \pi I_0$$

The view factor, F_{12}, is then

$$F_{12} = \frac{Q_{r_{12}}}{Q_{r_{\text{tot}}}} = \frac{1}{A_1} \int_{A_1} \int_{A_2} \frac{\cos\phi_1 \cos\phi_2}{\pi s^2} dA_1 dA_2 \qquad (11.15)$$

which can be rewritten more generally in the form

$$A_i F_{ij} = \int_{A_i} \int_{A_j} \frac{\cos\phi_i \cos\phi_j}{\pi s^2} dA_i dA_j \qquad (11.16)$$

From an examination of the symmetry of this equation, an important reciprocity relationship can be deduced

$$A_i F_{ij} = A_j F_{ji} \qquad (11.17)$$

The calculation of view factors is a tedious process and many examples of standard configurations are available in the literature [6] to make the task easier. For simple spacecraft designs, view factors can also be obtained experimentally, for example, by observing shadows cast by scale models [8] or using a device known as a *factometer* [9]. Although these techniques still have a place in the thermal engineer's arsenal, sophisticated software tools, such as ESARAD [10, 11], capable of handling large and complex spacecraft configurations, have largely superseded them.

Effective emittance

The effective emittance between two surfaces has a complicated dependence on surface optical properties, mutual reflections and reflections via other nearby surfaces. Its detailed treatment is beyond the scope of this book but can be found in Reference [7]. In the simple case in which all surfaces are perfectly black (i.e. $\varepsilon = 1$ for all surfaces, not just surfaces i and j), the effective emittance reduces to the trivial result $\varepsilon_{ij} = 1$. Specular surfaces and/or emittance values less than unity give rise to more complicated expressions for ε_{ij}. For the relatively simple case of two parallel, diffuse surfaces, separated by a distance that is small compared with their surface areas, it can be shown that

$$\varepsilon_{ij} = \frac{\varepsilon_i \varepsilon_j}{\varepsilon_i + \varepsilon_j - \varepsilon_i \varepsilon_j} \tag{11.18}$$

In spite of the fact that this is a special case, seldom encountered in practice, errors introduced by its use are small, provided the surfaces involved are diffuse and have relatively high ε values. This expression is commonly used for the calculation of practical effective emittance values and is acceptable for normal purposes in which the spacecraft interior is painted ($\varepsilon_i = \varepsilon_j \approx 0.9, \Rightarrow \varepsilon_{ij} \approx 0.82$). The use of specular and/or low-emittance surfaces will involve complicated analysis, which will need to be repeated for every change in the spacecraft internal configuration, and this should be avoided where possible.

11.4.4 Calculation of nodal temperatures

Let us consider, at a particular time, the heat balance of the ith node of a TMM consisting of n nodes. The net heat absorbed by node i per unit time is given by

$$Q_{\text{external},i} + Q_i - \sigma \varepsilon_i A_{\text{space},i} T_i^4 - \sum_{j=1}^{n} h_{ij} (T_i - T_j) - \sigma \sum_{j=1}^{n} A_i F_{ij} \varepsilon_{ij} (T_i^4 - T_j^4)$$

where

$$Q_{\text{external},i} = J_s \alpha_i A_{\text{solar},i} + J_a \alpha_i A_{\text{albedo},i} + J_p \varepsilon_i A_{\text{planetary},i}$$

is the external heat input. $A_{\text{solar},i}, A_{\text{albedo},i}$ and $A_{\text{planetary},i}$ are the effective areas receiving, respectively, direct solar, albedo and planetary radiation, $A_{\text{space},i}$ is the effective area with an unobstructed view of space and Q_i is the internal heat dissipation. If the mass and specific heat of the node are m_i and C_i, respectively, the heat balance equation for node i can be written as

$$m_i C_i \frac{dT_i}{dt} = Q_{\text{external},i} + Q_i - \sigma \varepsilon_i A_{\text{space},i} T_i^4 - \sum_{j=1}^{n} h_{ij} (T_i - T_j)$$

$$- \sigma \sum_{j=1}^{n} A_i F_{ij} \varepsilon_{ij} (T_i^4 - T_j^4) \tag{11.19}$$

The thermal behaviour of the complete spacecraft is thus described by the set of n simultaneous non-linear differential equations as above, with i varying from 1 to n. It should be noted that, in most practical cases, $Q_{external,i}$ and Q_i will also vary with time.

Generally speaking, the above set of equations is not amenable to analytical solution, and numerical methods must be used. A common approach relies on first linearizing the equations as follows.

If $T_{i,0}$ is the temperature of the ith node at some arbitrary time, t_0, the temperature of that node a short while, δt, later is given by

$$T_i = T_{i,0} + \delta T_{i,0} \quad \text{where} \quad \delta T_{i,0} = \frac{dT_{i,0}}{dt}\delta t$$

From the first of these equations and assuming that $\delta T_{i,0}$ is small compared with $T_{i,0}$ we see that

$$T_i^4 = (T_{i,0} + \delta T_{i,0})^4 \approx T_{i,0}^4 + 4T_{i,0}^3\delta T_{i,0}$$

and hence that

$$T_i^4 \approx T_i(4T_{i,0}^3) - 3T_{i,0}^4 \tag{11.20}$$

Steady-state calculations

In the special case of steady-state calculations, where $Q_{external,i}$ and Q_i are constant and dT/dt is zero, use of equation (11.20) allows us to write equation (11.19) in the linearized form as follows:

$$T_i\left[\sum_{j=1}^{n} h_{ij} + 4\sigma T_{i,0}^3\left(A_{space,i}\varepsilon_i + \sum_{j=1}^{n} A_i F_{ij}\varepsilon_{ij}\right)\right] - \sum_{j=1}^{n} T_j[h_{ij} + 4\sigma T_{j,0}^3 A_i F_{ij}\varepsilon_{ij}]$$
$$= Q_{external,i} + Q_i + 3\sigma T_{i,0}^4 A_{space,i}\varepsilon_i + 3\sigma(T_{i,0}^4 - T_{j,0}^4)\sum_{j=1}^{n} A_i F_{ij}\varepsilon_{ij} \tag{11.21}$$

This is a set of linear equations in n unknowns T_i, where $i = 1$ to n, and can be solved by standard matrix inversion techniques. The calculated temperatures can then be substituted for the original values of $T_{i,0}$ and the calculation repeated until the difference between newly calculated and previously calculated temperatures is sufficiently small for steady-state conditions to be assumed, that is, dT_i/dt is sensibly zero. It should be noted that, although the criterion that $\delta T_{i,0}$ should be small may not be met initially, it will be so for the final iterations and hence the final calculated temperatures will be correct.

Transient calculations

A similar approach can be used to analyse non-steady-state conditions. However, the left-hand-side of equation (11.19) must be replaced by the term

$$m_i C_i \frac{(T_i - T_{i,0})}{\delta t}$$

and the temperatures and heat inputs by their average values over the time interval δt, given by $(T_{i,0} + T_i)/2, (T_{j,0} + T_j)/2, (Q_{\text{external},i,0} + Q_{\text{external},i})/2$ and $(Q_{i,0} + Q_i)/2$, respectively.

Replacing the values of $T_{i,0}$ by the newly calculated values T_i and repeating the calculation will provide the temperature history at successive intervals of time δt. It should be noted that, for this to be accurate, $\delta T_{i,0}$ must be small since, at a given time, the temperatures are the result of a single calculation and not the result of multiple iterations as in the case of the steady-state calculation. In practical terms, this means that δt should not be too large and that the initial temperature values $T_{i,0}$ should be as realistic as possible. In the latter context, the temperature distribution resulting from a steady-state calculation using orbital-average heat inputs is often used as the starting point for a transient calculation.

Many software codes have been developed over the years to perform the above calculations. The one most commonly used in Europe is the ESA-sponsored ESATAN package [12, 13]. For small TMMs, a spreadsheet-based programme is also available, THERMXL [14].

11.5 THERMAL DESIGN

11.5.1 Requirements and constraints

The first task is to examine the proposed spacecraft concept and assemble the thermal parameters for the on-board equipment. The equipment designer should provide upper and lower safe operating temperatures for his equipment and should also specify the operating power dissipation or dissipation ranges. Detailed drawings and materials lists will be required in order to calculate nodal thermal capacitances, conductance paths and view factors. Orbital and other operating constraints must also be defined. For example, an Earth observation spacecraft may well need to be in a near-polar orbit with the plane of the orbit set at a specified angle to the solar vector in order to emphasize ground features (shadow effects). There may also be constraints on allowable attitudes. For example, an Earth observation spacecraft must turn so that its instruments always face the Earth, a solar observatory will be oriented such that its instruments always face the Sun and an astronomical spacecraft must ensure that its telescopes never come too close to the solar vector.

As an example, let us look again at our spherical spacecraft. For simplicity, let us assume that its mission is to enable upper atmospheric density to be inferred from measurements of the orbit decay over a period of a few weeks. The example chosen is relevant, and yet simple enough to show thermal design methodology in action. The principal investigator requires the external geometry to be a clean spherical surface that is as inert as possible to try to minimize any possible chemical effects (the spacecraft will be flying through a residual atmosphere which is nearly 100% atomic oxygen—see Chapter 2). The ideal would be a gold-plated sphere with a diameter of 1 m. Launch site and launch vehicle constraints require that the spacecraft be spinning and that it will be in a polar orbit, but the angle between the orbit and solar vector is unknown. The angle between the spin axis and solar vector is also unknown and not controllable. The spacecraft systems engineers determine that the spacecraft will be tracked via a battery-powered beacon, located at the centre of the sphere, which dissipates 5 W. This is mounted through a circular panel across the spacecraft's 'equator' and supported by two tubes along the spin axis, as shown

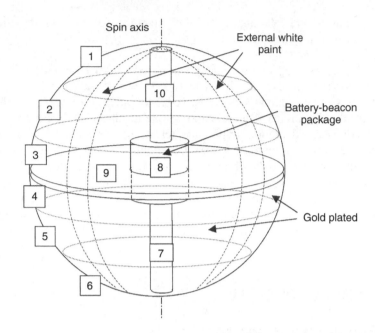

Figure 11.5 Spacecraft configuration and nodal breakdown

schematically in Figure 11.5. The two hemispheres on each side of the equatorial panel act as the beacon's antenna. The battery supplier confirms that, for the short lifetime of the spacecraft, the batteries can tolerate temperatures between −15°C and +60°C. We have already seen that the thermal analysis process involves approximations, particularly concerning the establishment of the TMM. We should, therefore, take an appropriate margin here, and design to stay within the range −5°C—+50°C.

11.5.2 Definition of worst-case conditions

The task of the thermal designer is not usually to achieve a specific temperature but rather to ensure that equipment stays within certain acceptable limits. To do this, *worst-case conditions* should be defined. These would typically be the orbits with maximum and minimum periods of sunlight, combined with certain extreme spacecraft attitudes and operational modes. In most cases, these worst-case conditions are established by inspection and experience. In the case of our air density spacecraft, we must design for both full-sun (hot case) and 59% sun (cold case) orbits, and for any attitude with respect to the solar vector.

From an inspection of Table 11.2, it is clear that our principal investigator will have to accept some compromise concerning the external surface of the sphere. A gold plated sphere will have an average temperature between 138°C and 176°C, neglecting the internal heat dissipation. It is evidently necessary to apply a low-α/ε surface coating, say a white paint, to part of the external surface. However, this should be the minimum area consistent with keeping the battery-beacon package below 50°C. A paint pattern consisting of several segments (see Figure 11.5) is relatively easy to apply and gives the necessary degree of surface uniformity.

11.5.3 The TMM

Figure 11.5 shows a simple nodal breakdown for the air density-measuring spacecraft. It consists essentially of a gold-plated aluminium sphere with a honeycomb panel across its equator, which supports the battery-beacon package at its centre. Two aluminium tubes are used to provide additional support along the spacecraft spin axis. Since the spacecraft is spinning and the only source of internal dissipation is at the centre, we may assume that temperatures vary only with spacecraft 'latitude'. Each hemisphere can therefore be modelled as a number of isothermal nodes (zones) generated by planes parallel to the spacecraft equator. For our example, each hemisphere is divided into three such nodes as shown in Figure 11.5. The various conductance terms, view factors and nodal thermal capacitances can now be calculated as indicated previously. Table 11.4 lists the resulting view-factor area products and conductance terms.

11.5.4 Results and discussion

For the analysis, the incident heat flux densities were the same as those used for the example worked in Section 11.3. That is,

$$J_s = 1371 \, \text{W/m}^2, \, J_p = 220 \, \text{W/m}^2,$$

$$J_a = 67.9 \, \text{W/m}^2 \, [\sim 0.33(0.15)J_s]$$

for the 100% sunlit orbit and 316.7 W/m² [∼ 0.33 (0.7) J_s] for the sunlit part of the 'cold' orbit.

The internal surfaces of the spacecraft were assumed painted with a matt white paint with $\varepsilon = 0.9$, whence $\varepsilon_{ij} = 0.82$. The paint pattern on the external surface consists of several (the number is not important) segments, each of which is defined by planes

Table 11.4 View factors and conductance terms

View-factor × area products (m²)

i\j	1	2	3	4	5	6	7	8	9	10
1		0.068	0.079	0	0	0	0	0.013	0.215	0.058
2	1.03		0.089	0	0	0	0	0.018	0.241	0.031
3	0	2.10		0	0	0	0	0.016	0.246	0.018
4	0	0	2.54		0.089	0.079	0.018	0.016	0.246	0
5	0	0	0	2.10		0.068	0.031	0.018	0.241	0
6	0	0	0	0	1.03		0.058	0.013	0.215	0
7	0	0	0	0	0	0.18		0.0027	0.031	0
8	0	0	0	0	0	0	0.16		0.054	0.0027
9	0	0	1.41	1.41	0	0	0	0.94		0.031
10	0.18	0	0	0	0	0	0	0.16	0	

Conductance terms (W/K)

containing the spin axis. This means that the percentage painted area of each node is the same. For this example, a silicate-based white paint with $\alpha/\varepsilon = 0.16$ has been used.

Table 11.5 and Figures 11.6 and 11.7 show the results (obtained using the ThermXL [14] software) for the case in which 15% of the external surface is covered with white paint, giving an effective $\alpha/\varepsilon = 1.38$ for the external surfaces of nodes 1 to 6.

It can be seen that the maximum predicted temperature of the battery-beacon package (Table 11.5) is comfortably inside the 50°C maximum. If required, the design can be trimmed, for example, by modifying the external paint coverage, and the analysis re-run until the desired maximum predicted temperature has been attained.

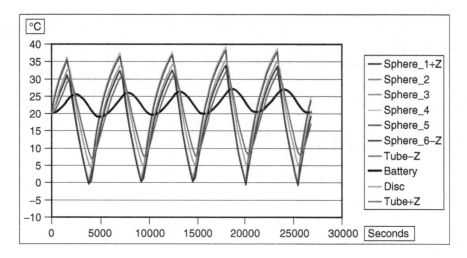

Figure 11.6   Temperature evolution, side-to-Sun, maximum eclipse

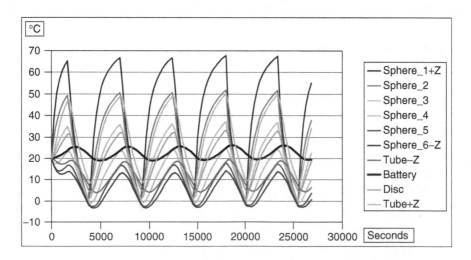

Figure 11.7   Temperature evolution, top-to-Sun, maximum eclipse

Table 11.5   Calculated temperatures for 100% sunlit orbit

Node Number	Location	Temperature (°C)	
		Side-to-Sun case	Top-to-Sun case
1	external surface	40.6	72.3
2	external surface	44.6	57.2
3	external surface	45.8	43.8
4	external surface	45.8	32.1
5	external surface	44.6	25.3
6	external surface	40.6	22.1
7	support tube	43.3	30.7
8	**battery-beacon**	**46.7**	**45.5**
9	Support platform	44.6	42.3
10	support tube	43.3	59.2

For the cold cases (Figures 11.6 and 11.7), the battery-beacon package remains well above the −5°C lower allowed limit. We may therefore conclude that our thermal design is adequate for this spacecraft and its mission.

More complex spacecraft with more complex missions and more variable environments will require more complicated thermal design. Reliance on thermal conduction, radiation exchange and insulation systems is known as *passive thermal control* and is the initial starting point for most spacecraft thermal design. However, where this proves inadequate, *active thermal control* techniques can be used. These are predominantly concerned with the active control of internal dissipation, for example, by the use of heaters, and of radiative, conductive and forced convective heat flows as a function of temperature. These active techniques are described in more detail in the next Section.

In the event that temperature requirements cannot be met, or can be met only at great expense or with significant risk, it will be necessary to revisit the requirements. This may involve re-qualifying hardware for different temperature ranges, or changes in spacecraft system design or mission profile. Project managers will expect such feedback earlier rather than later. Consequently, thermal design should start early with simplified TMMs, only moving to more detailed models once the overall concept has been shown to be sound.

11.6   THERMAL TECHNOLOGY

11.6.1   Passive control

Passive thermal control techniques available to the engineer consist essentially of the selection of surface properties, the control of conduction paths and thermal capacities and the use of insulation systems.

Surface finishes

Although a wide range of α/ε values is, in principle, available (see Table 11.3), the practical selection of surface finishes is often restricted by their ageing characteristics.

Contamination of low-α surfaces (white paint, polished or electroplated metal surfaces) will increase the α value and should be avoided. Paints generally consist of a binder that is transparent to visible light and a filler material that gives it its colour. Often the binder will be opaque in the infrared, which implies that most paints have high ε values. Unfortunately, many binders degrade and discolour under the influence of solar ultraviolet radiation, becoming less transparent to visible light. White paint on the outside of a spacecraft will suffer an increase in its α/ε value with time and care must be taken to select the most suitable formulation [5, 6]. The presence of contamination (grease, oil etc.) will exacerbate the situation. Reflectors such as *Second Surface Mirrors* (*SSMs*) (plastic film, aluminized or silver coated on the back side) or *Optical Solar Reflectors* (*OSRs*) (similar to SSM but using thin sheets of glass instead of plastic film) are less sensitive to solar radiation and are easier to clean. Compared to paints, however, SSMs and OSRs are more difficult to apply to spacecraft surfaces, particularly those with complicated geometry.

Suitable allowance must be made for any likely changes in surface optical properties during the design lifetime of the spacecraft.

Conduction paths

Solid conduction needs little discussion. Joints, however, need special consideration. The thin air gap between the non-contacting areas of a simple bolted joint contributes significantly to the effective conductance of the joint. Under vacuum conditions, this contribution disappears. The conductance of such a joint under vacuum conditions is strongly dependent on the quality of the mating surfaces and on the bolt tension. These uncertainties can be minimized by the use of interface fillers such as soft metals (e.g. indium foil) or loaded polymers (e.g. silver-loaded silicone) [5, 6].

Heat pipes and two-phase systems

Two-phase heat transport systems, such as *heat pipes, loop heat pipes* (*LHPs*) and *capillary-pumped loops* (*CPLs*), are devices which transfer heat in the form of the latent heat of vaporization, using a volatile working fluid that is circulated by capillary action in a porous wick structure. They are capable of large heat transport rates at very low temperature gradients (i.e. very large effective conductance).

A basic heat pipe is shown schematically in Figure 11.8. It consists essentially of a sealed tube possessing a porous structure (the wick) on its inside surface and containing a volatile liquid (the working fluid) in thermodynamic equilibrium with its vapour. The

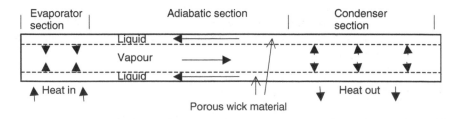

Figure 11.8 Schematic illustration of a basic heat pipe

quantity of liquid is just sufficient to saturate the wick structure. The working fluid and wick material are chosen to ensure mutual wetting (i.e. a small contact angle), which guarantees retention of the working fluid in the wick. If a temperature gradient exists between, say, one end of the heat pipe and the other, heat will be absorbed as latent heat at the hot end (the evaporator section), as the working fluid evaporates, which will then be given up again by condensation at the cold end (the condenser section). Continued evaporation causes the liquid-vapour interface to withdraw into the wick and it is the resulting curved meniscus that generates the pressure drop needed to drive the condensed working fluid back to the evaporator.

Assuming that the working fluid perfectly wets the wick material (i.e. contact angle is zero), the maximum capillary pressure that can be generated by the wick is given by

$$\Delta P_{\max} = \frac{2\sigma}{r_0}$$

where σ is the surface tension and r_0 is the effective pore radius of the wick.

With reference to Figure 11.9, the maximum rate dm/dt at which the working fluid can be recycled is then governed by

$$\frac{2\sigma}{r_0} = (l_{\text{eff}} \sin \alpha)\rho g + \frac{l_{\text{eff}}}{A_{\text{wick}}} \frac{\eta}{\phi \rho} \frac{dm}{dt} + \Delta P_{\text{vap}} \qquad (11.22)$$

where

ρ = density of the liquid phase, g = acceleration due to gravity,
η = dynamic viscosity of the liquid phase, ϕ = permeability of the wick structure,
l_{eff} = heat pipe effective length = $l_{\text{adia}} + 1/2(l_{\text{evap}} + l_{\text{cond}})$,
A_{wick} = cross-sectional area of wick, ΔP_{vap} = pressure drop in the vapour flow.

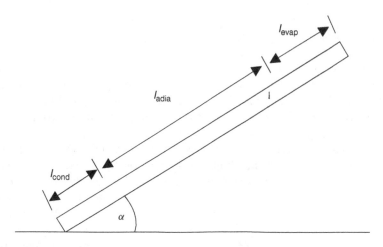

Figure 11.9 Heat pipe oriented with evaporator above condenser

In most practical applications, the pressure drop in the vapour is small compared to that in the liquid phase. Equation (11.22) can then be rearranged as follows:

$$\frac{dm}{dt} \approx \frac{A_{\text{wick}}}{l_{\text{eff}}} \frac{\phi\rho}{\eta} \left(\frac{2\sigma}{r_0} - \rho g l_{\text{eff}} \sin\alpha \right)$$

Hence, if H_v is the latent heat of vaporization, the maximum rate of heat transport Q_{max} is given by

$$Q_{\text{max}} \approx \frac{A_{\text{wick}}}{l_{\text{eff}}} \frac{\phi\rho H_v}{\eta} \left(\frac{2\sigma}{r_0} - \rho g l_{\text{eff}} \sin\alpha \right) \tag{11.23}$$

When operating in space ($g = 0$), it should be noted that the product $Q_{\text{max}}\, l_{eff}$ is a constant for a given heat pipe design—the longer the heat pipe, the smaller its heat transport capability. For this reason, heat pipe performance is often specified in units of Watt-metres. The second thing to note in equation (11.23) is the dependence on the angle of tilt α when operating on the ground. Capillary forces are relatively weak forces and a heat pipe with a performance of several hundreds of Watt-metres under zero-gravity conditions may cease to operate on the ground if its evaporator is raised more than a few mm above its condenser section. On the ground, a heat pipe only approximates to its zero-gravity performance when tested horizontally. The effect of gravity on performance must be taken into account fully during system design, particularly if a system of several heat pipes (e.g. in a radiator) is used or heat pipes are required to be bent into complex shapes. It is easy to design heat-pipe-based thermal control systems that prove to be unverifiable on the ground!

As noted above, the pressure drop in the vapour is relatively small. Since the vapour pressure is a strong (exponential) function of the temperature at the liquid-vapour interface (as described by the Clausius-Clapeyron equation), the temperature drop in the vapour will also be small. For the purposes of analysing the effect of a heat pipe in a TMM, little error is introduced by assuming the vapour space to be isothermal and an effective conductance h_{eff} can be defined, given by

$$\frac{1}{h_{\text{eff}}} = \frac{1}{h_{\text{evap}}} + \frac{1}{h_{\text{cond}}}$$

where h_{evap} and h_{cond} are the radial conductance values through the wall and wick in the evaporator and condenser zones, respectively. Unlike normal thermal conduction, the effective conductance of a heat pipe is not a function of its length. Figure 11.10 shows a constant conductance heat pipe developed for the European Automated Transfer Vehicles (ATVs) used to re-supply the ISS.

LHPs and CPLs are variations on the basic heat pipe, designed to improve ultimate performance and/or to improve flexibility of application. In a LHP, the working fluid is returned to the evaporator via an external pipe. The CPL takes the process a step further and several evaporators, operating in parallel, may be attached to the same liquid return line. This enables the thermal control of complex configurations that would otherwise require the use of active systems such as pumped liquid loops. Heat pipes, LHPs and CPLs often use anhydrous ammonia as the working fluid, since it permits operation over a wide temperature range from approximately $-50°C$ to about $+80°C$. Other working fluids are

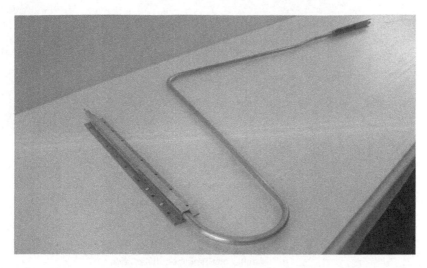

Figure 11.10 Typical constant conductance heat pipe used to conduct heat to a radiator. (Reproduced by permission of Euro Heat Pipes, Belgium)

available, however, allowing heat pipe devices to operate from cryogenic temperatures up to several hundred degrees.

Deeper discussion on the subject of two-phase heat transport systems is, unfortunately, outside the scope of this text. The interested reader is referred to References [15–19].

Phase change materials

Phase change materials (*PCMs*) can be used where increased thermal capacity is required, for example, to minimize temperature excursions during eclipse or other transient conditions such as occasional high power dissipation in equipment. They work by absorbing or releasing latent heat during solid-liquid phase changes. Care needs to be taken to ensure that there is sufficient free volume (ullage) to accommodate the volume changes accompanying phase change. There should also be sufficient internal structure to ensure adequate thermal conduction throughout the PCM material, especially under zero-gravity conditions. Typically, a PCM consists of a hydrocarbon wax but many substances are available [20] covering a wide range of temperatures.

Insulation systems

Extremely effective insulation systems can be designed to minimize radiative exchanges in the vacuum of space. These are normally referred to as *super-insulation systems* or multi-layer insulation (MLI) blankets. They consist typically of several layers of aluminized plastic film (e.g. Mylar of Kapton) acting as radiation shields, each separated by a low-conductance spacer (e.g. silk, Nylon or glass-fibre net). In practice, in many applications the spacers are omitted and instead alternate layers of aluminized film are crinkled or mechanically dimpled to minimize contact between adjacent layers. A typical insulation blanket might consist of 40 or more layers, each about $10\,\mu m$ thick and aluminized on

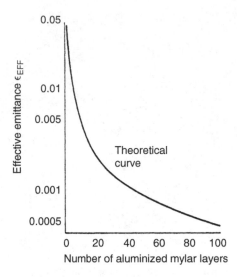

Figure 11.11   Effective emittance versus number of aluminized Mylar layers (theoretical)

both sides. This would have an effective emittance value of about 0.001. The theoretical values of effective emittance as a function of number of layers, neglecting the effects of conduction between layers, are shown in Figure 11.11.

It should be remembered that, in practical systems, full account must also be taken of edge effects, joints and fasteners. The design of MLI blankets and their method of installation must also ensure adequate venting, particularly during the rapid depressurization accompanying the launch phase. Further extensive information concerning insulation systems can be found in the literature [6].

11.6.2   Active control

Active thermal control systems are generally more complex than passive systems and often consume power and sometimes telemetry resources. Such systems are typically less reliable and often heavier. As a general rule, active systems should be used only when it has proved impossible to meet requirements by passive means alone. Active systems will typically be used for very temperature-sensitive equipment (telescopes, scientific instruments, atomic clocks etc.), missions in which environmental conditions are very variable (to Mercury, Mars, Europa etc.) or where heat dissipations are large/variable/random (e.g. in the cabin of a manned space vehicle). Active thermal control may also be required for specialized payload elements such as infrared sensors.

Heaters

Heaters constitute, probably, the simplest and most obvious active thermal-control device. A passive thermal-control design may lead to minimum predicted temperatures falling below permissible limits for a given equipment or subsystem. A thermostatically controlled heater can be used to prevent this. In view of the high reliability of modern thermostats, such a solution may well be considered competitive with alternative passive approaches, such as use of PCM units. Other typical applications for heaters include the propulsion subsystem (thrusters, fuel lines and valves, tanks etc.), batteries, experiment sensors and mechanical actuators. Heaters may be in the form of compact metal-mounted resistors, metal co-axial cables in which the heating element forms the core of the cable (thermo-coax), or adhesive sheets in which the element is configured rather like a Printed Circuit Board (PCB) between two layers of Kapton foil.

Variable conductance heat pipes (VCPHs) and diodes

A *variable conductance heat pipe* (*VCHP*), shown schematically in Figure 11.12, is a variant of the simple heat pipe described previously. A non-condensable gas, typically nitrogen, is used to progressively block the condenser section as a function of evaporator temperature.

There are several forms of VCHP—the one illustrated in Figure 11.12 is known as a cold, wicked reservoir design. During operation, the vapour always streams from evaporator to condenser, so driving the non-condensable gas always towards the reservoir. The reservoir is cold and hence the partial vapour pressure of the working fluid is low. Most of the gas in the reservoir is hence composed of the non-condensable gas. Since the vapour can only travel through this gas by diffusion, the vapour flow is effectively stopped when it meets the non-condensable gas. Hence, the heat flow to regions of the condenser section that are down-stream of the vapour-gas front is very much reduced. The reservoir is equipped with a wick structure that connects with the main heat pipe wick. This ensures that any vapour that does enter the reservoir and condenses will be removed by capillary action. In operation, any increase in evaporator temperature will cause the internal pressure to rise. This will compress the non-condensable gas, causing the vapour-gas front to retreat down the pipe and expose a greater length of condenser to the vapour flow. The device thus acts as a thermostat and, configured as described, operates completely passively. Such a VCHP, used to control the temperature of equipment in the avionics bays of the ATVs, is shown in Figure 11.13. Figure 11.14 shows the VCHPs integrated into the avionics bay of ATV-3 (left-hand-side of the picture). Also shown (hanging from the crane) is the avionics bay for ATV-2 (Johannes Kepler) in its final configuration, complete with radiators.

The ultimate temperature stability that can be obtained using a VCHP depends on the ratio of the reservoir and condenser volumes—the larger the reservoir, the better the

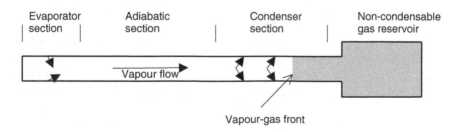

Figure 11.12 Schematic drawing of a cold, wicked VCHP

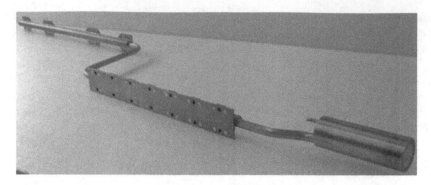

Figure 11.13 A VCHP designed for the European Automated Transfer vehicle (ATV). (Reproduced by permission of Euro Heat Pipes, Belgium)

Figure 11.14 VCHPs integrated in the ATV avionics bay radiators — lower left, during integration and, upper right, a completed avionics bay. (Reproduced by permission of EADS Astrium)

stability. If even greater temperature stability is required, a heater can be attached to the reservoir to control the partial vapour pressure of the working fluid and hence control the position of the vapour-gas front. This enables the vapour-gas front to be moved as a function of the temperature of the equipment being cooled, allowing temperature-control precision to fractions of a degree.

Consider now the effect of removing the connection between the heat pipe and reservoir wicks and omitting the non-condensable gas. Such a device, known as a *liquid trap heat pipe diode* [21], is shown schematically in Figure 11.15. When section 'A' is hotter than section 'B', liquid accumulates by condensation in the reservoir until the heat pipe dries out and stops conducting. When 'B' is hotter than 'A', the liquid evaporates from the reservoir, re-condenses in the heat pipe and the heat pipe works normally. Such a device can be useful, for example, for switching between radiators to prevent backflow of heat

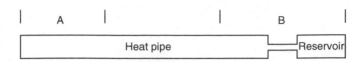

Figure 11.15   Schematic drawing of a liquid-trap heat pipe diode

from a temporarily sunlit radiator. Use of a heater on the reservoir enables the heat pipe to be operated as a thermal switch.

Mechanically-pumped two-phase loops

Mechanically-pumped two-phase loops are similar to CPLs with the addition of a mechanical pump in the liquid return line. This increases the heat transport capability and also renders the loop less sensitive to instabilities caused, for example, by gas bubbles in the liquid lines or nucleation within the evaporators. Such systems would normally be considered for complex thermal control applications involving large heat transport requirements such as might be encountered on a large space station.

Liquid loops

Perhaps the simplest concept for the active transport of heat is to use *liquid loops*, such as is used for cooling automobile engines or in the domestic central heating system. Liquid coolant is pumped between the various heat sources (dissipating equipment) and sinks (e.g. radiators or inter-loop heat exchangers) by, typically, an *electrically-driven centrifugal pump* and temperature control is effected by using by-pass valves. Gas-pressurized *accumulators*, in which flexible diaphragms separate gas and coolant, are used to control the pressure and *gas traps* are included to minimize the risk of damage from bubbles. Since the pump is both a single-point failure risk and the most vulnerable item in the loop, the pump package will usually consist of two pump units in cold redundancy.

Liquid loops are inherently rather massive. They would normally only be used where their particular attribute of system and operational flexibility is required. They find most application in manned spacecraft such as the Shuttle and the ISS. The liquid-loop-based active thermal-control system for the Columbus module on the ISS is shown schematically in Figure 11.16.

The choice of liquid should be made with some care. In order to minimize power consumption, the specific heat should be high (low mass flow rate) and dynamic viscosity low (low drag forces). In order to minimize system mass, the boiling point should be high (to minimize containment pressure). However, there must be no risk of freezing during any phase of operation, particularly if water is considered, which expands on freezing. For use in cabins containing a human crew, toxicity in the event of leaks must be taken into account. Consequently, water is used in all the manned volumes of the Shuttle and ISS. Because of its undesirable freezing characteristics and high freezing temperature, water is not used outside the pressurized modules. Instead, the water loops interface with the radiator assemblies via heat exchangers and separate external loops that use low freezing point fluids. The Shuttle uses Freon-21 in its external loop and the ISS uses ammonia.

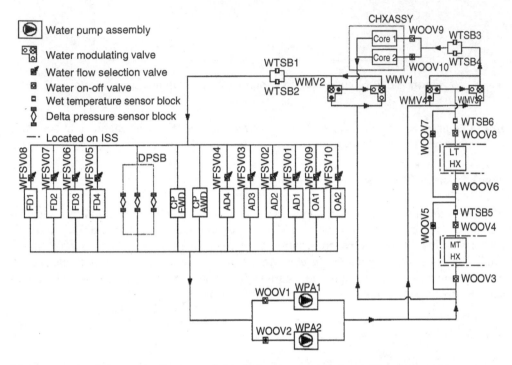

Figure 11.16 Schematic diagram of the Columbus water loop. (Reproduced by permission of EADS Astrium)

Louvres and shutters

A *louvre* [6] is a device that varies the effective emittance of a radiator in response to temperature. It is usually—but not always—mounted on the outside of a radiator panel and typically consists of an array of metallic blades rather like a venetian blind. The concept is shown schematically in Figure 11.17. When the blades are open (perpendicular to the radiator surface), the radiator has a good view of space and radiates accordingly. When they are closed, however, the radiator sees a low-emittance surface and most of its radiated heat is reflected back. The blades may be activated individually, for example, by bimetallic springs or by liquid-filled Bourden spiral actuators thermally connected to the radiator, or as a complete assembly by a single actuator. Actuators can be bimetallic, hydraulic or electrical, and operate in response to equipment temperatures, radiator temperatures or computer- or ground-generated commands. Figure 11.18 shows a photograph

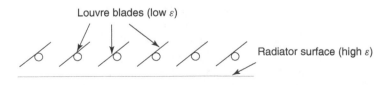

Figure 11.17   Schematic illustration of a louvre

Figure 11.18 Louvre assembly developed for ROSETTA. (Reproduced by permission of SENER)

of a lightweight louvre assembly driven by bimetallic springs, which was developed for the ROSETTA spacecraft. Louvres operate most effectively when mounted on radiators that are not illuminated by the Sun. In sunlit locations, complex reflections occur that make the prediction of louvre performance complicated and unreliable. Under these circumstances, an alternative approach, which was used on the GIOTTO spacecraft during its mission to Halley's Comet, is to partially cover the radiator surface with a motor-driven low-emittance roller-blind made from a sheet of metallized Kapton film.

Refrigerators and heat pumps

A *heat pump* may be used to increase the temperature of a radiator. Since the amount of heat radiated from a surface increases with the fourth power of its temperature, quite small temperature rises are sufficient for a significant increase in heat dissipation rate. The penalties for this are the mass and power consumption of the heat pump and any ancillary interfacing equipment such as liquid loops. Under normal circumstances, the engineer will find that it is usually better to try to find more radiator area. Exceptions would be where the amount of radiator area is fundamentally limited or where heat must be rejected in a relatively warm environment such as might occur on the lunar or Martian surface during daytime.

The most common applications are to cool sensors or other equipment in order to reduce signal-to-noise ratio and, in the specific case of manned space activities, to provide refrigeration for food and biological specimens resulting from scientific research.

Refrigeration requirements tend to fall within one of four categories, corresponding to temperature ranges.

1. $+4^oC$ — -20^oC. This range is basically concerned with food storage and is normally provided by *thermoelectric cooling* or mechanical two-phase heat pumps based on the *Rankine cycle*. Thermoelectric refrigerators are simple, rugged and reliable (no moving parts). They are not very efficient thermodynamically, however, and become very power-hungry when heat loads or the temperature-lift

(i.e. the difference between the cooled equipment and heat sink) become large. Two-phase heat pumps, such as those used in domestic refrigerators/freezers, are more efficient but much more complicated [22], a particular issue being to avoid damage to the vapour compressor due to accidental ingestion of liquid under zero-gravity conditions.

2. $-80\,°C$. This temperature is used for long-duration storage of biological specimens. It is too cold for thermoelectric cooling and a two-phase refrigerator would require several stages, involving substantial mass, power and reliability penalties. Instead, single-phase gas cycles such as the *Stirling* or *Brayton cycles* are used [23, 24]. The Brayton cycle is a continuous loop, whereas the Stirling cycle is reciprocating one. The Brayton cycle is inherently easier to interface with a large refrigerator, although it is not as thermodynamically efficient as the Stirling cycle. The MELFI specimen storage freezer (see Figure 11.19) on the ISS and the NICMOS cooler on the Hubble Space Telescope both use the Brayton cycle.

3. $80\,K\,(-193\,°)C$. This is an important temperature for Earth observation spacecraft since it is the temperature needed by the infrared detectors of instruments designed to look through, for example, the 8-μm atmospheric window (see Section 11.2.2). If heat loads are low and there is an unrestricted view of space, carefully designed radiators can achieve these temperatures. The METEOSAT spacecraft, orbiting at geosynchronous height, use such radiators. In LEO, however, the effect of the Earth's planetary radiation is difficult to avoid, and mechanical coolers using the Stirling cycle are now common [25]. Interest in these temperatures is not limited to Earth observation missions, however. Figure 11.20 shows the Stirling-cycle coolers developed for the gamma-ray spectrometer on board the Integral space observatory.

4. $<4\,K$. Temperatures in this range are predominantly of interest to astronomers wishing, for example, to measure the cosmic infrared background or detect the infrared emission from planets orbiting other stars. Depending on the application, cooling may be required down to the mK range, although it is an arguable point as to whether the achievement of such low ultimate temperatures is legitimately the responsibility of thermal control or more properly the province of the instrument developer. Techniques available for these temperatures [23, 24] include *Joule-Thomson, Brayton* and related *Claude cycles, sorption cooling, dilution refrigeration*, and *adiabatic demagnetization (ADR)*. The Planck spacecraft uses a cascade of radiators, a 20 K sorption cooler, a 4 K Joule-Thompson cooler (see Figure 11.21) and finally a Helium dilution refrigerator to achieve the 0.1 K temperature required by its microwave detectors.

11.7 THERMAL DESIGN VERIFICATION

Thermal design verification consists of two fundamentally different activities: (1) qualification of the hardware making up the thermal-control subsystem and (2) verification of the accuracy of the TMM. The latter enables the TMM to be used to make reliable temperature predictions for any particular mission scenario (orbit, attitude, payload operational time-lining etc.).

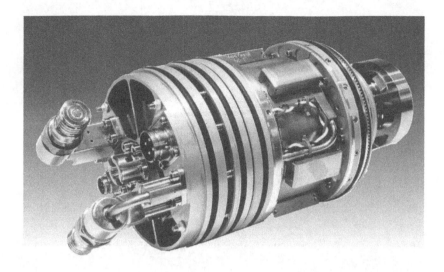

Figure 11.19 Reverse-Brayton-cycle cooler for the MELFI freezer. (Reproduced by permission of ESA/Air Liquide)

Figure 11.20 50–80 K Stirling-cycle cooler for the gamma-ray spectrometer on board the Integral spacecraft. (Reproduced by permission of EADS Astrium)

Figure 11.21 Cold stage of the 4 K Joule-Thompson cooler mounted on the Planck spacecraft. (Reproduced by permission of ESA. Image by Alphons Arts)

11.7.1 Hardware verification

As with any other spacecraft components, thermal-control hardware must be shown to be suitable for its purpose. The optical properties of thermal-control surfaces, the effective conductance of joints, and the performance of insulation systems, must be measured. The performance of heat transport loops, heat pipes, cryogenic systems, and so on, must be verified. It is also necessary to verify that the hardware will operate once it is in space (launch vibration tests, vacuum tests, temperature exposure tests) and will continue to operate correctly for the life of the mission (life tests). This is achieved by exposing qualification samples or units to conditions more severe than will be encountered in flight, to verify that the design is suitably robust. The flight hardware will then be tested to limits that also exceed expected flight conditions but are less severe than the qualification values. In many cases the thermal control engineer will be using hardware that has a proven track record in space (paints, insulation etc.), in which case qualification can be established by similarity with past applications. However, care should be taken on a case-by-case basis to ensure that previously established qualification limits remain valid for the current case. For example, external thermal-control paints that have good performance in an Earth orbit will probably prove unacceptable in the vicinity of Mercury, because of the high temperatures and intense solar ionizing radiation encountered. It should also be noted that, although the atmosphere is extremely thin at LEO altitudes, it is composed almost entirely of atomic oxygen with a very high kinetic temperature (because of the satellite velocity). This environment, particularly when combined with solar UV radiation, can be very damaging for some thin film materials [5].

11.7.2 Thermal model verification

It will be recalled that the TMM is an approximation to the real situation, devised by the thermal control engineer to enable temperature predictions to be made. It is essential to verify the accuracy of these models and, where inaccuracies are found, to amend the TMM accordingly. This is done by performing thermal balance tests that make use of sub-scale or full-size models of spacecraft or parts of spacecraft (e.g. a payload module may be tested on its own if the spacecraft service module to which it is to be attached is already a well-established design).

A spacecraft thermal balance test requires high vacuum conditions to minimize air conduction/convection, a heat sink to simulate the cold radiative environment of space and heat source(s) to simulate the external heat inputs (Sun, albedo, planetary infrared radiation). Facilities to test even small spacecraft are extremely expensive, both to manufacture and to operate. There are therefore relatively few of them and, in Europe, their use is co-ordinated by ESA/ESTEC [26] via the web-based European Environmental Test Facility Inventory (EETFI) (http://eetfi.esa.int/). The thermal engineer needs to consider carefully just how accurately the environment needs to be simulated, in order to verify his TMM, since the higher the required accuracy the higher will be the cost. It may be necessary to use a well-collimated and calibrated solar simulator to verify the external thermal interface between spacecraft and environment (including shadowing effects and multiple reflections). The absolute value of the simulated solar input is not so important, however, provided it is known. The same applies to the surrounding heat sink—most facilities use matt-black painted shrouds cooled by liquid nitrogen for this purpose. The shroud temperatures are monitored and suitable allowance made in subsequent analysis. Albedo and planetary infrared radiation are seldom simulated, although a slightly higher solar constant may be used to ensure that the test article (particularly if it is a real spacecraft) does not run too cold. A typical test sequence will consist of several steady-state tests at different spacecraft attitudes, together with a transient test (e.g. simulating eclipses). Such a test on a full-size development model of a spacecraft (or, indeed, on the proto-flight spacecraft) should provide sufficient information to enable the TMM to be verified and, if necessary, 'trimmed'.

These full-size solar simulation tests are expensive, however, and for large modern spacecraft problematic owing to the difficulty of getting a very large diameter solar simulation beam of the required quality. In such cases, adequate verification of the TMM can usually be obtained by a combination of (1) a full-size spacecraft test using infrared heaters instead of a solar simulator [27] and (2) verification of the external thermal interface using a sub-scale model specially designed for the purpose and sized to fit within the beam of available solar simulators.

As a general rule, the more testing that can be done, the better. However, in practice the amount of testing which a project manager will sanction will be based on compromise and will depend on such parameters as criticality of the thermal design, size of current temperature margins, complexity of external geometry, mission operational complexity and the project budget. However, the availability of reliable active thermal-control techniques and increasingly sophisticated thermal analysis tools, and the tendency to base new spacecraft on past well-proven designs, is steadily reducing the need for detailed extensive thermal balance testing. Experience indicates that the thermal control engineer will always request more testing than the project manager is prepared to pay for.

11.8   EXAMPLE OF SATELLITE THERMAL DESIGN – XMM/NEWTON

In this chapter we have attempted to describe the techniques that are used to devise, analyse and verify a spacecraft thermal design. In practice, every spacecraft is different and each thermal design tends to be unique. It will be a compromise between the thermal engineer's desire for perfection and other constraints such as mass, cost and available power and telemetry resources. Although designers will attempt to make use of the legacy of past successful designs, there is no 'standard thermal-control subsystem'. As a rule of thumb, the thermal control system will usually constitute between 2 and 5% both of spacecraft mass and development cost [28]. An example helps to illustrate the way in which these design techniques are used on modern spacecraft.

The X-ray Multi-mirror Mission (XMM) spacecraft [29], subsequently renamed 'Newton' and shown schematically in Figure 11.22, is an advanced X-ray observatory that uses three grazing-incidence mirror modules, each containing 58 concentric mirror assemblies,

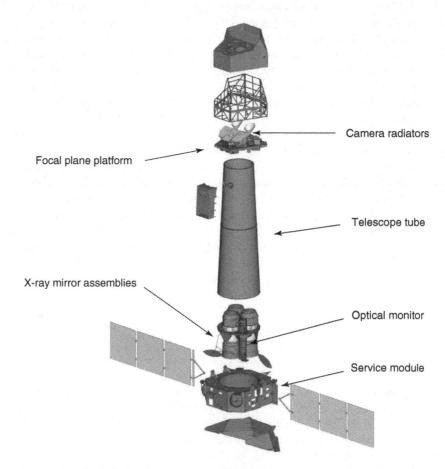

Figure 11.22 'Exploded view' of the XMM/Newton spacecraft. (Reproduced by permission of European Space Agency)

to focus X-ray images on the charge coupled device (CCD) detector arrays located in the focal plane. The spacecraft 'housekeeping' subsystems are located in the service module that surrounds the mirror modules. The solar arrays are attached to the service module. The mirror modules and focal plane instrumentation are separated by the telescope tube that is about 6.8 m in length. XMM has an overall length of around 10 m and a mass of about 10 tonnes.

In order to avoid interference from the Earth's trapped radiation, XMM is in a highly elliptical orbit and only performs its observations when well outside the radiation belts. Specifically, it is a 48 h orbit with apogee at 114 000 km and perigee at 7000 km. Observations are made during the 40 h of each orbit when it is above 40 000 km. Spacecraft attitude is controlled to within ±20° of 'side-to-Sun'. The inclination of the orbit is 40°, so it suffers from periodic eclipse seasons. The eclipses always occur well below the altitude at which observations are made, however, and never exceed 1.7 h. Under these circumstances, the external thermal environment is rather stable. There is little influence from the Earth, even during an eclipse season, and none during observing periods, and this has been taken advantage of in the thermal design.

The most critical requirement of the thermal design [30] was to ensure that thermal distortions would not jeopardize the alignment of the X-ray telescope (which has a 7.5 m focal length) or the relative alignment between this telescope and, for example, the star trackers. This translated into a requirement to control the temperature of the mirror modules and mirror support platform (which also supports the star trackers) at an average temperature of 20°C, and to limit temperature gradients to less than 2°C. This was achieved by equipping each mirror module, the mirror support platform and the entry and exit baffles with heaters controlled by the mirror thermal control unit (MTCU). The heaters are pulse-modulated with a duty cycle that is updated periodically by command from the ground. This approach was made possible by the large thermal mass and stable external environment. The surrounding service module is maintained at an average temperature of 15°C by a combination of equipment dissipation and thermostatically controlled heaters. The service module (Figure 11.23) is insulated from the environment by multi-layer super-insulation blankets, except for the panel radiators. At the other end of the spacecraft, the focal plane instrumentation compartment is controlled passively under normal operating conditions. If an instrument is turned off, its thermal dissipation is automatically replaced by switching on equivalent heaters. Contingency thermostats are also provided, which will switch these heaters on or off if an instrument approaches its non-operating temperature limit for any reason. The sunlit side of the compartment is fully insulated—thermal balance is achieved by adjusting the heat rejected from the shadowed sides. Special cone-shaped radiators are provided to cool the camera detectors to cryogenic temperatures. As shown in Figure 11.24, these are inclined away from the sun at an angle of 20° to avoid any risk of sunlight entering them, as well as being located behind the fully insulated sun-shield. The telescope tube is constructed from carbon fibre (high-ε surface) and is fully insulated externally. This results in a very stable environment with gradients across the tube diameter not exceeding 3°C. No heaters were implemented on the telescope tube.

As can be seen from the photographs, the thermal design of XMM involves extensive use of MLI blankets. These were typically constructed from 20 double-sided aluminized layers, separated by Dacron net spacers, together with an outer layer of carbon-impregnated Kapton. This black outer layer, which gives XMM its rather sinister black appearance, is electrically conducting and is intended to prevent the build-up of static

Figure 11.23   XMM/Newton lower module showing mirror entry baffles and surrounding service module. (Reproduced by permission of European Space Agency)

Figure 11.24   XMM/Newton focal plane instrumentation compartment. (Reproduced by permission of European Space Agency)

electricity and consequent electrostatic discharges. It has the added advantage that its thermo-optical properties will not change during the 10 year life of the mission, so helping to maintain spacecraft temperature stability.

ACKNOWLEDGEMENT

The assistance of the author's ESTEC colleagues is gratefully acknowledged. Particular thanks are due to Mr. Olivier Pin who ran the ThermXL analysis for the example in Section 11.5.

REFERENCES

[1] ESA (2008) *Space Engineering—Thermal Control General Requirements*, ECSS-E-ST-31C, November 2008.

[2] Lide, D. R. and Frederikse, H. P. R. (eds) (1995) *CRC Handbook of Chemistry and Physics*, CRC Press, Boca Raton, FL.

[3] NASA (1983) *Space and Planetary Environment Criteria Guidelines for Use in Space Vehicle Development* (1982 revision), TM82478, Volume 1.

[4] NASA (1983) *Space and Planetary Environment Criteria Guidelines for Use in Space Vehicle Development* (1982 revision), TM82501, Volume 2.

[5] ESA (2004) *Data for the Selection of Space Materials and Processes*, ECSS-Q-70-71A rev. 1, June 2004.

[6] ESA (1989) *Spacecraft Thermal Control Design Data*, ESA PSS-03-108, Issue 1.

[7] Redor, J-F., *Introduction to Spacecraft Thermal Control*, ESA EWP1599, Version 1.10, 1990.

[8] Eckert, E. R. G. (1959) *Heat and Mass Transfer*, McGraw-Hill, New York.

[9] Semple, E. C. (1967) *Principles and Techniques in the Passive Thermal Control of Spacecraft*, Royal Aircraft Establishment Technical Report No. TR 67100.

[10] ITP–ESATAN-TMS r3 Workbench Getting Started Guide [December 2010].

[11] ITP–ESATAN-TMS r3 Workbench User Manual [December 2010].

[12] ITP–ESATAN-TMS r3 Thermal Training Manual [December 2010].

[13] ITP–ESATAN-TMS r3 Thermal User Manual [December 2010].

[14] Knight, R., Pin, O. and Thomas, J. (2000) ThermXL: a Thermal Modelling Tool Integrated Within Microsoft Excel, *30th International Conference on Environmental Systems and 7th European Symposium on Space Environmental Control Systems*, Toulouse (France), 9-13 July, 2000.

[15] Dunn, P. D. and Reay, D. A. (1994) *Heat Pipes*, (4th edn), Pergamon Press, Elmsford, New York.

[16] Peterson, G. P. (1994) *An Introduction to Heat Pipes—Modelling, Testing and Applications*, John Wiley & Sons, Inc., New York.

[17] Faghri, A. (1995) *Heat Pipe Science and Technology*, Taylor and Francis, Washington, DC.

[18] Ku, J. (1995) Operating Characteristics of Loop Heat Pipes, Paper No. 1999-10-2007, *29th International Conference on Environmental Systems (ICES)*, Denver, CO (USA), 12-15 July 1999.

[19] Bodendieck, F., Schlitt, R., Brand, O., Labuhn, D., Goncharov, K., Kalesnikov, V. and Golovin, O. (2002) Emerging Loop Heat Pipe Applications for Small-Sat, MARS Mission and ISS, Paper No. 2002-01-2464, *32nd International Conference on Environmental Systems (ICES)*, San Antonio, TX (USA), 15-18 July 2002.

[20] Humphries, W. R. and Grieggs, E. I. (1977) *A Design Handbook for Phase Change Thermal Control and Energy Storage Devices*, NASA TP-1074, November, 1977.

[21] Groll, M., Muenzel, W. D., Supper, W. and Savage, C. J. (1980) *Transient Behaviour of Liquid Trap Heat-pipe Thermal Diodes*, in Heat Transfer, Thermal Control and Heat Pipes, Progress in Astronautics and Aeronautics, Vol. **70**, Olstad, W. B. (ed.), published by AIAA.

[22] Berner, F. and Savage, C. J. (1984) *Design and Characteristics of a Dynamic Cooler for Space Use*, Paper No. 840964, *14th Intersociety Conference on Environmental Systems*, San Diego, CA. (USA), 16–19 June 1984.

[23] Walker, G. (1983) *Cryocoolers, Part 1: Fundamentals and Part 2: Applications*, Plenum Press, New York (USA).

[24] Ros, R. G. (ed.) (1994), Cryocoolers 8, *Proceedings of the 8th International Cryocooler Conference*, Vail, Colorado (USA), 28–30 June 1994, published in 1995 by Plenum Press, New York (USA).

[25] Ross, R. G. and Boyle, R. F. (2006) An Overview of NASA Space Cryocooler Programs—2006, 14th *International Cryocooler Conference*, Annapolis, MD (USA), 14–16 June 2006.

[26] ESA (1993) *Europe Qualifies for Space—The European Coordinated Test Centres*, ESA BR-46, ISBN 92-9092-035-1, 2nd edition, December 1993.

[27] Tan, G. B. T. and Walker, J. B. (1982), Spacecraft Thermal Balance Testing using Infrared Sources, 12th *Space Simulation Conference*, NASA CP 2229.

[28] Wertz, J. R. and Larson, W. J. (1999) *Space Mission Analysis and Design*, (3rd edn), Microcosm Press and Kluwer Academic Publishers.

[29] Barré, H., Nye, H. and Janin, G. (1999) *An Overview of the XMM Observatory System*, ESA Bulletin No. 1000, December 1999.

[30] Van Katwijk, K., Van Der Laan, T. and Stramaccioni, D. (1999) *Mechanical and Thermal Design of XMM*, ESA Bulletin No. 100, December 1999.

12 TELECOMMUNICATIONS

Ray E. Sheriff[1] and Adrian R. L. Tatnall[2]

[1] School of Engineering, Design and Technology, University of Bradford
[2] Aeronautics and Astronautics, Faculty of Engineering and the Environment, University of Southampton

12.1 INTRODUCTION

12.1.1 The development of telecommunications satellites

Long before artificial Earth satellites became a reality, their potential in the field of telecommunications had been appreciated by visionaries such as Arthur C. Clarke. He described the concept of using geostationary satellites for communications in a paper entitled *Extra-Terrestrial Relays—Can Rocket Stations Give World-wide Radio Coverage?*, published in Wireless World in October 1945 [1]. In the years following the launch of Sputnik 1, the demand for global communications systems was one of the main driving forces—along with military and political considerations—in the rapid development of space technology. Today the need for communication satellites, boosted by the increasing requirement for communications and entertainment services, is at the heart of the space industry growth.

The birth of telecommunications by satellite can be seen, perhaps, in the launch of Telstar I in 1962. Echo 1, NASA's first passive communications satellite, was launched in August 1960 but Telstar I permitted, for the first time, transoceanic communications by satellite. By today's standards it was, of course, a very modest affair. Just under a metre in diameter and weighing 77 kg at launch, the satellite had one channel with a 50 MHz bandwidth, providing about 12 telephony circuits. But it was the start of a revolution in international communications.

The low elliptical orbit of Telstar I limited usage for transatlantic communications to three or four half-hour periods in each day. The first successful geosynchronous communications satellite was Syncom II, used for experimental transmissions between America and Japan. This again had a rather limited capacity, but the design after further improvements, became the basis for the first commercial communications satellite, Early Bird or, as it

Spacecraft Systems Engineering, Fourth Edition.
Edited by Peter W. Fortescue, Graham G. Swinerd and John P. W. Stark.
© 2011 John Wiley & Sons, Ltd. Published 2011 by John Wiley & Sons, Ltd.

was later called, Intelsat I. This satellite, launched in 1965, was similar in size and mass to Telstar I but was capable of providing 240 telephony circuits or a high-quality television channel.

The first worldwide satellite communications system was not established, however, until 1969 when Intelsat III satellites were in position over the Atlantic, Pacific and Indian Oceans, giving coverage of all parts of the globe other than the polar regions.

The subsequent development of satellite communications has been a response to a dramatic increase in the demand for international telephone, video and Internet traffic. This extraordinary increase in demand has been matched by a rapid increase in the size, power capability and traffic capacity of the spacecraft, as well as in the number of satellites in simultaneous use. There has also been a large increase in the number of spacecraft operators, although the global leaders are Intelsat Ltd and SES. In 2001, Intelsat changed from a treaty-based organization to a privately held company. As of 2010 it operates a fleet of over 50 satellites. SES, is a network of satellite operators across the world, with SES Astra operating in Europe and SES World Skies in North and South America.

Geostationary satellites were first used in 1981 to provide mobile communications to the maritime sector. The initial service provided by the International Maritime Satellite Organisation (*INMARSAT*) was based on frequency modulation (FM) telephony and was known as the Standard-A service, which was later renamed Inmarsat-A. Over the last thirty years, Inmarsat has evolved its family of services such that aeronautical and land environments are also served by its network of eleven satellites. Inmarsat won the 2010 MacRobert award for its Broadband Global Area Network (BGAN) service. This provides internet data connectivity globally and is used for remote connectivity for business, government, broadcasters, aid and relief agencies, and emergency services.

Driven by the huge potential market offered by the introduction of mobile telephones, the end of the twentieth century was notable for a new direction in satellite communications, with the introduction of non-geostationary satellites in Low Earth Orbit (LEO). Unfortunately, by the time these multi-satellite constellation systems (see also Chapter 5) were introduced, terrestrial cellular networks had already substantially eroded the anticipated markets. The existing Iridium, Globalstar and Orbcomm networks all filed for 'Chapter 11 bankruptcy' protection shortly after becoming operational but have since been restructured and continue to provide a service. Further evidence for the growing interest in the mobile communications market was the award in 2010 for a contract for 81 satellites to Thales Alenia Space for the Iridium NEXT constellation. The first launches are scheduled for 2015.

The success of satellite navigation has demonstrated the viability of non-geostationary satellite technology. In particular, the US Global Positioning System (*GPS*) and the Russian Global Navigation Satellite System (*GLONASS*) constellations, both of which have their origin in military applications, have successfully been applied to civilian use. The growing reliance of many terrestrial systems on navigation led to the decision by the European Union (EU) to develop its own satellite navigation system. With the backing of the European Space Agency (ESA) and the EU, Europe will deploy the *GALILEO* network of non-geostationary satellites in 2014. A contract was awarded in 2010 to OHB-System and Surrey Satellite Technology Limited (SSTL) to build the first 14 operational satellites.

12.1.2 The role of the communications payload

Some early experimental systems, such as Echo 1, used orbiting objects as passive reflectors or scatterers in order to achieve intercontinental communications. It quickly became apparent, however, that in order to meet the demand for high traffic capacities it would be necessary to use active satellites containing transponders that receive the signals transmitted from the ground, change their frequencies and amplify them before re-transmitting them to Earth. Figure 12.1 is a very much simplified block diagram of such a link. The power amplification factor (or *gain*) provided by the satellite is typically in the region of $10^{12}-10^{15}$ (120–150 dB).*

The most striking difference between a satellite link and a terrestrial link is in the length of the transmission path. This is typically about 50 km for a terrestrial microwave link, but 36 000 km from a ground station to a satellite in geostationary orbit (GEO), or 10 000 km to 20 000 km for a Medium Earth Orbit (MEO) or 700 km to 2000 km for a LEO.

The immediate result of this enormous increase in path length is that the most critical parameter in a satellite system is normally the available transmitter power—particularly on the downlink, in which the transmitter power is in turn one of the major demands on the primary satellite power. For broadcast or multicast services, the huge coverage area offered by a geostationary satellite, combined with its fixed position in the sky, makes it particularly suitable for this mode of service delivery. For mobile services, the selection of one particular orbit over another is not so straightforward. The smaller coverage area offered by a non-geostationary satellite, which continuously moves over the Earth, necessitates the need for multi-satellite constellations to guarantee continuous coverage on a regional or global basis. This, in turn, increases the complexity of the network infrastructure. However, the reduced transmission distance in comparison with a GEO, allows the power requirements to be relaxed and results in a reduction in the latency of the link.

Satellite communication is not restricted to the use of circular orbits. For instance, satellites of the Russian Molniya series are in elliptical, high-inclination, zero-drift orbits that allow good coverage of high latitudes. During the slow apogee transit, near GEO altitude, the satellite appears almost stationary in the sky, allowing continuous operation for up to about two-thirds of the orbital period (see Chapter 5).

The system illustrated in Figure 12.1 is one of a large variety of telecommunications applications of artificial satellites. The main types of systems are as follows:

- trunk telephone and television services;
- broadcast services;
- videoconferencing and other business services;
- communication with mobiles (hand-held, ships/aircraft/land vehicles);
- data transmission (between ground stations);
- data relay (between other satellites and ground stations);
- internet access;
- navigation;

---

* In telecommunications engineering, a power ratio (in this case the ratio of output power to input power) is often expressed on a logarithmic scale with decibels (dB) as units. On this scale the ratio $P1/P2$ is represented by $10 \log_{10}(P1/P2)$ dB. Thus, for instance, a power ratio of 10 is 10 dB and a factor of 2 is about 3 dB.

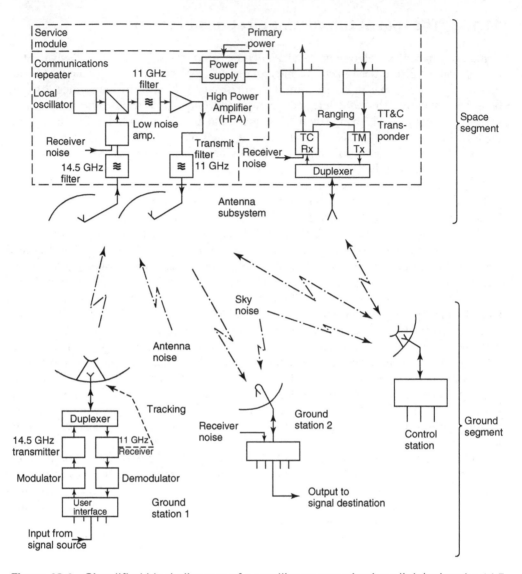

Figure 12.1 Simplified block diagram of a satellite communications link (using the 14.5 and 11 GHz bands)

- remote monitoring and meter reading;
- fleet management and tracking.

A single satellite may include transponders for more than one of these functions.

12.1.3   System constraints

The design of a satellite communications system involves considerations not normally required for a terrestrial system. In general these design constraints can be divided into three broad categories: customer requirements, technical requirements and international regulations.

Customer requirements form the starting point of the design, and are as many and as varied as the customers themselves. Among the parameters that one might expect the customer to specify would be:

- type of signals (voice, television, data and so on);
- capacity (i.e. number of channels of each type)—or bandwidths and frequencies;
- coverage area served by the satellite and the site(s) of the control station(s);
- uplink and downlink signal strength and quality—or types of ground terminals to be served;
- connectivity between different channels and traffic routes;
- availability (i.e. times of day/year, permitted outage times and so on);
- lifetime (typically seven years for LEO, 12 years for MEO and 12–15 years for GEO satellites currently in service, but likely to be about 20 years or more for future designs).

There may be a further requirement that some of these parameters can be altered in orbit by command from a ground station and there may also be other limitations such as a need for a high level of security, both against illicit interception and decoding of signals and against interference, jamming and illegal telecommands.

As with all major engineering projects, however, the most important constraints imposed by the customer and market are usually those of cost and timescale.

The *technical constraints* (such as transmitter power, receiver sensitivity, interference, environment, available components etc.), which apply to a terrestrial link generally apply also to a satellite link, but with some additional restrictions. The most significant are those imposed by the payload mass, the available power and the need for equipment to survive the launch phase and operate unattended in the space environment.

The interplay between technical and cost factors is often crucial. In most terrestrial systems, it is possible to guarantee adequate overall performance by allowing generous margins in the design. Such extravagance in the case of a satellite system will result in an increase in mass and power consumption and hence increased launcher costs. On the other hand, the cost of developing and space-qualifying new tailor-made equipment or of elaborate integration and test procedures may mean that the best technical design is not the most cost effective.

International regulations for communications systems are required to control possible interference between different systems and to ensure compatibility between the various national systems that may need to be connected end to end.

The International Telecommunications Union (ITU), which is now part of the United Nations Organization exists for the purpose of defining and enforcing international standards. More than 190 nations are signatories to its Convention that has the status of a formal international treaty. Within the ITU, there are two consultative bodies that collate information and formulate policy in relation to telephony and radio communications, respectively. These are the *ITU—Radiocommunication Sector (ITU-R)* and the *ITU—Telecommunication Standardization Sector (ITU-T)*. A third body, the *Telecommunication Development Sector (ITU-D)*, is concerned with the technical requirements of the developing world.

The recommendations of the ITU-R are in turn considered by the World Radio Conference (WRC) [2] which may then be incorporated into the ITU Radio Regulations [3]. These are binding on signatories of the ITU Convention.

Among the important system parameters subject to international regulation are the available frequency bands, the orbital location and the maximum permitted power flux density (PFD) at the Earth's surface. The bulk of telephony and television traffic has been in the C-band (4/6 GHz) and Ku-band (11 GHz). Due to overcrowding there is increasing use of the K- and Ka-band between 26 and 40 GHz. New Ka-band satellite broadband services have already started in the US and Canada (e.g. *WildBlue* — a high speed satellite internet provider), and are expected shortly in Europe with the launch of the *HYLAS* and *KaSaT* satellite systems.

The position of a satellite in the GEO is negotiated through the Space Service Department. The main requirement is that there should be sufficient separation between locations to allow a ground-station antenna of reasonable size (and hence beamwidth) to discriminate between adjacent satellites. For C- and Ku-band satellites this implies a separation of at least $2°$ of longitude between satellites using the same frequency. Some parts of the orbit are relatively empty, but in others, such as the Atlantic region, where traffic density is very high, the choice of location may be very restricted.

In what follows, we will be concerned primarily with the technical aspects of the system design. Later we will look at some of the units that make up the telecommunications payload but first we must examine some of the principles underlying the operation of the system as a whole.

12.2   TECHNIQUES OF RADIO COMMUNICATIONS

12.2.1   Introduction

Although most of the important elements of a satellite communications link are shown in Figure 12.1, this is a very much simplified picture. In particular, only two communications ground stations are shown and the system is a simplex (i.e. one-way) link. Nevertheless, the block diagram illustrates some of the features that are common to practically all systems. In fact, the extension to two-way (duplex) operation is often trivial. The same satellite transponder may carry both outgoing and return traffic provided that the same uplink and downlink frequencies are available at the two ground stations. However, this is not always the case. For instance, in a system serving mobiles, the satellite–mobile link may operate at L- or S-bands (1.5–1.7 GHz), and the satellite-fixed Earth station link at C/Ku/X (4–18 GHz) or Ka-band (26–40 GHz) In this case the satellite must carry separate forward and return transponders.

12.2.2   Modulation

Types of modulation

The signals to be transmitted by a communications system normally consist of a band of rather low frequencies, ranging, for instance, from a few tens of Hz to a few kHz in the case of speech or from a few tens of Hz to a few MHz in the case of television.

These *baseband* frequencies—those which constitute the original signal—are unsuitable for direct transmission as radio waves.

For transmission purposes, the signal is imposed on a *carrier wave* of much higher frequency—a process known as *modulation*. This is represented in Figure 12.1 by the *modulator* block in ground station 1. The reverse process—recovery of the baseband signal from the received signal—is known as *demodulation*. If we represent the high frequency signal by a cosine wave,

$$V = V_c \cos(\omega_c t + \phi_c) = V_c \cos(2\pi f_c t + \phi_c), \tag{12.1}$$

then the baseband signal may then be represented by a variation with time of either the carrier amplitude (V_c), its frequency (f_c) or its phase (ϕ_c).

Figure 12.2 illustrates these three basic types of modulation for the simple case where the baseband signal is itself a cosine wave. As shown in the algebraic expressions representing the waveforms in Figure 12.2, the magnitude of the modulation is represented by a *modulation index*, m (also called *modulation depth*) in the case of amplitude modulation (AM) and β in the case of phase modulation (PM) or frequency modulation (FM). PM and FM are different forms of 'angle modulation'. Any FM waveform can be represented as phase modulation (albeit by a slightly different baseband signal) and *vice versa*. For a sinusoidal modulating waveform, β, the peak phase deviation (in radians), is equal to $\Delta f / f_m$ where Δf is the peak frequency deviation and f_m the modulating frequency.

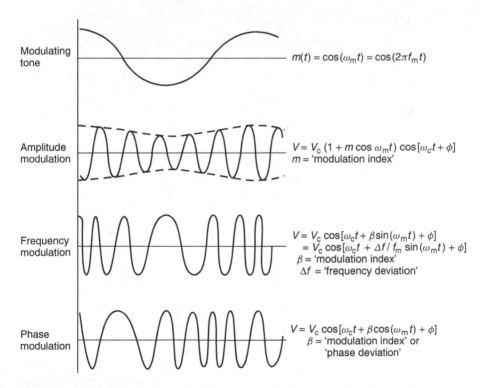

Figure 12.2 Types of analogue modulation. (In the interest of clarity the frequency of the carrier is shown very much reduced compared to that of the modulating signal)

Analogue and digital signals

The modulated carrier waves illustrated in Figure 12.2 are examples of analogue signals. They are characterized by the fact that the instantaneous value of the baseband signal may lie anywhere within a certain range.

The majority of telecommunications traffic consists of digital signals, that is, signals that can take only a finite number of discrete values—often only two values, corresponding to the binary digits 0 and 1. When the signals are in this form, the three types of modulation described above are known as *amplitude-shift keying (ASK), frequency-shift keying (FSK)* and *phase-shift keying (PSK)*.

Within these general categories there are many variants whose different characteristics make them suitable for differing applications. Some of the more commonly used techniques are illustrated in Figure 12.3 where, for instance, ASK is shown as 'on–off keying', with a binary 1 represented by full amplitude and a 0 by zero amplitude. FSK is illustrated by continuous-phase *fast frequency-shift keying (FFSK)* in which there are no discontinuous changes in phase at the frequency transitions and the two frequencies are separated by half the bit rate (i.e. number of binary digits per second).

This choice of frequency separation has certain advantages concerned with the ease of detection of signals in the presence of noise (see Sections 12.2.4 and 12.2.5).

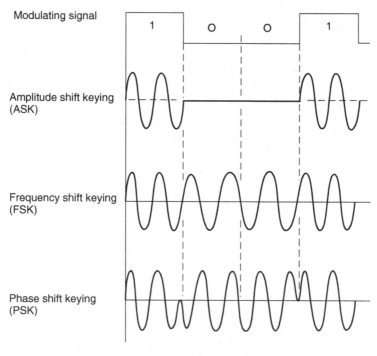

Figure 12.3 Examples of digital modulation. The examples illustrated are: ASK — on–off keying; FSK — fast frequency-shift keying; PSK — phase reversal keying (In the interest of clarity, the frequency of the carrier is shown to be very much reduced compared to that of the modulating signal)

PSK often uses two phases separated by π radians (as illustrated in Figure 12.3) and is then known as *phase-reversal keying (PRK)* or *Binary Phase-Shift Keying (BPSK)*. However, another important type of PSK uses four phases separated by $\pi/2$ radians. This is *quadrature phase-shift keying (QPSK)*. More complicated multi-phase or multi-phase plus multi-amplitude schemes may also be used, particularly where there is a need to achieve high data rates in a limited bandwidth. The *bandwidth* is the width of the range of frequencies included in the signal, and transmitted by the system.

Digital signals occur naturally, for instance, in the transmission of numerical data or in communication between computers. It is also possible to convert analogue signals into digital form. Most telephone traffic is transmitted by digital techniques, and digitized television and audio broadcasts are now becoming the norm.

The conversion of analogue signals into digital form may be described in terms of three distinct steps—*sampling, quantization* and *encoding*.

The sampling process is illustrated in Figure 12.4. Clearly, if it is known that the signal varies with sufficient smoothness between the sample times, the entire signal can be reconstructed with considerable accuracy from the sample values. It can be shown that this is guaranteed if the signal to be sampled contains no frequencies greater than half

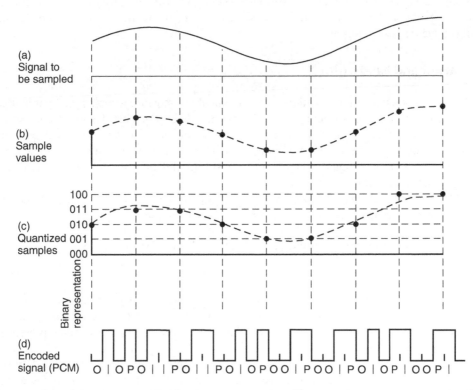

Figure 12.4 Sampling and digitization. The process illustrated here is an example of pulse code modulation (PCM). Each sample value is represented by a binary number with, in this case, three digits and one extra bit for parity checking. (a) Signal to be sampled, (b) sample values, (c) quantized samples and (d) encoded signal (PCM). The symbols marked 'P' are parity bits

the sampling frequency. This is known as the *Nyquist criterion*. If the sampling speed is insufficient an effect called aliasing is observed. This leads to spurious frequencies being generated.

The process of quantization involves rounding off each sample to one of a finite number of allowed values. This introduces errors that cannot be compensated in subsequent processing of the signal. In order to reduce this *quantization noise* to an acceptable degree it may be necessary to increase the number of allowable signal levels. However, the greater the number of levels, the greater is the number of digits that are necessary to identify a particular level and hence the greater the required bit rate. This reduces the transmitted energy associated with any one bit, which tends to enhance the signal degradation caused by mechanisms other than quantization (see Section 12.2.5). The system designer must seek a balance that results in best overall signal quality.

The final step, that of encoding, may take many forms. The simplest is to express each sample value directly as a binary number and then combine the numbers (normally with some extra digits for synchronization—that is, to allow the receiver to identify the start of each binary word) to form a long sequence of binary digits. This is the system usually understood by the term *pulse code modulation (PCM)*. However, a number of more subtle techniques may be introduced at this stage with a view, for instance, to making the signals less vulnerable to transmission impairments, providing encryption or modifying the spectrum. Thus digital modulation is in many ways much more flexible than its analogue counterpart.

Spectrum and bandwidth

In the design of a telecommunications system, one of the most important signal parameters is the bandwidth. It is this (among other things) that determines the design of the various filters shown in Figure 12.1. The effect of modulation of a carrier wave is to produce frequencies other than that of the unmodulated carrier, and the resultant spectrum depends both on the baseband signal and on the type of modulation in use. Some particular examples are considered below.

In the case of *amplitude modulation* by a co-sinusoidal signal, the spectrum may be inferred from standard trigonometrical identities, for example,

$$V_c(1 + m \cos \omega_m t) \cos \omega_c t = V_c[\cos \omega_c t + (m/2) \cos(\omega_c + \omega_m)t$$
$$+ (m/2) \cos(\omega_c - \omega_m)t] \qquad (12.2)$$

The components at frequencies $(\omega_c + \omega_m)$ and $(\omega_c - \omega_m)$ are known as *sidebands* and the term $V_c \cos \omega_c t$ is the *carrier component*. An increase in the modulation index m results in increased sideband levels but no change in the range of frequencies present in the signal. This is illustrated in Figure 12.5(a).

A simple development of equation (12.2) shows that if the baseband signal, $m(t)$, contains more than one frequency, each component of the baseband produces a pair of sidebands. In general, if $m(t)$ is not composed of discrete frequencies, its spectrum may be represented by its Fourier transform, $M(\omega) = \int m(t) \exp(-j\omega t) \, dt$. Whatever the baseband spectrum, the spectrum of the modulated carrier has the same form but in bands located symmetrically above and below the carrier frequency. The required channel bandwidth is twice the base bandwidth (irrespective of modulation depth).

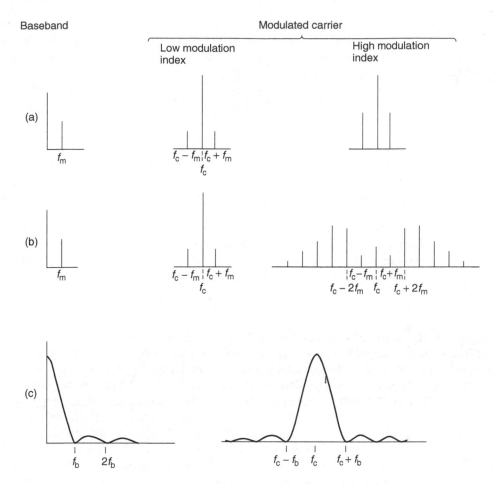

Figure 12.5 Spectra of modulated signals: (a) amplitude modulation (amplitude spectrum); (b) frequency modulation (amplitude spectrum) and (c) phase-reversal keying (power spectrum). The AM and FM spectra show modulation by a single tone of frequency f_m. The PRK spectrum is for a random binary sequence at a bit rate of f_b bits/s

The spectrum of a *frequency (or phase)-modulated signal* is considerably more compli-cated. In the case of co-sinusoidal modulation with a small modulation index, β, the pres-ence of sidebands similar to those for AM may again be demonstrated by standard trigono-metrical identities. If, however, the index is not small, the spectrum of FM is in striking contrast to that of AM. Fourier analysis of expressions, such as $\cos(\omega_c t + \beta \sin \omega_m t)$, yields an infinite set of components at frequencies $(f_c \pm nf_m)$ (where n takes all integer values) and with amplitudes that can be evaluated from Bessel functions of the first kind. The number of side frequencies having significant amplitude (and hence the effective bandwidth) increases with increasing β (Figure 12.5b).

When the baseband is not a single sine wave, the side frequencies are no longer related in a simple way to the baseband frequencies. For instance, if the baseband contains two frequencies f_1 and f_2, the modulated signal contains all frequencies of the form $f_c \pm nf_1 \pm mf_2$ (where n, m are integers). In strictly mathematical terms, the bandwidth is infinite but there is a useful semi-empirical expression (*Carson's formula*) for the bandwidth containing about 98% of the power:

$$\text{Carson bandwidth} = 2(\Delta f + f_m) \tag{12.3}$$

where Δf is the peak frequency deviation and f_m is normally taken to be the highest frequency present in the baseband. Frequency modulation with index sufficiently small so that only the first-order sidebands are significant (β less than about 0.5) is known as *narrow-band FM (NBFM)*, while for larger β the term *wide-band FM (WBFM)* is used.

In the case of *phase-shift keying* the spectrum depends both on the type of modulation (PRK, QPSK etc.) and on the encoding procedure used in producing the bit stream. As an example, we may consider PRK by a random binary signal (i.e. by a random sequence of 0s and 1s occurring with equal probability). A signal of this type does not have a Fourier transform but we may define a power-density spectrum $S(f)$ that is the power per Hz of bandwidth at the frequency f. For the random binary signal, we have

$$S(f) = V^2 T (\sin \pi f T)^2 / (\pi f T)^2 \tag{12.4}$$

where V is the signal voltage and T is the length (in time) of one bit.

Since PRK may be regarded as a type of ASK (the amplitude is switched between $+A$ and $-A$), the spectrum of the modulated signal is similar to the baseband spectrum but centred on the carrier frequency. This is illustrated in Figure 12.5(c). Strictly speaking, the overall bandwidth is infinite because, in this case, the baseband has infinite bandwidth. However, the power in the 'tails' of the spectrum is not very significant, the first subsidiary maxima already being some 13 dB below the central peak. In practice, therefore, it is usual to transmit at most, that part of the spectrum that lies between the first zeros. In many cases less will suffice. This has the effect of partially smoothing out the phase transitions, though without destroying the baseband information completely. Since $1/T$ is the number of bits per second, the effective bandwidth for the random binary signal is between one and two times the bit rate.

If the binary signal results from digitizing an analogue signal, the bit rate depends both on the base bandwidth and on the required level of quantization errors. For instance, for speech (bandwidth 4 kHz) quantized to 256 levels, the bit rate is 64 kbit/s since there must be at least 8000 samples per second with 8 binary bits per sample.

Much effort has been concentrated on devising digital techniques that make more efficient use of bandwidth in particular applications. An important example is speech encoding, when used in mobile communications. PCM is an unnecessarily powerful and general technique for this purpose, since it can reproduce all waveforms satisfying the Nyquist criterion—including waveforms that are not produced by the human voice and make no contribution to the intelligibility of speech. Encoding techniques modelled on the mechanisms of voice production or sensitive to the statistics of the speech waveform allow speech transmission of adequate quality at a rate of 2.4 kbit/s. Digital television, by using the Moving Pictures Expert Group (MPEG-2) coding standard, requires bit rates of the order of 5–7 Mbit/s, and about 25 Mbits/s for High Definition Digital Television (HDTV). In both cases, this is a significant reduction compared to the 100 Mbit/s that would be required if a simple PCM approach were to be adopted.

12.2.3   Multiple access

It is a requirement of most systems that several users (in some cases very many users) can pass signals through the satellite simultaneously. There are three main techniques for doing this. In *frequency-division multiple access (FDMA)* each user transmits a signal on a different carrier frequency, and at the receiving station the signals are separated by frequency-selective filters. The signals in this case are often analogue in nature.

In the case of digital signals, *time-division multiple access (TDMA)* or *code-division multiple access (CDMA)* may be used. In TDMA, an individual user transmits short bursts of digits in a particular time-slot within a repeating time-frame. Other users occupy different time-slots within the same time-frame. Synchronization signals are included, which allow the receiver to identify the start of the frame, and select those time-slots that contain a particular 'message'. The concentration into short bursts has the result that each signal occupies the whole of the channel bandwidth. However, no two signals occur at the same time. The Global System for Mobile Communications (GSM) digital cellular system, currently the most popular mobile system in the world uses this technique.

In CDMA, each signal uses the full channel bandwidth but the signals may well all be present at the same time. In Direct Sequence CDMA (DS-CDMA), the carrier signal is modulated twice, first by the required signal and then by a pseudo-random sequence of binary digits at a much higher bit rate. A receiver using the correct pseudo-random code can undo the effect of the second modulation and so recover the original signal. It does not, however, recover the signals of other users who have employed different codes. An alternative to DS-CDMA is frequency hopping. This involves changing the transmission frequency pseudo-randomly across the available bandwidth. At the receiver, the same pseudo-random sequence is used to synchronize with the transmitted frequency, to allow demodulation of the original signal. Most satellite applications employing CDMA use DS-CDMA. In both cases the net result is that the signal transmitted is spread over a much larger bandwidth than the original signal and these techniques are known as *spread spectrum systems*. Hybrid multiple access schemes, comprising combinations of the three techniques, are also employed. For example, a FDMA–TDMA scheme involves the division of the available bandwidth into channels (FDMA), each of which contains a TDMA frame.

12.2.4   Noise

Telecommunications would present few problems were it not for the presence of 'noise' in all electrical systems. In radio communications, electrical noise is a result of the random thermal motions of atoms and electrons in matter, which reveal themselves as small randomly varying electromotive forces and currents. Each resistive element in a circuit is a source of thermally generated electrical power of kT watts per Hz of bandwidth in the radio frequency (RF) range, where $k = 1.38 \times 10^{-23}$ J/K (Boltzmann's constant), and T is the temperature (in Kelvins) of the resistor (throughout this section it should be assumed that temperatures are measured on the absolute scale, that is, at 0° Celsius, $T = 273$ K).

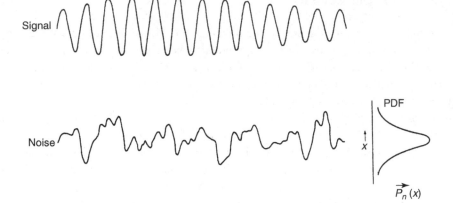

Figure 12.6 Signal and noise voltage waveforms

Quantum mechanics predicts a power spectral density given by the expression,

$$P_0(f) = \frac{hf}{\exp(hf/kT) - 1} + \frac{hf}{2} \tag{12.5}$$

where $h = 6.625 \times 10^{-34}$ J s is Planck's constant. If $hf << kT$ this reduces to $P_0(f) = kT$.

The essential differences between a noise voltage and a typical signal are illustrated in Figure 12.6. The signal has a regular quasi-periodic structure whereas noise is essentially irregular and unpredictable. The root-mean-square (r.m.s.) value may be used as a measure of noise magnitude, but a complete specification must include the probability density function (PDF), that is, the probability $p_n(x)\delta x$ that the instantaneous voltage lies in the range from x to $x + \delta x$. For many sources of noise the PDF is approximately Gaussian, in which case $p_n(x)$ takes the form

$$p_n(x) = \frac{1}{\sigma_n \sqrt{(2\pi)}} \exp\left(\frac{-x^2}{2\sigma_n^2}\right) \tag{12.6}$$

where σ_n is the r.m.s. value.

When two independent sources of noise are combined, it is their powers (or mean squared values) that must be summed. In particular, the noise powers contained in different frequency ranges are additive, so that for a constant power spectral density, $P_0(f) = kT$ W/Hz, the total power in bandwidth B is given by

$$P = kTB \tag{12.7}$$

The importance of noise is that it sets a fundamental limit to the sensitivity of a telecom-munications receiver. The crucial factor that determines the overall performance of the system is not just the signal power itself, but rather the ratio of signal power to noise power, that is, the *signal-to-noise ratio (SNR)*.

In any communications system there are many possible sources of noise. However, the signal power reaches its lowest level at the end of the transmission path and it is therefore the noise generated in the input circuit of the receiver which is most critical in

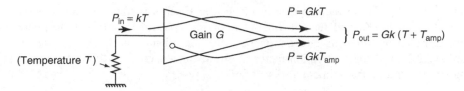

Figure 12.7   Definition of noise temperature

determining the system performance. For this reason, noise is shown in Figure 12.1 as an extra input to each of the receivers—a convenient fiction that has much the same effect as the reality.

There are two commonly used ways of expressing the 'noisiness' of an amplifier (or radio receiver). In the first, the expression for thermal noise power available from a resistor, $P_0(f) = kT$, is used to express the noise output as an equivalent temperature. The concept is illustrated in Figure 12.7.

If the amplifier were an ideal noise-free device, the noise power density kT delivered by the source resistance would give an output power density GkT, where G is the gain. In practice, the output is greater than this and we can write

$$P_{out}(f) = Gk(T + T_{amp}) \tag{12.8}$$

where T_{amp} is known as the *noise temperature* of the amplifier. The noise temperature is thus the additional noise power originating in the amplifier expressed as an equivalent increase in the source temperature. Note that noise temperatures, like noise powers, are additive.

In the second method of expressing the noise performance of the amplifier, we assume that (in the arrangement of Figure 12.7) the source resistance is at a standard temperature T_0. Were the amplifier noise-free, the output power spectral density would then be kT_0G but in reality it is greater than this by a factor F, the *noise factor* (or *noise figure*):

$$P_{out}(f) = FGkT_0 \tag{12.9}$$

From equations (12.8) and (12.9) it can be seen that

$$F = 1 + (T_{amp}/T_0) \tag{12.10}$$

The standard choice of T_0 is 290 K—a typical 'room temperature'.

A receiver may consist of a chain of amplifiers with different gains and noise temperatures. As might be expected, the total noise temperature can never be better than that of the first stage, and moreover for a given first stage noise temperature, the total noise temperature is minimized by making the first stage gain, G_1, as large as possible.

In addition to the noise originating in successive amplifier stages, the overall system temperature T_{sys} will include background noise received by the antenna (usually represented by an *antenna temperature*) and a (usually small) contribution transmitted with the signal.

Antenna noise results mainly from thermal radiation from various surfaces, which may be either in the 'main beam' of the antenna or in the less sensitive 'sidelobes' (see

Section 12.2.8). These include, for instance, the Earth's surface (at about 300 K), the Moon (about 200 K) and the Sun (about 6000 K). Also, the Earth's atmosphere emits thermal radiation when its attenuation becomes significant, as during a rainstorm (see Section 12.2.7). If the attenuation factor in a rainstorm is α (>1), the contribution of the atmosphere to antenna temperature is

$$T_{\text{atmos}} = T(1 - 1/\alpha) \qquad\qquad (12.11)$$

where the temperature, T, of rain is about 275 K. In the majority of geographical locations, in order to achieve the required limits on outage times, it is necessary to design the system for operation with significant atmospheric attenuation (see Section 12.2.7). Under these conditions the atmosphere may sometimes be one of the main sources of antenna noise (and indeed of system noise) for a ground station, where the receiving antenna is pointing towards the normally 'cold' sky. For an uplink, however, in which the satellite antenna points towards the Earth's surface, the antenna temperature is always in the region of 290 K and is little affected by the atmospheric attenuation (though of course the signal level is affected).

In the case of a satellite downlink, in which a transparent satellite transponder is employed, the transmitted signal is contaminated by noise originating on the uplink. The system is often designed so that the contribution of the uplink to the system noise is about 10 dB or more below that of the downlink. Under these conditions the transmitted noise causes a 0.4 dB degradation of downlink SNR.

12.2.5   Output signal-to-noise ratio

In *digital systems*, the mechanism by which system noise affects the output of a communications link differs markedly from the analogue case where noise always degrades the signal. The demodulator must contain one or more threshold detectors, which allocate to each received symbol one of the permitted values. If the symbol is correctly identified, the noise has no effect whatsoever. Occasionally, however, the noise voltage is large enough to cause the receiver output to lie on the wrong side of the detector threshold and so result in an error in interpretation.

The probability of error in any one bit—known as the *bit error rate (BER)*—depends on the product of the received signal power and the length of the bit interval—that is, on the received energy per bit. For typical transmission systems the BER, P_e, may lie in the range 10^{-3}–10^{-9}. The dependence of BER on the ratio of bit energy to noise power density (E_b/N_0) is shown for two types of binary PSK demodulation in Figure 12.8. These are ideal theoretical curves. In practice the value of E_b/N_0 needed for a given error rate may be 1–2 dB higher than shown.

The BER Pe can be significantly improved by the use of encoding techniques that permit error detection and correction at the receiver (this is *forward error correction—FEC*). The simplest form of error detection is illustrated in Figure 12.4(d). The values of the parity bits are assigned so that the number of 1s in each binary 'word' is (in this case) even. If at the system output a word is received with an odd number of 1 s, it is known that an error has occurred (it is assumed in this case that the probability of two errors in the same word is negligible). The same principle—that of transmitting more bits than are strictly required by the data—is used in more sophisticated coding techniques, which

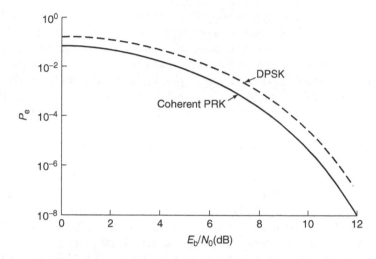

Figure 12.8   The effect of noise on digital systems. (a) BER versus E_b/N_0 for two types of system. In coherent PRK, the phase of each received bit is compared with that of a reference carrier derived separately from a long sample of the signal. In differential PSK (DPSK), the phase of each bit is compared with that of the preceding bit. White Gaussian noise, perfect synchronization and optimum detection are assumed

not only reveal the presence of errors but also (except in very improbable circumstances) make it possible to identify and so correct the offending bits. The consequent reduction in error rate more than offsets the effect of the required increase in bit rate.

There are two main types of FEC codes; *block* and *convolutional* codes. If a block code is used the input data is divided into pre-determined fixed length data symbol blocks and the code added converts them into fixed-length code blocks. This type of code is particularly appropriate if bursts of errors are expected. Convolutional codes work on bit or symbol streams of arbitrary length and work particularly well if the errors occur randomly. The two types of code are often used together in a concatenated code that is typically used for interplanetary missions where power is very limited.

12.2.6   Choice of modulation and access systems

The output SNR is one of the main factors in determining the choice of modulation technique. Other factors of primary importance are the equipment complexity, the payload electrical efficiency, the spectral occupancy and the general system efficiency, flexibility and adaptability.

If simplicity of the equipment and efficient use of spectrum were the only significant requirements, the natural choice would be AM. However, currently, this is rarely, if ever, the final choice.

Wide-band FM allows a specified S/N_0 to be achieved with considerably less power than is required for AM. Moreover, unlike in AM, the transmitter may be operated in a more efficient non-linear mode. Wide-band FM has been used extensively in the past for transcontinental television and *multi-channel per carrier (MCPC)* telephony, but today the approach is to employ digital methods. When many small Earth stations or mobile stations are involved, it is usual to employ PSK with a *single channel per carrier (SCPC)*. As with FM, the use of PSK results in a constant amplitude signal, hence it does not rely on strict linearity to avoid signal distortion, as would be the case in AM. The composite signal formed by adding several individual carriers is not itself of constant amplitude. Distortion of the amplitude waveform by a non-linear circuit results in the generation of *intermodulation (IM) products* the frequencies of which are related to (but not the same as) the input frequencies (see Section 12.3.2). It is usually necessary to operate the transmitter at reduced efficiency in order to keep the in-channel carrier-to-intermodulation power ratio (C/I) above a specified value—typically 15–20 dB—when the transponder is fully loaded.

The most effective way of avoiding IM products is to use TDMA. In this system no two signals are amplified at the same time, so there is no IM even when the transmitter is operated in a highly non-linear mode. The increased transmitter efficiency is partially offset by the need to transmit extra pulses for synchronization, for channel identification and to give the demodulators time to lock on to each signal in turn.

When the signals originate in analogue form, TDMA suffers from the fact that PCM requires more bandwidth than analogue modulation. However, there is compensation in the much greater flexibility of TDMA in switching and routing signals and in multiplexing signals of differing bandwidths and differing types in a variety of configurations.

12.2.7 Radio propagation

Latency

Radio waves consist of a system of electric and magnetic fields that travel through free space at a velocity of about 3×10^8 m/s. The transit time for a round trip to a satellite in GEO and back is thus about 0.24 s. If other normal delays from network sources are included this gives a typical one-way connection latency of 500–700 ms from the user to the internet service provider (ISP) for an internet link—much greater than the propagation delays normally encountered in communication links. A number of techniques have been used to reduce the overall latency encountered. Whilst the speed of light is constant, the protocols used can be modified to reduce the number of round trips to the satellite that are required. In addition recent technology developments have led to Transparent Transmission Control Protocol (TCP) acceleration being used to increase throughput without requiring any changes in end-system TCP implementations.

Polarization

In propagation through uniform isotropic media, the electric and magnetic fields are at right angles both to each other and to the direction of propagation. In a *plane-polarized* wave, the direction of the electric field lies in a fixed plane as shown in Figure 12.9(a). If a second wave travelling in the same direction has its plane of polarization at right angles to the first, then the two waves will propagate independently and indeed may carry quite different signals. A suitable (ideal) receive antenna could absorb all the power from one polarization while completely rejecting the other. The two polarizations are said to be orthogonal.

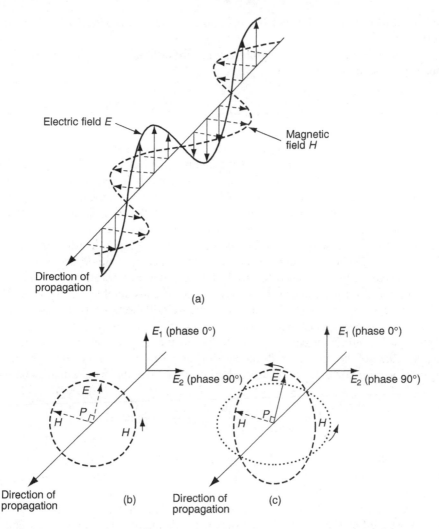

Figure 12.9   Field vectors in electromagnetic waves: (a) a 'snapshot' of a linearly polarized wave at a particular instant in time. The electric and magnetic fields are represented here by geometrical vectors; (b) representation of the E- and H-fields at a particular point, P, in space in a circularly polarized wave and of the E-fields of component linearly polarized waves with a 90° phase difference and (c) the same for elliptical polarization

If the two orthogonal plane-polarized waves are identical, apart from a phase difference of $\pi/2$ radians, then the combination can be regarded as a different type of polarization. In this case the electric field vector at any point in the propagation path rotates, the tip (in a geometrical representation) following a circular path if the two plane-polarized waves are of equal amplitude. This is known as *circular polarization*. It may be either right- or left-handed, depending on whether the vector rotates clockwise or anticlockwise, when viewed by an observer looking in the direction of propagation. Figure 12.9(b) shows the right-handed version. More generally, if the two plane-polarized waves are not of equal amplitude, orthogonal and in quadrature, the result is *elliptical polarization* (see Figure 12.9c).

Left- and right-handed circular polarizations are orthogonal and may be received independently by suitable antennas (see Section 12.3.3).

Propagation in the Earth's atmosphere

For most of their journey the signals from a satellite propagate through what is essentially free space, and they travel without change, apart from the steady diminution in intensity in proportion to $1/r^2$ as the distance r from their source increases. In the last few kilometres, however, as they pass through the Earth's atmosphere, they encounter phenomena of an unpredictable nature that may significantly affect the system performance. These atmospheric effects can be divided into two categories, those that occur in the ionosphere and those that occur in the troposphere. Ionospheric effects are very important at low and medium frequencies but in general their magnitude varies as $1/f^2$ and at the microwave frequencies used for satellite links, they can normally be ignored. Tropospheric propagation phenomena, however, can be of considerable significance. There are three main effects—refraction, attenuation and scintillation—but of these it is usually only attenuation that can cause serious problems to the system designer.

Atmospheric refraction causes a slight shift in the apparent elevation of the satellite. The magnitude of the shift depends on the elevation as well as on the atmospheric pressure and water vapour content. For a standard atmosphere, the shift is given approximately by

$$R = \frac{0.02}{\tan[e + 0.14 + 7.32/(e + 4)]} \text{degrees} \qquad (12.12)$$

where e is the true elevation in degrees. Minor variations caused by differing climatic conditions can readily be accommodated by causing the ground-station antenna to track the satellite.

Attenuation in the troposphere has two causes. The first is molecular absorption by gases, illustrated by Figure 12.10. This shows the attenuation as functions of frequency for oxygen and water vapour, which together account for nearly all of the gaseous absorption in the atmosphere. Both curves display strong absorption bands, and although these occur at higher frequencies than have so far found extensive use in satellite communications, the water vapour band centred on 22.2 GHz will certainly be a significant effect at 20 and 30 GHz. It is also of some interest that the strong absorption by oxygen at 60 GHz makes this frequency very suitable for inter-satellite links, operating above the atmosphere that provides a screen against interference from terrestrial sources. Apart from the microwave absorption bands, the background attenuation shows a steady rise with frequency owing to the tails of absorption bands in the far infrared. This clear-air attenuation is generally negligible at frequencies below 10 GHz but is not entirely insignificant at Ku-Band and above.

Much more dramatic attenuation effects are caused by rain. Water droplets both scatter and absorb radiation, the effect being strongest when the drop size is of the order of a wavelength. Thus in the microwave range, for which the wavelength is larger than a raindrop, the attenuation increases with frequency. Semi-empirical curves may be obtained showing attenuation as a function of frequency for various rainfall rates

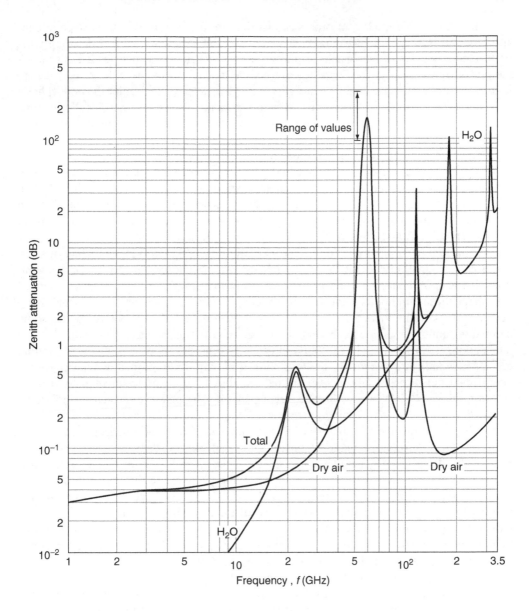

Figure 12.10 Total dry air and water vapour attenuation at the zenith from sea level [4] (Reproduced by permission of International Telecommunication Union)

assuming reasonable models of drop-size distribution and of rain-cell size. Since attenuation by rain is very variable, the system designer must seek some way of deciding what performance margin is reasonable to allow for the occasional deep fade. It is usual for the customer to specify an allowable outage time and of this some will be allocated to loss of signal because of rain. The designer must then attempt to predict the atmospheric attenuation that will not be exceeded for more than this allocated time. Ideally, experimentally determined statistics of attenuation may be available for the site of the ground station. Otherwise, it will be necessary to use an approach such as that provided by the ITU-R, which has adopted a general method for the conversion of rainfall statistics into predictions of attenuation [5]. The method provides attenuation values as a function of probability of occurrence, RF, ground station location and satellite elevation. When local rainfall statistics are not available, use may be made of maps of rainfall rate given in Reference [6]. For temperate climates, typical fade margins are in the region of 2 dB at 11 GHz and 3.5 dB at 14.5 GHz, for a link reliability of about 99.9%.

Scintillation, or rapid fluctuation in signal amplitude analogous to the twinkling of stars, is normally a small effect that for most purposes can be ignored. However, at low elevations, because of the longer atmospheric path, all propagation effects are greatly enhanced and under these conditions scintillation depths of several decibels are not unusual. As a general rule, satellite links are not operated at elevations less than about 10° but in some cases this cannot be avoided. The importance of scintillation is that, along with clear-air attenuation, it causes a slight reduction in capacity (and hence, in a commercial system, loss of revenue) at all times. In the long term this may be as serious as the occasional deep signal fade—especially in systems that are able to compensate for short-term fades in some of the signals by, for instance, redistributing the power allocated to individual channels or by changes in coding parameters.

Another consequence of rain is signal *depolarization*. Because raindrops are slightly flattened, they absorb one plane of polarization rather more than the other. Since the axis of the drops is in general inclined to the plane of polarization of the signal, this differential absorption causes a slight rotation of the plane, or in the case of a circularly polarized signal, a slight ellipticity of polarization. The resulting power loss due to mismatch between the signal polarization and that of the receiving antenna is not significant. However, the presence of the orthogonal polarization, albeit at a level well below that of the wanted polarization, can have a serious impact on systems employing frequency reuse by polarization discrimination (see Section 12.2.9).

One further propagation effect that requires mention is multi-path—that is, the simultaneous presence of two or more transmission paths between the satellite and Earth terminal. This may happen occasionally because of abnormal tropospheric conditions but it is more commonly caused by reflections from the ground or buildings. Signals arriving by the indirect path may interfere either constructively or destructively with the main signal, resulting in signal enhancement or fading. Since the conditions of interference depend on the difference *in wavelengths* between the path lengths, the transmission loss may vary rapidly with frequency across the channel bandwidth, causing signal distortion. Multi-path does not usually present any problems in fixed-satellite services where the Earth stations are well sited and employ narrow beam antennas (which can select just one of the arrival directions). However, in the mobile services using low gain Earth terminal antennas, possibly in an urban environment, additional link margins are required to allow for multi-path (as well as for shadowing by buildings etc.)

12.2.8 Antennas

This section covers the general properties of antennas as they affect the telecommunications system. Various types of on-board antenna will be discussed later (Section 12.3.3). The description will, in general, be expressed in the terminology of transmission, but it will apply equally well to reception. The main properties of an antenna—such as gain and beamwidth—are the same for the two functions.

Radiation pattern and beamwidth

No antenna is strictly *isotropic*. That is, no antenna radiates or receives signals with equal intensity (or sensitivity) in all directions. Indeed, most antennas are designed specifically so that they radiate very strongly in just one direction.

The directional properties can be represented by a polar plot of the radiated field intensity or power as a function of direction. A two-dimensional section through this pattern is known as a polar diagram (see Figure 12.11). It is seen that most of the radiation is emitted within a narrow range of directions known as the *main beam*. There are other directions in which radiation emerges, albeit at considerably lower level, known as *sidelobes*. Their importance lies in the fact that they may contribute significantly to antenna noise and to interference from, and to, other systems. Figure 12.11(b) also illustrates the concept of beamwidth, which is usually measured between $-3\,$dB points (as shown) but it is sometimes specified between other limits such as $-1\,$dB, $-4\,$dB or between the first minima in the radiation pattern. Beamwidth is related to antenna size. Many microwave antennas consist of a large physical area, or aperture, illuminated by a single primary feed or by many small radiating elements. The most common arrangement is a paraboloidal reflector illuminated by a horn (see Section 12.3.3). The radiation pattern measured at a large distance from such an antenna is the Fourier transform of the distribution of the electromagnetic field in the aperture. It follows directly from the scaling property of the Fourier transform that beamwidth is inversely proportional to antenna size. A useful rule of thumb for calculating the beamwidth is

$$3 \text{ dB beamwidth} = 70(\lambda/D)(\text{degrees}) \qquad (12.13)$$

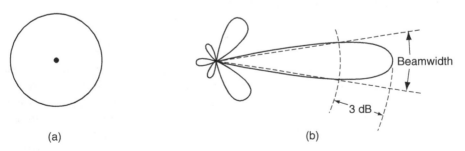

(a)   (b)

Figure 12.11 Antenna radiation patterns: (a) isotropic antenna and (b) directional antenna (large aperture)

where λ is the wavelength and D the antenna diameter. This formula applies to a circular aperture with a typical distribution of illumination but it may be used as a rough guideline for most antennas of the 'aperture' type.

It is not usual to design for uniform illumination. By 'tapering' the illumination at the edges of the antenna, it is possible to achieve lower sidelobe levels as well as reduced 'spillover' of radiation from the primary feed. The consequent slight increase in beamwidth and decrease in gain (see below) is often a price worth paying.

Antenna gain

The importance of using highly directional antennas is that they provide signal *power gain* as compared with the same system using (hypothetical) isotropic antennas. The relevant expressions for gain are given by

$$\text{directive gain} = \frac{\text{actual power flux density}}{\substack{\text{power flux density from an isotropic} \\ \text{radiator with the same total radiated power}}}$$

$$\text{power gain} = \frac{\text{actual power flux density}}{\substack{\text{power flux density from a loss–free isotropic} \\ \text{radiator with the same input power from the generator}}}$$

It is assumed that the flux densities are measured at a fixed large distance from the antenna. The power gain is slightly less than the directive gain because of the power dissipated in losses in the antenna.

For a uniformly illuminated antenna with physical area A, the directive gain at the centre of the main beam is given by

$$G = 4\pi A/\lambda^2 \tag{12.14}$$

We may use the same relation to define an *effective aperture* A_e for any antenna. Thus,

$$A_e = \lambda^2 G/4\pi \tag{12.15}$$

Non-uniform illumination results in the effective aperture being less than the physical area. We may define an *aperture efficiency* by

$$\eta = A_e/A \tag{12.16}$$

For typical microwave antennas η is in the range $0.5–0.7$. An overall antenna efficiency may be defined in the same way by using the power gain in place of the directive gain.

In a satellite system, where the coverage area is specified, the optimum antenna size depends on a trade-off between antenna gain and beamwidth. In order to optimize performance at the edge of coverage (which is usually the critical requirement in the system design), the gain in this direction should normally be about 4.2 dB less than at beam centre.

Polarization axial ratio

Most antennas are designed either for pure circular or for plane polarization, but in practice the radiation is always elliptically polarized. In either case the antenna quality, from this point of view, can be expressed as an *axial ratio*. This is defined as the ratio (dB) of the powers radiated in two orthogonal planes of polarization, in which the planes are chosen to maximize the ratio.

For circular polarization, the axial ratio should ideally be 0 dB. In a ground-station antenna, which points directly at the satellite, 1 dB should be achievable, but for an antenna on-board a satellite the axial ratio may be as high as 2–3 dB at the edge of coverage.

Because the transmitting and receiving antennas may not have their planes of maximum gain aligned, allowance must be made for possible *polarization loss*.

In the case of a plane-polarized antenna, the axial ratio should be as large as possible in order to discriminate against signals that may be present on the other polarization. An axial ratio (or *cross-polar discrimination: XPD*, as it is usually called in this case) as high as 30 dB is not unreasonable.

12.2.9   Frequency reuse

It has already been pointed out that overcrowding of the spectrum has led to a gradual shift towards higher frequencies. However, since higher frequency usually means less transmitter power and increased atmospheric loss, there are considerable advantages in making maximum use of the lower-frequency bands. Many satellites now use the same frequency for two or more signals. This may be achieved either by spatial discrimination, in which two signals at the same frequency may be transmitted on separate narrow beams to different regions on the Earth's surface, or by polarization discrimination, in which signals at the same frequency may be transmitted over the same path on orthogonal polarizations. In the former case, spot beam frequency reuse is similar to the frequency reuse of cellular radio towers. In the latter case, antenna imperfections and atmospheric phenomena may give rise to cross-polar interference. Although this can be troublesome, special receivers can use the signal in one channel to 'null out' the unwanted signal in the other, thus making the technique viable. The saving of space in the radio spectrum can be considerable.

12.2.10   The link budget

We are now in a position to discuss the calculation of the transmitter power requirement. We have seen that, for a given type of modulation, the overall performance depends upon the RF carrier-to-noise-power-density ratio, C/N_0, at the receiver. It remains to show how C/N_0 is related to the transmitter power.

The definition of antenna gain implies that a transmitter with output power P_T associated with an antenna of gain G_T can be replaced, for the purpose of this calculation, by an isotropic radiator with output power $P_T G_T$. The quantity $P_T G_T$ is known as the *equivalent isotropic radiated power (EIRP)*. From an isotropic radiator, this power would spread out uniformly so that the PFD (i.e. the power flowing through a unit area) at a distance r from the source is

$$S = P_T G_T / (4\pi r^2) \tag{12.17}$$

If atmospheric attenuation results in power loss by a factor L_A, then the flux density at the receiver is

$$S = P_T G_T / (4\pi r^2 L_A) \tag{12.18}$$

The effective area of a receiving antenna with gain G_R is

$$A_R = \lambda^2 G_R / 4\pi \tag{12.19}$$

Thus the signal power at the input to the receiver is

$$C = P_T G_T G_R (\lambda/4\pi r)^2 (1/L_A) \tag{12.20}$$

Finally for a system temperature T_{sys} the noise power density referred to the receiver input is kT_{sys}, giving a signal-to-noise-power-density ratio of

$$C/N_0 = P_T G_T (\lambda/4\pi r)^2 (1/L_A)(G_R/T_{sys})(1/k) \tag{12.21}$$

In this expression, which is known as the *telecommunications link budget equation*, the factor $P_T G_T$ can be regarded as a figure of merit for the transmitter and the term G_R/T_{sys} as a figure of merit for the receiving system. Leaving aside the constant $1/k$, the remaining factors refer to the propagation path. The quantity

$$L_S = (4\pi r/\lambda)^2 \tag{12.22}$$

is known as the *free-space loss*. We may think of it as the attenuation between two isotropic antennas separated by a distance r. It may seem curious that the space loss should depend on frequency but this is an artefact of the (slightly arbitrary) way in which we have separated out the factors related to the transmitter and to the receiver.

For a satellite in geostationary orbit (for which r is 3.6×10^7 m), the space loss from the transmitter to the sub-satellite point is given in decibels by

$$L_S(\text{dB}) = 183.6 + 20 \log_{10} f \text{ (GHz)} \tag{12.23}$$

If the ground station is not at the sub-satellite point, the increased path length results in additional space loss, depending on the elevation of the satellite as viewed from the ground station.

In terms of the space loss, C/N_0 is given by

$$C/N_0 = P_T G_T (1/L_S)(1/L_A)(G_R/T_{sys})(1/k) \tag{12.24}$$

Since the required C/N_0 can be determined from the system specification, this expression allows us to calculate the required transmitter power, P_T. A typical link budget is shown

Table 12.1 Example of a link budget[1]

Transmitter output power (per carrier)		−14.4	dBW
Multiple carrier loss[2,3]	0.2		dB
Transmitting circuit loss[2]	0.9		dB
Transmitted carrier power		−15.5	dBW
Transmitting antenna gain		18.0	dBi
EIRP		2.5	dBW
Space loss[2,4]	206.1		dB
Polarization loss[2,5]	0.1		dB
Atmospheric and multi-path losses[2,4]	4.0		dB
Total transmission loss[2,4]		210.2	dB
Ground terminal G/T		35.0	dBi/K
Boltzmann's constant		−228.6	dBJ/K
Received C/N_0		55.9	dBHz
Transmitted C/N_0[6]		69.2	dBHz
Resultant C/N_0[7]		55.7	dBHz
Carrier-to-intermod-density ratio, C/I_0[8]		67.0	dBHz
Overall C/N_0 (including intermods)		55.4	dBHz
Required C/N_0[9]		55.3	dBHz
Margin		0.1	dB

[1] On the basis of the shore-to-shore link in a study of a maritime satellite system. The budget given here is for the 11.7 GHz downlink carrying high-quality voice traffic (without forward error correction (FEC)) on a single carrier per channel basis. Global coverage is assumed.

[2] Loss factors are shown here as numbers greater than unity (i.e. >0 dB). Thus power is *divided* by the loss factor. In dB, the loss factor is subtracted.

[3] This correction allows for robbing of transmitter power by noise and IM products.

[4] For the minimum elevation of 5° at the ground terminal.

[5] Antenna axial ratios: satellite 3 dB; ground terminal 1 dB.

[6] From the uplink budget.

[7] The resultant C/N_0 is calculated by adding the noise powers, for example, $(N_0/C)_{tot} = (N_0/C)_{up} + (N_0/C)_{down}$.

[8] The ratio of carrier power to the power spectral density of IM products (regarded here as a type of 'noise').

[9] From the customer's requirements.

in Table 12.1. In this table the calculation starts from P_T and derives a value for C/N_0 in order to demonstrate compliance with the customer's requirement.

The signal-to-noise-power-density ratio is related to the bit energy to noise power density (E_b/N_0) described in section 12.2.5 by the following

$$E_b/N_0 = C/RN_0 \qquad (12.25)$$

where R is the channel data rate.

It is usually convenient to work in decibels and for this purpose in Table 12.1 all quantities that are proportional to power are expressed in decibels relative to the appropriate unit. For instance, a power of 1 W is 0 dBW (decibels relative to one watt) or 30 dBm (dB relative to one milliwatt), a bandwidth of 1 MHz is 60 dBHz (decibels relative to one Hz) and Boltzmann's constant is −228.6 dB J/K (decibels relative to one Joule per Kelvin). Note also that C/N_0, being power/(power per Hz), has the dimensions of frequency and may properly be expressed as dBHz.

12.3   THE COMMUNICATIONS PAYLOAD

12.3.1   The transponder system

Figure 12.12 is a simplified block diagram of a typical satellite repeater, which together with its associated antenna subsystem would make up a complete on-board transponder. Depending on the purpose of the satellite, there may be just one or perhaps many such repeaters within the same payload.

Before considering any of the details of the units, we shall trace the signal's path through the transponder, listing the main subsystems that it encounters and outlining their functions. The scheme adopted is a dual-conversion payload. Some alternative arrangements will be mentioned later.

The signal path

1. The *antenna subsystem's* function is to collect the incident signal power. Clearly the main requirements are that the aperture should be of sufficient size to receive a signal of adequate strength, and that it should be pointed in the right direction (an obvious need but this is often the source of many non-trivial problems in implementation). In a system relying on frequency reuse, the antenna subsystem must distinguish between signals on the basis of direction of arrival and polarization. Different signals on the same frequency must be presented at different output ports.

2. The *low-noise amplifier (LNA)* must amplify the weak signals arriving at the antenna to a level at which they can be processed without causing degradation of signal quality by noise in the electronic circuits. In addition, the LNA usually incorporates some preliminary filtering with the main purpose of attenuating any strong signals that may be present at adjacent frequencies.

3. The *down-converter* converts the signals to a lower frequency (the *intermediate frequency, IF*) at which much of the amplification takes place. This enables the creation of narrow-band channels that are suitable for many types of communication and may be essential for the implementation of on-board routing and processing functions. Frequency conversion also effectively eliminates the possibility of the amplifiers breaking into oscillation because of coupling between the output and the input of the repeater. The down-converter includes filters at both its input and its output. In Figure 12.1 the input filter is shown as part of the LNA. Its purpose is to give a more sharply defined RF passband and to reject the 'image channel' (see Section 12.3.5). The output filter rejects the many unwanted frequencies that are generated in the process of down-conversion, in favour of the one wanted at the IF.

4. The *IF processor*. The first part of the processor is normally a *demultiplexer* or set of filters that divide the broadband output from the down-converter into a number of separate channels. Although the proliferation of equipments in a multi-channel transponder carries a penalty in mass and cost, it also has advantages such as a reduction in IM products (see below) in the high-power stages, greater flexibility in routing signals between different antenna beams and different frequency bands, and the provision of *graceful degradation* of the system (meaning a gradual reduction in performance as equipment failures occur, rather than a sudden and total loss of capacity).

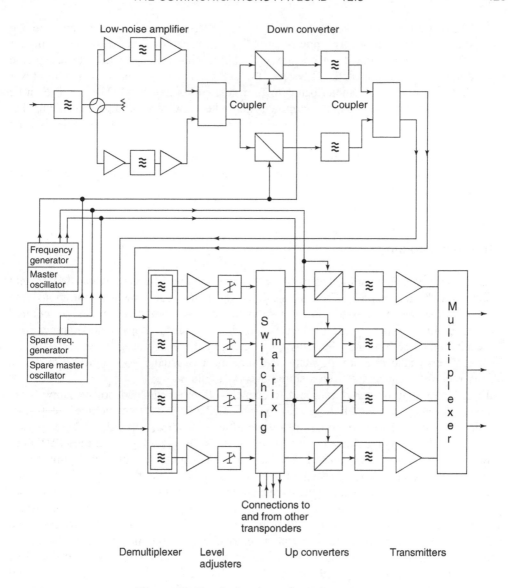

Figure 12.12 A simple on-board repeater

The second part of the processor is a switching matrix to perform routing operations and, in the case of equipment failures, to select channels that are still working. Finally there are the IF amplifiers, one for each channel, that provide most of the transponder power gain.

5. The *up-converter* reverses the function of the down-converter by translating the amplified IF signals to the higher frequency at which they will be transmitted. Like the down-converter, it produces many unwanted outputs as well as the one at the right frequency. There is consequently a need for further filtering before the signals pass on to the final amplifiers.

6. The *transmitters* (or *high-power amplifiers, HPAs*) raise the signal power to the required output level. The power amplifiers must achieve this with the maximum efficiency (minimum drain on primary satellite power) consistent with acceptable distortion. In a channelized system, the signals must then pass to a multiplexer that re-combines the signal bands to form a single output band. There will then be some filtering to remove the harmonics and other unwanted out-of-band signals that are always generated in a non-linear power amplifier. If the same antenna is used for both transmitting and receiving, the transmitter output filter and the receiver input filter may be combined in a single unit, the *duplexer*, which has the added function of ensuring that the incoming and outgoing signals are routed to their correct destinations.

Other payload units

There are several other units that form part of the communications payload but are not directly on the signal path. There is the local oscillator that provides continuous sine wave (CW) signals used by the down- and up-converters in order to provide the required frequency shifts. Typically the unit would provide several (perhaps many) outputs, one for each frequency converter. Possibly some or all of these will need to be harmonically related. Even where it is not a system requirement, it is usually good practice to derive all the required frequencies from a single master oscillator.

In order to achieve stable gain and frequencies, highly regulated power supplies are needed. Some power conditioning may be provided in the service module but further regulation is normally provided by units within the payload. In particular, where specialized power conditioning is needed, as for instance, for a travelling wave tube (TWT) or solid-state power amplifier (SSPA), there will be a dedicated unit for each equipment.

Finally the communications payload may include part of the Tracking, Telemetry and Command (TT&C) equipment. In many cases, once the satellite is on station, the telemetry and telecommand signals are transmitted at frequencies within the communications bands. The telecommand signals are extracted at the input demultiplexer or the IF processor and may be further down-converted or partially demodulated for the interface with the TT&C subsystem. Similarly, the telemetry signals are injected into the communications downlink either in the IF processor or in the output multiplexer. During the launch and early orbit phase (i.e. prior to switching on the communications equipment) and also as a back-up while on station, the TT&C system uses a dedicated transponder operating at one of the standard space operations frequencies. This is normally part of the satellite bus.

Other types of transponder

A few years ago, the dual-conversion repeater was in most cases the cheapest and most effective way of providing the necessary transponder gain and implementing on-board routing and processing. However, there is a useful reduction in complexity if the upconverter is eliminated, leaving a single conversion from the receive frequency to the transmit frequency. The bulk of the amplification and the other functions of the IF processor must in that case be performed at one of the radio frequencies (usually the downlink frequency, since this is normally lower). For frequencies of Ka-band and above the dual-conversion repeater can still be found.

The system described above is known as a *transparent* repeater. An alternative scheme is on-board demodulation of the signals followed by modulation of a separate downlink carrier. This arrangement is known as a *regenerative* transponder. Its use can lead to an improvement in SNR in digital systems where uplink noise is significant. This might be the case, for instance, if very small ground stations are in use so that uplink power is limited. In a regenerative transponder, digital signals can be 'cleaned up' at baseband so that only those noise peaks that are large enough to cause bit errors will be transferred to the output. Since the BER is normally very sensitive to the SNR, there is an advantage in a system where error rates rather than noise powers are cumulative. A further advantage of a regenerative transponder is that it allows sophisticated signal processing at baseband. In addition to routing on an individual signal basis this may include, for instance, changes in the coding structure and adaptation of error correcting codes in response to changing signal transmission conditions.

The use of redundancy

As is usual in all payload systems, the communications payload includes cold spares of the most critical units. The redundancy inherent in a channelized system has already been mentioned as a source of graceful degradation. Units that are common to all or many of the signal paths, such as (in the payload illustrated in Figure 12.12) the LNAs, down-converters and local oscillators, are usually provided with one or more spares, but where there are several parallel signal paths a lower proportion of spares may be appropriate. Wherever possible the spares are connected into the repeater using passive power splitters and combiners or hybrid couplers so that the selection of the operating unit can be effected simply by switching the power supplies on or off. The use of passive components rather than RF switches leads to greater reliability. However, since the use of any of these components involves some loss of signal (in the case of a power splitter or hybrid coupler, at least 3 dB loss) they cannot be used at the input to the receiver where the noise level is critical or at the output of the HPAs where the loss of transmitter power cannot be tolerated. In these positions low-loss switches must be used.

Mass and power budgets

One of the most important aspects of the payload system design, and a determining factor in the choice of payload architecture, is the estimation of the total mass and the power requirements. The way these quantities are apportioned between equipments shows substantial variations from one payload to another. The illustrative budgets shown in Tables 12.2 and 12.3 should not therefore be considered of general application. They are, however, broadly representative of fairly large satellites providing a mix of fixed-satellite and direct broadcast services.

The budgets given in the tables are for a 20-channel single down-conversion payload providing regional coverage. Differences between the numbers of equipments assumed in the mass budgets and those used for calculating power are due to the inclusion of redundant spares. The power budget shows the maximum demand when all 20 channels are driven at their highest power level. Particularly striking is the dominant role of the

Table 12.2 Example of a spacecraft power budget

	Quantity	Consumption (W)		Dissipation (W)
		Equipment	Total	
Receivers (LNAs, down-converters, local oscillators)	2	20	20	20
Channel amplifiers	20	70	70	70
TWTs	20	3310	3310	1250
Power conditioning units (for channel amplifiers and TWTs)	20	3710	330	330
Total payload consumption			**3730**	
Spacecraft bus DC power (equinox)			340	
Losses/battery charge (equinox)			670	
Demand at power source			**4740**	

Table 12.3 Example of a spacecraft mass budget

	Quantity	Total (kg)
Receivers (LNAs, down-converters, local oscillators)	4	8.5
Demultiplexers	4	16.0
Switch matrix	2	11.0
Channel amplifiers	28	8.5
TWTAs (including isolators, output switches etc.)	28	85.0
Multiplexer	2	9.0
RF cables/waveguides	—	10.0
DC harness	—	10.0
Fixing/mounting plates etc.	—	7.0
Total repeater mass		**165.0**
Antenna subsystem		60
Total payload mass		**225**
Structure/propulsion		290
Thermal control		60
Attitude determination and control		50
Telemetry, telecommand, ranging		50
Solar array		125
Batteries		180
Power conditioning and harness		80
Balance		3
Total platform mass		**838**
Spacecraft dry mass		**1063**

Note: TWTA. Travelling wave tube amplifier.

HPA in the mass of the payload and in the power consumption not only of the payload but of the entire spacecraft.

12.3.2   Payload system performance

Antenna coverage area

It has already been noted (Section 12.2.8) that the antenna coverage area and gain are closely related to the antenna dimensions (which are often determined by the available accommodation). Thus in calculating other key aspects of performance (such as EIRP and G/T) the coverage area and gain are usually regarded as predetermined parameters. Since performance requirements usually apply to the worst case, it is normally the antenna gain at the edge of the service area that is critical.

G/T

For a given antenna gain, improvement in G/T can be achieved only by reducing the system noise temperature, T_{sys}. The main contributions to T_{sys} are noise from the first amplifier stage and antenna noise due to thermal radiation from the Earth. Assuming that a 'state-of-the-art' amplifier device is used, there is little that can be done to reduce these contributions. Other contributions more amenable to the control of the designer come from later stages in the repeater, from noise generated in the transmitter which is not fully removed by the output filter and from the receiver input losses.

Input losses include attenuation in the antenna cable or waveguide, loss in the input filter, loss in a test coupler (where fitted) and mismatch losses. In designing the layout of the spacecraft, the LNAs should generally be placed as close as possible to the antennas in order to minimize cable losses. Filter loss depends on the filtering requirements and on the type of construction (see Section 12.3.8). Mismatch losses are caused by reflection of signal power at interfaces between equipments whose electrical impedance differs slightly from the correct design value. Reflections from a single interface can usually be kept as low as 1 or 2% of the power but the combination of several interfaces can result in a small but not negligible loss.

If the LNA has sufficient gain, the contribution to the overall system noise temperature from later stages can be made small. Despite this, however, the IF processor can still sometimes make a significant contribution. This is because the first IF amplifier is often preceded by power splitters, mixers and filters, which introduce attenuation and so undo some of the effect of the high gain LNA.

Detailed calculations of the system noise temperature are used to demonstrate compliance with the G/T requirement and to apportion the required gain between the various equipments.

12.3.3   The antenna subsystem

The antenna subsystem is often a critical factor in the spacecraft design because of its impact on total mass and stability, the possible need for stowage during launch and deployment in orbit and the requirement for Earth pointing, if necessary by the provision

Table 12.4 Spacecraft antenna diameters from geostationary orbit

	Earth coverage	UK coverage
Beamwidth	17°	1.5°
Gain (at edge of coverage)	18.5 dBi	39.8 dBi
	Diameter (m)	
Frequency		
1.5 GHz (L-band)	1.0	11.0
4.0 GHz (C-band)	0.37	4.1
12.5 GHz (ku-band)	0.11	1.3

of a de-spun platform. From the point of view of communications, the first constraint on antenna design is the required coverage area that determines the beamwidth and hence the antenna size (Section 12.2.8). Table 12.4 gives antenna diameters for three down-link frequencies and for two extremes of service area: Earth coverage, corresponding to beamwidth of $17°$ from GEO and a (directive) gain of 18.5 dBi (at the edge of coverage), and a *spot beam* covering, say, the British Isles, with a beamwidth of about $1.5°$ and an edge-of-coverage gain of 40 dBi. It is seen that for large coverage areas (comparable with Earth coverage) and especially at the higher frequencies, antennas of quite modest size are sufficient.

There are several possible advantages in the use of spot beams:

- There may be a requirement for a strictly limited coverage area in order, for instance, to avoid interference with other services.
- High antenna gain may be needed in order to serve very small ground terminals. The coverage area may in this case be served by many spot beams.
- More efficient use can be made of the available transmitter power if areas of high traffic density are served by high gain antennas (requiring relatively little transmitter power per carrier).
- Spot beams may enable frequency reuse by spatial discrimination.

The trade-off to decide on the number and size of the spot beams is clearly very complicated and involves many different aspects of the overall system design—the spacecraft structural design, antenna deployment, attitude control, power supplies, thermal control, and so on, as well as telecommunications capacity and market prediction.

Antenna types

1. The *horn antenna* can readily provide the small aperture needed for Earth coverage at 4 GHz or higher frequencies. In its simplest form it is a section of rectangular or circular waveguide spread outwards at the end to give the required aperture dimension (Figure 12.13a). In order to give improved radiation patterns, this simple structure is often modified by, for instance, the inclusion of steps or corrugations in the flared section.
2. *Patch antennas* (Figure 12.13b) consist mainly of a conductor mounted on a dielectric over a ground plane. Their use is generally limited to rather wide beams and gains less than about 14 dB. They can be used in phase arrays to provide a directive

radiation pattern that can be scanned and altered by changing the relative phases of the feed signals.

3. *Reflectors*, such as a paraboloid illuminated by a horn, are usually the most satisfactory solution when a narrow beam is required. The most usual configurations are the front-fed arrangement where a waveguide runs through or round the reflector to the focus of the paraboloid (Figure 12.13c), and the offset feed in which the reflector is a segment of a paraboloid taken from one side of the axis (Figure 12.13d). The front-fed arrangement has the disadvantage that the feed and its support structure cause blockage of the beam, reducing the gain and scattering power into the sidelobes. Moreover, the offset arrangement is often mechanically better, since the feed horn can be mounted rigidly on a face of the spacecraft. On the other hand its lack of symmetry results in poorer XPD when linear polarization is used, and in the case of circular polarization it can cause beams of opposite polarizations to point in slightly different directions. The Cassegrain system (Figure 12.13e), which is common on ground stations, is not often used on satellites because blockage by the sub-reflector usually makes it unsatisfactory for a small antenna. In some cases, where the service region is irregular in shape, there may be an advantage in forming an appropriately 'shaped beam' which, in a reflector antenna, is achieved by replacing the single primary feed by a cluster of horns. The reflector must be large enough to form the smallest features of the beam pattern.

4. *Phased arrays* are based upon the principle illustrated in Figure 12.14. The aperture is excited by many separate radiating elements that individually have only very weakly directive properties. Their combination may, however, have a very narrow beam because the radiation in some directions interferes constructively and in others destructively. The figure shows a one-dimensional array but the principle can readily be extended to the two-dimensional case.

   The advantages of this arrangement are that one array can produce a large number of beams simultaneously and that these can be steered electronically over a rather large angular range without the need for mechanical pointing systems. The distribution of transmitter power between many output paths reduces the demand on any one power amplifier and facilitates the use of solid-state devices. Finally, the array can be mounted rigidly and can often be made to conform to some convenient surface.

   The major disadvantage is the complexity of the associated equipment. Each of the equipment in a signal path to or from an array element must have a precisely controlled phase and amplitude characteristic and this must be maintained over a wide range of temperature, signal level, and so on. Since the number of elements may be large (possibly 100 or more) the programmes for manufacture, alignment and testing can be considerable.

12.3.4   The low-noise amplifier (LNA)

The primary requirements for the LNA are a low-noise temperature and sufficient gain to ensure that the noise contributions from the succeeding stages are small. A variety of amplifier types have been used in the past, including travelling wave tubes (TWT), tunnel diodes, parametric amplifiers and transistors.

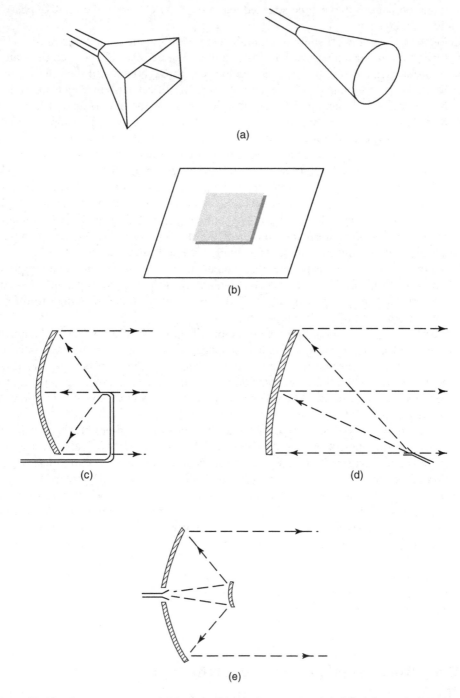

Figure 12.13   Antenna types: (a) horns; (b) patch antenna; (c), (d), (e) paraboloidal reflector antennas; (c) front-fed; (d) offset feed and (e) Cassegrain

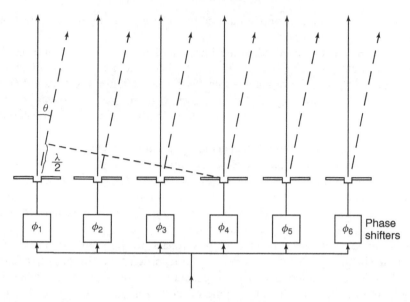

Figure 12.14   Phased array antenna. If all of the phase-shifts, $\phi_1, \phi_2, \ldots$ are equal, there is constructive interference in the broadside direction. At a small angle θ from this there is no emission since radiation from each element interferes destructively with that from an element three spacings away (for the six element array shown here). By changing the ϕ_i, the beam can be steered electronically without any movement of the antenna

In most current designs a transistor amplifier is chosen for the 'front-end'. In the first stage of amplification, in which the highest possible performance is required, an individual transistor would normally be used. For the later, generally less critical stages, monolithic microwave integrated circuits (MMICs) are widely used. These have an advantage both in reduced size and mass and in improved reliability and reproducibility.

To minimize input losses, the LNAs are accommodated as close as possible to the receiver antenna feeds. Cooling the LNAs in order to decrease their noise figure is likely to become increasingly common.

One further requirement for the LNA is good linearity over a wide dynamic range. Since the amplifier handles a multi-carrier signal (it is placed ahead of the demultiplexer that separates the signals into channels) any departure from linearity can give rise to IM products which cannot be removed by subsequent filtering.

12.3.5   Frequency converters

Any non-linear device will, in principle, serve as a frequency converter or mixer since, in response to an input containing frequencies f_1 and f_2, the output will contain every frequency of the form $nf_1 \pm mf_2$ (see Section 12.3.2). Usually the required output frequency is $f_1 + f_2$ or $f_1 - f_2$. Most conventional mixers depend on resistive diodes and these are usually arranged in a symmetric configuration, either singly or doubly balanced, which suppresses some of the unwanted outputs and provides better isolation between ports. In a balanced mixer the signal is applied at one point in the circuit and the local oscillator at another, ideally isolated, port.

Best performance is achieved when the local oscillator waveform is at a much higher level than the signal, so that the signal is in effect switched on and off at the local oscillator frequency, f_{LO}. As a general rule the local oscillator level should be at least 10 dB above the maximum signal level in order to minimize the *conversion loss* (the reduction in signal level from input frequency to output frequency), and to keep the level of IM products as low as possible.

A mixer produces many output frequencies in addition to the wanted one. It is also sensitive to frequencies other than the expected input frequency. In particular, in a down-converter, if the input frequency is $(f_{LO} + f_{IF})$, then the mixer also responds to $(f_{LO} - f_{IF})$, which is down-converted to the same IF (and *vice versa*). This is known as the *image response* of the down-converter. Noise and interfering signals in the image channel must be filtered out prior to down-conversion.

The noise temperature of a mixer is closely related to the conversion loss. To a reasonable approximation, a down-converter can be regarded as an IF amplifier fed through a resistive attenuator. Semiconductor noise generated by the diodes can usually be made relatively insignificant. Thus the noise figure of the down-converter (in decibels) is given roughly by the sum of the IF amplifier noise figure and the conversion loss. Noise figures in the region of 2 dB are possible for mixers operating at 14.5 GHz.

12.3.6 Local oscillators

Apart from the frequencies and signal levels, the two most significant aspects of a local oscillator's performance are the frequency stability and the phase noise. The specification of long-term stability usually limits the frequency drift to less than about 1 ppm/year, and this requires a crystal controlled source. Since crystal oscillators operate at less than about 150 MHz, frequency multiplication is necessary. A non-linear device such as a varactor diode or step-recovery diode may be used to generate harmonics, the appropriate microwave frequency being extracted either by filtering or by phase-locking a high-frequency oscillator to the correct harmonic.

12.3.7 IF processors

Digital signal processing (DSP) enables a number of payload functions to be performed including signal modulation/demodulation, channelization, channel power control, TWTA linearization, antenna control and autonomous packet switching. This results in a flexible approach to the service provision to be employed, a key requirement in the digital era, in which Internet access and digital audio and video transmissions are likely to vie for satellite resources.

Where DSP payloads are not employed, the main functions of the IF processor are to provide most of the transponder gain, to define the frequency response of each channel and, where necessary, to perform beam-to-beam routing functions. Thus the principal components are filters, transistor amplifiers and possibly electronic switching circuits. Where the routing requirements call for the division of the spectrum into many narrow-band channels, a rather low IF would be chosen. At IFs less than about 200 MHz, the best

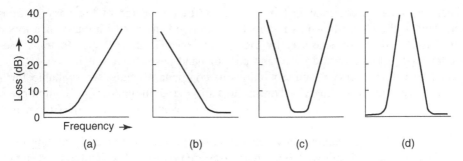

Figure 12.15   Filter types: (a) low-pass; (b) high-pass; (c) band-pass and (d) band-stop

choice of filter in terms of bandwidth, frequency stability and weight is almost always some type of crystal or surface acoustic wave (SAW) device. For broader channels and at higher frequencies, helical resonators provide satisfactory performance while at frequencies above 2 GHz, metal rod (coaxial) resonators or a thin-film technique would probably be used (this is a method of forming microwave circuits by etching a metal film deposited on a substrate of high dielectric constant).

12.3.8   Filters

The need for RF filters at various points in the transponder has already been noted. Most of these can be described by one of the labels, 'low-pass', 'high-pass', 'band-pass' or 'band-stop' (see Figure 12.15).

A typical RF filter consists of a series of resonant elements (or reactive elements for low- and high-pass filters) with precisely controlled electrical coupling between them. The most important electrical characteristics are:

- the attenuation within the passband,
- the rate of roll-off (increase of attenuation with frequency outside the passband),
- the flatness (lack of amplitude variation) of the in-band response,
- the phase linearity or group delay characteristic.

Some of the more important RF filters in the transponder are listed below.

- *The input filter.* A band-pass filter with main requirements: low in-band loss with adequate attenuation out of band.
- *The channel filter.* A band-pass filter with main requirements: rapid roll-off together with moderate loss and a sufficiently flat in-band amplitude and group delay response.
- *The output (harmonic) filter.* A low-pass or band-pass filter with main requirements: low in-band loss with adequate attenuation at specified higher frequencies.
- *The multiplexers and demultiplexer.* In general these can be regarded as combinations of band-pass and band-stop filters. There may be a requirement for very rapid roll-off if the channels are closely spaced. The multiplexer, coming after the HPA, must have low in-band loss but this requirement is usually less stringent for the demultiplexer.

There is much in common in the procedures for the design of the various types of filter. Any of the types shown in Figure 12.15 can be synthesized initially as a low-pass design, followed where necessary by the application of standard rules to transpose the impedances to those required for a high-pass or band-pass characteristic.

Standard response curves and tabulations of component values are available for low-pass prototype filters optimized according to various requirements. Two commonly used criteria are maximally flat in-band response (*Butterworth filter*), and maximum roll-off for a given in-band amplitude variation (*Chebyshev filter*). These characteristics can be obtained by including coupling only between adjacent filter sections. If coupling between non-adjacent sections is allowed, other responses are possible of which the most important is the elliptic filter response. This has a very sharp roll-off to attenuation maxima just outside the passband.

The theoretical response curves can be used to determine the number of sections (resonators) needed to achieve the required roll-off while maintaining adequate flatness in-band. It is then possible to predict other filter properties such as group delay response and in-band attenuation.

In designing a microwave filter for a space application, it is important to allow adequate margins for temperature variations. The main effect is a shift of centre frequency that for aluminium filters may be as much as 1 part in 10^3 over a 50°C temperature range. For Invar construction the corresponding variation is about 1 in 10^4 but there is a significant mass penalty. Alternatively the filter design may incorporate temperature compensation based on the use of two materials with different coefficients of expansion. Many filters are now made of carbon fibre based materials. These have both low mass and low thermal expansion but their use in some applications is limited by difficulties in manufacture.

12.3.9   Transmitters

There are only two generally accepted means of amplifying the signals to the level required for downlink transmission. These are the TWT and power transistors. The TWT has been the most commonly used but as transistor technology improves, solid-state amplifiers are finding an increasing number of applications, mainly because of their higher reliability and lower mass.

Travelling wave tube amplifiers (TWTA)

In a TWT, amplification is achieved by interaction between an electron beam and a signal in the form of an electromagnetic field travelling along an elongated guiding structure. For a significant effect, the electrons and the signal must travel at almost the same velocity, which implies that the signal must be slowed down very considerably by the guiding structure. In low- and medium-power tubes the 'slow-wave structure' is a helix of wire held in place by ceramic rods. The signal modulates the density of the beam and the beam in turn transfers energy to the signal, which increases exponentially in amplitude as it travels along the tube to the output port. Helix tubes are capable of up to about 200 W output. For higher powers a more robust slow-wave structure may be required, such as a series of electrically coupled cavities. However, for most space applications, including direct broadcast television, a helix tube is preferred.

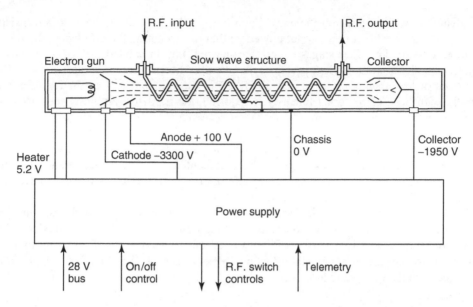

Figure 12.16   Simplified TWTA configuration

A disadvantage of a TWTA is the requirement for a complicated high-voltage power supply (see Figure 12.16). The accelerating voltage is several kilovolts and must be very well regulated. The efficiency of a TWTA can be high—as much as 70% for the tube itself or about 60–65% including the power supplies.

TWTs can now be made sufficiently reliable for most missions but they do suffer from a gradual deterioration in performance due to loss of cathode emission during their lifetime. In some cases a facility is provided for a telecommandable increase in heater current in order to restore performance. Alternatively, the correct cathode current may be maintained by a control loop that varies the anode or control grid voltage.

Though attractive from the point of view of gain, efficiency and power output, a TWT is a rather non-linear amplifier. When amplifying a multi-carrier signal, it both generates IM products and converts signal amplitude variations into spurious phase modulation. More linear operation can be achieved by 'backing-off' the tube to a lower power level, but this also results in a loss of efficiency.

It is becoming usual where a multi-carrier operation is required to incorporate a *linearizer* in the TWTA. This is a non-linear driver amplifier that pre-distorts the signal in such a way that the non-linear TWT characteristic is partially compensated. The tube may be operated closer to its maximum power (saturation) level (keeping the efficiency high) whilst IM product levels remain low.

Transistor power amplifiers

When compared with the equivalent TWTA, a SSPA has lower mass, higher reliability and lower cost. There is also the significant advantage of eliminating the need for the TWT's complicated high-voltage power supply. The simpler, low voltage supply required for an SSPA affords a further improvement in mass and reliability. Transistors are, however, at a disadvantage with respect to efficiency. The microwave power at the input to a transistor is a significant fraction of the output power, so that in computing the efficiency, the power requirements of the driver stages cannot be ignored. For an L-band (1.5 GHz) transistor amplifier under a multi-carrier operation, the overall efficiency might be around 30–35%, appreciably less than for a saturated TWTA. But the efficiency of a TWTA decreases rapidly with back-off. If good linearity is required, the transistor amplifier may in some circumstances prove to be more efficient than a TWTA.

The employment of phased arrays allows the use of distributed power amplifiers, where each individual SSPA has a fairly modest output. As in the case of an LNA, the critical stages (in this case the output stages) are implemented using individual transistors but, in order to reduce size and mass, the pre-amplifier and driver stages and any necessary control functions may be realized in MMIC technology.

TWTs continue to improve in performance, reliability, adaptability and cost but there is, particularly at low powers, considerable competition from solid state devices. The output per transponder is typically a few watts to a few tens of watts. Powers up to about 300 W are within the present capabilities of SSPAs at L-band. SSPAs are available in all frequency bands up to Ka-band.

12.4 CONCLUSION

A modern communications satellite is a far cry from the simple transponders of Telstar and Early Bird. What seemed in the 1960s, a bold and imaginative step is now a mature technology and fuels the continuing growth of the multi-billion pound satellite industry. Telecommunication satellites have become a key part of everyday life. Nevertheless, it is a technology that continues to develop both under the stimulus of advances in component technology and in response to new markets brought about by the digital era. Geostationary satellites continue to increase in power and moves to higher frequencies have led to a proliferation of telephony and TV transmissions, mobile and Internet services. Technology advances have not been limited to the geostationary satellite. After a period of over optimistic estimates of market potential, the use of LEO constellations of satellites to provide specialist services has stabilized. However, in terms of non-geostationary satellites, it is the navigation services that are, at present, the most successful. A further stimulus is expected with the introduction of the European *GALILEO* system.

As the service requirements of the twenty-first century citizen continue to evolve, the unique opportunities offered by satellites will ensure that this form of delivery will carry on playing an important role in the provision of telecommunication services for the foreseeable future.

ACKNOWLEDGEMENT

The authors wish to acknowledge the considerable contribution of Howard Smith to this chapter.

REFERENCES

[1] Clarke, A. C. (1945) Extra-terrestrial relays, *Wireless World*, **51**, 10.
[2] FINAL ACTS WRC-07, World Radiocommunication Conference, 2007, ITU, Geneva.

[3] Radio Regulations (2008) ITU, Geneva.

[4] Recommendation ITU-R P.676-7 Attenuation by Atmospheric Gases (02-2007) ITU.

[5] Recommendation ITU-R P.618-9, Propagation Data and Prediction Methods Required for the Design of Earth-space Telecommunication Systems (08-2007) ITU.

[6] Recommendation ITU-R P.837-5, Characteristics of Precipitation for Propagation Modelling (08-2007) ITU.

The reproduction of Figure 12.10 is made with the prior authorization of the International Telecommunication Union as copyright holder. The complete publication of the CCIR from which the figure reproduced in the present publication is extracted can be obtained from:

ITU General Secretariat

Sales and Marketing Division

Place Des Nations CH-1211 Geneva 20

Switzerland

13 TELEMETRY, COMMAND, DATA HANDLING AND PROCESSING

Nigel P. Fillery[1] and David Stanton[2]

[1]*EADS Astrium, Portsmouth*
[2]*Keltik Ltd, Hampton Hill*

13.1 INTRODUCTION

The *telemetry, command, data handling and processing* functions provide for the two-way flow of information between a spacecraft and its ground control station(s). There are transmission (downlink) and reception (uplink) functions to perform, as well as the tasks of gathering and processing data ready for transmission, and the processing and routing of command data from the ground station. In addition, there will be a transponder for ranging, and support services for the payload.

The mission, orbit, type of payload and the selected ground control station(s) play a considerable part in determining the nature of the design. This is illustrated by studying the *Intelsat* and *Inmarsat* series of satellites, in Geostationary Earth Orbit (GEO) with communication payloads; Earth remote sensing satellites such as *Envisat, METOP, TerraSAR-X* and *Sentinel*, in a near-polar Low Earth Orbit (LEO) with passive and active sensors to monitor the Earth; and the *International Space Station* (ISS), in a non-sun-synchronous LEO.

The *spacecraft/ground station link* from GEO can be a continuous direct one, with the spacecraft visible at all times from an appropriate ground station. This allows for a relaxed transmission link compared with that from a non-geosynchronous orbit. Envisat, for example, is visible to its ground station for only a few minutes at each ground pass. It therefore needs on-board data storage and a rapid data-transfer link. The ISS requires a continuous link, and this can be provided by a two-way link with any ground station in sight, and/or by a relay system [Tracking and Data Relay Satellite System (TDRSS)], which makes the ground link via special-purpose spacecraft in GEO.

Spacecraft Systems Engineering, Fourth Edition.
Edited by Peter W. Fortescue, Graham G. Swinerd and John P. W. Stark.
© 2011 John Wiley & Sons, Ltd. Published 2011 by John Wiley & Sons, Ltd.

Spacecraft destined for GEO will be non-synchronous in their early phases, such as during launch and intermediate orbits, and will require special ground support before they are handed over to their dedicated ground controllers. The Intelsat spacecraft use a worldwide Launch Support Network operating in C-band, both during launch and subsequently. Other spacecraft may use the ESA network operating in S-band for initial orbits, with control being transferred to a dedicated station operating in Ku-band when it is on-station.

The *telemetry downlink* must provide the ground control team with information about the functioning of the subsystems in the craft, so that they may detect whether it has its correct orientation, or whether any fault has developed, and if so, they must be able to diagnose its cause. It may also be the channel for passing information from the payload to the ground when the mission is scientific or Earth-observation. When the payload comprises communications equipment, that equipment will serve as an alternative route for the telemetry data, once it is deployed.

The *command uplink* must enable the ground controller to change the role of the spacecraft, such as to reorientate it, to correct a fault, to operate a mechanism, or for other reasons. It must do so in a highly reliable way, giving confirmation that the instructions have been carried out.

The *ranging transponder* forms part of the system by which the ground controller tracks the spacecraft and determines its orbit.

The *payload* may require significant control, data handling, data storage and processing functions. As spacecraft designs evolve towards autonomous operation, the bus itself may also require extensive data processing and storage functions.

13.2   SYSTEM ARCHITECTURE

All spacecraft need most of the functions described in this chapter. However, the physical distribution and implementation vary considerably.

13.2.1   Relationship with other subsystems

The context of the telemetry, command and data-handling subsystem with respect to the other main subsystems is illustrated in Figure 13.1. Whilst it is usual to consider any subsystem under discussion as the centre of the system, in this case it is true, for it is the digital system that spacecraft operators and users 'see' and interact with.

Commands are received via, typically, an S-band link, decoded and placed in a queue for either internal distribution or distribution to the other subsystems. The verification of each command is achieved by the feedback of telemetry, usually from each stage in the command distribution and after the operation of the command. This telemetry is checked to verify that all is well at each stage. The operation of all of the other subsystems is achieved in this way. Some subsystems are not demanding in terms of processing requirements but are still critical to the mission, for example, deployment. Others are very demanding but not so critical, for example, instruments. Separate downlinks, at X- or Ka-band, may be provided for the payload data, particularly when the data rate is too great for the normal telemetry link. The system protocol interacts with an on-ground command centre, and sometimes a separate data-collection centre. The system usually

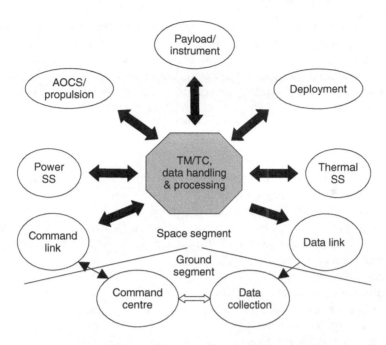

Figure 13.1   System context

operates in a number of different modes—ranging from a basic 'safe' mode, in which the spacecraft is kept alive from a command, thermal and power point of view, up to a fully performing platform and payload.

13.2.2   Avionics

The term *Avionics* is often loosely used. It can apply just to the electronic part of the Attitude and Orbit Control System (AOCS) subsystem but sometimes it is used to cover all of the platform electronics hardware and software, that is, elements for the AOCS, spacecraft-level telemetry/telecommand (TM/TC), data communications, data processing, data storage and the transponder. In a modest spacecraft, all of the processing for these functions may be in a single on-board computer. In large spacecraft, a number of distributed and sometimes different processors and devices are used. Avionics does not usually include the payload electronics such as that required by instruments and experiments, or the processor embedded in some communications payloads for RF beam forming.

13.2.3   On-board data-handling functions

The emphasis on the data-handling functions required varies according to the mission type. A typical communications satellite only has requirements for basic telemetry and telecommand functions in order to maintain the health and position of the platform on which the payloads are situated. Communications satellites with signal processing capability may need a little more interaction with the ground.

For all other mission types, the primary product of the mission is in the data, which is sent to the ground via telemetry. These missions also tend to operate in a number of modes according to mission phase or data product required, and thus involve more interaction with ground operations. The data-handling systems are therefore more directly involved with the mission product generation and tend to be more complex than for communications satellites.

Typical functions that can be required of the on-board data-handling (OBDH) system include the following:

- Enabling the flow of housekeeping and science data.
- Receiving and distributing commands.
- Performing telemetry and telecommand protocols.
- Time distribution around the spacecraft—required for synchronization, and the time stamping of information (datation). This is usually derived from an ultra-stable oscillator and a mission-specific epoch.
- Providing data storage.
- Executing commands and schedules.
- Controlling payloads and subsystems.
- Monitoring spacecraft health.
- Making autonomous decisions.
- Performing data compression.

13.3   TELEMETRY DATA FORMATTING

13.3.1   Classification of data

Telemetry data fall into three basic categories: housekeeping, attitude and payload.

Housekeeping data, sometimes known as engineering parameter data, needs to be monitored to keep a check on the health and operating status of the on-board equipment. Such data can arise in many forms and some typical examples are as follows:

- *Temperatures* of equipment boxes, solar arrays, attitude-control thrusters and plenum chambers, parts of the structure and so forth. Thermistors are used to convert temperature into an analogue voltage. For high temperatures, thermocouples are used and the output of a few mV d.c. is amplified to a level suitable for the telemetry encoder.
- *Pressure* in fuel tanks, plenum chambers and so forth. Various forms of pressure transducers are used.
- *Voltages and currents* of equipment power supplies. The rail voltages are scaled to a common full-scale range, which is often 0 to +5.12 V. Current monitoring may involve a variety of circuit techniques.
- Individual digital bits representing the *operational status* of equipment, each indicating that a particular functional mode is either selected or deselected. For proportional status information, such as amplifier gain settings, a number of bits are grouped together into words of appropriate length.

- *Redundancy status* that furnishes each relevant relay or switch with a set of contacts to provide a status bit provides information on whether the 'main side' or the 'cold redundant side' of equipment is in use.
- *Deployment of mechanisms*, separation from the launcher and so forth. A microswitch is fitted in order to provide an appropriate status bit.

On a modern, large communications satellite there will be several hundred to about 1000 engineering parameters to be monitored, and the result of every command is usually checked via the telemetry. The great majority of these will only need sampling at infrequent intervals of typically 30 s to 2 min, so the bandwidth required is quite small. A bit rate of a few hundred bits per second is sufficient to transmit the total information.

Attitude data arises from a variety of sensors such as Sun, Earth and star sensors, gyroscopes and accelerometers (see Chapter 9). The data can be analogue, digital or a mixture of both.

During transfer and intermediate orbit phase, the attitude and velocity will change rapidly, and frequent sampling is needed, typically from once to four times per second. For GEO operations, a reduced rate may be provided, selected by command.

Although only a few channels of attitude information are required, the high sampling rate needed during some mission phases may lead to a bandwidth that exceeds that which can be provided conveniently by a standard pulse code modulation (PCM) data system, and a separate wideband high rate system may have to be provided.

Payload data is very variable and each case needs to be considered individually. Science and Earth-observation missions may generate very large data volumes, for example, imagery is typically very data-intensive. Often only a few channels of data are required, but their rates may be as high as many Mbits or Gbits per second. An entirely separate high-rate system may then have to be provided, such as on Envisat, and data compression may also be used to reduce the rate. Whilst high-rate downlinks can be made available for LEO missions, deep space images may take hours or even days to transmit to Earth. The capacity to acquire such data may be limited by the amount of on-board storage as well as the bandwidth of the communication links.

A communications payload gives rise to many channels of engineering parameter data in addition to that generated in the service module. Because of its complexity and the large amount of redundancy employed, it is likely to demand considerably more channels than the service module.

Typical monitoring requirements of a communications payload are as follows:

- Temperatures of travelling wave tube amplifiers and other repeater equipment.
- Power supply voltages and currents for each of the main equipment.
- Operating and redundancy status monitors for the many waveguide and coaxial switches.
- Analogue power monitors for the signal levels at each main RF equipment interface.
- Telemetry monitoring of the digital gain settings telecommanded to the various channel amplifiers.

13.3.2   Influence of mission phase

The phases of a mission run from ground testing through the launch phase to the orbit phases. In each of these, the telemetry plays a part.

During *ground testing*, there is access to inter-unit connections, and test and diagnostic connectors, in addition to the normal telemetry data. On satellites with an avionics data bus, direct access to the bus may be possible. This allows higher data rates for telemetry and command, with the benefit of faster ground testing. However, RF links are still used for overall performance testing.

During the *launch phase*, satellite data is usually minimal and is restricted to a few housekeeping parameters such as battery condition, some key temperatures, reaction control equipment (RCE) pressures and deployed item status. Payloads are not usually switched on until in orbit and until out-gassing has been completed as necessary, but some telemetry activities may occur at separation. This data is sent down via the launch vehicle.

The *in-orbit* phases include the transfer and intermediate orbit phases of geostationary missions. In these cases, telemetry contact is not continuous unless relay satellites are used. Spacecraft operation must be autonomous as far as possible in order to avoid the need for intervention by ground control. For example, majority-voting techniques may be used instead of cold redundancy.

13.3.3   Telemetry data encoding

All the data considered so far arises in three basic forms: analogue, digital bi-level, and digital serial. The data is time-division multiplexed (TDM) into a PCM bit stream, which modulates the downlink RF carrier (see Chapter 12).

The first step in conditioning *analogue data* is to scale the data to a common full-scale range, usually 0 to $+5.12\,$V. This is done at the source of the data. Frequency components greater than half the sampling frequency need to be removed by a low-pass filter to prevent aliasing errors. It is good practice to include a simple low-pass filter in each analogue line.

After filtering, the channels are sampled in turn by analogue switches and each sample is converted to a digital word, which is mixed in with the digital data as described later. For most data, an overall accuracy of about 1% is sufficient, and an 8-bit analogue-to-digital converter is used to achieve this.

Analogue commutation is invariably carried out in two stages: by a mainframe multiplexer sampling at a relatively fast rate and a slower sub-multiplexer sampling at a binary sub-multiple of the mainframe rate. The larger number of housekeeping channels that need sampling only once or twice per minute use the sub-multiplexer, and the few channels requiring fast sampling go straight into the main multiplexer.

This arrangement is very convenient because some of the sub-multiplexers can be remotely located in subsystems or payloads, reducing the complexity and mass of the harness.

Digital bi-level data arising from relay contact closures and so forth, is first conditioned to appropriate logic levels in which the 'off' state is represented by nominally zero voltage and the 'on' state by a positive voltage suitable for the integrated circuit (IC) logic family used. Individual bits are then grouped together into 8- or 16-bit words and sampled by logic gates whose outputs are serialized in a parallel-to-serial converter and mixed in with the main PCM data stream.

Digital data is usually acquired in serial form, thereby simplifying the cable harness. Such data is initially stored as an 8- or 16-bit word in a shift register located in the equipment generating the data. In equipment using processors, blocks of words in random access memory are also used.

In normal PCM mode, the 8-bit parallel words from the analogue-to-digital converter and the serial digital multiplexer are loaded, together with bi-level status data, into a shift register in the parallel-to-serial converter and clocked out at the telemetry bit rate as a continuous PCM bit stream. Unique synchronization and identification codes are inserted in the bit stream so that all data can be identified when it reaches the ground. Finally, the bit stream is bi-phase modulated on to a coherent sub-carrier at an integral multiple of the bit rate before routing it to the two TM transmitters.

In many encoders, an alternative to the above allows sampling to be concentrated on a particular word, to the exclusion of the other data. A 'dwell' mode is achieved by loading the address of the desired channel into the control and timing block, which then sets up the input multiplexers permanently to that channel instead of cycling round all the channels.

Wideband analogue signals bypass the PCM section of the encoder altogether, and frequency modulates a voltage-controlled oscillator (VCO) instead, using one of the US Inter-Range Instrumentation Group (IRIG) standard centre frequencies. Since the phase-shift keying (PSK) sub-carrier and VCO frequencies are different, the PCM and FM channels can operate simultaneously.

13.3.4 Telemetry list and data format

The first step in designing the telemetry system is to draw up a telemetry list for the spacecraft. For each item, this will contain the signal identification, the type of data (analogue, digital bi-level, or digital serial), the required accuracy and the sampling rate required. This list is first established during the initial 'Phase A' feasibility study and it evolves with the project. It is important to allow enough spare channels at the outset to cater for natural growth as the project matures.

The next stage is to lay out the format of the PCM message, and a simple format consisting of eight frames, each containing 64 8-bit words, may be used. The first two words of each frame contain a fixed synchronization code, 16 bit long, which the ground station equipment will recognize as the start of a frame. The next word is a frame-identification channel. The ground station de-commutation process is then able to identify every channel in the format and display the parameter contained in it on the 'quick look' facility at the station. Further processing by computer provides outputs in engineering/scientific form for users.

The rest of the format consists of data channels. A channel is typically sampled once in every frame, and such a channel is known as a mainframe or prime channel, and could be used, for example, for a nutation sensor or other data source needing fast sampling. When a still higher rate is required, the same data can be put into more than one mainframe word—such a channel is said to be super-commutated.

Housekeeping parameters requiring infrequent sampling can be sub-commutated by sampling only once per format.

By extending the principles of sub- and super-commutation over a longer sequence of frames, a wide variety of sampling rate needs can be accommodated, and once this is done, the bit rate needed to provide every parameter with at least its minimum sampling rate can be worked out. The sub-commutation depth of eight frames used in the above example would not be sufficient for a typical communications satellite.

These PCM systems are used with spacecraft in which there may be existing infrastructure and/or equipment, or where there is no need for the additional utility of a Consultative Committee for Space Data Systems (CCSDS) packet system.

13.3.5 Packet telemetry

The *CCSDS* has produced a series of recommendations for data systems such as 'Packet Telemetry' [1] and 'Space Packet Protocol' [2], which have been adopted by ESA through the European Cooperation for Space Standardization (ECSS-E-ST-50 series). These supersede the PCM Telemetry Standard. They represent the evolution from the time-division multiplex methods used in the past and come into their own on multiagency projects such as the ISS, where increased interoperability and standardization are essential. The data flow in a packet telemetry system is illustrated in Figure 13.2.

Application data is first encapsulated by the source into a 'source packet' by prefacing the data with a standard label known as the 'packet header'. This is used to route the data through the system and must therefore contain identification of the source and its particular applications process, the number of the packet in the sequence of packets produced by the source (so that the packets can be delivered in the right order at the data sink end of the link) and the length of the data field attached to the header. The provision to enable source packets exceeding a prescribed length to be segmented into several shorter 'telemetry packets' that can then be interleaved with packets from other sources has been eliminated from the CCSDS standards.

The next stage is to assign each group of sources to a so-called 'virtual channel'. The packets are inserted into fixed-length frames known as virtual channel data units (VCDU), which are then multiplexed into a single transfer frame for transmission on the downlink. If necessary, each virtual channel can be sampled at a different rate, appropriate to the bandwidth of that channel.

After transmission, the transfer frame can be de-multiplexed in the normal way, and the telemetry and source packets can be reconstructed and routed to their destinations using the headers. Labelling the source packet with all the necessary information about its routing and interpretation is the key to the flexibility of the packet system.

A number of extensions to the packet system have been developed, providing even more utility and higher levels of protocol. The need is for reliable file transfer and internet-type operation. The CCSDS 'Advanced Orbiting Systems (AOS)' [3], and 'Space Communications Protocol Specification' [4] are two examples (see 13.5).

The advantage of using packet telemetry is its inherent adaptability to changing data-transmission requirements. This is due to the variable-length nature of the packets and the fact that the packet headers include fields that can be used to identify and interpret the contents of the packet. The data length and periodicity are not constrained by having to occupy a particular position within a TDM frame. Bandwidth is not wasted in sending periodic data to the ground, which is then discarded. Data packets can be issued asynchronously by any data source, and the packets can be of whatever length to accommodate the data requirements at that instant. The service provided by packet telemetry

CCSDS recommendation for packet telemetry

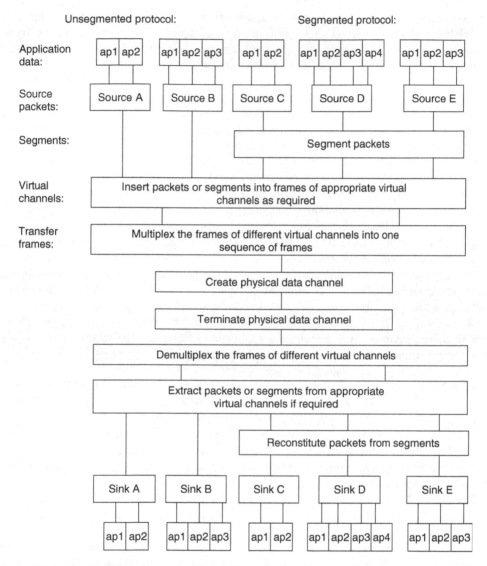

Figure 13.2 Packet telemetry data flow. (Reproduced by permission of the European Space Agency)

is also directly compatible with higher layer application services such as networking and Internet applications.

CCSDS packet telemetry is now the dominant technology in use for missions all around the world. Space-qualified chip sets, software libraries, and global networks of earth stations are available, and there is a great body of expertise within the space industry.

13.3.6   Error-control coding

On communications spacecraft, there is usually enough transmitter power for a bit error rate (BER) of better than 1 in 10^5 on the downlink. The loss of even a complete frame is not catastrophic, so there is no great need for correction of such errors—nevertheless an error-checking code is sometimes included in the frame.

In a typical scheme, this is the 16-bit remainder that results from dividing the data portion of the telemetry frame by a specified polynomial such as $x^{16} + x^{12} + x^5 + 1$. The remainder is placed in the last two words of the telemetry frame and transmitted to the ground with the rest of the data. On the ground, the complete frame including the check bits is again divided by the same polynomial, and if a non-zero result is given, then the frame is flagged as being in error. This scheme is not able to correct errors and offers no saving in transmitter power since the actual error rate on the link is unaffected.

When *forward error correction* is needed, a scheme based on convolutional coding provides good correction capability in a Gaussian noise channel and is simple to implement in hardware. Significant further improvement (particularly with bursts of errors) can be obtained by concatenating a Reed–Solomon (RS) block code with the convolutional code. The RS code is used as the outer code while the convolutional code is the inner one.

One application of forward error correction is in packet telemetry. This needs nearly error-free channels for the successful routing and control of the data packets. The coding techniques are used either singly or together and are an integral part of the standards.

Missions involving very long transmission path lengths, such as the *Huygens Probe* sent to Saturn's moon Titan, require the application of coding. The telemetry transmitter may then dominate the power budget, and the coding gain of several dB can be exploited to reduce the RF power needed for a given BER.

Joint NASA/ESA telemetry coding guidelines have been issued to standardize on particular convolutional and block codes to facilitate network cross-support on joint missions. The convolutional code recommended is the constraint length $k = 7$, rate $r = \frac{1}{2}$ code that is in common use by ESA, and within NASA by the Goddard Space Flight Centre and the Jet Propulsion Laboratory. CCSDS recommendations have been adopted by ESA. The ECSS Telemetry synchronization and channel coding standard [5] provides convolution, RS, concatenated and turbo code standards.

The basic RS code is a (255, 223) code, which can be used either non-interleaved ($I = 1$) or interleaved to a depth of $I = 5$. This means that 223 I bytes of uncoded information are fed into the RS encoder and emerge unaltered with 32 I check symbol bytes appended. The whole code block of 255 I bytes is then transmitted, followed by a further 223 I information bytes, and the cycle is repeated indefinitely. The RS code is capable of correcting bursts of errors in up to 16 I consecutive bytes, and it can be seen that the interleaved form is considerably more effective in combating long burst errors.

The RS code generator is a more complex proposition than the convolutional coder as it involves binary multiplication and the intermediate storage and processing of up to 1275 bits in a shift register or equivalent RAM. However, custom chips, designed by both NASA and ESA, reduce the hardware to manageable proportions. Software solutions are also used where the data rate is relatively modest and will be used increasingly as the performance of space-qualified processors increase.

Using soft-decision *Viterbi decoding*, convolutional coding typically gives a 5.1 dB increase in link margin (reducing the power requirement for a given rate to nearly a quarter). RS encoding can give a further 2.5 dB increase.

13.3.7 Downlink frequencies and modulation

A spacecraft uses one or more of the worldwide ground station networks for command, telemetry and ranging, at least in the early mission phases. The frequency bands are constrained to be those that are supported by the chosen ground stations.

The first step in choosing the band(s) to use is to contact the Frequency Management Office of the networks concerned, with a request for the favoured frequency band. The second step is the selection of discrete frequencies within the allocated bands. This choice is a very complex and lengthy process because of the need for coordination with other space agencies and national authorities.

Many networks provide telemetry support in the ultra high-frequency (UHF) band, an exception being the Earth stations covering Intelsat missions.

Phase modulation of the RF downlink is universally used, and there is a degree of compatibility between the ESA and NASA standards, which allows cross-support, provided certain limitations are observed.

At low bit rates, a PSK-modulated sub-carrier is used to prevent modulation sidebands occurring too near to the RF carrier and upsetting the phase-lock loops in the ground station receivers. At higher bit rates a preferred alternative, achieving the same result, is to modulate the RF carrier directly with PCM-SPL data as defined in Figure 13.3.

Pre-modulation filtering of the square-wave PCM bit stream is usually provided to attenuate high-order modulation sidebands.

13.4 TELECOMMAND

13.4.1 Telecommand user interface

For relatively simple missions, there are three basic types of commands, as follows:

- *Low-level on–off commands*. These are logic-level pulses used to set or reset bi-stable logic.
- *High-level on–off commands*. These are higher-powered pulses, capable of operating a latching relay or RF waveguide switch directly. Typically these may be 12 to 28 V pulses lasting several tens of milliseconds and capable of supplying 90 mA for a relay drive, or up to 600 mA to drive an electromechanical RF switch. Separate 'on' and 'off' pulses are supplied on two different lines to drive the two coils of the latching device.
- *Proportional commands*. These are complete digital words, which may be used for purposes such as the reprogramming of memory locations in an on-board computer, or for setting up registers in the attitude control subsystem.

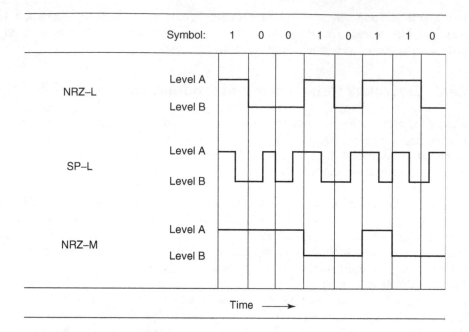

Figure 13.3   PCM waveforms

Enough spare channels of each type need to be provided to allow for natural growth, otherwise a redesign may be necessary at a later stage. This is controlled through the generation and maintenance of command lists.

More advanced missions require more sophisticated services, such as the ESA Packet Utilization Standard (PUS). This provides standard facilities for the above functions as well as features for loading timelines, loading, checking and dumping memory, software patching and so forth. The PUS also includes standard telemetry services.

13.4.2 PCM telecommand standards — Intelsat and the US Air Force Satellite Control Facility (SCF)

The agencies responsible for the ground station networks publish standards covering space–ground interface requirements and procedures to be adhered to for the transmission of commands. Their purpose is to ensure compatibility between the spacecraft and the equipment at the ground station. Conformance to these standards is mandatory and any deviations for a particular mission have to be justified and approved by the agency before being included in the design.

In GEO, with continuous visibility from several Earth stations, it is very seldom that a sequence of commands has to be sent within a short period of time. The Intelsat command standards are therefore based upon a command-verify-execute strategy in which each command is held in the satellite decoder and verified through telemetry before it is executed.

Figure 13.4 shows a simplified block diagram of a typical decoder for an Intelsat spacecraft. The explanation relates primarily to the Intelsat V system, but more recent spacecraft in the series have used fundamentally similar principles.

The command uplink signal consists of a sequence of tones that represent data 0, data 1 and execute information 'bits'. These are frequency-modulated on the 6-GHz carrier and demodulated by the two command receivers. The ground operator is able to choose which receiver is to be used by selecting either of two alternative data 0 tone frequencies.

Each digital command message includes an introductory series of zeros to select the receiver and to synchronize the bit detector clock, and this is followed by a decoder address word, a command vector, and finally an on–off command word or a proportional data command word. The 0 and 1 bits comprising the message are stored in a shift register.

Each command unit has a unique address, which is hardwired, and only messages containing this address will be accepted. The data held in the command message store are telemetered to the ground for verification, and assuming this is satisfactory, an execute tone

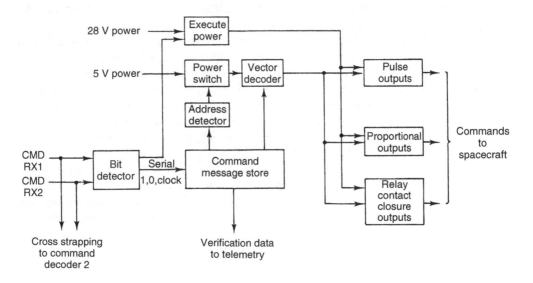

Figure 13.4 Telecommand decoder

burst is sent. This turns on the execute power switch, which then outputs the command to the specified user channel. The combination of power switching and the use of diode isolation and redundant components in the output circuits allows the outputs of the two command decoders to be combined in a fail-safe manner without loss of reliability due to single-point failure modes. The US Air Force SCF tracking network used a basically similar ternary (three-tone) modulation scheme using frequency-shift keyed tones.

13.4.3   PCM telecommand standards — ESA and NASA

Agencies supporting scientific and LEO missions often need to send rapid sequences of commands in a short time period, such as during a brief ground station pass. Although the ESA and NASA systems do make use of ground verification, the structure of the telecommand message allows the checking and correction to take place inside the satellite's decoder itself.

The ESA telecommand message starts with a fixed number of acquisition bits to initialize the message, usually fifteen 'zeros' followed by a single 'one'. The message sequence proper then consists of an unbroken sequence of one or more 96-bit frames similar to that shown in Figure 13.5. A single 16-bit spacecraft address and synchronization word (ASW) is used to terminate the message. The initial 16-bit ASW is used to identify individual spacecraft and to synchronize the decoder.

A 4-bit mode selection word using 2-out-of-4 redundant coding is included in each frame after the ASW. This word is repeated once and provides means for selecting on–off pulse commands, proportional commands or time-tagged commands for delayed execution.

Since a single-bit error will give an invalid mode word, the acceptance criterion in Figure 13.5 ensures the acceptance of a command even if a single-bit error has occurred, and its rejection if more than one error has taken place. In addition, a command receiver squelch cuts off the input to the bit detector if the signal-to-noise ratio is too low, thus ensuring a low bit error probability (normally better than 10^{-5}) and a correspondingly high probability of frame acceptance.

The three data words following the mode word are each of eight bits and represent respectively the address of the 16-bit memory to which the data are to be sent, the first eight bits of data and the final eight bits of data. Each data word is repeated once to increase the probability of acceptance, and four Hamming-code check bits are appended to each word to permit error detection and correction in the spacecraft. By comparing the received check bits with a set generated locally from the data word, it is possible to detect all single- and double-bit errors and, in the case of single-bit errors only, to determine which data bit is in error and correct it.

The end-to-end probability of command rejection can be reduced to less than 1 in 10^6, and the probability of an erroneous command to less than 1 in 10^8.

Spacecraft that use the ESA standards are compatible with the NASA ground network and procedures, whose modulation and coding techniques are to a large extent compatible.

13.4.4   Packet telecommand

The PCM telecommand standards have been superseded by a packet-based system, which has some similarity to the packet telemetry system. CCSDS recommendations have been

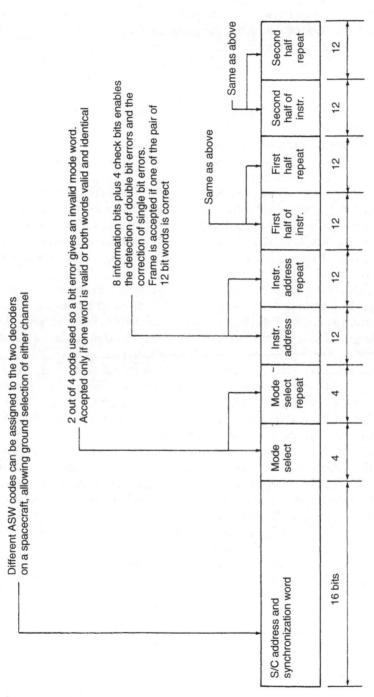

Figure 13.5 Memory load command frame structure

produced and are adopted by ESA through its 'Packet Telecommand Standard' [6] and ECSS standards.

The packet telecommand system is arranged in a number of layers, which provide the mechanism for the flow of data from the ground (source) to the spacecraft subsystem (sink). These layers are as follows:

- An *Application Process Layer* allows in-orbit payloads or subsystems to be controlled remotely by a user, providing the interface from humans and their computers. It can also be used to support advanced higher-level protocols such as file management.
- A *System Management Layer* provides the translation between the commanding language used in the Application Process and the detailed communications and control language used in the Packetization layer.
- A *Packetization Layer* allows the user to optimize his application data with minimum constraints and independently from other users. The application data is formed into TC packets consisting of a fixed-length packet header of 48 bits and a variable-length packet data field. The header provides identification, sequence control and packet length. Data field headers and packet error control may be added to the data field as options.
- A *Segmentation Layer* provides the mechanism for a number of variable-length TC packets to be formed into one virtual channel. It also enables the multiplexing of TC packets from a number of different sources and control of the data flow. The standard structure of a TC segment consists of an 8-bit segment header, a variable-length segment data field and an optional variable-length trailer.
- The *Transfer Layer* uses TC transfer frames to provide error-free transfer of data over the RF link. Two types of transfer frame provide for sequence control, and one for an expedited service typically used in recovery situations. The standard structure of a TC transfer frame consists of a 40-bit frame header, a variable-length data field (containing one TC segment) and a 16-bit frame error-control field.
- A *Coding Layer* provides the error correction and the synchronization required by the receiver. Each transfer frame is placed into a single command link transfer unit (CLTU). The CLTU consists of a 16-bit start sequence, n code blocks of up to 64 bits and a tail sequence the same length as a code block. Here n is an integer, sufficient to encompass one encoded transfer frame, with fill data being used when required. The code applied is a (63,56) using the polynomial $x^7 + x^6 + x^2 + x^0$ to produce a 7-bit parity covering up to 56 bits of data.
- The *Physical Layer* is the radio link itself and operates in one of four Carrier Modulation Modes. These are

1. unmodulated RF carrier,
2. modulated with Acquisition sequence,
3. modulated with one protocol data unit (CLTU),
4. modulated with Idle sequence.

The receiver, demodulator/command decoder and OBDH operate on the same layers but in reverse order. At each stage, telemetry reporting is fed back to ground, providing confirmation of good receipt and successful command operation when applicable.

As is the case for packet telemetry, packet telecommand protocols provide the user with greater flexibility than the earlier standards. It accommodates the dynamic commanding needed for advanced satellite operation. It makes possible complex interactions between ground and spacecraft such as software uploads, on-board file system manipulation and even access to web sites on the satellite from anywhere on the terrestrial internet.

13.4.5   Error-control coding

In general, error-control encoding is a less intensive processing task than decoding. For telemetry links, this places the more complex error-control tasks in the ground segment. It is difficult to implement powerful telecommand error-control decoding in the space segment because of the constraints on on-board electronic systems. For this reason, the error coding in telecommand standards is rather basic. The CCSDS telecommand recommendations choose a Bose–Chaudhuri–Hocquenchem (BCH) algorithm whose major strength lies in error detection. Indeed, the failure of this code is used to denote the end of a telecommand frame.

In any case, error control via coding still leaves a small but significant possibility of error, which may not be important for telemetry but could be disastrous in a mission critical command. Command links in general, therefore, use an automatic retransmission mechanism (the CCSDS protocol is known as COP-1) to guarantee command delivery by detecting errors and retransmitting failed frames. The protocol also ensures that telecommands are always received and executed in the correct sequence.

13.4.6   Uplink frequencies and modulation

Command uplink frequencies are typically in the S-band (2025–2120 MHz) or X-band (7145–7235 MHz) and support relatively modest data rates. The early ESA RF and Modulation scheme is based on a sub-carrier and a maximum data rate of 4 ksymbol/s. However, newer applications require higher rates and thus the CCSDS developed a modulation standard that supports rates of up to 256 ksymbol/s using Split Phase-Level/Phase Modulation (SP-L/PM) directly on the carrier. More advanced schemes are given in the ECCS standards [7].

Still higher rates are possible by using direct bi-phase-shift keying (BPSK) carrier modulation, although, unlike the lower rate systems, this will not support simultaneous ranging.

13.5   COMMUNICATION TECHNIQUES AND PROTOCOLS

13.5.1   Ranging

Ranging is achieved by means of a transponder, which is integrated into the telemetry-command subsystem. This retransmits ranging tones through the telemetry channels in response to tones received via the command route.

On Intelsat X, the output of a command receiver on a 6-GHz carrier can be connected to a beacon transmitter that phase-modulates the ranging tones on to the 4-GHz downlink carrier. On Earth-observation satellites such as Envisat these operations are carried out using the S-band links. The signals are tones as specified in the ESA ranging standards. These may be phase-modulated on to the uplink carrier in place of command signals and on to the downlink simultaneously with telemetry signals. Interference on the downlink is avoided by the choice of sub-carrier frequencies. Also, integrated Doppler measurements are made in order to provide spacecraft tracking—see Reference [8].

13.5.2 Advanced orbiting systems (AOS)

In the late 1980s, the requirements of the ISS, and space transportation systems such as Hermes and Earth-observation systems outgrew the capabilities of the CCSDS packet telemetry scheme.

The deficiencies were in the areas of

- support for isochronous (e.g. voice/video) services,
- support for high-volume downlinks (e.g. Earth-observation imagery, including synthetic aperture radar (SAR), video, etc.),
- support for interactive services (e.g. two-way fast file transfer).

The CCSDS recommendations addressed these issues and also took the opportunity to adopt a service-driven structure to the recommendation rather than the protocol/data format-oriented system of the packet TM recommendations. AOS is used in the downlinks of the ISS and Envisat. AOS also introduced networking to the CCSDS standards, with the incorporation of the efficient connection-oriented path protocol and support for the ISO 8473 connectionless network protocol—subsequently amended to support the Internet Protocol (IP).

13.5.3 Proximity links

CCSDS Proximity-1 is applicable to power-constrained links and is specifically designed for use in inter-satellite links, constellation missions and planetary orbiter/lander/rover configurations.

Proximity-1 operates symmetrically—that is, there is no distinction between telemetry and telecommand or forward and return links. Proximity-1 includes facilities for the following:

- Automatic detection of link availability and the negotiation of link parameters. A medium access control (MAC) sub-layer provides for the establishment of a link before data is transferred, its maintenance and termination. It also controls link characteristics, such as data rates, and the state of the data link and physical parameters.
- Reliable, in-sequence, complete delivery in both directions through the use of automatic retransmission. This uses a process known as 'persistence' to verify correct transfer.
- Convolutional and RS coding in both directions.
- Use of fixed-length or variable-length frames.

Exploration missions to Mars are typical examples in which Proximity-1 may be used. It uses features adapted from both the CCSDS packet telemetry and telecommand standards.

13.5.4 Space communications protocol standards (SCPS)

NASA initiated the SCPS development in collaboration with the US DoD. It was driven by perceived requirements for military and civil satellite constellations and sensor networks. SCPS takes the existing protocols used in the networking, transport and application layers of the terrestrial Internet and extends them to cater for the more onerous requirements of space networking. The components of the SCPS protocol suite are the SCPS network protocol (SCPS-NP), security protocol (SCPS-SP), transport protocol (SCPS-TP) and file protocol (SCPS-FP).

The application of SCPS to space missions has been slow, and is particularly applicable to more complex space constellations. However, there has been much interest in its use in terrestrial applications that have constraints similar to the space environment.

13.5.5 CCSDS file delivery protocol (CFDP)

CFDP provides a reliable data-transfer service. It is an application process (see Section 13.4.4) that operates over CCSDS or other link and networking protocols. The major features of CFDP are as follows:

- *Store and forward networking* differs from real-time networking in that an end-to-end path does not have to be present at the same time to allow networked communications. The protocol takes account of a dynamic network configuration with disjointed links.
- *Continuous file delivery* supports the delivery of unbounded TM streams. It can deliver file segments as they arrive, as well as deliver what can be recovered from a failed file transfer.
- *Negative automatic report queuing (ARQ) scheme* guarantees data delivery and is optimized for large bandwidth/delay products and for a low overhead in the return path. The retransmission requests can be optimized for operations; for example, they can be issued as errors occur or can be bundled together until a suitable opportunity for retransmission occurs.
- *Proxy transfer facilities* provide for third-party control of data transfer.
- *Graceful suspend/resume* for predictable link outages.
- *Garbage clearance facilities* protocol recognizes that software may have faults or may be disrupted by single-event upsets (SEU) and ensures that systematic errors do not accumulate and require operator intervention.
- *File Manipulation* allows for the remote naming, deletion, copying and management of on-board file systems.

The CFDP protocol has emerged from requirements to support spacecraft with on-board file systems and to provide a reliable service for bulk data transfer. This is particularly the case for spacecraft telemetry in which advances in on-board autonomy and data compression, which reduced the data redundancy in the downlink, making uncorrected bit errors less intolerable.

13.5.6   The interplanetary internet (IPN)

The IPN is an initiative of the Internet Research Task Force (IRTF). The IRTF is an Internet-based group largely responsible for the future direction of Internet development. The IPN Research Group (IPNRG) carries out this development. Despite its name, the initiative is intended to reach out into all space networking applications and incorporates both near-Earth and deep space applications.

The IPN consists of the following:

- *In situ Internets* are deployed Internets such as the Earth's Internet, a Mars Internet or an Internet local to a satellite constellation. These are composed, in general, of dedicated sub-network protocols supporting common transmission control protocol/Internet protocol (TCP/IP) for the network and transport.
- *The Interplanetary backbone* provides a long-haul transport service between remotely deployed Internets. The long-haul transport protocol (LTP) is currently under development and uses some of the techniques developed for the CFDP.
- *Interplanetary gateways* connect deployed Internets to each other or to the interplanetary backbone.
- *Inter-Internet dialogues* provide the store and forward networking necessary to operate with either long delays or disjointed communications resulting from orbital or planetary motion. It provides the store and forward networking required to interconnect deployed Internets.

The inter-Internet dialogues are based on the concept of bundling. Just as the IP networking layer provides a common means of data transfer across heterogeneous subnets in real time, bundling provides a common means of data transfer across heterogeneous Internets in a store and forward environment.

It is expected that, in the interplanetary environment, special bundle applications will need to be developed to avoid an excess of short discrete transfers, as would be the case if some common email exchange protocols, such as SMTP, were used.

13.6   ON-BOARD DATA HANDLING (OBDH) AND PROCESSING

13.6.1   Platform and payloads

The OBDH functions reside both on the spacecraft platform and within the payloads. They provide both the command and data management associated with the telemetry and telecommand operations discussed earlier. Communications spacecraft such as Inmarsat have separate and very complex payloads for the receipt and transmission of the user data. Remote sensing and scientific spacecraft often have separate provision for the handling and transmission of instrument data. The latter may require special measures to cater for very high data rates. For example, synthetic aperture radar can generate data at rates from around 100 Mbps to over 1 Gbps.

Classical OBDH architectures are based upon a central processor, typically connected via a video or digital path to the RF communications subsystem. This central processor will communicate with the platform subsystems such as the AOCS and the payloads using a serial bus with high data integrity. ESA spacecraft typically used an ESA standard **OBDH** bus operating at 250 kbps with separate Interrogation and Response lines, transformer-coupled and Litton-encoded. A variety of terminal devices are used depending on the particular need.

More recent ESA and other agency spacecraft use a bus based on the Military Standard 1553B, as does the Ariane launch vehicle. The 1553 bus is a serial bus capable of operating at 765 kbps and more, using transformer coupling and Manchester encoding. The speed of such buses is gradually being increased. It is interesting to note that certain spacecraft have been designed with mixed bus architectures to allow flexibility in equipment procurement.

The *central processor* has two primary functions. These are to provide for telecommand decoding and to operate as the central terminal for the bus. Depending upon the spacecraft needs, these functions may be implemented using from one to five separate units. The precise arrangement will depend upon the mission. It may be necessary to be able to extract, prior to any processing, certain high-priority commands for configuration switching operations. The subsystem may be involved in the arming and firing of pyrotechnic devices. Also, expanded macrocommands and time-tagged commands are often required. Ultrastable oscillators may be incorporated for accurate timing of control functions and for datation purposes. Certain missions where ground contact is not continuous will require a degree of autonomous operation or at least a fail-safe survival mode. An example of a remote sensing platform is given in Figure 13.6, in which the central processor has become a data-handling subsystem.

Terminal devices vary considerably and can be either stand-alone units or embedded within an instrument or subsystem. Stand-alone units may be unintelligent, as in the case of a remote terminal unit (RTU), or sophisticated communications processors. The RTU caters for bus coupling and the Litton code. It provides the distribution of simple 16-bit serial memory load and single-bit discrete commands, and high-level commands with the

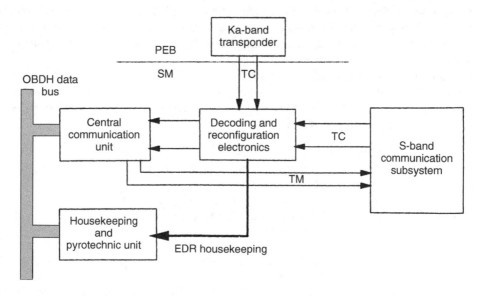

Figure 13.6 Data-handling subsystem (central processor)

capability to drive relays directly. It provides for the collection of 8/16-bit serial and discrete bi-level telemetry. The RTU may also include the conditioning and analogue-to-digital conversion of analogue telemetry. An RTU will typically use a switched matrix to provide for a high number of interfaces whilst minimizing the amount of internal circuits. It will include multiplexing and de-multiplexing functions.

Communications processors operate as Intelligent Terminal Units and can provide data-packetization services for the payloads. The interfaces between a Communications Processor and the payload may be either through dedicated ports or a bus.

Embedded terminals can operate as either mini-RTUs or more sophisticated remote bus interfaces (RBIs), which provide for direct memory access within the payload. In both these cases, separate units handle bus coupling and Litton-coded functions.

The more complex instruments may include their own processor-based intelligent control units (ICU) utilizing an embedded RBI. The processing power of some payloads sometimes exceeds the capability of the platform computers.

Payload processors can provide the functions of a communications processor, ICU and science data processor.

A *typical on-board computer* will use a 32-bit microprocessor, a microcontroller or digital signal processor (DSP)—built in radiation hard or tolerant technology. The embedded software will be produced to high standards controlling the software life cycle by addressing aspects such as documentation, review, structure, margins and verification. Secure methods and tools are used including object-orientated techniques and high-level languages such as ADA or C-ANSI. Optimized operating systems are an option. Normally, the system would be designed to allow the flight software to be patched and dumped, allowing modification and investigations to be carried out in orbit.

The production of on-board code is evolving from the situation where every line of on-board code was written for that particular mission, or at the very most, inherited from a very similar mission. The processing power and storage available on-board are following terrestrial industrial processors, and the mission requirements are also evolving towards those prevalent in terrestrial control applications. These trends manifest themselves in requirements for complex data organization, autonomous decision-making, intensive signal processing and multitasking, and the coordination of large distributed development teams.

The production of such software systems for highly reliable application has been addressed in terrestrial systems and has resulted in the availability of high integrity, real-time operating systems. These are now finding their way onto on-board processing systems and bring the advantages of

- faster, cheaper development,
- ease of maintenance and developer team coordination,
- off-the-shelf development systems,
- off-the-shelf processing and communications libraries,
- built-in file handling, multitasking and inter-task communication,
- large investment in software reliability.

The instrument data is often collected by a separate data-handling system, one example being known as the *payload data handling and transmission* (*PDHT*) system. These systems allow for the direct and independent collection of data at rates much higher than what can be handled by standard buses. They have their own dedicated RF ground link. Such systems will embrace high-speed multiplexing of data packets and provide data storage when direct ground contact is not available. Magnetic tape recorders were used for a long time with capacities of 10 Gbits and more. However, the availability of very high-density semiconductor memories has enabled modern spacecraft to use solid-state data stores. They offer the advantage of random access, plus the option of applying architectures capable of operating with peak rates of several Gbits per second and terrabit capacities.

13.6.2   Instruments, experiments and sensors

Figure 13.7 illustrates the main elements that might exist in a spacecraft generalized instrument, experiment or sensor. A payload may consist of one or more of these, some typical examples being

- active radars, such as Altimeters or the Advanced Synthetic Aperture Radar (ASAR) on the Envisat spacecraft;
- passive microwave sounders, such as the Micro Humidity Sounder (MHS) on the NOAA and METOP spacecraft;
- IR and visible spectrum radiometers, such as on the METEOSAT series of spacecraft;
- multi- or hyper-spectral instruments, such as the Compact High-Resolution Imaging Spectrometer (CHRIS) on the PROBA spacecraft;

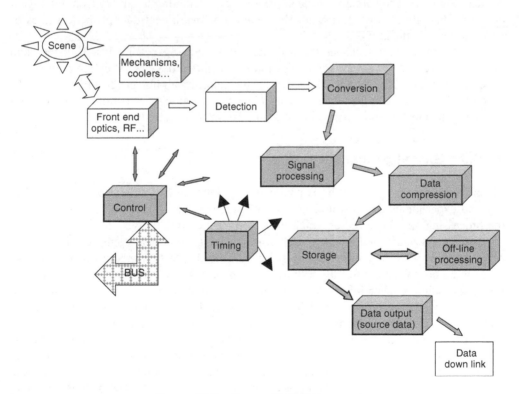

Figure 13.7   Generalized instrument

- laser-based instruments such as the ALADIN Lidar for the Aeolus mission;
- star sensors used for attitude control;
- science instruments, such as on the Rosetta Spacecraft.

Except for the simplest case, most systems will have a controller that provides the functional interface with the platform and provides local and typically real-time operation. In addition, there may be digital signal processing and data-handling functions to support. The data output from these systems will typically be in *source packet* format. Many instruments are multimode.

The *controller* will typically be a standard microprocessor, such as the SPARC based LEON. Alternatively, a DSP device, such as the TSC21020F operating at 20 MIPS and 60 MFLOPs, has been used where signal processing is required in addition to the instrument control function. At the time of writing, DSP developments for space applications requiring 1 GFLOPS products are in progress. In some applications standard processor designs such as the LEON, or high performance DSP functions are embedded within application-specific integrated circuits (ASICs), with appropriate intellectual property arrangements. These standard designs are known as 'IP cores'.

Digital communication within an instrument may be based on a multi-drop bus or dedicated parallel and point-to-point serial links.

Multi-drop buses, such as the ESA OBDH, the Mil Std 1553B and the CAN bus, are used where one device (master) needs to communicate with more than one other device (terminal). They are used for commanding and gathering telemetry from intelligent devices. They are also used for gathering science or instrument data at modest rates of up to around 1 Mbps.

Point-to-Point Links, such as the Mil Std 1355, ESA SpaceWire [9], HotLinks and GigaLink, are used for direct communication between two devices. They are particularly suited for the transfer of blocks of data and can support much higher rates than the multi-drop bus. Rates can vary from 10 Mbps to 1 Gbps plus. A number of parallel point-to-point links may be used to support higher rates and data-routing functions.

13.6.3 Signal processing

Signal processing in this context is applicable to platform sensors and operation, and to payload experiments and instruments, in both analogue and digital form. The nature of the processing and the speed at which it is done vary considerably.

Platform operation and sensors—control and navigation subsystems rely on a number of sensors providing information on the Sun, Earth, stars, spacecraft inertia, and so forth as inputs to the control algorithms (see Chapter 9). Sensing elements require signal conditioning, conversion to digital form, detection, and the manipulation of the data acquired to yield the information. Charge-coupled device (CCD) and Active Pixel detector arrays, as found in conventional digital cameras, require timing, read-out circuits, and analogue-to-digital conversion. Such sensors produce data at rates of typically 20 Mbps.

Payload experiments and instruments often require the processing of a baseband signal in either the time or frequency domain. Some examples are digital filtering, compression, feature extraction, data formatting, error control, and encryption. Operation with a number of single channels, each at up to 200 Mbps or more may be required.

13.6.4   Compression and storage

Data compression can be achieved by eliminating the redundant or duplicate content of the data. Image compression can be achieved by eliminating, for example, unwanted information or by reducing the resolution.

On-line Data Compression such as block-adaptive quantization may be used to reduce the data rate to more manageable proportions and would typically reduce the rate by factors from 2 to 8. They operate in real time on the data stream, perhaps taking advantage of first-in-first-out (FIFO) buffering.

Off-line Data Compression usually works in conjunction with a large data store and would run in dedicated processors or ASICs. There are many algorithms available that can give lossless compression with a reduction factor of perhaps 2, or lossy compression with reduction factors of up to and beyond 100. The selection of the algorithm really depends on the information required from the data and the data reduction needs. Compression algorithms are available to established standards such as Java Picture Experts Group (JPEG) and CCSDS [10].

Data storage is an important function for missions in which the data volume produced is greater than the link capacity and/or availability. Where tape recorders were once employed, solid-state memories using dynamic RAM are used, providing very large capacities. Typically, they do not have external addressing to each RAM location but operate on a block or file basis. The block may be of the order of 1000 bytes or sometimes it will correspond to one source packet. It is important to maintain the integrity of the data and so data coding and error correction is employed to protect against errors such as those caused by SEUs induced by cosmic radiation. The detailed arrangements of the input and output busses have to take into account the data rates and volumes produced by the instrument and the corresponding capability of the following downlink. To provide for flexible operation and degradation in the memory array, a data-routing function is usually included on both the input and output.

The large capacity of on-board mass memory has led to the same challenges in organizing stored data as is found on disk storage in terrestrial computers. The data is best arranged hierarchically, with data files identified by name rather than exact locations. This is leading to two trends—firstly, the adoption of industrial real-time operating systems, as mentioned earlier, and secondly, the requirement for remote file manipulation facilities as provided by the CFDP protocol.

13.6.5   Data downlinks

A typical *System Architecture* will consist of a data-handling function followed by a RF transmit chain or chains. The data handling needs to provide for data routing, the buffering of different input and output data rates (using a simple FIFO or a large data store), data formatting, carrier modulation, amplification and the generation of an RF beam.

Modulation techniques are normally quadrature phase-shift keying (QPSK), or of a higher order. In the case of QPSK, a 100-Mbps data stream would require an RF bandwidth of 55 MHz. The bandwidth available for LEO missions in X-band is 8025 to 8400 GHz, which with guard bands can support around 500 Mbps. However, permission has to be obtained from the International Telecommunications Union (ITU) and it is unlikely that all of the frequency range would be granted to one mission.

Amplification to achieve an equivalent isotropic radiated power (EIRP) of typically 15 dBW would normally require travelling wave tube or solid-state amplifiers and an antenna system with some gain. There are many *antenna* solutions ranging from fixed pseudo-omni-reflectors to very focused electronically steered beams. The beam has to be well controlled to achieve a reliable link over the required area of the earth and to limit the power flux density to satisfy World Administrative Radio Conference (WARC) regulations.

Link Layer data processing is needed to convert source packets into physical channel access protocol data units (PCA-PDU). This includes the formatting to establish VCDU, as required by the selected CCSDS standard; pseudo-randomization to smooth out the spectral content of the data and the addition of synchronization bits to enable the ground processing to recover the randomized data, forming channel access data units (CADU); differential encoding to resolve phase ambiguities; and the splitting of the data into in-phase and quadrature (IQ) channels prior to the modulator. Forward error correction may optionally be included to improve the link budget.

The *Link Availability* is an important system design consideration for LEO missions and it depends upon the spacecraft orbit and the location of the ground station. Typically, the link to a reasonably high latitude station will be available from 3 to 10 minutes during one orbit of 100 minutes. Low latitude stations will have a significant number of orbits with no availability.

13.7 TECHNOLOGY

Spacecraft data-handling systems are implemented using electronics technology. The space environment places major constraints on the use of this technology in a number of ways. The spacecraft resources available to electronic systems are limited. The three major parameters that are at a premium are mass, power and volume. The physical environment is very stressful for electronics components and subsystems. These must be chosen and designed to withstand the rigours of launch vibration and in-orbit temperature extremes. They must also be compatible with the vacuum environment in not having pressurized voids in the packaging (common in plastic packages) and having no outgassing materials that may contaminate other spacecraft equipment, for example, optical surfaces.

A major consideration when choosing appropriate components is the radiation environment. There are three effects to be considered (see also Chapter 2):

- *Single Event Upsets* (*SEU*) are temporary effects due to ionizing radiation changing the state of an electronics node on a semiconductor device. The effects are unpredictable. SEUs can cause noise in data, random switching events, changes in memory locations, random jumps in software code or illegal state-machine configurations. Data stored over a long period in on-board memory is subject to randomization by cumulative SEUs. The effects can be mitigated using error-correction codes, which are checked on a regular basis and the data is corrected if necessary. This is known as memory scrubbing. Other circuits may be protected using voting logic (e.g. triple module redundancy) and by incorporating software watchdog timers.

- *Total Dose* damage is due to the cumulative effect of ionizing radiation over time. The result is a catastrophic device failure. A device's tolerance to total dose is measured in krads, and the immunity required varies from 1 krad to around 100 krad depending on the spacecraft trajectory and lifetime. Devices can be shielded from total dose effects by the spacecraft structure, by the equipment housing, by specific device packaging or spot shielding.
- *Latch up* is another catastrophic condition and is caused by a single energetic ion initiating a runaway current flow in the device leading to failure. Again, shielding can help (although the emission of secondary ions can exacerbate the effect). An alternative strategy is to protect the device with current sensing and limiting circuitry that blocks the runaway effects and returns it to normal operation when the event has passed.

Note that the above catastrophic effects are significantly reduced when devices are unpowered. Another mitigation strategy is therefore to power-down electronic subsystems when not required.

Microprocessors that are specifically designed for use in the space environment emulate those available for terrestrial uses. However, they tend to lag terrestrial technology by a number of years. This is not necessarily a problem. Routine spacecraft processing tasks such as attitude control, housekeeping reporting, systems management, command execution and timeline sequencing require very little processing power. In this case, simple 8-bit microcontrollers have been adequate for the task.

It is worth remembering that many support devices such as RAM, ROM, buffers, and so forth come with limited width or serial interfaces. Choosing a microprocessor with a wide data bus causes a proportionate increase in the number of support devices, and a minimum bus width should always be selected.

For more demanding applications, 32-bit processors and signal processors are available in space-qualified technology. These are limited in processing power compared to their terrestrial counterparts, but find application in hosting complex spacecraft or payload software, or in performing intensive signal processing tasks.

Although some processors have been produced in space-qualified technology, the same cannot be said for peripheral devices. Universal asynchronous receiver and transmitter's (UART), dynamic RAM controllers, I/O controllers and communications devices usually need to be implemented on a custom basis using field programmable technology.

The electronics design cycle for spacecraft electronics used to be very costly. Past solutions used discrete digital devices with typically less than 50 gates per device. This resulted in large assemblies and a number of generations of costly and time-consuming printed circuit board (PCB) layout, before a satisfactory flight design was reached. The size of the electronics systems was greatly reduced by the advent of ASICs. The ASIC design and modification process is, however, costly and time consuming.

Radiation-hard field programmable gate arrays (FPGAs) are now in common usage. These can host a very large number of functions and can be designed and programmed using a desktop computer. This significantly decreases the time required to complete a design. Some devices can be continuously modified; others can only be programmed once. For critical or mass production applications, FPGAs are often used to prototype ASIC designs. The use of FPGAs can result in a single-board layout cycle in which the philosophy is to route all signals on the board into the FPGA and finalize the FPGA functionality later. In conjunction with desktop PCB design, FPGA technology is increasing the available functionality and shortening development timescales for on-board electronic systems.

13.8 TOOLS AND CONTROLLING DOCUMENTS

Because of the relatively high cost of equipment development and production, the criticality of the application, and the need to manage risk, it is essential to keep good control of the requirements and design. It is also essential to ensure that the solution is properly verified at each stage in its life cycle.

In the data handling and related fields, this is increasingly being achieved using computer tools such as very high-speed integrated circuit hardware description language (VHDL), mathematical modelling of digital signal processing, simulation of analogue and digital circuits, and auto layout and routing. There are a number of good commercial software packages available.

The definition and control process is also achieved through the use of formal documentation such as national and international standards, specification and user manuals, telecommand and telemetry directories, hardware/software interface control documents, and electrical and mechanical interface control documents. These are just some examples of the documents produced and maintained throughout a product's life cycle.

Some of the ESA PSS specifications mentioned have been replaced by the European Cooperation for Space Standardization (ECSS) documents or later versions of the CCSDS documents. Also, the reader needs to be aware that some terminology is evolving, for example, see Reference [2] annex C.

REFERENCES

[1] *CCSDS Packet Telemetry*, 102.0-B-5 Issue 5 (November 2000).

[2] *CCSDS Space Packet Protocol*, 133.0-B-1 Issue 1 (September 2003, TC September 2010).

[3] *CCSDS Advance Orbiting Systems*, Networks and Data links, 701.0-B-3 Issue 3 (June 2001).

[4] *CCSDS Space Communications Protocol Specification*, 713.0-B-1 Issue 1 (May 1999).

[5] *ECSS Telemetry Synchronisation and Channel Coding*, ECSS-E-ST-50-01C Issue 2 (July 2008).

[6] *Packet Telecommand Standard*, ESA PS-04-107 Issue 2 (April 1992).

[7] *ECSS Radio Frequency and Modulation*, ECSS-E-ST-50-05C Issue 1 rev 1 (March 2009).

[8] *ECSS Ranging and Doppler Tracking*, ECSS-E-ST-50-02C Issue 2 (July 2008).

[9] *ECSS SpaceWire Links, Nodes, Routers and Networks*, ECSS-E-ST-50-12C Issue 2 (July 2008).

[10] *CCSDS Lossless Data Compression*, 121.0-B-1 Issue 1 (May 1997, TC September 2007).

LINKS:

http://public.ccsds.org/publications/default.aspx
http://www.ecss.nl/
http://www.ipnsig.org
http://www.irtf.org

14 GROUND SEGMENT

Franck Chatel

German Space Operations Center (GSOC), Oberpfaffenhofen

14.1 INTRODUCTION

The ground segment is defined by the European Cooperation for Space Standardization (ECSS) institution in Reference [1] as composed of ground operations organizations, which employ the personnel involved in the mission, and ground systems, that group together all ground infrastructure required to support the mission from its preparatory stage to its completion. Spacecraft operations consist mainly of monitoring telemetry data received at the ground station and configuring the on-board equipment to execute the mission. The fulfilment of these two main tasks implies an intense preparatory phase before the launch. This comprises not only the design, implementation and test of the ground segment hardware and software, but also the planning of the operations and the training of the personnel involved in the mission execution.

Spacecraft operations can be compared to driving a racing car—that is, engineers design a highly complex product using state of the art technologies to meet the customer requirements. However it is the driver, and the way that they drive the car during the race, that determines the success or otherwise of the endeavour. A final product satisfying all of the requirements, but offering poor ergonomics to the driver, is not likely to win the race! Just as it is common practice in car racing to involve the driver at the design stage, so it is with spacecraft operations. Operations engineers should be involved in the design process of the spacecraft as early as possible. The design then benefits greatly from the previous experience of the operations engineers, who would in turn acquire in-depth knowledge of the spacecraft that they will operate subsequently.

This chapter presents the main features of spacecraft operations. It describes the tasks carried out by operations engineers in the preparatory phase and during the mission execution, as well as the hardware and software that are typically implemented. The organization and features described below are implemented in a flexible manner. That is, each spacecraft mission is unique and requires the ground segment to be tailored to its needs, sometimes leaving out some features, sometimes developing new functionalities. For example, what is useful for an Earth-observation mission like ENVISAT is not

Spacecraft Systems Engineering, Fourth Edition.
Edited by Peter W. Fortescue, Graham G. Swinerd and John P. W. Stark.
© 2011 John Wiley & Sons, Ltd. Published 2011 by John Wiley & Sons, Ltd.

necessarily appropriate to a project flown by students and professors at a university. The organization adopted depends also upon the available personnel resources. The chapter is structured around the four main systems usually involved in the ground segment:

- The ground station;
- The flight dynamics system;
- The ground data system and
- The flight operations system.

This structure focuses on the teams of experts which contribute to the mission, and Figure 14.6 provides an overview of how these groups are typically organized. It is a complementary to the ground segment breakdown used in Reference [1].

14.2   THE GROUND STATION

The ground station provides the communication interface with the spacecraft. It takes care of all the Radio-Frequency (RF) aspects of the ground segment. The central (and most visible) part of a ground station is the antenna (see Figure 14.1). However the processing of the RF-signals, either their generation for commanding the spacecraft, or their demodulation of telemetry (TM) data, plays an equally important role within a ground station.

Figure 14.1   Weilheim 30 meters antenna (Credit: DLR)

14.2.1   Station location and antennas

Antennas are the communication interface with the spacecraft. The characteristics of the missions which have to be supported determine the type of antenna that is required. Operational aspects should be considered in the ground station design.

The first issue is where the ground station should be located. Since it is supposed to exchange data with the control centre, it requires some communications infrastructure (e.g. telephone, network etc. - see Section 14.4.2), as well as access to a power supply system. Although such facilities are readily available in proximity to cities, ground stations located in remote places (desert, polar region etc.) might lack such infrastructure, thus implying greater cost. Some stations in extremely remote areas, such as the Antarctic, may be controlled remotely, and inspected only during summer time. However, in the event of an anomaly, a service outage of such a ground station is obviously a problem and the implementation of redundancy is all the more necessary. An important feature of the antenna characteristics is its horizon mask, which defines the region of the sky within which the antenna can communicate with the spacecraft. As well as mechanical limitations in the way the antenna is mounted, it is obstacles such as surrounding mountains, buildings and other antennas that determine the minimum elevation of the antenna, as illustrated in Figure 14.2. On the other hand, such topographic features also act as a protection against external RF interference, coming from airports or other similar radio emitters. Another example of a ground station location at Weilheim, Germany is shown on Figure 14.3 where tracking down to an elevation of 3° is possible.

The underlying geology at the site is of interest, as a robust and stable foundation is required when building the antenna, as this impacts the pointing accuracy. The weather conditions that prevail throughout the year are also an influential characteristic of the site as significant precipitation can affect RF reception, and cloudy skies can inhibit the use of Laser Communication Terminals (LCT). The use of the latter has increased in recent years because of the data rates they offer (inf the order of magnitude of one to hundreds of gigabits per second).

The antenna characteristics should also be examined carefully. The frequency at which the antenna operates is dictated by the spacecraft to be supported. This frequency is set by the up- and down-converters (see below) and impacts directly the antenna gain. Operations

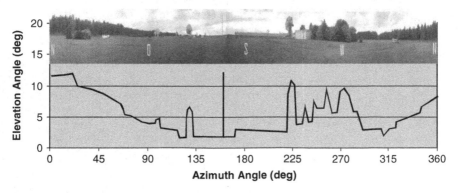

Figure 14.2   An example of a horizon mask, illustrating the constraints on minimum elevation at a ground station (Credit: DLR)

Figure 14.3 The Weilheim ground station is located on a plain, and is surrounded with forest, with the Bavarian Alps in distance. Tracking down to three degrees elevation is possible (Credit: DLR)

are often carried out in S-band during the *Launch and Early Orbit Phase* (LEOP) and in the case of an emergency (safe mode). However payload data are often transmitted at very high data rates, and therefore require a higher transmit bandwidth than is available in S-band. As a consequence, payload operations tend to use higher frequencies (X, Ku or Ka). In order to reduce the coordination effort at the ground station, payload and platform are operated nominally in the same frequency band during the routine phase. Table 14.1 below summarizes the most commonly used frequencies for satellite communications. A more detailed discussion on this topic can be found in Chapter 12.

Table 14.1 Typical RF frequencies used for satellite communications

Band	Frequency (GHz)	
	Downlink	Uplink
S (deep space)	2.29–2.30	2.11–2.12
S	2.20–2.29	2.025–2.11
C	3.4–4.2	5.925–6.425
X	7.25–7.75	7.9–8.4
	8.4–8.5	7.145–7.235
Ku	10.7–12.75	13.75–14.5
Ka	17.7–21.2	27.5–31
	25.5–27.5	

Precise information about frequency allocations for satellite communication can be found in the frequency allocation plan published by the International Telecommunication Union (ITU). The use of higher frequencies generally pose greater difficulties, which explains why critical operations like LEOP or emergency modes are still handled using S-band. However, it is likely that even these critical operations will be executed using higher frequencies in the near future. It is in the interest of the ground station to be able to support as many types of mission as possible. A ground station must apply for a transmission license from its national telecommunication administration agency, which in turn must comply with ITU regulations.

The diameter of the antenna is directly linked to the surface available to collect the signal emitted by the spacecraft and is a key contributor to the Equivalent Isotropic Radiated Power (EIRP) and the ratio of the antenna receive gain to the system noise temperature (G/T) in the link budget (cf. Section 12.2.10). The steering capabilities of the antenna mounting are also relevant. Although not important in case of a fixed geostationary satellite, otherwise these capabilities define the operational region of the sky where a spacecraft can be tracked. For example, an antenna implemented on an azimuth-elevation mounting cannot view a region of sky around the zenith due to mechanical limitations. Finally the antenna radiation pattern, showing the strength of the radiated signal in all directions around the antenna, is useful information in the determination of the size of the main lobe (defined by the maximum gain $-3\,\mathrm{dB}$).

The antenna motion during contact with the spacecraft is controlled by the *Antenna Control Unit* (ACU). Normally the antenna will track the spacecraft as it passes through the visibility area of the ground station. To do so, the antenna can simply point in a direction given by the spacecraft's ephemeris, describing the orbit, which is computed off-line (this is referred to as *program tracking*). This type of tracking can be extended by implementing a searching motion (for example a spiral) around the spacecraft's position indicated by the ephemeris. This feature is useful when no signal is being received, especially in the case of first acquisition contact. It is possible to improve the tracking by making the antenna point in the direction where the signal is the strongest. This feature is implemented as different modes in the ACU. For example, the *auto-tracking mode* is used with full motion mono-pulse antennas, whereas the *step tracking mode* is used with fixed position antennas pointed to geostationary satellites. The latter is achieved by deviating the antenna through a small angle and comparing the new signal strength with the previous one. This allows the direction in which the antenna should be moved to be determined, in order to get the strongest signal. This method is applied on both axes of the antenna separately. When an antenna is used in auto-tracking or step tracking mode, the measurement of its pointing angles (azimuth and elevation) can be used for orbit determination.

If an appropriate transponder is present on-board the spacecraft, the antenna can also be used for ranging and Doppler measurements. Ranging consists of sending a specific signal to the spacecraft receiver, which in turn sends it back as soon as it is received, thus allowing the computation of the range using the round-trip time. Doppler measurements are obtained by sending a signal with a definite frequency, which is also returned by the receiver at a frequency defined by a fixed ratio of the uplink frequency (the so-called *coherent mode* to avoid interference with the uplink signal). From the measurement of the downlink frequency, it is possible to compute the Doppler shift, giving access to the radial velocity (also called range-rate). Both range and range-rate measurements are inputs for orbit determination.

As already mentioned, first acquisition of the spacecraft by the ground station can be challenging. If program tracking is not successful in acquiring the spacecraft, an acquisition aid antenna could be added in the control loop. This acquisition aid antenna has a wide main lobe in its radiation pattern (cf. Section 12.2.8), which allows the scanning of a wider region of the sky. However, it results in a poorer communications link because the radiated energy is spread over a larger beam. This antenna is therefore used as an aid to controlling the pointing of the main dish, which is the dedicated means of communication with the spacecraft. After a stable contact has been established, auto-tracking can be initiated.

14.2.2   Data processing

After reception of the signal by the antenna, it must be processed further before being ready to be sent to the control centre. Due to the low received power, the signal needs first to be amplified by a *Low Noise Amplifier* (LNA). It is placed as close as possible to the antenna feed to avoid increasing the noise due to cabling. The frequency of the signal is then decreased by a down-converter, from the RF carrier level (between 2–3 GHz in S-band for example) to a so-called intermediate frequency level (usually 70 MHz). This down-converted signal is then processed by the *baseband unit*, which is the central component of the ground station. This unit first converts the signal from analogue to digital, and then performs the demodulation, depending on the coding used for communication (see Chapter 13). Finally the baseband unit performs the bit synchronization, frame synchronization and Cyclic Redundancy Check (CRC), if the signal is not encrypted. The result of these operations is a bit stream containing telemetry data that is sent to the control centre for further processing (see Section 14.5.1).

For the task of commanding the spacecraft, the processing is similar. The baseband unit receives a bit stream from the Monitoring and Control System (MCS) (see Section 14.5.1), which is first modulated and converted from a digital signal to an analogue one. This processing is done at the intermediate frequency, and the frequency is then increased to the required up-link frequency by the up-converter. Finally the signal is amplified using a *High Power Amplifier* (HPA), and then radiated by the antenna. These operations require a reference frequency, provided either by a reference oscillator or directly from a time source (e.g. GPS). The baseband unit usually provides the possibility to store and replay some commands generated at the control centre for testing purposes. It is also able to replay saved telemetry data, emulating a real-time contact. Figure 14.4 shows the relationship between these units.

The data processing chain of the ground station must be tested for compatibility with the spacecraft, which is the objective of the *RF-compatibility test*. This is executed either with the spacecraft itself or, more often, with the aid of a 'RF-suitcase' delivered by the spacecraft manufacturer, which simulates the characteristics of the spacecraft at RF level. This suitcase is then used at each ground station involved in the mission. It is connected immediately after the feed system of the ground station antenna. The atmospheric losses are simulated by attenuators and a telemetry signal which emulates the spacecraft is injected into the processing chain. This test signal is then processed as if it were a real signal coming from the spacecraft. Correct demodulation and Bit Error Rate (BER)

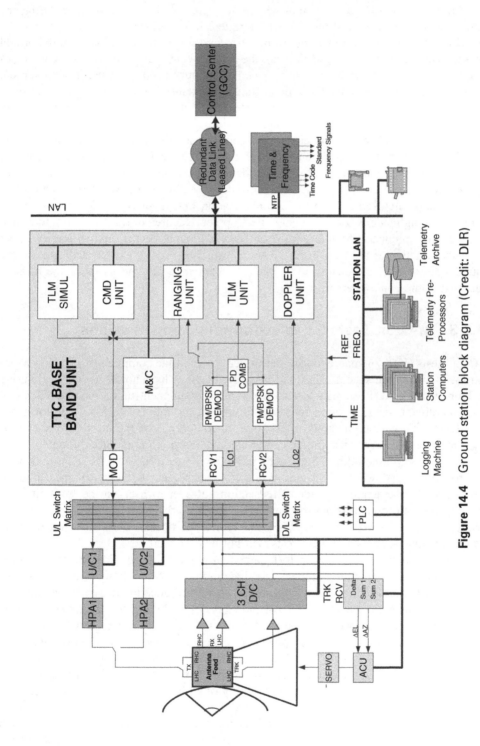

Figure 14.4   Ground station block diagram (Credit: DLR)

are verified. Commands are also sent which are representative of those used during the mission, and the reception of the correct bit pattern in the RF-suitcase is verified.

During the mission, prior to a pass of the spacecraft over the ground station, the orbital ephemeris of the spacecraft are loaded into the antenna control unit, the HPAs are warmed-up and a *data-flow test* is performed between the control centre and the ground station. Telemetry data are replayed at the ground station and transmitted to the control centre for processing. The control centre then sends some test commands to the antenna switched to a dummy load, dissipating the power into heat (in order to avoid actually radiating these commands). Both of these tests ensure the correct configuration of the communication link, the ground station equipment and the control centre for the up-coming contact. During the contact, the antenna exchanges data with the spacecraft from *Acquisition of Signal* (AOS) to *Loss of Signal* (LOS). These data are directly transmitted to the control centre but also archived in case the communication link is interrupted. Pointing angles (if auto-tracking mode is used), ranging and Doppler data are not transmitted in real-time but as off-line products. After LOS, the antenna is brought back to its parking position. All of these activities are overseen by a monitoring and control system dedicated to the ground station, the functionalities of which are similar to the one used for the mission execution at the control centre.

14.2.3   In-Orbit Tests for telecommunication missions

In-Orbit Tests (IOT) of spacecraft are performed to verify the compliance of the on-board instruments with their specification, and to ensure that they have not been degraded during the launch phase. IOT campaigns also allow the measurement of the characteristics of these instruments, and are often repeated throughout the mission to assess performance degradations. Of course, the kind of tests performed depends on the mission type (Earth-observation, astronomy, telecommunication etc.) and each requires different equipments. For example, one of the calibration tests for a Synthetic Aperture Radar (SAR) is done with reflectors on ground, which are arranged in a cross shape, thus allowing the measurement of distortion in the acquired image. This section however focuses on the In-Orbit Tests for telecommunication missions because these particularly involve the ground station. Chapter 12 gives a good overview of the typical performance factors measured on communication satellites and these tests will not be described here.

A ground station involved in an IOT campaign requires outstanding levels of performance to allow the verification of the payload requirements. Not only must absolute values be recorded, but also their stability in time. Measurement uncertainties must be compatible with the required accuracy. IOT measurements also make use of specific equipments, such as, for example, a spectrum analyser used in the characterization of Adjacent Satellite Interference (ASI). All these equipments are integrated together in portable racks, and a software program is used to automate the data acquisition and generation of reports. Due to these special features, such a ground station is normally not involved in the daily activities of spacecraft operations. The capability of an IOT ground station to support the test campaign adequately must be demonstrated during acceptance tests, and the measured performance is used to establish the IOT specification. For example, the determination of the Input Power Flux Density (IPFD) at saturation can be limited by the EIRP of the ground station.

The methodology used during IOT must be carefully planned. Many factors influence the measurements, and each of them must be taken into account to get a meaningful result. For example, the interference coming from the uplink and downlink signals of neighbouring satellites have to be characterized so that its influence can be taken into account in the test measurements. The IOT campaign is performed as far as possible at the final on-station position of the satellite, but such interference might impose the need to move the spacecraft to another orbital location to perform the testing. The influence of the Earth's atmosphere (cf. Section 12.2.7) also plays an important role, especially for Ku or Ka-band payloads, and it is necessary to get daily weather forecasts during the campaign. Some tests (e.g. gain measurements) might even be postponed if clear sky conditions are not available or wind speeds are too high. The pointing of the satellite platform is also relevant since it impacts the spacecraft's antenna pointing, and some measurements cannot be performed during station-keeping manoeuvres. This influence of the platform attitude is nevertheless useful in mapping the antenna pattern; instead of having many ground stations measuring the signal strength, the platform can be rotated and the entire pattern mapped with a single station. Finally coverage corrections for the projection of the beam onto the Earth's surface, and corrections due to degradation have also to be taken into account.

14.3 FLIGHT DYNAMICS

Flight dynamics experts deal with all aspects of the mission related to the spacecraft orbit and attitude. During the preparatory phase, they are mostly involved in the mission analysis. Based on this analysis, they supply experts from other systems (ground station, ground data and flight operations) with products containing predicted orbital information (called *operational products*) to support their preparation. During the mission, flight dynamics experts perform orbit and attitude determinations to control the actual execution and deliver updates of the operational products. Finally flight dynamics provides services related to space awareness and collision avoidance.

14.3.1 Mission analysis and operational products

Flight dynamics experts perform the *mission analysis* in close liaison with the satellite manufacturer. This relationship is essential to ensure that all mission requirements are met, and to allow negotiation concerning whether and how some requirements may be loosened if necessary. Mission analysis is discussed in detail in Chapter 5, so this chapter will only outline the main outputs required by the ground segment. These can be summarized as:

- The identification of the network of ground stations that satisfies the mission requirements in terms of ground coverage. This network is typically global and dense during the early orbit phase, the focus of which is to check that the spacecraft systems are operating normally, and to ensure that orbit manoeuvres (if necessary) are performed correctly to reach the final orbit. The choice of the first acquisition station must be considered carefully, as the first contact with the spacecraft is a critical part of the mission. The distance between the launcher separation point and the first acquisition station, the quality of the delivered tracking data and the availability of an acquisition aid antenna are some of the criteria to be taken into account when making this choice. After spacecraft commissioning and the commencement of routine operations, the ground station network is reduced to a few stations. This network forms the basis for the preparatory phase of the ground data system.

- The *launch window* determination taking into account requirements in terms of final orbit and manoeuvre planning. It is usually computed over many months around the planned launch date to cope with launch delays. This output is delivered to the launch provider, and discussions might arise if there is a co-passenger.
- The orbit manoeuvre planning (if required by the mission) to reach the final orbit from the transfer orbit. A tree of backup strategies is also computed in case one of the manoeuvres cannot be performed as planned.

The strategies for final orbit acquisition and station keeping also come under the umbrella of mission analysis and its main outputs. They are however very dependent on the mission (formation flying, rendezvous, interplanetary mission, GEO, and so on). In the case when orbit manoeuvres are required, the orbit acquisition strategy can only be finalized after completion of the orbit determination following the last manoeuvre. This is due to the dispersions in the main engine thrust, as well as modelling errors. Apart from these main outputs, several more specialized operational products are generated from the mission analysis. These operational products are of course mission specific but the following are typical:

- *A list of orbit related events*. This is the basis for the mission planning (see Section 14.5.4). It identifies events in chronological order. For example, such events are eclipse entry and exit, apoapsis and periapsis crossings and sensor blinding by the Sun or Moon.
- *Ground station events*. This is similar to the orbit related events but focused on the ground segment. For example, events of interest are acquisition and loss of signal (AOS/LOS) and RF sun interference (when the sun is close to the station-spacecraft line).
- *Ground station ephemeris*. This has to be generated for each ground station in the network. It provides information (the azimuth and elevation angles) to point the antenna during a spacecraft pass in program-track mode.

All these operational products are specified in *Interface Control Documents* (ICD). Of course, at this stage these products are only theoretical, being based on tools making use of gravitational field or engine performance models. The update of these products during the mission is the topic of the next section.

Another category of operational products to be specified are command and telemetry files that are exchanged between flight dynamics and flight operations during the mission execution. Telemetry files contain measurements related to the sensors and actuators (GPS position and velocity, attitude quaternion from star sensors, earth sensor roll and pitch angles, wheel speeds, etc) down-linked by the spacecraft and these are used as inputs for orbit and attitude determinations. Based on these files, flight dynamics also computes the values of some parameters that must be up-linked to the spacecraft. Such parameter values are delivered to the flight operations team in the form of command files ready to be loaded into the monitoring and control system. Typical examples of command files are on-board orbit propagator parameters, orbit manoeuvre parameters such as ΔVs, quaternions for spacecraft re-orientation into manoeuvre attitude, gyroscope drift after a calibration or equipment misalignment.

The tools used for the mission analysis are implemented in the flight dynamics system that will be used during the mission, together with the ones for processing telemetry and generating command files. The preparation of this system for the mission sometimes requires the development of specific tools (depending on the type of mission activities—formation flying, rendezvous, high accuracy orbit determination, etc.) and their testing.

14.3.2   Orbit and attitude determination

As pointed out in the previous section, the operational products generated during the preparatory phase are based on theoretical models. The actual mission performance differs either slightly when operations are nominal or dramatically in the case of an anomaly. An update of these operational products is then required.

The first opportunity for update comes during the launch phase or immediately after separation. Based on actual launcher performance (mostly the lift-off time and the computed orbital elements of the transfer orbit, which are both delivered by the launch provider), flight dynamics experts have to decide, or not, to update the ephemeris of the first acquisition ground station. It is a critical decision that has to be taken possibly before AOS (usually due within the next minutes) or worse still, during the first contact! This update will determine whether the spacecraft can be successfully tracked during the first contact, which is very important. It is usually the case that the tracking of the spacecraft based on the orbital elements provided by the launcher becomes increasingly difficult. If the first acquisition is to take place a significant time after the separation, tracking data using an aid antenna might help bridge the gap. In the last resort, the two-line orbital elements provided by the US Strategic Command (USSTRATCOM, formerly known as NORAD) can be used. Unfortunately the availability of the first such set of orbital elements can vary from a few hours to several days after launch!

Another opportunity for the update of operational products is a non-nominal sequence of the launcher, leading to the spacecraft being released into an orbit different to that targeted. In such an event, the operational products of all ground stations need to be updated immediately to ensure they are capable of tracking the spacecraft. When the launcher performance is not nominal, the orbit perigee is often too low, which may force the performance of an orbit manoeuvre as soon as possible to avoid the denser parts of the atmosphere. This requires the computation of a completely new manoeuvre planning strategy and, as a consequence, the re-planning of all subsequent operations.

Immediately after the first successful contact, the flight dynamics experts can begin with their major task during the mission, namely that of *orbit determination*. This may be based on a variety of data types, depending on the kind of measurements available on the mission. For example, down-linked GPS data if the spacecraft is equipped with a GPS receiver, range and range-rate data gathered at the ground station in the case of an on-board transponder supporting these measurements, or simply tracking data comprised of antenna angles if the dish is in auto-track mode. An overview of orbit determination methods can be found in References [2] and [3]. Measurements are required over at least one complete orbit revolution, with a good global distribution to get a reliable

Table 14.2 Typical accuracies from orbit determination methods

Orbit Type	Used measurement type	Position accuracy (m)
LEO	Angle data	< 100
	GPS navigation solution data	< 10
	GPS raw data (single frequency receiver)	< 2
	GPS raw data (dual frequency receiver)	< 0.01
GTO/GEO	Angle, range, range-rate data	$x, y : <100, z : <400$

determination. Typical accuracies of orbit determination using different measurements methods are given in Table 14.2.

Orbit determination is required after each orbit manoeuvre. The comparison of the orbital elements prior and after the manoeuvre allows the determination of the ΔV actually achieved during the burn. The difference between this and the planned ΔV is used to determine a calibration factor (usually applied to the thrust level) that can be taken into account in the planning of the next manoeuvre. Sometimes computing this calibration factor is complicated by the fact that spacecraft attitude is controlled by thrusters, which also contributes to the overall ΔV. Although it is possible to filter some of the thruster contribution by performing measurements at well chosen times (for example just before the thrusters firing), the thrust contribution resulting from the attitude control during the manoeuvre can only be separated by a more precise analysis. After orbit determination, an update of the operational products is delivered to all parties. The manoeuvre planning is eventually updated as well.

Similar to orbit determination, the *attitude determination* is also the responsibility of the flight dynamicists during the mission. Measurements depend on the sensors and actuators implemented on-board the spacecraft's attitude control system (see Section 9.2). Reference [4] presents a useful review of methodologies. The result of the attitude determination can be used in the preparation of attitude manoeuvres, in the same way that orbit determination can be used for orbit manoeuvre planning. It is also used to calibrate some equipment (gyroscopes for example) or verify the proper functioning of the on-board pointing algorithm.

14.3.3 Space situational awareness and collision avoidance

Operating a spacecraft safely requires knowledge of the threats posed by the space environment. The characteristics of the space environment is treated at Chapter 2 and shows that these threats originate either from the natural environment (Earth atmosphere, solar and cosmic radiations, meteoroids etc) or from the consequences of space utilization by humans (active spacecraft or space debris). Acknowledging the need to have reliable information in this domain, the European Space Agency (ESA) started the *Space Situational Awareness* (SSA) programme in November 2008. It aims to monitor all Earth orbiting objects (satellites and debris) as well as the space environment (so called *space weather*).

The ever increasing number of satellites and space debris in near-Earth space is causing growing concern about the risk of collision between orbiting objects. This risk is enhanced

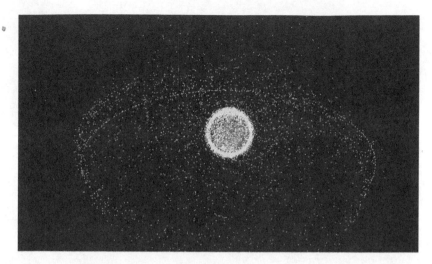

Figure 14.5 Earth-orbiting objects on 17th of February 2010 compiled from the entire USSTRATCOM TLE catalogue (14137 objects at the epoch mentioned) (Credit: DLR)

in certain popular operational orbits such as GEO (where most telecommunications assets can be found) and sun-synchronous LEOs which accommodate many Earth observation spacecraft. A schematic of the current debris environment is shown in Figure 14.5.

Although the cross-section of a single spacecraft is extremely small compared to the available volume, the collision of Iridium 33 with Kosmos 2251 on 10th February 2009 (the first ever between two intact catalogued objects), reminds us that the risk must not be underestimated. The accuracy with which close encounters of spacecraft with other objects (spacecraft or debris) can be predicted depends upon the accuracy of their orbital parameters. The USSTRATCOM catalogue of *Two-Lines Elements* (TLE) is the only public source of information available about orbiting objects. Unfortunately, their quality is limited and some military satellites are not included in the catalogue (which is why the ESA SSA programme has been established). Consequently, close encounter predictions, taking into account the data accuracy, might be too pessimistic and lead to a high number of warnings. Any unnecessary collision avoidance manoeuvre results in a loss of operational lifetime (increased fuel consumption), service interruption and a possible loss of science data. In the event of a potential collision warning, it becomes necessary to refine the orbit knowledge of the other object by implementing a tracking campaign based on ground radar stations. This allows the refinement of the necessary orbital information, and eventually the implementation of an evasive manoeuvre (if necessary) in due time.

Space weather must also be monitored carefully. The activity of the Sun is closely monitored by many satellites (SOHO, ACE, GOES and POES satellites, for example) and ground stations. Disturbance of the Earth magnetic field, radiation storms and radio disturbances can be forecast and the flight operations team can decide to suspend the spacecraft activities during such events. In some cases, the spacecraft can even be re-oriented to present a less-sensitive face to the Sun.

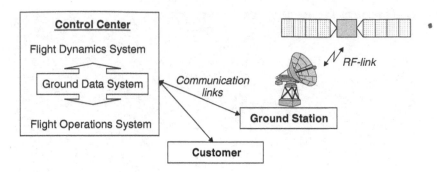

Figure 14.6 A simplified overview of the ground segment (Credit: DLR)

14.4   THE GROUND DATA SYSTEM

Ground data system personnel take care of the ground segment infrastructure required to support the operations. This infrastructure can be seen schematically as a spacecraft control centre, hosting the flight operations and flight dynamics systems, interfaced to the ground station network and the organization representing the customer, as pictured on Figure 14.6. The exchange of data within the control centre or with external sites is done via communication links. Spacecraft specific implementation and testing of the whole ground segment is necessary to ensure its proper functioning.

14.4.1   The control centre and network infrastructure

The control centre hosts all personnel and infrastructure involved in the mission. Sometimes the same infrastructure is shared by many different missions (a multi-mission control centre). The first obvious consequence is that many computers need to be accommodated together and combined into a network. As a consequence, a thermal control system is required within the centre to dissipate the heat generated. This problem is further exacerbated by the redundancy requirement of critical function hardware, such as the monitoring and control server or the data archive server. The redundancy switching is often performed autonomously to avoid human intervention and to optimize the system availability. All the data in the control centre is backed-up periodically (every day) to avoid any loss. For precise timing applications, all computers are synchronized to Coordinated Universal Time (UTC), which is distributed over the control centre network. The public power supply is often a single point of failure and it is necessary to install an Uninterrupted Power Supply (UPS) based on a separate source (for example, a generator driving a battery system) to bridge possible outages until public grid electricity is available again. The switching to the secondary power supply is controlled autonomously by software.

Running a control centre requires dedicated teams for network administration and infrastructure maintenance. Computers are usually running continuously, which puts a heavy load on the hardware. This requires carefully planned maintenance and replacement strategies, especially taking into account delivery delays and spare-parts management. Standardization of equipment throughout the control centre is certainly good practice. In general, maintenance activities should not impact running operations. However, long missions (20 years in GEO for example) are a challenge with respect to maintenance, because mission tools do not evolve much in the routine mission phase. Consequently the same computer hardware and software needs to be maintained over a long period of time, together with the system knowledge. The emergence of virtualization techniques, where the hardware and software of a virtual machine are entirely emulated at software level on a host machine, can help in solving such problems.

Consideration of security is becoming more important as a requirement. It aims at ensuring the confidentiality, the integrity and availability of mission data, both during the preparation and execution phases of the spacecraft mission. Security impacts the entire mission and, as such, needs to be taken into account from the very beginning. Each mission is different, but usually there are similarities. The security considerations begin with the installation of separated Local Area Networks (LAN), protected by firewalls to restrict the access to sensitive data. Personal login user names and passwords are required on each computer, and personnel must be cleared through governmental agencies to handle classified material. This may pose problems for system administrators, who must have access to all machines and data. Emails and exchange of data in general are encrypted, which supposes that the functionalities required to do this are available. Control of access to sensitive areas (server or control rooms) is required together, with logging functions (listings of who accesses which areas or machines) and video surveillance. The telemetry and commands are often encrypted, which impacts both the spacecraft and the MCS. This must be taken into account by dedicated processes on the ground. Special attention must be given to security methods because it could threaten the mission if not correctly implemented. Finally, it is important to make the control centre personnel aware of this, through the use of dedicated training sessions.

Documentation related to the control centre design is essential, especially for long-term missions when changes in personnel occur periodically. This documentation is reviewed at the Preliminary Design Review (PDR) and Critical Design Review (CDR).

14.4.2 Communication links

Operations require the smooth interaction of people and systems. Communication links between the ground stations and the control centre, communications within the control centre itself and with the customer, are the 'glue' of the mission, without which day-to-day functioning of the overall system would not be possible.

The most obvious need for a communication link is between the ground station and the control centre, through which telemetry data is forwarded from the antenna to the MCS. Command data obviously moves in the opposite sense, to be radiated to the spacecraft. All these data are usually carried by leased ISDN lines, although sometimes the internet is used as the medium. The process may include a data relay satellite. The communications lines are shared by different missions, unless the customer requires dedicated lines. Attention must be paid to the data format—so called 'gateways' are often required for protocol conversion and this implies extra effort in the development and testing. The recommendation of the Consultative Committee for Space Data Systems (CCSDS) concerning *Space Link Extension* (SLE) Services [5] acknowledges the need for standardization in this domain and proposes some new concepts to improve the interoperability between ground segments. The customer or the spacecraft manufacturer may also request to receive spacecraft telemetry from the launch pad to perform a last health check. Other common requests prior to launch are command tests, and the reception of the television of the launch by the launch agency (if available) or at least images of the launch.

Another use of the communication links is the delivery of operational products to the different partners involved in the operations. Together with the data discussed in Section 14.3.1, that are directly used to run the operations, scientific data or images may need to be extracted from the archive and transferred to the customer or to other control centres for further processing. Data transfer is also complicated by security measures. A specific tool is usually used to automate these transfers between well defined pick-up and delivery points.

Apart from all data exchanged between machines, people are always involved in the operations and there is no better way to exchange information than by using the medium we are all accustomed to in our day-to-day lives - voice. A voice conference system with different channels (called *voice loops*), is often implemented at the control centre and the ground stations. It allows the different groups of personnel involved in the operations to speak with each other and to hear what is currently going on. All people taking part in the operations must have permission to talk and/or listen on the different channels, depending upon their involvement. As a backup, regular phones are also used within the control centre and external phone lines, independent of the regular private communication network, might be available to bridge service outages. Finally fax machines are required to transmit signed documents. This is particularly important during the launch phase, when the launch provider needs to confirm the lift-off time, the separation time and the orbital elements of the transfer orbits by written means for clarity.

All these communication links must be documented in the ground segment description, but more importantly in interface control documents. Ground procedures, similar to flight operations procedures (see Section 14.5.3) need to be developed and validated for configuration tasks and redundancy switching. This documentation is reviewed during the PDR and CDR. A more detailed overview of the ground segment documentation can be found in Reference [6].

14.4.3   Configuration management and testing

The ground system required for the mission is complex and its preparation requires similar techniques to that used for satellite product assurance (see Chapter 19).

At the beginning of the mission, ground system requirements are defined in terms of facilities and software. At the same time, an overall test plan is issued, which describes the way in which these requirements will be verified. Different methods can be used for this verification which include inspection, analysis or testing. The overall test plan is then broken down into several test specifications focused on a particular sub-system (for example, acceptance testing performed on operational machines) or at system level (integration tests including the interfaces between subsystems). Test specifications are delivered with test procedures, detailing exactly the steps needed to verify the requirements. Prior to testing, a *Test Readiness Review* (TRR) is held with all persons involved in the test, under the supervision of a *Test Review Board* (TRB). The TRR ensures that the product is under configuration control, that its documentation is available, that the test is correctly documented and that a deviation reporting procedure is in place. Finally a test report is issued, summarizing all deviations, the 'as-run' test procedures (that is, with comments added during the test) and the conclusion of the TRB stating the outcome. A review of all the test results takes place at the Qualification Review (QR).

Managing such a complex system as a control centre cannot be done without a proper configuration management process of all operational parts (hardware, software, flight procedures, database, and so on). The tracing of the installation status is important. As discrepancies are reported during tests or simulations, an update of some tools is often necessary, prompting some new tests, and the integration of the new tools into the overall system. Special processes like a *Non-Conformance Report* (NCR) in the case of an anomaly or an *Engineering Change Request* (ECR) in the case of a required change ensure that each problem is documented and that all changes are traceable, thus allowing a mechanism for correction. As the preparation for the mission progresses, changes will always be more critical to implement, due to the risk of impacting other parts of the system. Finally, shortly before launch, the system is frozen and changes can only be implemented after the agreement of a special board involving all parties responsible. In a multi-mission control centre, this impacts the resources shared with other missions.

14.5   THE FLIGHT OPERATIONS SYSTEM

The flight operations team is in charge of conducting the operations, which consist mainly of commanding the spacecraft and monitoring its telemetry data. Both these tasks are performed with the help of the monitoring and control system, which processes telemetry and command data in near real time and archives them. Operations are based on flight operations procedures, which describe the way in which the spacecraft is to be commanded, and the mission planning describing when they are to be executed.

14.5.1   The monitoring and control system

The *monitoring and control system* is the heart of the operations. On the telemetry side, it processes the bit stream generated at the ground station and transmitted to the control centre. It also presents these processed data to the flight operations team in a meaningful way. On the command side, it translates the orders from the operators into a bit stream that the spacecraft can interpret. Details about telemetry and command encoding technologies are given in Chapter 13 and the term 'bit stream' is used in this chapter to avoid being too specific about the technology.

The processing of telemetry data involves the following steps:

- Extraction of the raw value of each telemetry parameter from the bit stream. This is done with the knowledge of the parameter's first bit location and its bit length.
- Translation of the *raw value* into an *engineering value*. Polynomials or linear interpolation between predefined points are often used as translational aids. A discrete status is calibrated using tables (0 = OFF, 1 = ON for example).
- *Verification of the validity of the parameter*. Parameter specific validity conditions (as opposed to the overall validity of the bit stream verified with the CRC) can be defined. For example, a reading associated with a particular item of equipment can be flagged invalid when this equipment is switched off. This is useful, for example, in situations where the last value received from the unit is down-linked although the unit itself is not delivering any more data.

- *Computation of out-of-limit status*. Equipments are tested on the ground, where their proper functioning is ensured under different conditions (qualification and acceptance testing). In order to avoid using the equipment in untested conditions, which may result in damage, limits are defined on the values delivered. The first limit stage, called *soft alarm* or *warning*, signals that the evolution of this value must be monitored closely and might reach potentially dangerous levels. The second limit stage, called *hard alarm*, signals that the value has reached a potentially dangerous level, where a corrective action must be undertaken. It is also possible to define so-called delta-limits on the difference between two consecutive values.
- *Derived parameters*. These are parameters computed from the values of other parameters (raw or engineering). The MCS usually proposes a set of mathematical functions for this purpose.

Once the telemetry data have been processed, a display system (sometimes integrated into the MCS) presents the data to the flight operations engineers. The basic types of display pages are *alphanumeric*, where parameters are listed together with their description and value, as well as *graphical*, where the value of a parameter is plotted against time or the value of another parameter. Apart from these types, alarm displays summarize the parameters with out-of-limit conditions and *synoptics* give a graphical overview of a subsystem. Three-dimensional animations controlled by telemetry parameters can also be used, although this is not yet commonly-used. The three basic types of display are illustrated in Figure 14.7.

The processing of command data is similar to that of telemetry data. The display function takes the form of command queues (an example is shown in Figure 14.8), which

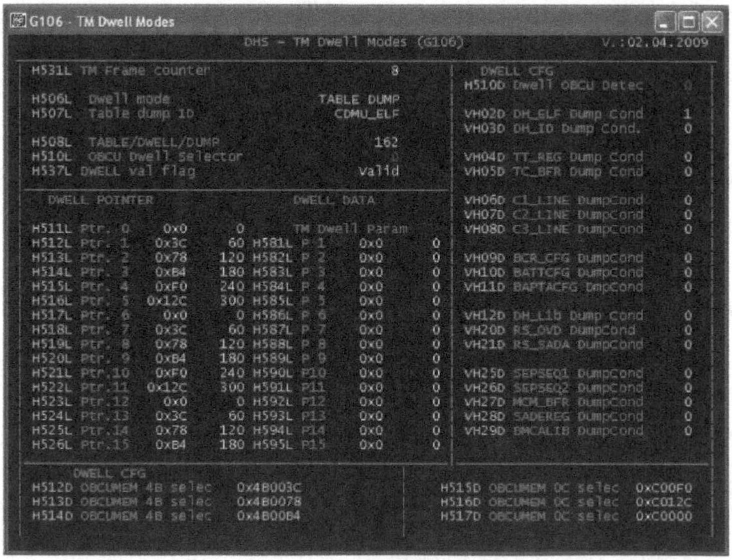

(a)

Figure 14.7 Example of display pages. Alphanumeric (a), graphical (b) and synoptic (c) (Credit: DLR)

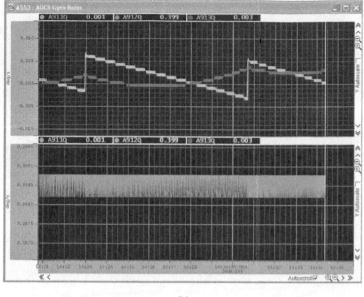

(b)

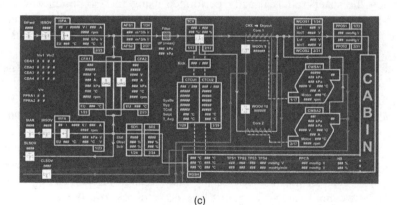

(c)

Figure 14.7   Continued

are lists of commands that can be executed manually (using functions like delete, load, move, insert, modify). The command process is performed directly from the command queue either manually by sending them one by one, batched together or automatically. Depending on the technology used at the ground station and on-board the spacecraft, some information about the command processing (radiation acknowledgement, on-board reception and execution) is returned to the MCS, which can display these different stages. There are two additional important functions in the command process. The first is the *Pre-Telemetry Verification* (PTV)—this ensures that the values of a list of telemetry parameters must be within a specified range before the command can be sent. The second is the *Command Execution Verification* (CEV), which checks that a list of telemetry parameters are within a specified range to indicate that the command execution has been

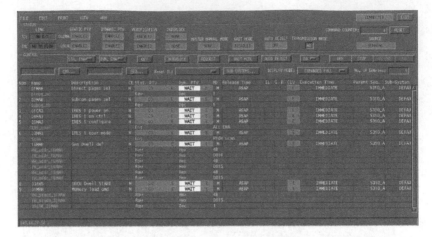

Figure 14.8 Example of a command queue as implemented in the MCS SCOS-2000 (SCOS-2000 is a monitoring and control system developed by ESA) (Credit: DLR)

successful. These two functions are safety measures that can be overridden if necessary. Commanding the spacecraft is usually performed using the Communications Operation Procedure-1 (COP-1) as recommended by the CSSDS in Reference [7]. When using this procedure, commands are sent either in *expedited service* (By-pass Data or BD service), where they are transmitted only once, or in *sequence-controlled service* (Acceptance check Data or AD service), which implements a retransmission mechanism ensuring that no command is lost, duplicated or delivered out of sequence.

All the information required by the MCS is stored in a database, which is delivered by the spacecraft manufacturer to the flight operations team with proper documentation. This is the source of a potential conflict that should not be underestimated. The spacecraft manufacturer uses the database with its own system for ground testing. On the other hand, the flight operations system might use another MCS with a different database structure. The decision to convert the database to the MCS format already in use at the control centre must be weighed carefully against the risk and cost of conversion or adaptation. However, using the MCS of the spacecraft manufacturer implies that the operations team will require additional training so that it acquires the required proficiency. Tools to manage the database (editing, version control, and so on) are also required. Some efforts are being made by the Object Management Group (OMG) to specify an information model for telemetry data and commands in order to standardize the exchange of databases. The resulting XML Telemetric and Command Exchange (XTCE) became the topic of a CCSDS recommendation (see Reference [8]) and may help in improving the situation.

14.5.2 Archiving and off-line processing

Telemetry data processing is usually performed in near real-time - that is to say that there are only a few seconds delay between the data generation on-board and its processing on the ground. This is the case for geostationary satellites or more generally other Earth-orbiting satellites when visible from a ground station. Near real-time processing is however not always possible in the case of an interplanetary mission or when the satellite has no immediate ground contact. Data are then usually stored on-board and dumped, together with current real-time data, during a contact with a ground station. These data are processed 'off-line' (as opposed to near real-time processing). The same MCS may be used to process both real-time and off-line telemetry data, especially if merging is required. However, science data can also be processed using a different system in a different control centre.

All data down-linked from, or up-linked to the spacecraft must be permanently archived, including off-line processed data that has to be merged with the real-time data. The temporal ordering of the data in such an archive is usually done using the on-board time (the time defined by the spacecraft's computer), especially if off-line data are present. Attention must be paid when a time reset occurs, for example after a reboot of the on-board computer, or when the on-board time is implemented as a counter (for example an integer increased every second and coded using a limited number of bits). In the latter case, the counter may exceed its maximum value and is then set back to zero ('wrap-up'). Sometimes, in the case of continuous ground contact, the *Earth-Received Time* (ERT) (the time given by the ground station baseband unit at the reception of the telemetry data) is used to avoid these problems. The commands are archived directly by the MCS, using the server time, usually synchronized with UTC.

The definition of an archive system also dictates at which stage data must be archived. The raw bit stream may be temporarily archived at the ground station to avoid losing it when the communication link between the station and the control centre is down. The ground station can usually replay these data, thus allowing the MCS to process them normally (but delayed in time of course). The raw bit stream can also be archived before being processed by the MCS. The data must then be reprocessed prior to further use, but this can be difficult if the satellite manufacturer uses another system or if cryptography is used to code the data between the MCS and the spacecraft. Another possibility is to archive only the engineering data after processing by the MCS. In this case, access is faster but data can only be accessed by the MCS. The solution adopted depends entirely on the mission, and several archives are usually used.

Archived data are only useful if they can be easily accessed. The first possibility is data replay within the MCS. Data are simply played-back continuously or step by step, and parameter values are displayed as if in real-time by the display system. This is particularly useful for precise analysis of an anomaly. The display of the command queue during the playback is an asset allowing a full picture of the situation. This function is also used to compile the mission report for the customer.

Another possibility to access the archived data is telemetry extraction. This functionality allows the extraction of values of parameters between a start and a stop time. It is used to generate products to be transferred to other groups, such as the flight dynamics team or the scientific community, who may use different data analysis tools.

Although the MCS offers the possibility to define derived parameters, this is often not sufficient to support complex analysis. Derived parameters are well adapted for simple computations performed in near real-time, and are used to support the operations, but they are not suitable for other purposes. Complex analysis requiring advanced visualization of parameters, frequency analysis (Fourier's transform), or complex orbital calculations are better done with external tools. These are referred to as *off-line tools*. They clearly depend on the mission and can be specified by the satellite manufacturer if they are to be used during the operations. A typical example of an off-line tool is the fuel mass calculation derived from the on-time and pulse count of the thrusters. Such tools will of course be specifically developed and tested.

14.5.3 Flight procedures and validation

Due to the high cost and complexity of a spacecraft, there is no place in spacecraft operations for improvization. Many constraints, either at sub-system level (equipment temperatures, warm-up times, fields of view, and so on) or at system level (power balance, available memory, angular momentum, and so on), need to be taken into account when commanding the spacecraft. All the relevant information must consequently be summarized in *flight operations procedures* that are executed during the mission.

A flight operations procedure can be seen as a list of steps in chronological order, itemizing the actions to be performed. The two basic functions are the verification of telemetry parameters and the verification of commands. The former comprises a check of their code in the MCS database and the expected value or range (e.g. 'verify AU50K<50'), and the latter a check of their code and the value of associated parameters (for example 'send 01MAH, Dirpg_nb = 2' would correspond to line 1 of the command stack shown on Figure 14.8). Other actions are typically ground segment configuration tasks ('configure MCS to use COP-1 AD service' for example - see Section 14.5.1) or calls to execute other flight procedures. These actions might be embedded in branching conditions (e.g. 'if-then-else'). An important characteristic is the timing of the steps with respect to main events (for example 'boost start minus 5 min'), which gives a guideline with respect to timing of the execution, providing a basis for mission planning. Based on these procedures, files containing the relevant commands are generated, and these files are loaded in the MCS when the procedure is to be executed.

Flight procedures need to cover not only all nominal activities expected during the entire mission, but also a set of well-chosen contingency cases. It is not realistic to develop a contingency procedure for all possible single and multiple failures. However, all cases where the mission could be endangered without a swift reaction need to be considered, and a contingency procedure developed. The advantage of having such procedures is that their proper functioning has been validated beforehand, and that command files have been prepared and are ready to be loaded in the MCS. The experience of the flight operations team is a valuable asset in defining these contingency cases, and design engineers tend to underestimate what could go wrong in a space mission, especially as telemetry data are the only means of diagnosing failure. Of course it is always possible to develop contingency procedures during the mission execution, which is fine when there are no time pressures that might put the spacecraft at risk.

Flight procedures are delivered by the spacecraft manufacturer, together with the documentation detailing the spacecraft design and operation (usually in the form of operations handbooks). It is the task of the flight operations team to rework and customize these procedures to make them relevant to the particular control centre, and to the MCS. This customization is especially required if the spacecraft manufacturer and operations team use a different MCS! All flight operations procedures must then be validated to ensure their execution is feasible in a timely manner as well as being 'correct'. This validation is performed either against a software simulator or an engineering model, using the delivered database. The software simulator is delivered by the spacecraft manufacturer and must be representative of the spacecraft in the way that telemetry values react to commands. The environment simulation also needs to be realistic, although a complete simulation of the thermal aspects is not usually required. Moreover it should offer telemetry and command interfaces to the MCS and the possibility to simulate failures. A software simulator is clearly the most flexible solution as it allows the operations team to work in parallel with the *Assembly, Integration and Verification* (AIV) team. Otherwise, time would need to be allocated for validation using the engineering model, thus blocking opportunities for ground testing or integration. The validation of flight procedures is an iterative process involving not only the flight procedures themselves, but also the spacecraft documentation, the MCS database, the simulator and the display pages. All these aspects are validated in the process. Finally an end-to-end *System Validation Test* (SVT) involving all the command and telemetry equipments that will be used for the mission, is performed to demonstrate the compatibility between the control centre and the spacecraft. The RF part of this is normally excluded from the SVT since it is included in the RF-compatibility tests.

During the validation process, the definition of a common set of display pages is agreed between the flight operations team and the spacecraft manufacturer team, which is important for the mission execution. Since the spacecraft manufacturer is often asked to support the operations, it is important that experts from both teams can interact easily. The validation status of the flight procedures is presented at the Operational Readiness Review (ORR).

14.5.4 Mission planning

The *Mission Planning System* (MPS) builds the interface between the customer, the space segment and the ground segment. It aims at using the space segment resources optimally to supply the customer with products (images, measurement data, bandwidth, and so on). Such a system is minimally used during the LEOP and fully employed during routine operations, when the payload on-board the spacecraft is fully operational.

Mission planning is first considered when establishing a timeline for the LEOP. Later, during the mission execution, mission planning also supports the spacecraft and ground resource management and as such contributes directly to operations.

The timeline produced for LEOP by the mission planning process is a detailed sequence of all the events foreseen during this early phase of the mission, timed accurately to the second. It is based on operational products delivered by flight dynamics, augmented by activities from flight operations and ground data systems. The *sequence of events* is the guideline for all personnel involved in the mission, and consequently takes all aspects of the mission into account, including:

- orbital events as defined in the operational product delivered by flight dynamics;
- ground station events as defined in the operational product delivered by flight dynamics—examples of items included in the planning are ground station handovers, stations used for uplink and downlink, backup stations during critical events and ranging periods;
- spacecraft and ground operations—this is the planned sequence of flight and ground procedures with their duration (these procedures are related to orbital events and the sequence shows what activities are planned, and at what time);

- shift plan—this schedules the presence of the various personnel required for the planned activities.

Of course, the establishment of a sequence of events is an iterative process and all these activities are interconnected. For example, the execution of an orbit manoeuvre, as computed by the flight dynamics team during the mission analysis, implies starting the operations in good time, so that the spacecraft is correctly oriented and configured at the computed boost start time. However in the process of planning, it may turn out that the specified flight procedures period is such that the required gyroscope calibration, based on Sun and Earth sensors, would take place in a collinearity region, when the Sun vector and Earth vector seen from the spacecraft are almost parallel. This would imply poor accuracy in the attitude determination computed from these two pieces of information. Depending on the other constraints, the flight operations team may have to start earlier, leaving a pause between the calibration and the boost. This may in turn lead to the requirement for an extension of the ground support by a backup station. The other possibility is that the manoeuvre may have to be postponed. The feasibility of any operational sequence needs to be demonstrated using the process of mission rehearsal.

Mission planning must be clearly supported by appropriate software to allow for rapid re-planning during mission execution. Re-planning is always necessary anyway to cope with the continuous updates of operational products in nominal cases, but of course re-planning will be necessary in the event of anomalies, or simply late user requests. Graphical and textual outputs are required. Usually the mission planning sequence is timed using UTC (since teams from all over the world may be involved in the mission), but it is tabulated using *Mission Elapsed Time* (MET) which is the time since the launch. Doing so allows the planning process to be independent of potential launch delays.

The main aspect of mission planning is, however, related to resource management. For complex missions, this function can even be dedicated to a specific system, which takes care of the entire process of planning, and the acquisition and dissemination of the data produced by the spacecraft. Each mission is different, but an example will help to highlight the key features of a mission planning system. Consider TerraSAR-X, a low-Earth orbiting satellite carrying an X-band synthetic aperture radar, and supplying a large community of commercial and scientific users. These users specify their requests by specifying a region of the Earth of which they require an image, and a time frame within which the image is to be acquired. Short-term requests are often made by commercial users or in the case of a disaster-monitoring situation. The construction of a feasible mission planning strategy includes constraints imposed by the space segment (memory size, orbit position, and so on) and by the ground segment (ground contact opportunities). In the case of the TerraSAR-X mission planning system, approximately one hundred constraints were modelled in the process. User requests are accepted up to six hours before the next ground contact (typically two per day) and planning is issued for the next three days. Depending on the user, priority levels (from 0–9) are applied to these requests to resolve potential conflicts. In the last resort, such conflicts are resolved on a first-come-first-served basis. Requests with the highest priority are scheduled first and the resulting planning is then checked against the constraints. If this planning is not feasible, the request is discarded. Otherwise it is accepted and the next request is processed until the process is complete. The planning is then converted into command files and uploaded for the next 24 hour period, thus covering an interval with two possible ground contacts, which is robust with respect to ground station failure. A command operator enters a feedback in the system about which requests were actually up-linked, and the mission planning system records an overview of the requests already scheduled on-board. However since requests are accepted up to six hours before the next contact, the mission planning system must be able to re-plan (deletion or addition of requests) some late high-priority requests. Orbit maintenance manoeuvres or contingency activities must also be taken into account in the planning process as no images can be acquired in these cases. The mission planning system operates continuously, and provides the users with up-to-date information about their requests. It is fully automated and integrated with the MCS.

14.5.5 Mission execution

The success and overall quality of the mission execution is dependent upon the thoroughness of activities undertaken in the preparatory phase—for example, untested equipment on the ground is likely to fail during critical support activities, or insufficiently trained people are more likely to fail to detect an error on-board the spacecraft. As such, great efforts must be made to get all personnel involved in the mission execution to a level of proficiency where no question remains unanswered, and no uncertainty remains about the roles and tasks to be performed.

The role of each person involved in the mission execution must be defined clearly. As well as the personnel of flight dynamics team, the ground data team and the flight operations team, the representatives of the customer are also involved and ultimately decide what is to be done in the case of non-nominal operations. The customer usually involves personnel from the spacecraft manufacturer (forming the Satellite Support Team, SST, or Engineering Support Team, EST) to support the early orbit phase (from launcher separation to final orbit) directly in the control room, as well as the commissioning phase (verification that all spacecraft systems survived the launch and are working nominally). All these people form the combined flight operations team, which is led by a flight director. The flight director conducts the operations and coordinates the action of the combined flight operations team (see Figure 14.9). Within this team, individuals are usually responsible for a spacecraft subsystem (such as power, attitude determination and control, orbit control, thermal control, data handling, and so on) either in a prime or a backup position. These backup positions are important to ensure that expertise is available, even in the case when the prime expert is missing due to illness or accident.

This organizational structure must be documented (in the mission operations plan), which is presented to the customer. Personnel are formally linked with the subsystems for which they are responsible, and the related tasks they are to perform during the mission. *Decision authorities* are clearly identified, as well as the decision process to be applied in the case of an anomaly. The list of flight operations procedures (including their version number and validation status), and the mission planning combine to complete the flight operations plan. From this, a training and simulation plan is developed to give the required proficiency to the combined team. All these documents must be reviewed throughout the entire preparatory phase especially at the operational readiness review, the goal of which is to state whether the ground segment is ready for the mission.

The training of personnel may begin with classroom sessions on spacecraft design. Such sessions can be based on documentation (e.g. operations handbooks) delivered by the spacecraft manufacturer. If possible, these sessions should be led by people from the spacecraft manufacturer to increase the quality of information (covering a broad range of

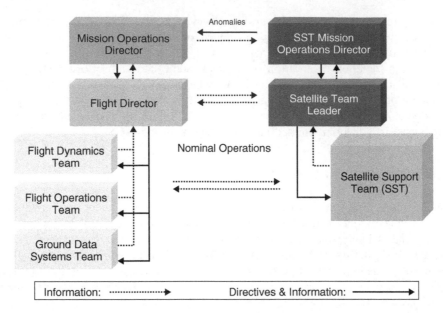

Figure 14.9 A typical example of a ground operations team structure, illustrating the decision flow (Credit: DLR)

topics) passed to operations personnel. If necessary, tools such as the software simulator can be covered as well.

As the preparation continues, the flight operations team members will acquire specialized training related to one or more subsystems of the spacecraft, so that they acquire proficiency in the operation of these subsystems. Each such subsystem expert takes care of the flight procedures related to their subsystem, especially validation. This knowledge about subsystem operations is then passed on to the other members of the team during dedicated training sessions. Tools specific to subsystems and developed for operations (off-line tools, derived parameters, and so on) are also covered.

In the meantime, simulations of different aspects of the mission are organized. They are performed with the satellite software simulator or the Electrical Ground Support Equipment (EGSE), using the validated flight procedures, the operational tools, the MCS and the delivered database. The sequence of events, defined by the mission planning process, is also used to verify that the activities are feasible as planned. Personnel from the spacecraft manufacturer and the customer that are involved in the operations participate in the process to become accustomed to the operations environment.

Finally one or two *mission rehearsals* are executed, covering the entire mission. These are performed in the flight environment (the final flight operations system with the MCS, the database, procedures and the sequence of events) by the combined flight operations team (which includes flight dynamics and the ground data systems, the spacecraft manufacturer, the customer and eventually the ground station team). The aim of these rehearsals is to demonstrate the proficiency of the combined flight operations team and the readiness of the ground segment. This includes the correct loading of the last parameters from the spacecraft parameter book, which is a document containing the numerical inputs for operations, such as sensor alignments or the final spacecraft mass. This is delivered in its final version shortly before launch. Contingency situations are also simulated to ensure that the decision process is well understood and that personnel are also proficient with these situations.

The mission execution itself is nominally only a repetition of the rehearsal. Once the RF-contact with the spacecraft is established, a checkout of the bus and platform is performed to ensure that no system has been damaged during the launch phase. Thereafter, orbit manoeuvres are performed to bring the spacecraft to its final orbit. The verification of instrument performance against design requirements follows, during the commissioning phase. After commissioning, the spacecraft is handed over to the customer for routine operations at the In-Orbit Acceptance Review (IOAR). On this occasion, all data (for example—orbital elements, next planned station keeping manoeuvre, loaded time-tagged commands, current warnings and alarms, and so on) required for further operation of the spacecraft (as defined in the hand-over plan) is communicated to the customer.

After hand-over, routine operations commence and the size of the all the teams (flight dynamics, ground data and flight operations) is usually reduced gradually. Some expert personnel leave the project, to be replaced by others, illustrating the necessity of the maintenance of up-to-date documentation and the provision of training sessions periodically to maintain the team proficiency. Spacecraft performance is monitored through trend analysis to detect the effect of aging or degradation. Periodic reviews are conducted with the customer to assess the quality of the delivered service. The last critical events at the end of the mission are passivation of the spacecraft and its de-orbiting. Last but not least, lessons learned and experience gained on the mission is documented to improve future missions.

ACKNOWLEDGMENTS

The author wishes to thank his colleagues from the German Space Operations Centre (GSOC) for their review and comments: R. Faller, M. Haeusler, W. Kruse, T. Kuch, M. Wickler and H. Wobbe.

Special thanks are offered the GSOC flight dynamics group (S. Aida, R. Kiehling, M. Kirschner and F. Meissner) for their inputs on space situational awareness and their compilation of the USSTRATCOM catalogue in Figure 14.5.

Finally the author thanks Graham Swinerd for his helpful support.

REFERENCES

[1] European Cooperation for Space Standardization (2008) *ECSS-E-ST-70C, Ground systems and operations*, ESA Publications Division.
[2] Centre National d'Etudes Spatiales (1995) *Spaceflight Dynamics Part 1*, Cépaduès Editions.
[3] Montenbruck, O. and Gill E. (2000) *Satellite Orbits*, Springer Verlag, Heidelberg.
[4] Wertz, J. R. (1980) *Spacecraft attitude determination and control*, Reidel Publishing Company, Dordrecht.
[5] Consultative Committee for Space Data Systems (2006) *CCSDS 910.3-G-3, Cross Support Concept - Part 1: Space Link Extension Services*, CCSDS Secretariat.
[6] European Cooperation for Space Standardization (2008) *ECSS-E-ST-50C, Communications*, ESA Publications Division.

[7] Consultative Committee for Space Data Systems (2003) *CCSDS 232.1-B-1, Communications Operation Procedure-1*, CCSDS Secretariat.
[8] Consultative Committee for Space Data Systems (2007) *CCSDS 660.0-B-1, XML Telemetric and Command Exchange (XTCE)*, CCSDS Secretariat.

BIBLIOGRAPHY

Hallmann, W., Ley, W. and Wittmann, K. (2009) *Handbook of Space Technology*, 3rd Edition, John Wiley & Sons, Ltd, Chichester.
European Cooperation for Space Standardization (2008) *ECSS-E-ST-50C, Communications*, ESA Publications Division.
European Cooperation for Space Standardization (2008) *ECSS-E-ST-50-02C, Ranging and Doppler tracking*, ESA Publications Division.
European Cooperation for Space Standardization (2009) *ECSS-E-ST-50-05C, Radio frequency and modulation*, ESA Publications Division.

RELEVANT INTERNET LINKS

- International Telecommunication Union (ITU): http://www.itu.int
- European Cooperation for Space Standardization (ECSS): http://www.ecss.nl
- Consultative Committee for Space Data Systems (CCSDS): http://www.ccsds.org
- USSTRATCOM: http://www.space-track.org
- Celestrak (TLE): http://www.celestrak.com
- XML Telemetric and Command Exchange (XTCE): http://www.omg.org/space/xtce
- Space Weather Prediction Centre (NOAA): http://www.swpc.noaa.gov
- ESA Space Weather Web Server: http://esa-spaceweather.net

15 SPACECRAFT MECHANISMS

Guglielmo S. Aglietti

Aeronautics & Astronautics, Faculty of Engineering and the Environment, University of Southampton

15.1 INTRODUCTION

Spacecraft usually contain mechanisms, the functions of which are indispensable to the craft's mission. Often, mechanisms will be part of one of the major subsystems, such as the attitude and orbit control system (AOCS) or the power supply system, where they perform essential tasks supporting the subsystem's operation. It follows that great care has to be taken to ensure their reliability as the effect of a mechanism's failure upon the mission is never less than serious—and is frequently catastrophic—if it occurs. In this sense, mechanisms are critical. This at once makes reliability a fundamental requirement for every mechanism design.

As is the case for other subsystems, the development of spacecraft mechanisms evolves from system requirements and specifications that cascade down from the system to subsystem-level. Hence the requirements applicable to spacecraft mechanisms cover various areas, ranging from the definition of the functions of the specific mechanism and its interfaces (physical envelope, mass, electrical interfaces, and so on) to more *general requirements* [1, 2], determined by the launch and space environments, which have been described in Chapters 2, 7, 8 and 11.

Based on these initial requirements, some concept designs are defined, which will provide further information to define more detailed requirements. Depending on the nature and heritage of the type of mechanism, various concepts may be developed in parallel, until a design is selected as the baseline for the hardware development (with possible back-up options) and the requirements are transformed into the mechanism specifications.

In this chapter, typical space mechanisms requirements will be briefly discussed, followed by the description of design applications divided into two broad categories:

Spacecraft Systems Engineering, Fourth Edition.
Edited by Peter W. Fortescue, Graham G. Swinerd and John P. W. Stark.
© 2011 John Wiley & Sons, Ltd. Published 2011 by John Wiley & Sons, Ltd.

- *One-shot devices*, which, as the name suggests, are mechanisms required to function only once during the spacecraft mission (e.g. satellite release mechanisms, deployment mechanisms for whip antennas, most of the deployment mechanisms for solar arrays, etc.).
- *Continuously (or intermittently) operating devices*, which include all those mechanisms that are required to run continuously (e.g. momentum and reaction wheels) or intermittently (e.g. antenna pointing mechanisms, solar array drives, etc.) throughout the life of the spacecraft.

The typical mechanism's components are then described, and the chapter is concluded by three sections about materials, tribology and testing.

15.1.1   Requirements for space mechanisms

Mechanisms have to be designed ensuring that their function(s) achieve performances that meet the system level requirements and specifications. This means that the parameters characteristic of the type of mechanism, e.g. the accuracy of a pointing mechanism, the angular momentum of a momentum wheel or the torque of a reaction wheel, have to be specified (with their tolerances) and verified to meet the system specifications. Also, electrical parameters such as power consumption (and its transients) must be within specified limits.

The physical and electrical interfaces of the mechanism are required to match those of the subsystems that interface with the mechanism. This will affect the specifications of mechanical mountings, total mass (and moments of inertia), position of the centre of mass and the variation of these parameters while the mechanism functions. Similarly, all the electrical connections will need to be specified.

During the functioning of the mechanism, it is very likely that the moving parts (due to their inertia and accelerations) will produce some mechanical reactions (forces and moments) at their supports. In some cases, these interface loads are generated intentionally to fulfil the mechanism's functions (e.g. reaction wheels). However, moving parts (e.g. a piston of a pump for a cooler, or a rotor that, due to manufacturing tolerances, is slightly unbalanced, or ball bearings that are not perfectly spherical and roll on a surface which is not perfectly flat., etc.) can also produce very low level mechanical disturbances (microvibrations) that are transmitted through the mechanism interface and propagate throughout the spacecraft structure. Certain types of space instruments or payloads (e.g. high resolution cameras and telescopes and interferometers) have very stringent stability requirements, and therefore the microvibrations produced by other on-board equipment (typically mechanisms) have to be controlled and minimized whenever possible. As a consequence, for some types of missions there will be a limit on the maximum level of microvibrations that can be emitted by the mechanisms on board.

As some mechanisms have a relatively large mounting interface (or multiple interfaces—such as an *antenna pointing mechanism* (APM) that interfaces on one side with the satellite structure and on the other, with an antenna dish supporting frame) the mechanism itself, in its stowed configuration, can act as a (secondary) load path. Hence, the stiffness of the mechanism (or its interfaces) may need to be maintained below a certain threshold to control the loads through the mechanism. On the other hand, when the mechanism is 'deployed', there will be requirements on the minimum stiffness to guarantee that the deployed appendage has a resonance above a specified limit, to avoid dynamic coupling with the satellite AOCS.

The design of mechanisms, including material selection, structural dimensions and sizing of mechanical and electrical components, should follow good design practice and meet sets of standards or requirements agreed between the organization carrying out the mechanism development and the final customer of the mechanism. For example, the Standard ECSS-E-ST-33-01C (issued by the European Space Agency) contains requirements, including design rules with specific values for applicable design margins, based on years of experience in space mechanisms development.

Clearly, the mechanisms have to be able to withstand the various environments that they will encounter during their life, and overall it should be demonstrated that the mechanism will be able to operate, meeting the specified performance, until its designated end of life. For mission critical mechanisms, conformance to the specified reliability figure should be rigorously demonstrated, whereas, for non-mission critical mechanisms, some simplified methods of reliability analysis can be applied. A *Failure Mode Effects and Criticality Analysis* (FMECA) (see for example ECSS-Q-ST-30-02) should always be carried out. All single point failure modes should be eliminated (e.g. using redundancy) and, if that is not viable, an appropriate justification should be provided (and accepted by the customer).

All the requirements applied to spacecraft structures are usually applicable to the mechanism structure, and here the launch conditions often provide the worst (i.e. the most demanding) mechanical environment.

The space environment is generally not very hostile to mechanisms, with the two important exceptions of tribology and temperature (especially thermal gradients). Nearly every mechanism failure in space is due to either poor choice of materials, an inadequate understanding of space tribology or poor estimation of thermal gradients, which can lead to high loads and high torques. The requirement for very long life in rotating machinery puts an even stronger emphasis on these three areas. Many failures of space mechanisms are due to poor engineering practice and an inability to assess risk. Every decision taken by the designer clearly involves an element of risk, and an ability to appreciate and assess that risk is, in the author's opinion, *condicio sine qua non* for success.

15.2 ONE-SHOT DEVICES

Typically, the function of one-shot devices is to change the structural configuration of the spacecraft. These changes can be dramatic events, such as stage separations of a launch vehicle, major changes in a satellite configuration, such as deployment of large solar arrays, or they may be relatively minor events such as the deployment of a small whip antenna. In all cases, the change in configuration is produced by allowing and then forcing the relative motion of some of the parts of the mechanism that were previously constrained by a *hold-down* device. Consequently, in most systems, it is possible to distinguish at least two separate functions carried out by two (or more) different type of devices.

The first function is to restrain firmly (hold-down) one or more parts of the mechanism for all the time up to the moment when the system is finally 'fired'. The actuation is generally carried out by pyrotechnic devices (e.g. explosive bolts or pyrocutters). However lower shock devices (e.g. non-explosive devices based on memory shape alloys) are also available. When fired, the moving part of the mechanism is released, directly or by activating/opening a more complex latching mechanism.

The second function is to enforce a predetermined movement (e.g. deployment) of particular parts of the mechanism. In order to perform this function, most mechanisms store some form of potential energy which, during the release/deployment, powers the movement. Since these devices are designed to work only once, one of the most commonly used methods is to store the potential energy in springs. The springs are loaded when the mechanism is 'charged', and they store the energy until the moment when mechanism is activated, which can be months or years later (as in the case of the release mechanisms for the *Rosetta* lander or for planetary probes such as *Huygens*). The release of the elastic energy stored in the spring is a relatively sudden phenomenon and the moving part(s) might acquire a relatively large amount of kinetic energy. Consequently, damping devices are often included, in particular to control the arrest of the mechanism when fully deployed. Devices for separation, for example launch vehicle stage separation or satellite release mechanisms, do not generally require the addition of dampers, since the kinetic energy stays with the parts that separate, causing them to drift apart. Other mechanisms are powered by the expansion of gas (e.g. nitrogen) previously compressed and stored in cylinders. In this case the release of the gas is controlled by valves and the expanding gas can be used to move linear piston-like devices, or to deploy inflatable structures. For spinning spacecraft, the deployment of radial appendages can by driven by 'centrifugal' force (e.g. Geos), but this is the exception rather than the rule. Some mechanisms, for example those to deploy booms, are sometimes required to be retractable, and in this case they are generally actuated by electric motors. However these mechanisms cannot be classified as truly one-shot devices.

15.2.1   Separation systems

In every mission, various separation systems are used, and classic examples are the mechanisms used to release satellites from the launch vehicle. Most of these mechanisms are based on the use of a *Marmon clampband*, which has been in use since the early days of space exploration. It remains one of the most reliable and commonly-used mechanism [3].

The basic version of this device, whose cross-section is shown in Figure 15.1, is a circular belt with shoes having a V-shape horizontal grove, secured to its internal surface (see also Figure 8.4). When the clampband is taut, the V-grove on the shoes engages two circular mating flanges (one flange is part of the launch vehicle structure and the other one is part of the satellite structure) holding them together. The mechanism is activated by a pyrocutter or similar device that cuts a retaining bolt. Once this bolt is cut, the belt opens up (usually helped by torsional springs) pulling out the shoes, and thus freeing the flanges. At this moment the satellite is pushed away by separation springs secured to the structure on the launch vehicle side. For flanges with a diameter larger than around 1 m, the clampband is usually divided into two or more segments of equal length, joined by the pyrotechnically-actuated bolts. Figure 15.2 shows a view of all the mechanical components of a large separation mechanism. In this case, these devices have to be perfectly synchronized to make sure that the belt opens up at all the junctions simultaneously. The main advantage of this type of mechanism, apart from its simplicity, is that the two flanges are held together along the whole perimeter, thus transferring the loads in a relatively uniform manner. This is particularly advantageous when the two structures to be constrained are cylindrical, which for launch vehicles is always the case.

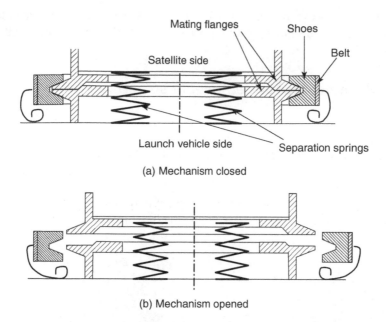

(a) Mechanism closed

(b) Mechanism opened

Figure 15.1   Cross section of Marmon clampband-type release mechanism

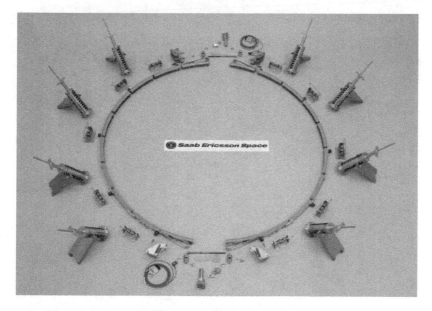

Figure 15.2   Components of the Saab Ericsson Space separation mechanism.
(Reproduced by permission of Saab Ericsson Space)

Other types of release mechanisms for spacecraft have been designed, tested and used successfully, but the marmon clamp band, in its various versions, is still the most widely used.

15.2.2 Mechanisms for deployable structures

Various types of mechanisms can be used in order to deploy elements of the spacecraft which are generally referred to as *appendages*.

The simplest method to deploy an appendage is to use a knuckle joint located at the root of the appendage. This joint can be a simple hinge coupled with a torsional spring. An example of this is shown in Figure 15.3, which is used to deploy a whip antenna. The torsional spring forces the rotation of the hinge, driving the deployment of the antenna and keeping the structure fully deployed once the hinge reaches its end-stop.

Figure 15.3 Hinge of whip antenna. (Reproduced by permission of AEC- ABLE Engineering Inc)

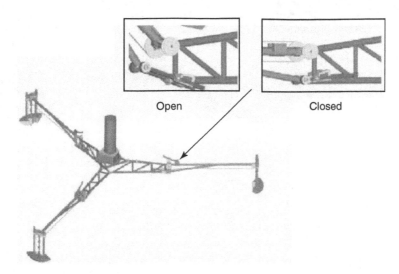

Open   Closed

Figure 15.4 Rosetta lander landing gear. (Reproduced by permission of Dr H. Rosenbauer)

In some cases, deployment mechanisms are required to be self-locking, in order to guarantee that the deployed part of the structure does not fold back when loaded. For example, this is the case for the joints used to deploy the legs of the Rosetta lander's landing gear (Figure 15.4). Here it is crucial that the legs, once deployed are kept in position, and therefore a device called a *mechanical diode* is used to guarantee that the deployment movement of the legs cannot be reversed by the forces applied during touch down. The hinges used for the deployable solar arrays on the satellite CFESat (see Figure 15.5) used a similar concept. At the end of the rotation, driven by the clock spring visible in Figure, when the dowel hits the hard stop a latching lever (pushed by a secondary torsion spring not visible in Figure) prevents the hinge from possible rotation backwards.

A very simple type of self-locking joint is the *tape spring hinge*. This flexible linkage is made by two tape springs parallel to each other, which connect two structural segments as is shown in Figure 15.6. The joint is folded by allowing the tape spring internal to the hinge to buckle and then bending the deployable part of the mechanism. Once folded,

Figure 15.5 Hinge for solar array used on CFESat. (Reproduced by permission of Surrey Satellite Technology Ltd)

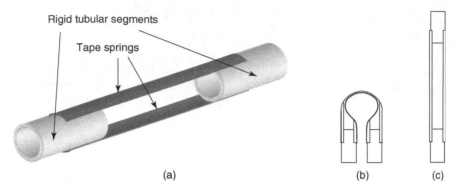

Rigid tubular segments

Tape springs

(a)   (b)   (c)

Figure 15.6 Tape spring joint (a) Three-dimensional view of the joint. (b) Side view of the joint (folded). (c) Side view of the joint (deployed)

strain energy is stored in the tape springs and, as soon as the constraint to the deployable part of the mechanism is removed, the tapes spring straight thus opening the joint. Once straight this joint has a high rotational/bending stiffness (produced by the high buckling load of the tape springs) but relatively low shear stiffness. This mechanism does not control and restrain the parts as precisely as a hinge, but it is much simpler and cheaper. Also, this mechanism can be exploited to deploy relatively long booms made up of shorter segments connected by tape spring hinges. This approach has been used for the design of a 2.5 m long boom on the micro-satellite *FedSat*. The boom, made of six rigid tubular segments connected by tape spring hinges, carries a magnetometer on its tip, and is used to stabilize the satellite, exploiting the gravity gradient torque.

A common problem in the design of these types of mechanisms is that the structure to be deployed is made up of one or more relatively long structural members, which have to be constrained against the external surface of the spacecraft prior to deployment. This problem can be overcome using telescopic booms, and in fact these mechanisms have been commonly employed on spacecraft. Various methods have been used to deploy telescopic booms. For example, a looped tape, held taut between pulleys and supported

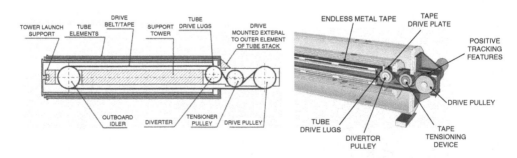

Figure 15.7 Telescopic boom (a) mechanism concept (b) detail of the tape drive assembly. (Reproduced by permission of Surrey Satellite Technology Ltd/SULA systems)

by a pillar inside the boom, was used in the SULA BOOM [4] (see Figure 15.7) to drive out the tubes one after the other, with a latching mechanism to control the release and latching of the tubes.

Other designs utilized a lead screw that rotates inside the interior tube of the telescopic assembly. However, telescopic booms still require a relatively large stowage volume, having its longest dimension in the direction of the boom.

A solution to this problem is to use devices similar to a carpenter's measuring tape, which is stowed rolled on a reel, and deployed by unrolling the reel. The concept is illustrated in Figure 15.8. The boom is elastically flattened into a strip (tape) and stowed rolled-up onto a drum. When the tape is unrolled, to extend the boom, it reacquires its original shape, thus gaining the bending stiffness necessary to keep the boom straight. This type of boom has good structural properties when extended. However, high shear

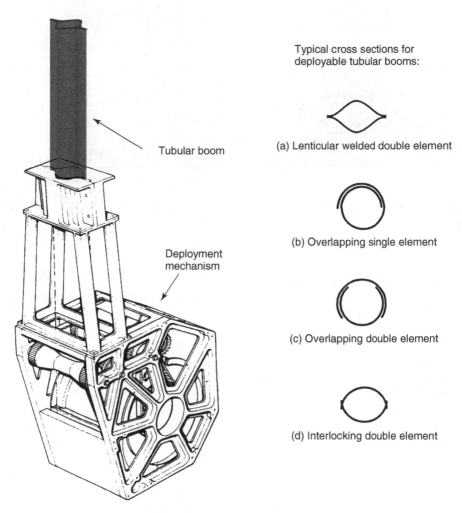

Tubular boom

Deployment mechanism

Typical cross sections for deployable tubular booms:

(a) Lenticular welded double element

(b) Overlapping single element

(c) Overlapping double element

(d) Interlocking double element

Figure 15.8 Collapsible Tube Mast Deployment mechanism. (Reproduced by permission of Sener Ingenieria y Sistemas, S.A)

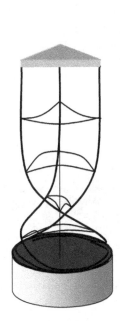

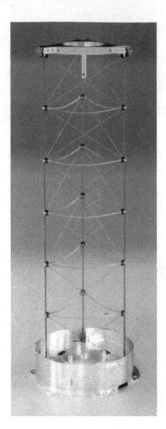

Figure 15.9 Mars lander deployable camera mast. (Reproduced by permission of AEC-ABLE Engineering Inc)

stresses arise in the welding between the two halves when the boom is rolled onto a small reel. Simpler devices can be built using just one element (Figure 15.8b). However this type of boom has inferior structural properties, and in particular it has a very low torsional stiffness. To improve the structural properties of the deployed booms there are several possibilities, as shown in Figure 15.8 (c) and (d). However these types of devices require two reels, and therefore add complexity to the mechanism.

Another class of deployment mechanism is the deployable lattice mast. A particularly successful example of this is the AEC- ABLE CoilABLE mast, shown in Figure 15.9. This mast is an integral part of the Imager for Mars Pathfinder (IMP) system, holding aloft a stereoscopic camera which provided panoramic views of the Red Planet. When fully deployed, the open lattice mast is about 0.7 m long. However, when stowed, it retracts to 10% of its extended length, coiling up inside a 76 mm long canister. The stored strain energy in the structure allows the CoilABLE mast to self-deploy without expensive motors that add weight and complexity. The lightweight steel and fibreglass construction provides a steady, rigid base for the IMP camera, with a mass of less than 0.7 kg. Alternatively lattice masts can be in the form of articulated trusses composed by rigid segments hinged together.

15.2.3   Surface deployment devices

Solar arrays

Typically, solar arrays can be assemblies composed of rigid panels, flexible surfaces in the form of a blanket rolled up on a drum or a semi-rigid surface folded like a concertina. Rigid arrays are composed of a number of panels, usually sandwich structures with an aluminium honeycomb core and CFRP skins, hinged together. The assembly is folded, and the resulting stack of panels is secured against the side of the satellite during launch, and then unfolded in space. Most often, the panels are also connected together and with the yoke (that connects the deployed array with the satellite) by a synchronization mechanism, generally composed of cables tensioned by springs and pulleys. The purpose of this mechanism is to control the deployment sequence, to avoid interference with other parts of the satellite. Sun pointing of the deployed array is then maintained with a solar array drive mechanism (SADM) that rotates the yoke with respect to the satellite. The hold-down and release mechanism must be stiff enough to prevent movement during launch. The possibility of friction welding at the load points is very real.

A typical solar array, held in the stowed position by a stack of cup-and-cone devices, is shown in Figure 15.10, and one of the main concerns in the design is to ensure appropriate preload to prevent gapping during the launch loads. The hinge system must accommodate out-of-plane movement, so spherical-bearings are commonly used. There have been cases of distortion due to thermal gradients producing torques high enough to stop deployment. The latch-up mechanism must stiffen the array and eliminate backlash. In

Figure 15.10   Stowed rigid solar array, showing detail of the cup-and-cone stack. (Reproduced by permission of Surrey Satellite Technology Ltd)

all spring-driven deployment systems, a problem needing careful resolution is the control of the stored energy. The torque to be provided should never be less than four times the estimated resisting torque, and must ensure final latch-up. To absorb the surplus energy, a wide variety of devices have been used, including clutter mechanisms, friction brakes, crushable honeycomb plugs, eddy current and fluid dampers. The balance between stored and absorbed energy is always one that requires very careful judgment.

Flexible arrays require a tensioning system, which can be an articulated mast or other type of deployable boom system. In either cases, this provides a framework to unroll or unfold the array and maintain it flat like a membrane supporting (tensioning) its edges.

Overall, even the simplest concepts can be full of hidden failure mechanisms, and there are too many cases of failures during deployment of solar arrays to encourage complacency.

Antennas

Antennas (see Section 12.3.3) vary from the elementary single rod whip antenna to the complexity of the TDRSS and Galileo unfolding umbrellas (Figure 15.11). In the latter design, the antenna surface is a wire mesh in which the tension of every wire is predetermined and set. The antenna is then folded as an umbrella and, when unfolded in space, the surface shape must be precisely recovered. Since the diameter of such an antenna is about 4 m the mechanical problems are formidable. The design, manufacture and assembly

Figure 15.11 An Umbrella-type antenna. The figure shows the Galileo 4.8 m diameter, 18-rib wire mesh high gain antenna in its deployed state. (Reproduced by permission of NASA/JPL - Caltech)

are driven by the need for extreme precision in the final shape of the antenna which in turn impacts on every mechanical item in the system. Once in Jovian orbit, unfortunately the Galileo antenna did not deploy completely, to the disappointment of scientists and engineers. The reason for this malfunction was most likely excessive vibration experienced by the mechanism during terrestrial transportation. In fact, the launch of the Galileo spacecraft was delayed a few years, during which it was transported back to JPL and then again to its launch site. This episode emphasizes how the life of a mechanism has to be followed by its designers, and how the impact of changes in the mission planning has to be assessed in order to avoid undesirable results. With the benefit of hindsight, the Galileo mission to Jupiter will be seen as one of the major achievements in the history of astronautics, despite this significant mechanism failure.

Inflatable structures

Inflatable technology has been used in aerospace applications over many decades, as the ability to deploy large structures with a low mass and very high packing efficiency has always been a requirement in the space industry. Flight heritage began in the 1950s with the *Echo Balloon* series, which were designed as passive space-based communications reflectors. *Echo Balloon 1* was approximately 33 m in diameter, weighed 61 kg and was constructed from 12 micron-thick Mylar [5]. Other flight demonstrations followed such as the Goodyear antennas in the 1960s, and the Contraves antennas and sunshades in the late 1970s to the mid 1980s. One of the most notable flight experiments to have occurred in recent decades was performed on the NASA shuttle mission STS-77 in 1996. This resulted in the successful deployment and test of the Inflatable Antenna Experiment (IAE), which demonstrated significant advances in inflatable technology with low construction costs. Inflatable antennas supported by telescopic booms (SULA BOOM) were also used on the satellite *CFESat*.

Applications of space inflatable structures are widespread, and include: solar arrays, communication systems, radar and reflector arrays, solar concentrators, solar shades, solar sails, planetary surface exploration infrastructure and impact protection, and human habitats. An inflatable module was proposed for the International Space Station (ISS), which was subsequently designed and developed by the NASA Johnson Spaceflight Centre. This concept of an inflatable transit habitat became known as 'Transhab'. Although cutbacks to the ISS program prevented the construction and flight test of the Transhab module, the concept was taken up by private companies and led to the flight test of Genesis I and II in 2006 and 2007 respectively, designed by Bigelow Aerospace. For further information concerning inflatable structures for space applications the reader is referred to References [6, 7 and 8].

15.3 CONTINUOUSLY AND INTERMITTENTLY OPERATING DEVICES

In general this covers a large variety of devices, but in limiting ourselves to spacecraft mechanisms it is useful to divide the category into:

- *mechanisms that operate throughout the life of the satellite* such as momentum wheels, solar array drives, de-spin mechanisms and horizon scanners;

- *mechanisms that operate intermittently* or on demand, such as antenna-pointing mechanisms and retractable appendages.

15.3.1 Continuously rotating mechanisms

Solar array drive (SAD)

These are needed to decouple the motion of the solar array from that of the satellite to maintain a Sun-pointing direction. The speed for a geostationary orbit will clearly be one revolution per day whilst in LEO it may be about 15 revolutions per day.

Figure 15.12 reveals the construction of a typical SAD mechanism. The shaft is mounted on two angular contact bearings lubricated by a film of lead and preloaded by a titanium diaphragm, which also accommodates thermal expansion and the movement required to off-load the bearings for launch. Allowing the bearing to slide on the shaft is discouraged in all spacecraft systems due to the risk of seizure. Both the shaft and the housing are made from beryllium, to take advantage of its high stiffness, good thermal conductivity and a coefficient of expansion matching that of steel. For off-loading, the bearings are displaced by a lever system onto a conical seating. Figure 15.12 shows the system in this state. The lever is retained by a pyrotechnically operated pin-puller. Two brushed DC motors are provided for redundancy, and are acceptable because the number of rotations required for a seven year life is under 5000. The slip-ring system is conventional with two dry lubricated brushes running on each ring. The drive system is totally redundant with

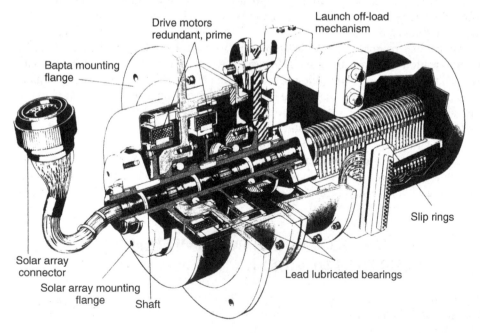

Figure 15.12 Schematic of a typical solar array drive mechanism. (Reproduced by permission of EADS Astrium Ltd)

control signals being generated by Sun sensors mounted on the solar array. Although an old design it has a very good operational history and shows the main features of a SAD.

Giotto de-spin mechanism

With the *Intelsat* series of dual spinners there is significant experience of the design and operation of de-spin mechanisms in the USA. However, in Europe such experience is relatively limited—a good example of a European design is the *Giotto* de-spin mechanism. As a spin-stabilized spacecraft Giotto needed such a mechanism to de-spin the high gain antenna from 14 rpm, with a further requirement to nullify the speed with no jitter. A cross-section of the mechanism is shown in Figure 15.13. Normally, a very stiff servo loop needing some 180 updates per revolution would be needed. However, to simplify the system, the designer decided on one update per revolution, thus throwing the problem back to the bearing designer to achieve a low and consistent torque throughout the life of Giotto. Tests at the European Space Tribology Laboratory (ESTL) demonstrated that lead-lubricated bearings could achieve the desired performance and they were adopted. The off-load mechanism, released by a pyrotechnic, can be seen together with the two drive motors. These were the same 1200 steps-per-revolution motors as used in the Hubble primary deployment system, but here they run as synchronous motors to achieve absolute speed control. The success of the Giotto mission is now space history.

Momentum and reaction wheels

The different roles for these two types of wheel in the AOCS are made clear in Section 9.4.7 of Chapter 9. Momentum wheels have large momentum (around 50–200 Nm s) and a maximum speed of up to 10 000 rpm. Reaction wheels, meanwhile, have a

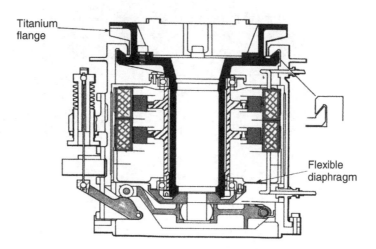

Figure 15.13 Giotto de-spin mechanism. (Reproduced by permission of EADS Astrium Ltd.)

smaller capacity (about 2 Nm s) and a speed up to 4000 rpm in both directions. Electro-mechanically, however, they are similar devices and use the same technology. The speeds of both types of wheel are limited by the bearing system and by power consumption, and although, from a mass point of view, a very small wheel rotating very fast would appear attractive, the power needed to produce the same torque from it would be very large since the energy in the wheel is a function of the speed squared. The bearing system is critical to success and almost all ball-bearing mounted wheels follow a similar approach. The angular contact bearings are preloaded by carefully designed springs to avoid variation of torque with temperature. Ball bearing lubrication remains the principal life-limiting factor for momentum and reaction wheels, as it is necessary to maintain a small and constant (to avoid torque variation) quantity of oil lubrication. To prevent loss of oil and to maintain extreme cleanliness, the wheels can be encased in a hermetic canisters filled with an appropriate gas, e.g. helium. Figure 15.14 shows an example of reaction wheel assembly.

To avoid some of these problems wheels supported by magnetic bearings have been the subject of intense development for more than 40 years in France, Germany and the USA. To support a wheel magnetically in five degrees of freedom (DOF) a choice must be made between a passive permanent magnet and an actively controlled electro-magnet. In the nineteenth century it was shown to be impossible to support all five degrees of freedom with permanent magnets alone; the system is fundamentally unstable and one degree at least *must* be actively controlled. Permanent magnets are attractive in their simplicity and almost-zero power consumption - they do in fact have a very small drag, but low stiffness [9].

It has become common practice to identify space wheels by the number of actively-controlled DOF. The first magnetic-bearing wheel to fly in Space was a French 1 DOF wheel, in *SPOT* 1 (three wheels) in 1986. In this type of wheel the actively controlled DOF

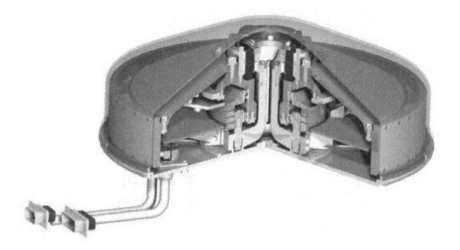

Figure 15.14 SSTL 200SP Reaction wheel assembly. (Reproduced by permission of Surrey Satellite Technology Ltd)

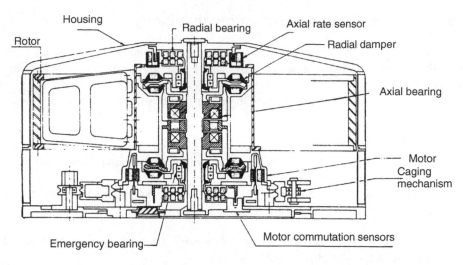

Figure 15.15 SPOT magnetically suspended momentum wheel (prototype).
(Reproduced by permission of Centre National d'Études Spatiales, France)

was the axial displacement of the rotor, which led to a relatively large axial dimension
of the wheel (see Figure 15.15).

Subsequently, actively controlled 2 DOF wheels were developed, where the wheel is
controlled along two orthogonal radial directions. This design results in a flatter geometry
and higher momentum-to-mass ratios. 2 DOF wheels have flown successfully on *SPOT* 4,
SPOT 5, *HELIOS, ENVISAT* and *METOP*. These types of reaction wheels, developed by
Aerospatiale, have a maximum angular speed of 2500 rpm. However, the new generations
of 2 DOF wheels can reach up to 10 000 rpm [10].

The 5 DOF wheel, for example, the MWI series wheels developed by Rockwell Collins,
Inc. in Germany, can be regarded as the most advanced on the market and embodies all
the capabilities of magnetic suspension. All magnetic bearing wheels include ball-bearings
as safety touch-down elements in case of failure.

Each type of wheel has advantages and drawbacks. The ball-bearing wheels are relatively simple and therefore cheap, and give reliable lives up to 10 years or more.
They are, however, mechanically noisy due to the bearings; furthermore, there is difficulty in achieving perfect balance, and they cannot act effectively as nutation dampers.
Magnetically-suspended wheels eliminate some of these problems. By removing bearing
wear, the speed limitation is now governed by power consumption and the root stress in
the wheel. The need for a hermetically sealed case disappears [11].

The 5 DOF wheel adds further sophistication [12]. With the availability of full control
of each DOF the wheel can be fully balanced, and the damping capability in each loop can
be used to control nutation of the spacecraft. Yet more significant is the ability to tilt the
wheel axis through a small angle, currently $\pm 1.7°$ (MWI series), allowing either three-axis
control or cross-momentum storage with only one wheel. Further extension of the system,
by adding mechanical gimbals for a larger angle of tilt, is feasible or, alternatively, two
skewed contra-rotating wheels could be used to achieve full attitude control.

15.3.2 Intermittently operated mechanisms

As examples of this type of mechanism, we look at *antenna pointing mechanisms* (APMs), although in some modes of operation the APM may be continuously trimming the antenna direction.

All APMs will be designed to operate in one or more of three modes as follows:

- *steady-state pointing*, maintaining alignment with any predefined angle on both axes to an accuracy of ±0.01°;
- *re-pointing*, when it shall be capable of re-pointing the antenna through any required angle to an accuracy of ±0.01°;
- *tracking*, when it shall be capable of maintaining the antenna on a station, following a ground-based RF beam in closed loop mode to ±0.015°.

In addition, in the event of complete failure of a drive it must be possible to return the antenna to a nominal zero position and lock it.

APMs are required to rotate the antenna in the direction of a specific 'target'. In general, to perform this function, it is necessary to control independently two rotational degrees of freedom. These are controlled by gimbal-drives and, at least in theory, the axes of these drives could have any orientation. In practice it is desirable for the gimbals to have axes along mutually orthogonal directions, one of which is either 'horizontal' or 'vertical'. This leads to two possible configurations [13].

In the first configuration, called *az-el*, the elevation of the antenna is controlled by a horizontal-axis gimbal-drive, which in turn is supported by a vertical-axis gimbal-drive. This configuration is very common in the driving mechanisms for ground station antennas, but is also used for APM on board satellites, see Figure 15.16.

The second type of configuration, called $x - y$, is shown in Figure 15.17 This mechanism allows the deployment of large antennas situated on the side of the body of the spacecraft and closed in a clam-shell fashion for launch. By displacing the two axes of the

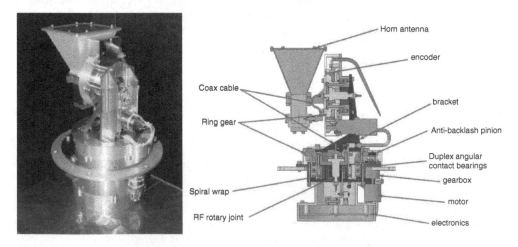

Figure 15.16 X-Band Antenna Pointing Mechanism. (Reproduced by permission of Surrey Satellite Technology Ltd)

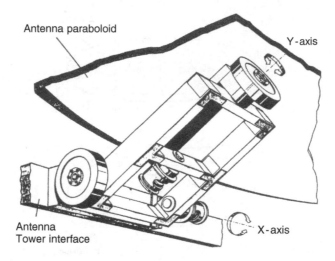

Antenna paraboloid

Y-axis

Antenna
Tower interface

X-axis

Figure 15.17 Dornier Antenna Pointing Mechanism: television satellite version (schematic). (Reproduced by permission of EADS Astrium Ltd)

APM, as shown in Figure 15.17, the outer becomes also the deployment axis whilst the inner remains a pointing axis of limited amplitude. This development required enlargement of the outer axis bearings, a reconsideration of the lubrication system and stiffening of the bearing mountings, none of which changed the fundamental features of the concept.

An example of an alternative design is the Olympus APM. This mechanism utilizes the swash-plate or rotating-wedge principle. The structure, shown in Figure 15.18, consists of four cylindrical and co-axial elements. The two end sections act as the spacecraft and payload interfaces and are prevented from mutual rotation by a bellows. The two centre sections are wedge-shaped and are able to rotate independently about the common axis by means of a bearing system. The two swash-plates may therefore be considered to have pointing vectors, equal to their wedge angles, which may be rotated independently. Their vector sum is the resultant pointing angle of the mechanism. By driving these wedges, any pointing angle may be achieved within a conical pointing range with a semi-cone angle equal to twice the swash angle; for Olympus the range is 8.6°.

The excellent mechanical strength and stiffness of the mechanism removes all need for an off-load device during launch. Its relatively small size, its substantial load capacity of 50 kg, and its advantage of being a closed and self-contained unit makes for easy integration in the spacecraft and flexibility of use for other pointing applications [14].

15.4 COMPONENTS

15.4.1 Electric motors

Since these are the only source of continuous mechanical power on spacecraft they are perhaps the most important mechanism component. Reliability is paramount and in critical systems they will be either double-wound or fully duplicated. In space applications, the

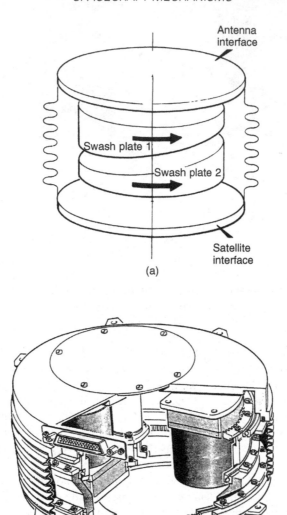

Figure 15.18 Olympus Antenna Pointing Mechanism (a) principle of operation and (b) the mechanical arrangement. (Reproduced by permission of EADS Astrium Ltd)

most commonly used electric motors are *DC Motors* (brushed or brushless), or *Stepper Motors* [15].

DC motors used in space are usually permanent magnet machines but may be either brushed or brushless. As they give maximum torque at zero speed and, in the brushed configuration, can be used with minimum electronics, making them attractive for many space applications. Dissipation of heat is a major problem and space motors may have to be de-rated by as much as 70%. There are cases where the motor is actually housed in a sealed compartment (e.g. reaction and momentum wheels produced by Satellite Services Ltd, UK) containing gas at 1 atm pressure, and in this case the full power potential of the motor can be exploited. Brush wear is of course the life-limiting parameter (see

Section 15.4.2). Brushless motors have the advantage of being very controllable at the expense of considerable electronic complexity. They are now widely used both in space and on the ground to take full advantage of the DC motor's performance linked to excellent controllability.

In general the design of brushless DC motors tends to achieve a particular type of performance profile. Where high torque is required (up and beyond 1000 Nm), the motor tends to have a relatively large diameter and short axial length. If higher speeds are required, the ratio between axial length and diameter of the motor will be higher. Also, if specifically smooth speed/torque characteristics are required, so-called 'toothless' brushless DC motors are employed. However, it should be remembered that mass, size and power consumption for the electronics necessary in a brushless motor are larger than that of a brush assembly. The use of brushes in some applications in space may also have unacceptable drawbacks in terms of the brush's behaviour in vacuum (wear, maintainability etc.) therefore making a brushless motor more attractive.

In the solar array drive (Figure 15.12), the motors are brushed only because at one revolution per day the lifetime total will not exceed about 5000 revolutions.

Stepper motors have simple driving electronics, which together with the incremental stepping motion of the shaft, makes them ideal for several applications, particularly open loop positioning/pointing devices. These types of motors can be divided into three main groups:

- Hybrid stepper motors
- Reluctance stepper motor
- Permanent magnet stepper motors

However, in space applications, hybrid stepper motors are by far the most commonly used.

This type of motor has a high power/torque ratio (especially at low speed), maximum torque in the region of 5–10 Nm, number of steps per turn between 50 - 1 200 (for motors with an external diameter between 15–150 mm) and a speed between ten and a few hundred steps per second. Reluctance stepper motors have a poor torque/power ratio. They are, however, much simpler devices, which makes them suitable for applications working in cryogenics conditions.

Stepper motors are very versatile as both drive and servo prime movers and find frequent application in spacecraft systems. For the *Giotto* de-spin application, the motors are 1200 step-per-revolution steppers but run in the synchronous mode at 14 rpm.

All the motors considered so far are rotational devices. Spacecraft mechanisms, however, often require linear motors. These often use the same principles exploited in rotational motors, and they are usually grouped into two categories:

- *Short stroke linear motors*, with a stroke length usually shorter than 50 mm, generally have a single phase winding which makes the electronic control relatively easy.
- *Long stroke linear motors*, with a stroke longer than approximately 50 mm, which is achieved by exploiting multi-phase windings.

Motors for space are often designed specifically for each application since performance, power consumption, heat dissipation and materials must all be closely controlled. Adapting commercial motors for space is acceptable only in conjunction with a very stringent evaluation program. See References [16, 17].

15.4.2 Slip rings

There are a number of space mechanisms where an electric current must be transmitted across a rotating joint, the solar array drive being a typical example, and in this case 'slip rings' are needed. These may be axial, as in Figure 15.12, or radial in the form of a disc, but the material requirements are the same. The brush material must have good electrical conductivity, a low wear-rate and a self-contained lubrication system. Graphite cannot be used as it becomes an abrasive in vacuum. Several compact materials based on copper, silver and molybdenum disulfide with a wear-rate between 10^{-14} and 10^{-15} m^3/ Nm have been widely used, but a lower wear-rate would be preferable for long-life mechanisms since the control of 'wear debris' in space is a growing problem. The ring material is normally copper, plated with gold or silver, although the latter suffers too readily from sulfide contamination in air.

Oil can be used as a lubricant but must be only vestigial. In one case of dry-lubricated slip-rings in a 60 rpm de-spin mechanism, the wear-rate of the brushes was markedly reduced by oil vapor from the liquid-lubricated bearings. The mechanism ran on test for over six years.

Honeywell have developed their Roll-Ring design to the stage at which it can be used with confidence in low speed applications. The concept is based upon a flexible hoop rolling between the stationary and rotating elements to carry the current.

15.4.3 Gears and bearings

These two machine elements may be discussed together. They are both essential components of a variety of space mechanisms and their technologies are both critically dependent on tribology and materials selection.

Space gearboxes differ from industrial units of similar size in their much reduced permissible tooth-loading and their limited choice of both lubricant and material pairing. The tooth module, defined as the pitch circle diameter divided by the number of teeth, will define the tooth size. As a general rule the minimum module is 0.08 but the choice is a balance between tooth load, accuracy of transmission and gear ratio.

The two types of tooth stress governing performance are the contact, or Hertzian, stress and the bending root stress. In practice the former has been identified as the more important in failures because it also controls the sub-surface shear stress and, by implication, the fatigue failure. A useful rule of thumb is to limit the tooth load of metal gears to a maximum of 10 N per mm tooth width, and to reduce this figure by a factor of 10 for plastic wheels. Metallic materials for gears include carbon, stainless and maraging steels, cast iron, bronze, aluminium and titanium, with suitable surface treatments. Favoured plastics are polyimide and polyacetal, often loaded with molybdenum disulfide or carbon fibre. Plastic gears should always run against metal gears, usually stainless steel, although aluminium and titanium are sometimes used.

The lubrication of lightly-loaded plastic gears is usually achieved by the incorporation of a solid lubricant such as polytetrafluoroethylene (PTFE) or molybdenum disulfide in the plastic. For heavily-loaded metal gears the choice is between thin films of soft metal, e.g. gold or lead, or molybdenum disulfide. Grease and oil may also be used but control against migratory loss will be needed.

Sometimes it is possible to achieve redundancy in a gear drive by incorporating a differential gear, one half of which is locked and released only if the other half should seize. Such complexity can only be justified in very critical systems. It is well to remember that a gearbox is nearly always reversible, and achieving true irreversibility, even with a worm gear, can be very difficult.

Harmonic Drives are often used in space. These are characterized by very high transmission ratios with a high torque capability, together with high positioning accuracy (virtually zero backlash) and repeatability [18]. The first major space application of Harmonic Drives was in 1971, on the Lunar Roving Vehicle of the Apollo 15 mission, and since then they have been used quite frequently where compact and powerful positioning drives are required.

The principle of operation of these devices, illustrated in Figure 15.19, is relatively simple and it is based on three mechanical components: the wave generator, the flexspline and the circular spline which are shown in Figure 15.20. The flexspline is basically a flexible gear that is deformed into an elliptical shape by the wave generator. The wave generator rotates inside the flexspline forcing the teeth of the latter to mesh with those of the circular spine which is fixed (thus acting like the case of a gear box). The manner in which this functions is illustrated in Figure 15.19, and highlighted by the movement of the black dot which is fixed to the flexspline. In this case, after half a turn of the wave generator the flexspline has rotated only 1/48 of a circumference which corresponds to a

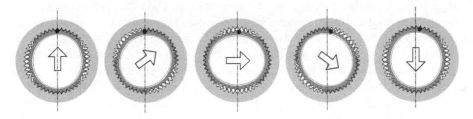

Figure 15.19 Harmonic Drive principle of operation

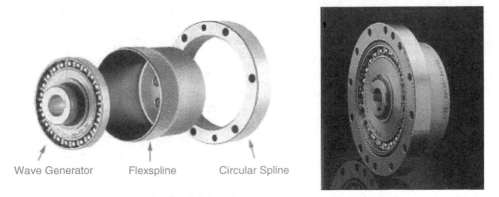

Wave Generator   Flexspline   Circular Spline

Figure 15.20 Harmonic Drive basic components and final assembly. (Reproduced by permission of Harmonic Drive AG)

transmission ratio of 24:1. In practice the number of teeth of real devices is much higher, and it is possible to reach transmission ratios from 30 up to 320:1.

For *bearings*, the application of the apparently simple ball-bearing to space mechanisms might be thought to be a straightforward adaptation, but it has taken more than thirty years of research and more than two million hours of testing in vacuum to reach the present state of knowledge, and there is still more to be learnt. Speed, load, reverse rotation, type of cage, material and both wet and dry lubrication exert major influences on the performance of this machine element. The duty cycle and life required of the bearing will be important in the choice of these factors. Temperature, particularly thermal gradients, is most important; the thermal conductivity of a liquid-lubricated bearing can be ten times that of one with dry lubrication. The warning must be never to use a ball-bearing in a space mechanism without the guidance of a space tribology expert (see Section 15.6). Plain bearings are also used in space but almost exclusively for one-shot systems. They are always dry-lubricated and polymer-metal pair based. Spherical rod end bearings are widely used in deployable elements to avoid high friction from thermal distortion. They must always be validated for space since some commercial materials are unacceptable. Reference [19] is a good, general survey and [20] is important reading too.

15.4.4   Sensors

Unless operating completely open loop, which is rare, an electro-mechanical system will embody a sensor, and the choice of this device will be crucial to the success of the system. The angle over which it must operate will limit the choice. Some sensors, such as the pick-offs used in gyros, are extremely sensitive but operate over very small angles. Others such as synchros, resolvers, stepper-motors, and encoders can be used over 360°. Optical encoders are commonly used in space, their development commencing in the 1950s. In its simplest form an optical encoder consists of a collimated light source (often a light emitting diode, LED), a rotating code disk (with transparent and opaque radial segments), a grating element (or slit plate) and one or more light detectors. The code disk is attached to the element whose rotation has to be determined, whilst all the other elements are fixed. The light beam propagates from the LED through the code disk, to the grating device and finally to the detector. The rotation of the code disk produces interruptions of the light beam, and from the pattern of these interruptions it is possible to determine the position of the disk. The optical encoders used on the Hubble Space Telescope have an accuracy better than 1 arcsec.

There are many factors to be considered in the choice of a sensor since many sensors are very sensitive to mechanical stability and thermal distortion, particularly over the longer lives now being demanded. Mechanically robust devices such as resolvers can be very attractive for this reason. Wear can ruin a mechanical contact encoder and optical units are dependent on the reliability of the optical elements.

15.4.5   Release devices

Pyrotechnics have always been, and still are, the principal elements for releasing space mechanisms [21, 22]. They are very energy efficient (better than 10 kJ/g), operate very rapidly (less than 10 ms) and, when well engineered, are very reliable. Space pyrotechnics always have two initiators for each charge and fully redundant firing circuits. The pressure generated by combustion is used to operate a simple mechanical device such as a pin-puller or pin-pusher, cable-cutter or valve-actuator. They can, of course, only be operated once and are then discarded, which seriously limits ground and pre-flight testing. They are also difficult to seal hermetically, although there have been a number of successful designs using deformable capsules or bellows. Safety is a very important aspect and in the early days there were incidents in which stray currents and electrostatic discharge, even from the human operator, caused untimely ignition. This led to the need for elaborate protection systems to absorb electrostatic discharge, which are now built into the initiators of every space pyrotechnic.

One of the drawbacks of pyrotechnics is the mechanical shock produced by the explosion of the charge. Pyrotechnics are in fact forbidden in some spacecraft (e.g. spacecraft with 'delicate' optical elements such as the Hubble Space Telescope), and *memory metal actuators* are often used alternatively as activation devices. Here, the metal, an alloy of nickel and titanium, can be deformed into a new shape (basically the material undergoes a phase transformation between austenitic and martensitic states) below a critical transformation temperature. On heating above this temperature (which is less than $100°$), the shape changes back to the original. If this change of shape on re-heating is restrained, then the alloy can be used to generate a significant force capable of doing work which, in turn, can be used to operate a mechanism. On the HST it was used to release the off-loading system of the SAD.

Practical examples of how this phenomenon is exploited in real mechanisms are the Frangibolts produced by the TiNi Alloy Company, which are commonly used in spacecraft. The mechanism, schematically represented in Figure 15.21, exploits the expansion of the shape memory alloy to produce an extra tension in the already loaded bolt, which consequently breaks the bolt across the area identified by the circular notch around its shank. A possible issue with this type of device is that usually there is a few seconds uncertainty on the exact time when the bolt will actually snap.

High Energy Paraffin actuators are another possible release initiator. They rely for their action on the large expansion of paraffin wax when changing from the solid to the

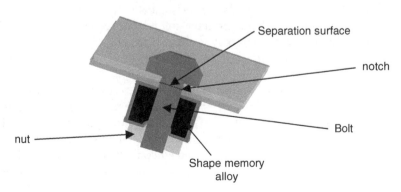

Figure 15.21   Schematic representation of the three-dimensional cross section of a Frangibolt

liquid phase. They can generate high forces and the action is indefinitely repeatable when thermally cycled - an important advantage in their favour. In another device, a heated knife is used to cut a tensioned Kevlar cable to release a solar array.

These devices take time to operate - of the order of one minute - so that simultaneous release of multiple elements cannot be guaranteed.

Burn wire mechanisms have also been used, due to their simplicity, to trigger release mechanisms. Basically, the separate parts of the mechanism are pushed together, overcoming the force of some preloading device (e.g. a spring) that tries to push them apart. These are then tied together by a wire (made of high strength and high resistivity material). To release the mechanism, a current is passed through the wire, heating it up and ultimately melting it, thus allowing the parts of the mechanism to separate. This mechanism can be actuated in fractions of a second, depending on the characteristics of the wire.

15.5 MATERIALS

The selection of the materials for a space mechanism is possibly the single most important decision that the designer will have to make, only equalled by the selection of the lubricant. The requirements follow very closely those specified for space structures and the reader is referred to Chapter 8. Only materials that have been validated for use in space should be selected. If a new material is chosen, then the cost of validation must be accounted. See [23], which is an essential guide, and the more recently published [24].

Besides relatively standard aluminum alloys, aluminum beryllium alloys are very interesting for some mechanism applications due to their very high specific stiffness and low specific mass. However, the toxicity of this material during machining often discourages its use.

When high specific stiffness and low mass are required, carbon fibre reinforced plastic (CFRP) is also used, and here an appropriate lay up can also achieve a near-zero thermal expansion coefficient. For some high accuracy applications where thermal stability is required, more exotic materials such as invar can offer very good performance.

Of particular importance to mechanisms is *stress-corrosion cracking* (SCC). It is sometimes necessary to use materials that are not generally recommended because of this factor. For example, the bearing steel 440C, although more susceptible to SCC, is preferred over the widely used 52 100 because of its better resistance to corrosion.

Most mechanism parts are machined from the solid, which is very expensive. The small number required for a project, perhaps only four or six, usually precludes other production methods. However, some space firms have economically used the lost wax casting process for short runs.

Polymers are as widely used in mechanisms as in structures. Polyacetal and polyimide are used for small components and, with appropriate fillers, are self-lubricating. But all polymers have high thermal coefficients of expansion and low thermal conductivity which may be unacceptable, particularly when used with metals. Other materials such as ceramics are now finding application in space. Hot-pressed silicon nitride has been used for bearing balls and indeed complete bearings. Titanium carbide and titanium nitride are frequently applied to surfaces as hard coatings.

All *plating processes* must be fully evaluated before being applied to a space mechanism. Cadmium, for example, is widely used for earthbound mechanisms but is totally forbidden in space due to sublimation in vacuum.

In LEO, attack by *atomic oxygen* (see also Chapter 2) is an environmental hazard. It is particularly damaging to exposed polymers and can also attack the surfaces of metals which are sensitive to oxidation, such as silver.

15.6 TRIBOLOGY

Tribology is the science of interacting surfaces in relative motion, and encompasses lubrication, wear, surface physics and the physics and chemistry of lubricants. Thus, a good understanding of this science is essential for anyone pursuing the design of space mechanisms.

In most cases, parts of a mechanism will move relative to each other, and to facilitate this motion, and protect the surfaces against damage and wear and tear, a material (lubricant) is introduced in the 'gap' between the parts in relative motion.

Lubricants can be liquid (wet) or solid (dry), with greases placed in an intermediate category that is, in any case, subject to the same concerns as liquid lubricants and which, therefore will not be explicitly discussed here.

Dry and wet lubricants are both used in space applications and each have advantages and disadvantages. For example, wet lubricants will tend to evaporate and creep (with loss of lubricant over time) and will require a seal - however they deliver low mechanical noise. On the other hand, dry lubricants will certainly have fewer problems of evaporation and creeping, but debris in the lubricant is likely to produce noise. In addition, the generally poor thermal characteristics of dry lubricants can be disadvantageous with respect to the high thermal conductance of wet lubricants.

Wet lubricants

In practice, two types of wet lubricants are used today in space applications:

- Synthetic hydrocarbons, that are essentially multiply alkylated cyclopentanes (MACs).
- Perfluoropolyethers (PFPE).

One of the main parameters describing wet lubricants is viscosity, and currently available on the market is the MAC Pennzane (synthesized hydrocarbon fluid) which comes in different levels of viscosity (from low viscosity for low temperature applications to high viscosity for high temperature applications). Perfluoropolyethers (PFPE) e.g. Fomblin Z, Krytox, Demnum and Brayco, are particularly suitable for elastohydrodynamic lubrication, due to their high density.

During operation of the mechanism there will be loss/degradation of the lubricant and generally the mechanism will fail when all the lubricant is gone. Although various studies have been conducted on the possibility to replenish the lubricant, generally this is not an option.

Dry lubricants

There are essentially three options:

- Dichalcenogides, of which molybdenum disulfide is the most useful. It performs better in vacuum than in air because it is degraded by the presence of water.
- Solid lubricant composites based on polymers loaded with molybdenum disulfide or PTFE.
- Soft metals of which lead and gold are the most effective.

Dry lubricants are generally applied like a thin film on the surface that needs lubrication, or bonded on the surface. However, there are also cases where components can be manufactured using some self-lubricating material. For dry lubricants, it is necessary to establish the rate of wear during operation, as this will determine the life of the mechanism.

For both solid and dry lubrication, accurate life prediction using ground test data and/or mathematical modelling is generally difficult and, as the rate of consumption of the lubricant depends on the operating conditions, the extrapolation of data from accelerated life tests in the lab to the space environment can be quite inaccurate.

The European Space Agency (ESA) has recognized the importance of tribology to the reliability and integrity of space mechanisms, and has established a special facility - the European Space Tribology Laboratory (ESTL). Its main remit is to develop space tribology, and at the same time provide meticulously controlled facilities for testing space mechanisms.

The following five precepts summarize the principles to be applied for successful lubrication in space:

1. The optimum lubrication system is an integral part of the mechanism design and not a process to be added when the design is complete.
2. Avoid making the operation of the mechanism dependent upon close control of the coefficient of friction.
3. The lubrication process must be fully codified and documented to ensure consistent repeatability of performance.
4. The lubricant must be approved and validated for space to a recognized specification, and must be source traceable.
5. The test programme to determine the performance of the lubrication system in the mechanism must reproduce all the operational conditions of duty cycle, life, and both space and ground environments that it will experience in application.

The apparently excessive requirements of the fifth precept still represent the only effective means to ensure that the lubrication process will work as required in terms of performance and duration.

Good general references on this subject are [25, 26], whereas Reference [27] covers in detail all the relevant areas.

15.7   TESTING AND VERIFICATION

For a mechanism, meticulous testing is the only method of demonstrating the integrity of both the design and the build of the machine, and verifying its ability to operate over its designed life. Many faults can be revealed only by testing in an environment which simulates the conditions that it will encounter in operation. There are no short cuts. Taking an APM or a SAD as examples, vacuum chambers with the ability to create thermal cycles and thermal gradients in a clean room environment must be provided. Perhaps one of the most difficult problems during testing is to recreate the microgravity environment in which the mechanism will operate. For deployable structures, a relatively standard technique is to suspend each segment of the structure from its center of mass (using light-weight wires) which compensates for the 'one-*g*' environment. Life testing, qualification testing and testing of individual builds are also essential and will add significant cost to the programme; they could absorb as much as 25% of the budget.

For one-shot systems of the single hinge variety, it is all too easy to accept a few deployments in the laboratory as verification and then be mightily surprised when it fails to deploy in space due to temperature differentials. The remarks above apply equally to such systems. Failures of this type are still much too frequent.

Reference [28] sets out the rationale to be applied to formal testing of mechanisms, and covers also the mechanical environment tests. Information concerning thermal vacuum testing can be found in [29, 30].

15.7.1   Microvibrations characterization

Although the microvibrations emitted by a mechanism operating on the ground will be different from those emitted during operation in the typical space microgravity environment, it is usually acceptable to perform microvibration tests under lab conditions, taking appropriate measures to reduce background noise.

Most often, the experimental characterization of the microvibrations emitted by mechanisms such as reaction wheels, APMs or other 'sources' is carried out with the equipment rigidly grounded (on a multi-axis dynamometer—e.g. a *Kistler table*), by measuring the forces and moments transmitted to the support/ground.

However, to completely characterize a source to allow it to be precisely represented when mounted on a supporting structure (of unknown stiffness), the accelerations (or velocities, or displacements) at the interface should also be measured during the characterization.

Alternatively, the measurement of the forces/moments transmitted and the resulting accelerations can be carried out directly at the interface between the equipment and the spacecraft structure. This method clearly requires the availability of the spacecraft structure (or a dynamically representative model) and the results clearly will include precisely the effects of the dynamic interaction between source and mounting structure. The issue here is that the experimental results obtained can be used only in situations where the source will be mounted on a structure with characteristics similar to that used in the characterization tests. If the test results have to be used for other dissimilar structures, and if accurate results are required, then the effect of the structure that was used as support during the tests has to be analytically removed from the test results.

At low frequency ($< 50\,Hz$) the facility's noise (i.e. vibrations of the building) is typically of the same order of magnitude as the signal to be measured. Therefore the methods above require that the test equipment is put on an appropriate suspension system to isolate it from the ground noise (i.e. ground vibrations). For multi-axis dynamometers (force tables/Kistler platforms), a vibration isolation table (i.e. a typical modern optical table implementing active/passive vibration isolation) is usually a good support.

15.8 CONCLUSION

Mechanisms are fundamental spacecraft subsystems, necessary to perform operations that are crucial to the success of the mission. Currently, there are a relatively large number of mechanisms (e.g. spacecraft release mechanisms, reaction and momentum wheels, solar array drives etc.) that have reached a certain maturity in their development, and thus have relatively standard designs. However, progress in engineering that produces continuous improvements and upgrades in the mechanisms design is always necessary. Furthermore, there will always be missions that require mechanisms with completely new requirements, and their design is a challenging and exciting endeavour. This very brief look at mechanisms shows the range of skills needed for their design. Specialist knowledge in such diverse areas as motors, sensors, pyrotechnics, tribology and structures will be needed, together with a wide knowledge of materials. Testing is yet another area where specialist expertise will be required.

The application of Product Assurance (PA) to mechanisms has not been addressed in this chapter and the reader is directed to Chapter 19. The reliability and avoidance of failure in space mechanisms depends, in no small measure, upon the strict application of the principles and practice of PA throughout every stage of the design process. Meticulous attention to detail, the refusal to take anything for granted, and a carefully planned and executed test programme are fundamental ingredients for success. References [31–33] are important.

For the reader interested in expanding his or her knowledge of spacecraft mechanisms, References [34–36] are excellent sources.

ACKNOWLEDGEMENT

The author gratefully acknowledges the considerable contribution of H. Mervin Briscoe, who died in 2001, to this chapter.

REFERENCES

[1] General Specifications for Assemblies Moving Mechanical for Space and Launch Vehicle (1988) MIL-A-83577B (USAF).

[2] Labruyere, G. and Urmston, P. (1996) ESA Mechanisms requirements, *Proceedings of the 6th European Conference on Spacecraft Structures, Materials and Mechanical Testing*, Paris.

[3] Anckarman, B. and Thuswaldner, M. (1998) The Development and Testing of the ACU2624 Separation System, *Proceedings of the European Conference on Spacecraft Structures, Materials and Mechanical Testing*, Braunschweig, Germany.

[4] Humphries, M., Haslehurst, A. and Forster, D. (2007) The Design and Development of a Multi-Element Telescopic Boom (known as the SULA BOOM), ESMATS.

[5] Jenkins, C. H. M. (2001) *Gossamer Spacecraft: Membrane and Inflatable Structures Technology for Space Applications*, AIAA.

[6] Lou, M., Fang, H. and Hsia, L. (2000) A Combined Analytical and Experimental Study on Space Inflatable Booms, *IEEE Aerospace Conference Proceedings*, **2**, 503–512.

[7] Huang, J. (2001), The Development of Inflatable Array Antennas, *IEEE Antennas and Propagation Magazine*, **43**, No. 4, 44–50.

[8] Cassapakis, C. G., Love, A. W. and Palisoc, A. L. (1998) Inflatable Space Antennas—A Brief Overview, *IEEE Aerospace Applications Conference Proceedings*, **3**, 453–459.

[9] Scharfe, M. (1996) Development of a Magnetic-Bearing Momentum Wheel for the AMSAT Phase 3-D Small Satellite, *International Symposium on Small Satellites*.

[10] Scharfe, M. (2001) The challenges of miniaturisation for a magnetic bearing wheel, *Proceedings of the 9th European Space Mechanisms and Tribology Symposium*, 19–21.

[11] Samuel, A. (1996) An Overview on Aerospatiale Magnetic Bearing Products for Spacecraft Attitude Control and for Industry, *Spacecraft Guidance, Navigation and Control Systems, Proceedings of the 3rd ESA International Conference*.

[12] Gerlach, B. (2005) Gimballing magnetic bearing reaction wheel with digital controller, *Proceedings of the 11th European Space Mechanisms and Tribology Symposium, ESMATS 2005*.

[13] Pritchard, W., Suyderhoud, H. and Nelson R. (1993) *Satellite Communications Systems Engineering*, -Prentice Hall, New York.

[14] Brunnen, A. J. D. and Bentall, R. H. (1982) Development of a high stability pointing mechanism for wide application, *16th Aerospace Mechanism Symposium*, NASA CP2221.

[15] Favre, E, Brunner, C. and Piaget, D. (1999) European Electric Space Rated Motors Handbook, *8th European Space Mechanisms and Tribology Symposium*, Toulouse.

[16] Slemon, G. R. and Straughan, A. (1980) *Electric Machines*, Addison Wesley, New York.

[17] *Second International Conference on Small and Special Electrical Machines*, September 1981, Institution of Electrical Engineers.

[18] Uera, K. and Slatter, R. (1999) Development of the Harmonic Drive Gear for Space Application, *8th European Space Mechanisms and Tribology Symposium*, Toulouse.

[19] Todd, M. J. (1987) Models for steady and non-steady Coulomb torque in ball bearings, *3rd Space Mechanisms and Tribology Symposium*, ESA SP279.

[20] Gill, S., Forster, D. J. and Rowntree, R. A. (1992) Thermal vacuum performance of cycloid and harmonic gearboxes with molybdenum disulphide and liquid lubrication, *5th European Space Mechanisms and Tribology Symposium*, ESA SP334.

[21] Catalogue of European Pyrotechnic Devices, (2nd edn), (1982), ESA SP1021.

[22] Explosifs et Pyrotechnie, *Proceedings of Symposium* (1979), ESA SP144.

[23] Data for Selection of Space Materials, ESA PSS-01-70.

[24] ECSS-E-ST-32-08.

[25] Tribology International-Special Issue on Space Tribology, **23**, No.2, April 1990.

[26] Jones, W. R. and Jansen, M. J. (2008) Tribology for space applications, *Proc. IMechE Part J: Journal of Engineering Tribology*, **222**, No. 8, 997–1004.

[27] Roberts, E. W. (2002), *Space Tribology Handbook* (3rd edn) AEA Technology plc, ESTL.

[28] Briscoe, H. M. (1983) A rationale for the testing of space mechanisms, *First European Space Mechanisms and Tribology Symposium*, Act. 1983, ESA SP196.

[29] Robbins, E. J. (1975) Tribology tests for satellite application: simulation of a space environment, *First European Space Tribology Symposium*, ESA SP111.

[30] Parker, K. and Duvall, J. A. (1983) Achievement of reliability by thermal vacuum testing of spacecraft mechanisms, *First European Space Mechanisms and Tribology Symposium*, Act. 1983, ESA SP196.

[31] Basic Requirements for Product Assurance of ESA Spacecraft and Associated Equipment, ESA PSS-01-0.

[32] Failure Rates, ESA PSS-01-302.

[33] Failure Mode Effects and Criticality Analysis, ESA PSS-01-303.

[34] Conley, P. L. (1998) *Space Vehicle Mechanisms - Elements of Successful Design*, John Wiley & Sons, Inc., New York.

[35] Sarafin, T. P. (ed) and Larson, W. J. (1995) *Spacecraft Structures and Mechanisms: From Concept to Launch* (Space Technology Library), Microcosm, California.

[36] Fusaro, R. L. (1995) Space Mechanisms: Lessons Learned Study, Vol. 1 & 2, NASA-TM-107046 and 107047.

16 SPACECRAFT ELECTROMAGNETIC COMPATIBILITY ENGINEERING

Ken M. Redford

British Aerospace, Bristol

16.1 INTRODUCTION

Electromagnetic Compatibility (EMC) for a system or equipment requires that

1. It does not cause interference with other systems or equipment.
2. It is not susceptible to emissions from other systems, equipment or electrical environments.
3. It does not cause interference within itself that can cause the system or equipment to malfunction or behave in an undesirable manner.

EMC problems were first noticed on domestic radio receivers in the early part of the twentieth century. These early vacuum-tube receivers were sensitive and able to pick up interference clicks and buzzes generated by electric motors, overhead wires and connections used in electric transport vehicles.

EMC has become much more important as technology progresses. More complex electronics performing many new functions are packed into smaller and smaller enclosures.

This trend is particularly true for space vehicles. The wide variety of communications, weather, robotic, remote sensing and scientific spacecraft, and more recently, small satellites launched as constellations, all pose complex EMC problems.

The trend has caused spacecraft contractors to impose stringent EMC requirements on both spacecraft and ground support equipment.

Spacecraft Systems Engineering, Fourth Edition.
Edited by Peter W. Fortescue, Graham G. Swinerd and John P. W. Stark.
© 2011 John Wiley & Sons, Ltd. Published 2011 by John Wiley & Sons, Ltd.

EMC provisions for all electrical and electronic equipment have now become law in all EEC member countries. It is illegal to sell any electrical or electronic product without a 'CE' mark certifying that the equipment meets the required European EMC standards.

Further reading on EMC can be found in References [1] and [2].

16.2 EXAMPLES OF EMC PROBLEMS

In domestic situations, most people have experienced interference from car ignition systems and on TV and radio reception. The recent discussions on the safety of operating mobile phones close to the brain and whether close proximity to overhead power lines affects health are all examples of EMC problems where one system could cause another to malfunction or behave in an undesirable manner.

EMC is fundamental to the military sector since interference can directly affect the operation of military equipment and is indeed used as a weapon in Electronic Warfare.

Fortunately, EMC problems in spacecraft are relatively rare since major contractors are careful to make early EMC provisions on all projects. Most EMC problems encountered on spacecraft result in a slightly degraded performance but some have had more serious effects. Electrostatic Discharge (ESD) problems between thermal blankets and the spacecraft structure on early maritime communications satellites were solved by careful grounding. The interference from these tiny spark discharges was sufficient to cause telemetry status latches monitoring the power subsystem to flip over. This initiated shedding of the payload communications power until reset by ground.

These examples illustrate the importance of EMC provisions in all electronic applications.

16.3 EMC SPECIFICATIONS

EMC Requirements Specifications are derived and written for each spacecraft depending upon its mission and they generally contain

- Spacecraft level requirements and derived subsystem/unit requirements broken down into the EMC categories described later.
- Details of how particular requirements are to be verified are also specified. This verification is done either by Inspection/Analysis or Test, and an indication is given of which tests are to be carried out at spacecraft/subsystem or individual electronic unit levels.
- Design guidelines are sometimes included to address particular EMC problems. For instance, if a scientific spacecraft is attempting to measure the magnetic field environment of a planet with a sensitive magnetometer, then there may be limitations on the use of magnets or magnetically permeable material in the spacecraft build.

EMC test methods are generally based on the American Military Standard MIL—STD-461 [3]. This is widely used in both Europe and the USA as a military standard for both ground-based military and spacecraft electronic hardware. The standard covers EMC requirements and test limits for electrical, electronic and electromechanical equipment, subsystems and systems and establishes techniques to be used for measurements.

EMC requirements for ground support equipment used for testing at spacecraft/subsystem and unit level are sometimes covered in separate requirements documents.

16.4   ELECTROMAGNETIC COMPATIBILITY – TERMS AND DEFINITIONS

16.4.1   Common terms used

Commonly used terminology in the EMC area is now introduced.

EMC – Electromagnetic Compatibility

This is a general term addressing all categories of electromagnetic emissions and susceptibility. It is generally split into four areas:

- *Radiated emissions* measures all electric or magnetic fields emitted from the equipment.
- *Conducted emissions* measures all power, analogue and digital signals that are conducted *via* cables, harnesses and structural parts of the spacecraft.
- *Radiated susceptibility* measures the ability of the spacecraft to operate satisfactorily when placed in an environment that contains externally generated electric or magnetic fields.
- *Conducted susceptibility* measures the ability of the spacecraft to operate satisfactorily when externally generated conducted interference signals are directly injected into the spacecraft's harness cables or structure.

EMI – Electromagnetic Interference

This is a general term referring to conducted and radiated EMI produced by a system.

RFI – Radio Frequency Interference

A more specific version of EMI referring to emitted radio frequency electric fields—generally in the Radio frequency (RF) bands from 3 KHz up to 3000 GHz.

ESD – Electrostatic discharge

An ESD will occur if two equipments or systems that are electrostatically charged at different potentials are brought together.

A lightning strike on a spacecraft launch vehicle or a spark ignition on a car engine are both examples of ESDs. ESDs to the human body can sometimes be felt when touching electrostatically charged car bodies.

Two main effects have to be considered when an ESD occurs:

1. The effect of the spark discharge that generates *radiated* electric and magnetic fields.

2. The effect of the resultant *conducted* discharge current through the equipment or system. This discharge can rise to hundreds or sometimes thousands of amperes flowing at the ESD point in a very short period.

EMP — Electromagnetic Pulse

This is the intense electromagnetic wave produced when a nuclear detonation occurs. It is characterized by extremely high electric and magnetic fields occurring in an extremely short period.

It was first noticed about 1943 when electronic equipment used to monitor the first atomic bomb explosions were destroyed by the EMP.

Nuclear 'hardening' has since become a significant requirement for all strategic military equipment, government communications equipment and critical electronic systems.

16.5   EMC FUNDAMENTALS

16.5.1   The basic EMC problem and its solution

All EMC problems have these aspects in common:

1. a source or transmitter that produces emissions,
2. a receiver that receives the emissions,
3. a transfer or coupling path between transmitter and receiver.

Interference occurs if the received signal causes the receiver to misbehave in some way.

There are three possible means to reduce the interaction between the transmitter and receiver.

1. Reduce the transmitted emissions.
2. Alter the coupling path between the transmitter of interference and the receiver by physical separation.
3. Make the receiver less susceptible to the interfering signal.

Clearly, reducing the transmitted emission is the most desirable since this is the cause of the problem. However, this is not always possible.

For example, the prime function of a telemetry transmitter on a spacecraft is to generate an RF signal and to send it to Earth *via* an antenna on the spacecraft. The transmitted power is governed by the link budget and the performance of the spacecraft and ground systems. Primary RF emissions cannot, therefore, be reduced for EMC reasons.

In contrast, a Switch Mode Power Converter also generates RF emissions. However, its primary purpose is usually to convert direct current (DC) power to different regulated voltages. It is, therefore, very important, and possible, to try to reduce the secondary RF emissions from the converter, even if this makes the converter slightly less efficient.

Altering the coupling path between transmitter and receiver by physical separation can be effective in some cases. An example of this can be found on the Ulysses Spacecraft. One of its functions is to measure the magnetic environment around the polar regions of the Sun. The Magnetometer sensor is mounted on a 5.6 m radial boom to minimize magnetic interference from the spacecraft body.

Finally, the third option of making the receiver less susceptible to the transmitted interference is almost always possible. Examples of this are filters on spacecraft electronic units' interfaces to eliminate conducted interference from pulses on power and signal lines on the spacecraft.

16.6   THE SYSTEMS APPROACH TO EMC

One of the most important aspects of EMC is that it must be considered as an integral part of the specification, design, manufacturing and testing phases of any spacecraft.

A 'top down' systems approach to EMC must be adopted at the same time that the basic requirements of the spacecraft are defined and specified. This involves examining each aspect of the spacecraft mission requirements, considering how each EMC category is influenced and deriving detailed EMC requirements and specifications.

System margins and budgets are attached to EMC performance in the same way as power budgets are applied to Power Systems or pointing budgets are allocated to Attitude and Orbit Control Systems.

16.6.1   Safety margins for EMC

These margins are defined as the difference between system susceptibility levels and the level of system emissions.

For instance, the electronics units mounted on a spacecraft platform will be required to perform their required function in the presence of RF emissions back scattered from the telemetry transmitter antenna sending signals to Earth. The systems designer may have calculated the worst case field strength *emission* from the transmitter at any location on the spacecraft as, say, 1 V/m. However, in the *susceptibility* specification for other electronic units the designer may impose a requirement that the 'units shall not be susceptible to a field strength of 10 V/m'. The designer has then assigned a 20 dB ($\times$10) safety margin between the *emission* and *susceptibility* specifications.

The size of this margin may reflect

1. the uncertainty in the designer's calculations,
2. that there are perhaps several 'sources' of interference from several equipments that can 'add up' at system level,
3. the criticality of the subsystem or equipment.

Generally, the size of the margin between susceptibility and emission is at least 6 dB ($\times$2) but it could be as high as 20 dB for safety critical systems such as pyrotechnic release mechanisms or military systems.

16.7   EMC CATEGORIES

As introduced in Section 16.4.1, EMC categories can be split into *radiated* and *conducted* main areas, each of which can be further subdivided into *emissions* and *susceptibility* as shown in Figure 16.1.

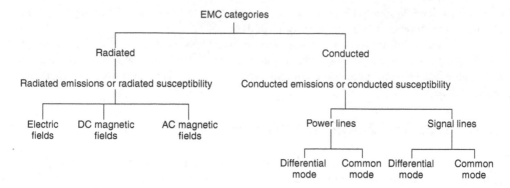

Figure 16.1   EMC categories

16.7.1   Radiated emissions

Radiated emissions are electric fields and AC or DC magnetic fields, which are emitted from either the subsystem units, or from the harnesses and cables that interconnect the subsystems.

Electric fields

These are radio waves generated by RF transmitters or signals radiated from the wires in electronic units and associated interconnecting harnesses acting as small antennae. They are formally defined as sinusoidal RF radiations in volts per metre over a frequency range 3 KHz to 3000 GHz. On most spacecraft, they are usually of concern only in the 15 KHz to 30 GHz region.

RF emissions are measured on a Spectrum Analyser using a variety of different antennae to pick up emissions. The equipment under test and the antennae are enclosed in a screened room to avoid external EMC influences.

Electric field emissions can be reduced as follows:

- Adopting a good grounding and screening philosophy for the equipment or subsystem under test.
- Choosing the slowest digital and analogue technologies consistent with the mission objectives—particularly on interface circuits between subsystems and electronic units.
- Effective shielding of all cables, harnesses and connectors both inside units but, more importantly between units. This is done by the use of screened wires and connector shrouds, coaxial cable, twisted wire, overall shielded braids on wires, wire bundles and so on. Almost all Electric Field emission problems can be minimized by adequate shielding and grounding of harnesses, cables and connectors between units and subsystems.
- Effective shielding and grounding of all electronic units by encasing all units in metal screened boxes or enclosures. Holes or apertures in units should be minimized and lid-fixing screws should be close together to avoid apertures. In extreme cases, RF

gasketing material can be used between adjacent metallic parts (e.g. lids/access parts) although this is only effective in the higher frequency ranges above a few MHz.

- Metal parts/panels should be electrically bonded together—giving typically less than 10 mohms resistance between adjacent parts.

DC magnetic fields

These are fixed DC magnetic fields, usually measured in units of *PicoTeslas* at 1 m distance. They do not vary with time and are produced by permanent magnets or DC currents flowing in any circuit. Their magnitude is important, and should be minimized on scientific spacecraft that use magnetometers to measure magnetic field environments. It is impossible to construct a spacecraft that does not have some level of DC magnetic field, and so if sensitive measurements are required, the magnetometer sensors are generally mounted on booms several metres in length, away from the spacecraft body.

DC Magnetic Fields can be reduced by

- minimizing the use of ferromagnetic or permeable materials,
- reducing DC currents and minimizing the loop area around which they flow.

In practice, this cannot be perfectly achieved. For instance, transformers, inductors and relays rely on ferromagnetic material. Flow control valves and latch valves controlling gas jets for attitude control purposes contain magnetic material. Even the mounting cans and connecting wires on integrated circuits and transistors are made of magnetically permeable nickel alloy.

Compensating magnets, or magnetic screening using 'Mu-metal' alloy material, can be used with some success in critical applications. DC magnetic fields can be further reduced by 'de-perming' the electronic units. This generally involves placing the unit in a coil that is energized to produce an AC magnetic flux density and slowly reducing this field to zero.

AC magnetic fields or B fields

These are alternating magnetic fields that vary with time (AC) and are produced by inductive circuits such as transformers, inductors, and, indeed, any wire loops or circuits in which alternating current flows. They are also measured in *PicoTeslas* at 1 m distance. AC magnetic fields are mostly of concern in the 50 Hz to 100 KHz frequency region. Above 100 KHz, electric fields tend to become more important than magnetic fields.

AC magnetic fields can be reduced by adopting the following methods:

- Reducing the magnitude of AC currents in the equipment and harnesses.
- Reducing the loop area around which these AC currents flow. If this is not possible, then cancellation or screening techniques can be used—although usually only with limited success. Some reduction can also occur if cancellation techniques or 'shorted turns' are used. A copper strap around the outside of a transformer body is an example of 'shorted turns' that do not affect basic transformer action. However, any stray external magnetic field present will induce currents in the strap that will, in turn, produce a compensating magnetic field.

Magnetic Screening with Mu-metal can sometimes be effective but only at frequencies less than about 1 KHz. Magnetic Fields tend to be very directional and some EMC problems can be minimized by simply physically moving or rotating the offending source of interference.

16.7.2   Radiated susceptibility

Units or subsystems can be caused to malfunction or to be *susceptible* to the presence of radiated AC or DC magnetic or electric fields.

These externally applied fields can be picked up on the harnesses and cables between subsystems or in internal wiring of the units themselves. These induced signals can produce conducted signals in the units and cables that can cause circuit functions to fail momentarily. In extreme cases, electrical interfaces can be permanently damaged.

Susceptibility testing is essentially the reverse of testing for emissions. RF power, from generators that can be swept over the required frequency range, is fed to a variety of antennae to radiate the equipment or spacecraft under test. Field strengths at the equipment under test are generally measured first. Then the equipment is switched on in its most sensitive mode to detect any susceptibility.

16.7.3   Conducted emissions

These are generally split into emissions from power lines and signal lines and are further categorized into *Differential Mode* or *Common Mode* signals as shown in Figure 16.1.

Noisy circuits and components inside a subsystem can cause conducted emissions to be present on both signal and power lines between units on the spacecraft. These emissions can also be conducted to ground *via* the chassis, case or screens in cables between units.

Measurements are made directly using voltage or current probes monitoring conducted signals, and these can be displayed on an oscilloscope or spectrum analyser.

The difference between a *differential mode* and a *common mode* signal is now explained. Figure 16.2 shows a typical signal interface between subsystems in which, for example, digital data is being sent from Subsystem 1 to the data-handling subsystem (Subsystem 2) for subsequent transmission to ground.

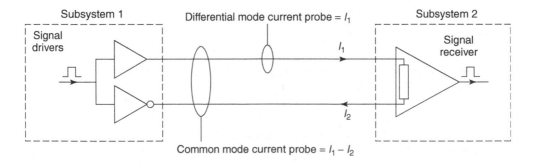

Figure 16.2   Differential mode and common mode signals

Most power, analogue signals and digital signals are transmitted between subsystems on 'two wire' interfaces since any common interference present is equally induced on both wires. This is termed *'common mode noise'*. The signal sent on one wire is designed to be complementary to the other, that is, if a logic '1' digital signal level is sent on one line then a logic '0' is transmitted on the second line. The interface circuit receiving the signal or power is designed only to look at the relatively interference-free signal *between* the two wires to detect the logic state or signal transmitted. The *common mode current* between the two subsystems is the algebraic sum of the currents flowing in the two wires. Two-wire interfaces are also commonly used for analogue signals and for distributing power around the spacecraft, using the same principles.

In the diagram, the *differential current* can be measured using a current probe around either wire and measures either I_1 or I_2. *Common mode current* is measured with a current probe around both wires so that the net forward current measured is $(I_1 - I_2)$. Similar definitions apply to *differential and common mode voltage* signals between interfaces.

16.7.4   Conducted susceptibility

Units or subsystems can be caused to malfunction or to be susceptible to the presence of conducted signals on the spacecraft harness or structure.

These externally applied voltages and currents are injected directly into the harnesses and cables between subsystems or directly on the unit's connector pins. These signals can cause a temporary malfunction, commonly called a 'glitch', or in extreme cases, electrical interfaces can be permanently damaged.

Susceptibility testing is essentially the reverse of testing for conducted emissions.

Sinusoidal and pulse signal generators, which can be swept over the required frequency range, are used as signal sources. These signals are power amplified and injected *via* current probes, transformers or voltage probes directly into the equipment or spacecraft under test. Applied signals are generally measured with an oscilloscope before the equipment is switched on, in its most sensitive mode, to detect any susceptibility.

16.8   ELECTROSTATIC DISCHARGE

This is similarly categorized into two areas, radiated and conducted.

- When an ESD occurs, there is always a 'spark', for example, lightning. The spark discharge occurs between two bodies or parts of a spacecraft—for example, a human body and a car door handle. The occurrence of the spark can generate a very large but extremely short duration burst of electric and magnetic *radiated emissions* over a wide frequency band. It is easy to confirm this by listening to any radio set tuned to the Medium or Long Wave bands during a thunderstorm. If any equipment, harnesses, wires or cables are close to this spark discharge, these fields will induce currents and voltages in the wires and equipment. This can cause either a temporary malfunction or, in some cases, permanent damage to equipment interfaces.

Testing for ESD radiated susceptibility involves generating a spark discharge of known energy usually 30 cm from the equipment and its connecting harnesses, and monitoring the equipment for any temporary or permanent malfunction.

- The secondary effect of an ESD is the *conducted current* and path of the ESD. The magnitude of this current can be up to 50 000 A with rise times in the order of nanoseconds to microseconds. If this conducted current is applied at equipment interfaces, or even discharged through an electronic unit's case, then this can cause either a temporary malfunction or permanent damage to the equipment.

Testing for ESD conducted susceptibility generally involves directly injecting current pulses from spark discharges, or directly from amplified pulse generators, into the equipment structure, harnesses and connector bodies and monitoring for any susceptibility.

The proximity of charged particles in the environment around any spacecraft can cause charge to build up on any isolated conductive surface. An account of this on-orbit charging mechanism can be found in Reference [4]. ESD problems can be eliminated by avoiding this charge build-up by grounding and bonding all parts of the spacecraft to the structure.

The importance of ESDs has increased as technology progresses. Vacuum tubes had an extremely high immunity to ESDs. Semiconductors used in spacecraft, particularly the high impedance metal oxide silicon (MOS) devices, can be very sensitive to even the smallest ESDs caused simply by handling the device without the appropriate precautions. Discharges of about 3.5 kV can occur from the fingers of an engineer and they are of such a short duration that they cannot be seen or felt. However, these discharges can easily destroy sensitive semiconductor devices, some of which are susceptible to voltages as low as 50 V.

For this reason, spacecraft clean rooms and manufacturing areas take special precautions and use grounded conducting floors, benches and chairs to minimize any possible charge build-up. Wrist straps are also used to connect personnel to ground during handling operations.

16.9   SPACECRAFT GROUNDING SCHEMES

Several fundamental questions need to be asked during the initial design phases of any spacecraft:

- What types of digital and analogue signals pass between the subsystems of the spacecraft, between subsystems and payloads, and what are the data transmission rates and bandwidth requirements of each of these signals?
- How much power is supplied by the spacecraft power subsystem to different parts of the spacecraft and payload and in what form? For example, are there direct supplies at main bus voltage levels in either regulated or unregulated form, or converted and regulated supplies distributed at lower or higher voltage levels?

Answers to these questions may perhaps already be known, since they may be customer-specified, or perhaps earlier studies may have answered them. For instance, data rates between the scientific payload of an (earth resources satellite) and the data-handling subsystem telemetering data to ground may be hundreds of Megabits per second. In contrast, communications satellites may only be telemetering a small amount of background housekeeping data.

Similarly, spacecraft with active microwave payloads such as a synthetic aperture radar (SAR) may consume kilowatts of power and would therefore be *directly* supplied by main bus power at voltages generally in the range of 28 to 120 V. Small satellites may supply payloads and subsystems with common *converted* and regulated supplies at, perhaps 5 V and 12 or 15 V levels.

A *Spacecraft grounding scheme* defines how all analogue and digital data signals and power are distributed and referenced on a spacecraft. All systems have a reference point or points to which all electrical signals and power is referred. This is usually called '0 V', 'earth', 'common', 'ground' or 'chassis'.

On any spacecraft, all these common points would be at the same potential, in an ideal world, regardless of whether signals in the system are 'Direct Current' (DC), (e.g. power buses), or Alternating in nature (AC) such as digital data signals between subsystems. In practice this is not true. Two points 'grounded together' with a direct wire connection, or perhaps *via* a conducting spacecraft structure, will not be at the same '0 V' potential. It depends on the currents that pass along the wire or structure, and their frequencies.

For example, if a DC current of 1 A is flowing in a 24 gauge ground wire of 10 cm length, there will be a very small potential difference between the two wire ends of, perhaps, 10 mV—the wire resistance would be 10 mohm from Ohm's Law. However, if AC signals are flowing in the ground wire at a frequency of 10 MHz, the wire will have a significant impedance—perhaps 10 ohms—due to the *inductance of the wire*. A much larger potential difference will exist if 1 amp flows in this wire at 10 MHz, in this case 10 V. Therefore, the two points apparently 'grounded together' with this wire for DC currents are not at the same potential at this frequency due to the wire inductance.

This example shows how essential a grounding scheme is to define all interfacing and grounding, and to pass only defined signals and power along defined paths or interfaces. This will minimize cross talk between signals (or power supply voltages) by minimizing common paths for the signals and power.

Computer calculations or estimations can also be made of the potential (DC) or noise (AC) at any point in the system with respect to a spacecraft ground point. A grounding scheme diagram is an essential aid to achieving these objectives.

There are three common ground schemes used on Spacecraft:

- Single-Point Grounding (SPG) Scheme.
- Multipoint Ground (MPG) Scheme.
- A Hybrid SPG/MPG Scheme.

16.9.1 Single-point ground (SPG) scheme

The example of an SPG scheme showing power and signal connections between two subsystems on a spacecraft is shown in Figure 16.3.

An SPG scheme has the following features:

- There is only one 0 V reference point to which all power and signals on the spacecraft are referred.
- Main Bus Power is fed to each subsystem by a 'twisted pair' set of wires.
- All signal and return wire currents pass down wires and therefore, no currents appear in the spacecraft structure. This is important particularly on some scientific satellites

(e.g. the European GEOS and Ulysses spacecraft), which contained plasma wave detecting instruments sensitive to interference from spacecraft structure noise.

- Note that the 0 V lines are grounded to the 0 V reference point at only one point on the spacecraft. This connection is usually made close to the Power Subsystem but can sometimes be located near the Data-Handling Subsystem.
- The single-ended digital signal transmitted between subsystems will be on slightly different logic '1' and logic '0' levels since points A and B will be at slightly different potentials. The signal current path shown indicates that both points A and B, in Figure 16.3, will have 'AC' signal frequency noise present because of the signal current flowing, and the 'AC' impedance between the 0 V point and A, and between the 0 V point and B. In practice, this is not a serious problem since logic chip integrated circuits have built-in 'noise immunity'. However, if many signals at very high frequencies are passing between the two subsystems, then the noise at these points with respect to the reference 0 V level could be sufficient to cause problems. This noise can cause distortions in the shape of the rise and fall time edges of digital signals. This gives rise to 'glitches' or interference pulses on signals between subsystems as the noisy signals cross and re-cross the logic '0' and '1' threshold levels. On modern spacecraft, these problems are invariably overcome by the use of differential drivers and receivers as shown in Figure 16.2 above, or by the increasing use of opto-coupled interface circuits. These pass information by switching infrared signals on and off. Opto-couplers, therefore, eliminate the flow of *electrical* signal currents between subsystems and are immune to noise on ground points between subsystems. However, they are usually not as fast for data-transfer purposes as conventional interface circuits.

A similar situation exists when analogue signals are passed between subsystems—for example, analogue telemetry signals monitoring currents and voltages, or those monitoring temperatures using thermistor sensors. Two wire interfaces are commonly used although

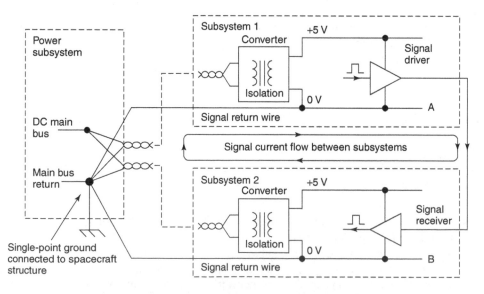

Figure 16.3   Single-point ground (SPG) scheme

they are often grouped with common return lines to minimize wires between subsystems and reduce harness mass.

In general, screened/twisted pair cables are used in the harness between units for both analogue and digital signals to minimize radiated emissions and susceptibility problems.

The main disadvantage of the SPG scheme is that, for large systems, the ground leads can be numerous and long, making the harness quite heavy. Although ideal for low frequency systems or distributing DC power, this can sometimes create problems since at higher frequencies, signal currents flowing in long wires can cause significant radiated emissions. These problems can be solved by using a MPG Scheme.

16.9.2 Multipoint grounding (MPG) scheme

An MPG scheme grounds all signal and power wire returns locally to a common ground plane. This ground plane has very low inductance between any two points on its surface and so ground currents flowing, even at very high frequencies, will not produce significant noise above ground potential on any part of this plane.

The ground plane is a flat, thin and wide conductor preferably made of plated copper. However, it could be any large, flat, and thin conductive surface—like the aluminium honeycomb structure used for many spacecraft platforms. This geometry gives a very low inductance between any part of the ground plane.

Separate flat, wide and thin conductive strips made from plated copper are used as a ground plane for those spacecraft that have carbon fibre structures, since carbon fibre has a comparatively high electrical resistance compared to aluminium.

If many signal interfaces are grounded locally to this ground plane, then many signal currents will flow in the ground plane but the effect of these is negligibly small since the inductance of the plane is so low—that is, 'Almost ideal' ground points are created throughout the spacecraft.

Figure 16.4 shows an example of an MPG scheme.

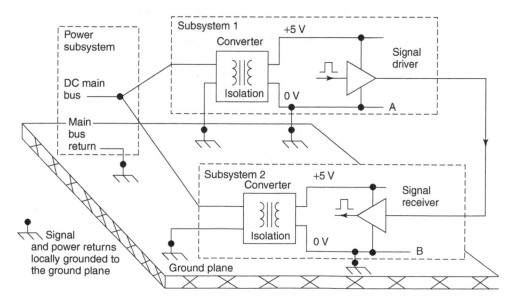

Figure 16.4 Multipoint ground (MPG) scheme

The grounds of logic chips on Subsystem 1 are closely coupled to those of Subsystem 2 by direct connection locally to the ground plane. This will allow interface circuits to operate at much higher frequencies (>30 MHz) using faster logic families. Note that it is still good practice to use differential drivers, receivers and screened cables and, if possible, opto-coupled interfaces between subsystems.

There may be several ground planes in a spacecraft, some inside the individual electronic units in each subsystem. For example, there may be separate planes for logic signals, sensitive analogue signals and high current digital drives for motors and solenoids. These precautions are taken to minimize cross coupling between radically different systems handling different signals. For example, the logic family ground plane may be handling 0 to 5 V logic signals with several milliamps of current in the ground plane. In contrast, the separate ground plane for motor drive or switch mode power converter circuits may be handling 50 V signals with 1 amp spikes induced in the ground plane.

16.9.3 Hybrid SPG/MPG scheme

It is common on spacecraft to adopt a hybrid-grounding scheme to take advantage of the best of both systems. For example, a typical strategy may be an MPG scheme for all digital signal transfers with separate ground planes for, say, high current digital drives, and an SPG scheme for the distribution of power and analogue signals. The SPG point and the many MPG points are shown in Figure 16.5.

This gives the advantage of the good high frequency performance of the MPG scheme together with the excellent isolation and elimination of ground loops for low frequency/analogue or power supply lines of the SPG system.

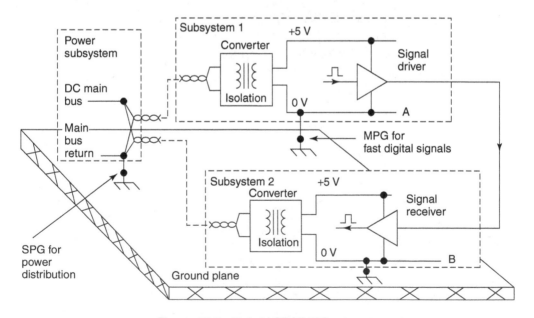

Figure 16.5 Hybrid SPG/MPG scheme

16.10   MAJOR CAUSES OF SPACECRAFT EMC PROBLEMS

16.10.1   Power supplies

Power supplies, particularly Switch Mode Power Converters, are usually major causes of EMC problems on any spacecraft. These generally convert main bus DC supplies down to other regulated voltages for use in subsystems, payloads and individual electronic units. They operate by first converting the DC input voltage into a rectangular AC signal using transistor power switches operating at frequencies up to several MHz. The AC signal produced is then applied to a transformer to produce several different AC voltages, before converting it back to new DC levels. A typical converter in a Data-Handling Subsystem unit would generate one 5 V and two 12 V supplies from one 28 V DC main bus input.

The speed of operation of the switching transistors and the use of magnetic components, combined with high voltages and circulating currents, gives rise to electric and magnetic field emissions at the fundamental switching frequency, and most of its harmonics. RF radiation can extend into the tens or even hundreds of MHz regions if adequate precautions are not taken.

The most effective way of reducing radiated interference from these converters is to reduce the radiations at source by slowing down transistor switching speeds. Reduction of transient voltages and currents, which are inevitably associated with the conversion process, is also necessary. In addition, close attention must be paid to the printed circuit board, unit layout and screening provisions as described in Section 16.7. Early testing on prototype models will identify any problems, since circuit layout problems are very expensive to fix after flight models of equipment have been manufactured and delivered.

Conducted interference currents, either to chassis or to other wires on the spacecraft, are also significant in these converters. Fast transistor switching times and large voltage handling can cause currents to be coupled, *via* stray capacitance effects, into the spacecraft structure or other wires. Since capacitance is inversely proportional to proximity, these problems are exacerbated by the trend towards smaller and smaller power supplies. Close coupling of heat generating components to chassis, to minimize thermal problems, can also cause these problems. In addition, stray capacitance between transformer windings can cause interference to pass back along main bus power leads or out to the converter loads.

Copper foil shields between windings can reduce these problems. Also, a careful choice of inductive and transformer components can reduce magnetic field leakage problems.

Diodes are used to convert the high frequency AC signals back to DC. The fast on-off switching of these rectifier diodes can also cause some problems by producing radiated emissions at harmonics of the converter switching frequency. Special circuits called 'snubbers' (usually a capacitor and resistor across each diode) can slow down the switching times of diodes to reduce this problem.

16.10.2   DC motors and actuators

The inductive nature of motors and actuators, the pulse width modulated nature and fast rise/fall times of the drive signals mean that these components have very similar EMC problems to switch mode power supplies. The resultant voltage and current waveforms and their harmonics can cause both radiated and conducted interference. Similar avoidance techniques are required.

16.10.3 Harnesses and cables

A significant percentage of all radiated and conducted emissions and susceptibility problems appear to be caused by the spacecraft harness. However, the harness is completely passive and cannot, by itself, either generate or be susceptible to conducted or radiated signals. It can radiate emissions and conduct electrical signals that are placed on the harness by electrical transmitters and receivers located at either end of the cable.

Electric and magnetic field emissions from harnesses and also, susceptibility to these signals can be minimized by

- Adopting a suitable grounding and bonding scheme for the systems.
- Partitioning and physically separating harnesses into power, pyrotechnic, noisy (digital) and quiet (analogue) signals.
- Paying careful attention to interface circuit designs, and where necessary, using matched drivers, transmitters and cables such that cables look like transmission lines.
- Matching the required electrical characteristics of each transmitted signal with the capabilities of the associated driver circuit, the harness wire characteristics, and receiver interface circuit properties. For example, for an interface transmitting digital signals between subsystems, this means slowing down logic signal rise and fall times as logic levels change from '0' to '1' and *vice versa*. It also means designing the slowest driver and receiver interface circuits, which, together with the harness wire characteristics, will successfully transmit these signals. Slowing down these signals, and also reducing the voltage and current levels used, prevents unwanted noise and associated harmonics appearing on the harness wires for subsequent electric and magnetic field radiation. Special filter connectors can also be used to slow down or filter signal and power lines.
- Using individual screened or twisted cables between signal interfaces, and making electrical connection to any screen ends *inside* metal backshells on the connectors at either end of the interface. In critical cases, overall shielding or braid can be fitted and electrically bonded with conductive tape to conductive backshells on connectors.

Conducted emission on harnesses depends critically on interface circuit designs and the overall grounding scheme adopted for the system or equipment.

16.11 ANALYSIS METHODS FOR SPACECRAFT EMC ENGINEERING

Computer analysis of radiated electric and magnetic fields is notoriously difficult since it relies heavily on accurate physical models of the environment. It is difficult also to predict with precision the nature of the frequency and time-domain interference that occurs. This is the reason most spacecraft contractors rely on preventative measures during the design phases, and early EMC testing and analysis of test results to characterize and identify problems early in the spacecraft development.

Analysis of conducted interference, using well-tried analysis software such as PSpice^R [5], is much more successful and is widely used. The input and output response of filters on power lines, interface circuits on both digital and analogue lines, and grounding diagrams are easily and accurately modelled and predicted using these computer programs. In addition, it is easy to generate computer simulations of a wide variety of interfering sinusoids and pulse signals to test the conducted susceptibility of all conducted power and digital/analogue interfaces. Differential responses to interfering signals can also be accurately predicted.

However, care must be taken when trying to predict the common mode responses of interfaces. Correct simulation relies heavily on correct models of stray capacitances and couplings of harnesses to each other and to the chassis.

Care must also be taken at frequencies above the 10 to 100 MHz region, since computer models of particular components may not be representative at these frequencies and beyond. A good example of this is the response of low frequency analogue circuits to RF radiation in the GHz region. Unless adequately screened, these RF fields can make small changes to the DC characteristics and offsets of the integrated circuits used. This produces unexpected changes in analogue circuit performance.

REFERENCES

[1] Clayton R. Paul (1992) *Introduction to Electromagnetic Compatibility*, John Wiley & Sons, Inc., New York.
[2] Morgan, D. and Peregrinus, P. (1995) *A Handbook of EMC Testing and Measurement*, IEEE Press.
[3] Requirements for the Control of Electromagnetic Interference Characteristics of Subsystems and Equipment, MIL-STD-461F, December 2007, USA Department of Defense Interface Standard.
[4] Hasting, D. and Garrett, H. (1996) *Spacecraft-Environment Interactions*, Cambridge University Press, New York.
[5] PSpice^R, Cadence Design Systems, Inc., San Jose, CA, USA, Web address: www.cadence.com.

17 ASSEMBLY, INTEGRATION AND VERIFICATION

Terry Ransome

EADS Astrium, Stevenage (Retired)

17.1 INTRODUCTION

The preceding chapters show that many engineering disciplines work together to produce a spacecraft design that is capable of fulfilling specified mission objectives. Here we discuss *AIV* (*Assembly, Integration and Verification*)—the engineering discipline that verifies to a very high level of confidence and probability that the hardware will perform the desired mission. The AIV process includes the assembly of many components, parts and manufactured subsystems, followed by the integration and test of the flight spacecraft. This is the most visible part of the hardware stage of the programme and is generally known as *AIT* (*Assembly, Integration and Test*).

There are however many other processes and tasks that contribute to the verification objective. Verification processes are at work from the beginning of the design stages. First we will look at the concepts and terminology used in the field of AIV. We will then show how, by careful planning, the verification (and test) activities are incorporated into the overall spacecraft programme—not just between manufacture and launch, but from the beginning of the project development phase. We will see how Verification Planning determines or influences programme logistics—the number of hardware items needed to fulfil verification objectives, the provision of test facilities and test equipment, and how verification tasks occupy a significant proportion of the project schedule.

17.2 SOME DEFINITIONS

An understanding of the following terms and concepts will help the understanding of the verification process.

Assembly—the process of mechanically bringing together hardware components, large and small, from many suppliers to build up the deliverable spacecraft. They range

Spacecraft Systems Engineering, Fourth Edition.
Edited by Peter W. Fortescue, Graham G. Swinerd and John P. W. Stark.
© 2011 John Wiley & Sons, Ltd. Published 2011 by John Wiley & Sons, Ltd.

from spacecraft structures and propulsion modules, spacecraft subsystem electronics and payload sensors and data processors, down to cables, thermal blankets, bolts and washers.

Approved and proven processes and specifications are used in the assembly phase—for example, use of adhesives, bolt torquing, and application of paints and surface finishes.

Integration—the process of physically and functionally combining component equipments, in a controlled and measured sequence. Put simply this means making sure it all works together. Each and every interface between two components needs to be shown to be correct. Does the power go to the right place? Is the output voltage of unit A what unit B expects? Does unit B draw the anticipated current? Are the individual data measurement points encoded and sampled into the correct telemetry packets?

Integration covers more than electrical signals though. We need to check, for example, that fields of view of sensors and deployment envelopes are free of obstruction (with some margin). Furthermore, integration includes verifying the system-level software/hardware interface. These might appear to be simple issues that should have been resolved early in the design stage. However, when components are delivered from a number of different contractors, errors and misunderstandings can, and do, occur. This is the time to discover interface discrepancies.

The integration phase ends with a known *functional* configuration.

Verification—the total process by which conformance to all applicable *performance* requirements are demonstrated. Verification should be seen as the sum of two main objectives:

- The first is *Qualification*—demonstrating that the spacecraft design is fully capable of meeting all the applicable requirements, i.e. that it is suitable and adequate for the planned mission with proper margins. That means that *qualification tests* (see later) include an element of 'over test' relative to the flight environment.

  Qualification does not necessarily need the flight hardware—any hardware properly built to the flight design standard can be used in the qualification process. Qualification can be transferable. That is, a design that has been previously qualified, especially if it has been flown, is automatically qualified for a future mission—provided the new mission environments are no more severe than the earlier mission. Conversely, the same design proposed for a new environment must be re-qualified.

- The second objective is *Acceptance*—demonstration that the end product, the *flight hardware*, is free from workmanship and materials defects, that no errors have been introduced during assembly and integration. A most important prerequisite to acceptance is confirmation that the flight hardware has been built to the qualified design.

Different methodologies may be applied to different disciplines and requirements, at different stages in the build programme. Often they are used in combination with each other. There are four principal *verification methods*:

- *Test*—the preferred method of verification which involves (a) the stimulation of hardware with inputs representative of an outside influence, or simulation of an outside environment (e.g. vibration, temperature, light or radiation sources) and (b) the measurement of the response in the form of an electrical signal (from the spacecraft or from non-flight sensors fitted temporarily), physical motion or telemetry data. Each test result is expected to be within a specified range of values. Often we are looking for the absence of a response to a stimulus (things that should *not* happen).

The test phase also includes the taking of *measurements*—for example determining the mass properties of the spacecraft (mass, centre of gravity), and pointing directions by measuring alignment parameters.

- *Analysis*—if verification by test or other means is not physically possible or prohibitively expensive, analytical techniques—mathematical modelling and computer simulation—are used. A subtype of analysis is *similarity*—where a requirement can be verified by direct, but detailed, parameter by parameter comparison with another mission, item of hardware or an earlier test.
- *Inspection*—a method of verification that determines conformance to specified constructional features, engineering drawings, physical features and workmanship standards. Standard laboratory metrology equipment can be used. Inspection includes witnessing by Quality Assurance personnel that specified processes are used.
- *Review of Design*—a method of verification that looks at approved design reports, technical descriptions and engineering drawings, along with evidence of the validation of the processes that contributed to those documents, to show unambiguously that the requirement is met. One application of review of design is to show that a previously-used equipment design is shown to be qualified and that no further qualification analysis or testing is required.

Verification level—the level within the hardware architecture at which the relevant verification is performed. Subject to detailed planning, this can be at spacecraft level, module or subsystem level, unit, equipment or component level.

In this chapter we will generally describe the AIV of a spacecraft. The spacecraft may however comprise a Service Module and a Payload Module, and each of these will be an assembly of various subsystems and units ('black boxes') and equipments (deployable booms, antennas and solar panels) mounted on the main structural component. The AIV processes will be allocated to module, subsystem, unit or component level for a number of reasons:

- It is generally better to detect anomalies, faults and failures as early in the programme as possible for easier rectification.
- Contractual arrangements require the manufacturer of a unit or equipment to prove, demonstrate and document the performance of their hardware at the time of delivery.
- Testing of smaller items is easier and cheaper and often more practical—because specialized test facilities large enough to accommodate the whole spacecraft may not be available.

There is always a need to frequently re-verify throughout the programme to ensure that performance is maintained, and that failures have not occurred. There is nothing better, at the end of the day, than a final, comprehensive test that demonstrates that the flight assembly (the spacecraft that will be launched) is 'ready to go'.

17.3   THE VERIFICATION PLAN

The flow diagram shown in Figure 17.1 is derived from ECSS-E-ST-10-02C [1], the appropriate standard for Verification Engineering, of the *European Cooperation for Space Standardization* (ECSS). Other space agencies have similar standards. The diagram shows

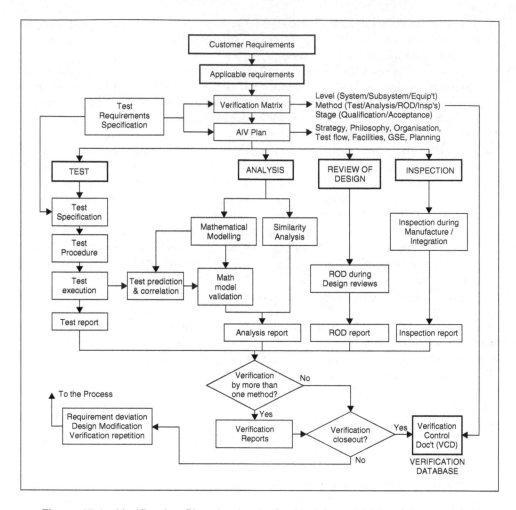

Figure 17.1 Verification Planning Logic, Derived from ECSS-E-ST-10-02C [1]

the relationship of the planning processes and the documentation needed to finalize a Verification Plan. At the top are the *customer requirements*, comprising not only the operational phase mission-specific requirements, but many *applicable requirements* too. The latter include customer-specified suppliers, test facilities or launcher systems, the interface to the ground segment, range safety requirements, international protocols (e.g. ITU), standards for manufacturing processes—and many more.

Right from the start of the project, the AIV engineering team will begin to analyse the requirements to identify programmatic and cost drivers (facility procurement costs or long test durations, for example) and to plan the verification programme. The first task is to prepare the *Verification Matrix*, within which all requirements are listed. For each and every requirement, a number of decisions are made.

Firstly, *how* will the requirement be verified? This is the verification method - analysis, test, review of design or inspection.

Secondly, *at what stage* will the requirement be verified? This is the verification level-spacecraft level, module, subsystem or equipment level.

At the same time, *where* will the verification take place? We need to ensure that appropriate test facilities will be available when needed, especially if they must be developed or upgraded.

Verification by test is chosen wherever possible for safety-critical and mission-critical features. It is a direct and visible demonstration or measurement of performance—as long as it can produce clear go/no-go results. Does each test measurement fall within an acceptable range of results, or not? But verification by test alone is not always practical, nor can it cover all possible mission scenarios within a reasonable timescale. Taking the example of a planetary lander, a representative landing site (a sandy or rocky surface) might be found or constructed somewhere on the Earth's surface. It is not however feasible to completely verify a parachute landing system from atmospheric entry to soft touchdown in a continuous 'end-to-end' test sequence if gravity and atmospheric parameters are not Earth-like.

Hence, alternative, or complementary, methods of *verification by analysis* are needed. Analysis will start early and will initially distinguish the good designs from those that will not work. The whole spacecraft and mission is analysed before any hardware is manufactured. More detailed mathematical modelling and simulations are then developed with the aim of representing the many environments in which the spacecraft must operate - the external stimuli such as mechanical induced loads or thermal radiation and conduction (and convection if in a planetary atmosphere). Models are improved and re-run as the detailed design of the spacecraft progresses to incorporate the known characteristics of chosen materials, subsystem designs and configurations, and the outcomes of technology developments. Thus the behaviour of the spacecraft is predicted. Out-of-specification results are investigated to determine if the modelling is wrong or if the design is inadequate.

Just as testing is not always a practical option, so analysis may be too complex, or the modelling may be uncertain so that confidence in the results may not be high enough. Therefore, analysis and testing usually work together to contribute to verification. For example:

- Analysis of spacecraft performance can be supported by the test results of lower-level equipments.
- Analysis predicts test results, and a comparison between predicted and actual measurements or observations can be made. If they do not agree, maybe the analytical methods need changing, or perhaps the as-built design has not been adequately represented in the model. Or, ultimately, the design is not good enough.
- Once modelling has been developed, it can be re-run at moderate cost. Tests can be expensive and time-consuming to repeat. On the one hand a test should only be run when sufficient confidence in the modelling has achieved. On the other, there does come a point when it is vital to 'test something' to get physical measurements and results, rather than to perform analysis after analysis.

A basic requirement is that any analysis must itself be *validated*. Initially this may be through previous application and experience and by rigorous testing of the analysis itself. Ultimately it is the agreement of predicted results and actual test results that clinches it—at which point the mathematical model can be declared to be 'validated'.

It is worth remembering here that the validated modelling will continue to be used throughout the mission at the control centre to predict in-orbit performance. Validating the model is as important as verifying the spacecraft.

Then, when all the requirements to be verified by analysis or test (or both) have been identified, *Review of Design* or *Inspection* are assigned to the remaining appropriate requirements.

The first draft of the verification matrix is thus completed, but it is hardly ever straightforward. A further distinction needs to be made in the verification matrix between activities that contribute to either qualification or acceptance, or both. Qualification will draw heavily on test and analysis methods, which explore all aspects of the planned mission, using a range of test environments. Since the primary aim of acceptance testing is to detect workmanship and materials faults in the flight hardware (and to eliminate them), there are fewer activities. They are primarily tests and inspections, and the tests need only look for go/no-go results rather than gathering extensive data. That said, sufficient data needs to be recorded to characterize flight performance. Then, a number of particular aspects of the programme should be looked at before the matrix is finalized. They may arise as a result of on-going equipment level design. These include:

- Design peculiarities—what's different or new? The design can drive the verification process. Size is important. Are test facilities large enough and capable enough for a large, heavy spacecraft? Does a large spacecraft (e.g. Envisat or Newton) need to be separated into modules for test and transportation? Are facilities clean enough for a biologically-clean planetary lander? Maintaining cleanliness of large optical mirrors was a driver for assembly and test of the Herschel spacecraft (see Figure 17.2).

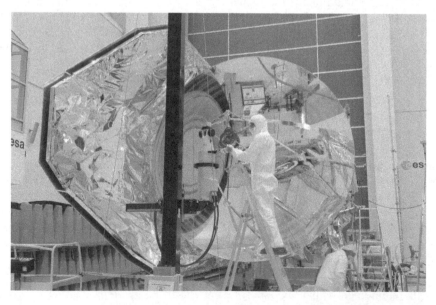

Figure 17.2 The Herschel spacecraft during alignment measurements on its mirror system. Optical and thermal surfaces demand strict control of cleanliness. (Photo: ESA, Anneke Le Floc'h)

- Are the necessary verification tools and techniques available? These might be hardware (test and metrology equipment) or analytical tools. If they do not exist, they must be developed (and verified in their own right), or alternative methods sought. Can we achieve the cryogenic temperatures needed to verify noise levels in the detectors of an infrared telescope? How can we simulate (in test or by modelling) a microgravity environment?
- What is the qualification status of the hardware? Has it undergone, and passed, a qualification process before? For the same environments? If it has, then no further qualification is needed. If it has been qualified but flown in a different environment, then it needs to be re-qualified for the new environment. The term 'delta-qualification' is used in intermediate cases—perhaps the hardware needs to be qualified by test for the mechanical environment but not for the thermal environment (perhaps because a different launcher is to be used to place a spacecraft into the same orbit as a previous, similar mission).

Answering questions such as these may modify the initial verification matrix; but there is still more planning to do!

17.4 RELATIONSHIP BETWEEN ANALYSIS AND TEST

It is recognized that there is no test that truly and completely represents the mission environments. For many years mechanical testing was performed one linear axis at a time. But test facilities now test in three axes concurrently, and even with six degrees of freedom—closer to reality, but never an exact simulation of the launch regime.

In the case of thermal control, there are many mission phases with considerably different thermal conditions to consider—launch, transfer orbit, apogee and perigee, and eclipse and sunlight modes. In elliptical orbits the thermal environment changes slowly throughout the orbit period, but the transition between sunlight and eclipse is sudden. Thermal conditions change between beginning of life and end of life as the thermal control surfaces degrade. A planetary probe will experience many more thermal scenarios—en route and at the destination planet.

It is not practical to test the thermal control performance in each and every flight case, and certainly not for the long durations of some phases. Test facilities use lamps to simulate the Sun, and nitrogen-cooled shrouds to represent cold views to space—they are not the 'real thing'. Thus, the logic depicted in Figure 17.3 applies:

- Mission scenarios are listed and modelling used to predict the performance of the thermal control subsystem for all mission cases.
- The flight cases that are to be tested (i.e. the *test cases*) are selected. They are likely to be the more critical or complex flight cases, and certainly any for which there is uncertainty in the modelling. Representative transition modes between test cases are also specified. Where there is a high level of confidence in the modelling, testing can be minimized.
- The flight modelling is adjusted for the test cases by including the calibrated performance and characteristics of the thermal balance test facility—such as the lamp-simulated 'solar' input, and the view factors of surfaces that 'see' chamber shrouds instead of deep space (which in some directions may be obscured by test fixtures).

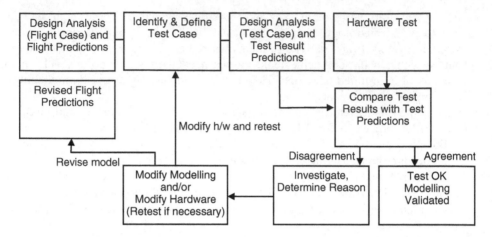

Figure 17.3   Relationship between analysis and test

The effects of conducted heat 'leaks' through test fixtures and test cables is factored in. The performance of the thermal control subsystem in the test cases is predicted.

- The hardware thermal test cases are run to provide actual test results. Test results are compared against the test case predictions.
- If they agree, the modelling is considered to be validated. If not, anomalies and disparities will be investigated. If the test results are within specification, but not as predicted, the modelling will be investigated until the anomaly is explained, and new flight predictions obtained from the revised model. If a test result is outside its specified range, either (1) the specification can be revised to accommodate minor discrepancies or (2) the hardware will need to be changed by redesign in some way. Then the modelling will be changed to reflect the new design, and re-run to revise flight predictions. Depending on the severity of discrepancies and design modification, some retesting may be required.

17.5   THE AIV PLAN

In practice, the spacecraft prime contractor will employ a number of subcontractors, each contributing hardware or software to the programme. The prime contractor is responsible for the planning and execution of an efficient but effective AIV programme across the project to:

- Ensure all partners and subcontractors are aware of their responsibilities to contribute, at their particular levels and using agreed methods, to the verification objectives.
- Assemble and integrate the hardware/software in a logical sequence. The planning will determine required delivery schedules for incoming equipments as well as the delivery of analytical results and modelling. It should also include some contingency to be able to react to delays and problems at equipment level.
- Schedule and perform a programme of tests, inspections and measurements of the hardware to support the verification programme. Test programme planning includes

provision of suitable test facilities, test and servicing equipment, test instrumentation, transportation planning and staff training.

- Perform sufficient 'health checks' on the product—moving it around, subjecting it to transport loads or spurious conditions can damage the hardware and induce faults. Has a sensor or thruster been knocked out of alignment during movement or test? Has the propulsion system 'sprung a leak'?
- Detect adverse 'trends' in performance—a gradual decline in battery capacity with use, or calibration changes in temperature sensors over time?
- Clearly demonstrate—to the customer, launch service providers and others—the successful completion of the key stages of qualification and acceptance, at all verification levels.
- Ensure that all necessary activities can be completed in order to meet launch dates. Launch dates may be driven by scientific needs—interplanetary launch windows for example. A commercial telecommunications satellite customer will suffer loss of revenue if the spacecraft operational status is delayed. The longer a spacecraft is under test, the more the cost increases.
- Record and document all verification results, analyses, measurements and conclusions. Acceptance test data are needed not only to demonstrate flight readiness but are also used by the mission operations centre to characterize the in-flight hardware.
- Implement precautions in test planning to ensure, at all times, the safety of the spacecraft hardware, and of the personnel who are working with it. A spacecraft may contain a number of pyrotechnic devices, deployment mechanisms, charged batteries, high voltages and high current sources, RF radiation and radioactive isotopes. Propulsion systems are pressurized, the fluids are toxic (and even nitrogen is an asphyxiant). Wrongly designed or badly maintained lifting equipment can cause hardware to drop—possibly onto someone.
- Ensure and continually confirm that activities similar to the above are performed by lower-level contractors on deliverable hardware, and that all test, analysis and inspection records are delivered with the hardware to the spacecraft assembly stage.
- Ultimately, deliver the product to the customer on time and within cost constraints. It is important for the planning process to understand at which point contractual delivery will take place. A scientific or Earth observation spacecraft is, in Europe, generally delivered to an institutional customer prior to shipment to the launch site. There may well be some verification activities that have not been completed at that stage, so they will carry forward into the launch preparation phase, and/or to in-orbit verification phases with the agreement of the customer. More usual for a commercial programme and a commercial customer, contractual delivery is with the spacecraft in a fully operational state, after an in-orbit commissioning phase.

17.6   TESTING: GENERAL

In summary, then, we have seen that the Verification Plan defines how each requirement will be verified. It is now necessary to plan the AIT programme to include all activities that wholly or partly require some aspect of testing.

Looking back at Figure 17.1 we can see that the Verification Plan leads to test planning that includes:

- *Test specifications*—exact definitions of how each of many different tests will be performed and what measurements are required from them.
- *Facility planning*—specifying and verifying the performance required of the facility and all interfaces between the test hardware and the facility.
- *Transportation planning*—how the test hardware will be moved between test locations.

At the same time, requirements will be developed for a range of specialized test equipment, transportation, handling and storage equipment—everything needed for the test stage. There may be specific requirements for facilities and equipment to maintain cleanliness, to protect optical, thermal and other sensitive surfaces and sensors or to implement planetary protection requirements. Are there specific hazards? Are special test rigs needed for deployment testing of large antennas and solar panels in a gravity field?

It is worth noting that the spacecraft has to 'testable'. It is quite acceptable for the AIT engineer to request that the spacecraft design incorporates provision for efficient test set-ups. It may be necessary to include 'hard' lifting and handling points into the structure design, or to wire test connections from units to the outside skin of the spacecraft, so that it is not necessary to remove access panels.

Requirements for the testing of space hardware are comprehensively specified in Europe by the standard ECSS-E-10-03A—Space Engineering, Testing [2], and similar standards within other agencies. That document is applicable at all levels of testing—down to equipment level. All levels of testing contribute to the whole, and the principles are similar at different levels. Here however we will concentrate on spacecraft-level hardware test activities.

17.6.1 Assembly and integration

Eventually, hardware arrives in the AIT facility, and assembly and integration can begin. The first task is to assemble and integrate equipments and subsystems into an operational system according to the design build standard and test objectives.

As part of this activity, sensors, thrusters and antennas will be aligned relative to the spacecraft axes, and the alignment parameters recorded. Functionality of deployable equipment will be verified after installation on the spacecraft structure.

17.6.2 Functional and performance testing

The key test following the integration stage is the *Integrated System Test* (IST)—a test that verifies the performance of all the elements working together at spacecraft level, in all operational modes. It includes redundant elements, back-up modes and foreseen transitions between operational states, and is generally performed in a sequence that reflects the mission sequence—launch, transfer orbit, final orbit, and so on. The IST is sometimes known as the *System Functional Test* (SFT). The test will be repeated again at the end of the test campaign. Often it will be performed part-way through the test programme, after a series of mechanical tests, but before the start of thermal testing.

A related test is the *Integrated System Check* (ISC), or *Abbreviated Functional Test* (AFT)— a subset of the IST that aims to test all major electrical and electronic functions in order to verify, within a reasonably short duration, that the spacecraft has survived a given test environment. It is therefore performed after each environmental test, (e.g. between individual mechanical tests) and when appropriate *during* an environmental test (while the environment is being imposed). Similarly *Mechanical Functional Tests* are performed on mechanisms, deployables, valves, and other moving parts at similar stages in the environmental test sequence.

The first IST sets the baseline against which the results of later ISTs and ISCs are compared, to identify any trends in performance that might indicate wear or aging effects, or to detect failures induced during the test programme (and to clearly identify the particular activity that caused the failure).

In the same way that the IST and ISC tests are repeated at intervals, so *alignment measurements* and *leak checks* are performed at the same stages to detect damage or degradation.

17.6.3   Test Reviews

It is important that the *build standard* of the spacecraft during formal tests is accurately known and recorded for every formal test. This is checked at a *Test Readiness Review* (TRR). The TRR also checks that the test facility and test equipment, procedures and personnel are ready and gives the go-ahead to start testing. The build standard is important. No matter what efforts are made to ensure that test hardware is flight representative, some deviation from the ideal is likely. Thermal blankets or close-out panels may be absent during some tests to allow access to test connectors. Test accelerometers, strain gauges, thermocouples and other instrumentation and cables will be fitted. These are predictable and acceptable differences. They are specified in the test specifications, incorporated into the *Design Build Standard* for the test and reflected in test planning.

Sometimes the Design Build Standard cannot be achieved. A piece of equipment may not be available because of a schedule delay, or because of failure. It may remain absent, or it may be replaced by an equivalent item. Provided the deviation from the required build standard does not prevent test objectives being achieved, the test can continue. The key task then is to record in fine detail, for every stage in the test programme, the *Actual Build Standard* for comparison against the Design Build Standard. Every item present in the spacecraft build is recorded, by part and serial number, so that they can be represented in test-case modelling activity and post-test analysis.

After each test, a (post-)*Test Review Board* (TRB) convenes to review the results and to ensure that any anomalies are properly assessed. It is the TRB that authorizes the dismantling of the test set-up—or orders a retest.

17.6.4   Environmental test

When the first IST has been satisfactorily completed, the hardware is ready for the *environmental test phase*—in which (a) overall system survival and performance is demonstrated during and after exposure to simulated flight conditions, (b) analysis and tests combine to validate modelling and (c) workmanship and material defects are identified.

How are workmanship or materials faults detected? For example, vibration or acoustic noise tests quickly identify loose bolts and connectors, and stress points in wiring and structural items—put simply, something will break or come loose and audibly rattle. Otherwise resonances and high amplitude displacements will be recorded by test instrumentation. Large surface areas (e.g. sunshields, shrouds, antenna dishes) are particularly prone to breaking loose or 'flapping' in the presence of acoustic noise. Video recordings during test runs are useful tools for observing the effects.

Thermal cycling tests (repeated cycling between hot and cold extremes) cause thermal expansion and contraction and will disclose problems such as dry solder joints and bad grounding connections. Throughout the environmental test phase, *measurements* are compiled into a spacecraft database—how does telemetry data change according the external environment, or in response to a telecommanded change of state?

This collection of data characterizes and calibrates the hardware, and is delivered to the ground segment as a reference data set during the operational life of the mission.

17.6.5 Margins

A number of *margins* are applied throughout design and testing, to arrive at the worse case extremes of the test environment (including the qualification levels). We can use equipment temperature levels as an example. The specification of test margins (see Figure 17.4) begins with the predicted temperature range during service life. This is bounded by the upper *predicted* temperature (worst case) and the lowest *predicted* temperature (worst case).

There will be uncertainties in the thermal control performance since materials from different manufacturing batches may have different characteristics, and there will be some uncertainty in calculations, for example. Hence a *thermal control uncertainty* factor is added to the predicted cases. This may be 10°C (if verification by test) or 15°C (if

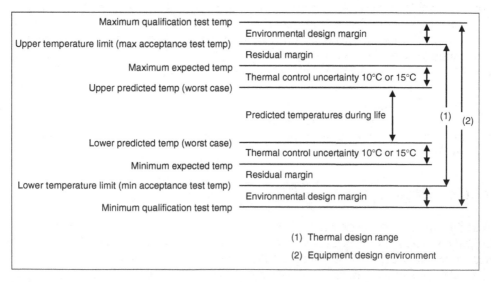

Figure 17.4 Temperature limits and definition of margins — derived from ECSS-E-10-03A [2]

verification by analysis), and it sets the 'maximum *expected* temperature' and the 'lowest *expected* temperature'.

The specified flight *acceptance* test levels are based on these expected temperature extremes, and a *residual margin* of up to 5°C set above and below them. The *qualification* test temperatures are then defined as 10°C above and below the acceptance levels. It is important to recognize that the thermal control subsystem has then to be designed to withstand and to perform at the qualification test levels—the greatest extremes to which the hardware will be exposed.

17.7 TEST TYPES

In this section, we look in some detail at the majority of the tests and measurements that must be included in a qualification (Q) or acceptance (A) test programme. Test methods, test arrangements, measurement and instrumentation methods, and the tolerances applicable to test inputs and to test measurements are described and specified in detail in the standard ECSS-E-10-03A. The list below is not necessarily complete—it does not include tests applicable to manned missions, aerodynamic environments and planetary landing, which require specific considerations on a case-by-case basis and specific tests to be developed.

Static Strength (Static Load) tests (Q) determine whether the design of load-bearing structures will sustain quasi-static and dynamic accelerations, induced by the launcher, boost motors and, if appropriate, its own spin. The static load test is an early structure-level qualification test. A flight representative structure is held in a very rigid test stand representing the launch vehicle adapter and subjected to limit, yield and ultimate loads through hydraulic jacks. Structural strains and displacements are measured. *A Spin Test* (Q) is required if loads of 2g or more are predicted in a spinning spacecraft.

Sinusoidal Vibration tests (Q,A) primarily validate mechanical modelling and verify that the spacecraft can withstand the physical vibration environment induced during launch. A vibrator, or 'shaker', subjects the spacecraft to a 'sweep' of sinusoidal frequencies rising from 4 to 100 Hz or 150 Hz at prescribed levels and a sweep rate specified by the launch vehicle manual, times a factor of 1.25 for qualification. The qualification test run compares the response (natural frequencies and mode shapes) to the sine input, and is used to assess the accuracy of the mass and stiffness model.

If run on the flight model, then the sine test can compare responses of the flight hardware to qualification results. Different natural frequencies or responses may indicate problems with materials or workmanship.

Modal Survey testing (Q) determines by experimental methods the natural frequencies, mode shapes and damping factors of the spacecraft. The modal survey is investigative in nature, and to some extent a trial and error approach is adopted. There are not necessarily pass/fail criteria, except when it is an objective to avoid certain natural frequencies. It is useful when similar sine vibration-induced natural modes combine into single modes, and the isolation of modelling errors becomes impossible.

The spacecraft structure is attached to an isolated seismic block. The natural modes are excited by single or multipoint actuators with transient, random or sinusoidal characteristics.

Random Vibration and Acoustic Noise tests (Q,A). The acoustic noise environment is greatest at lift-off when noise is reflected from the launch pad, and this can be of particular

Figure 17.5　The Herschel spacecraft on the Estec DUAL Multishaker for a vibration test. Note the protective canopy to prevent contamination from above. (Photo: ESA, Anneke Le Floc'h)

concern for large area, lightweight structures. The structural response to acoustic noise is predicted in terms of a random vibration. Random vibration tests using a vibrator/shaker are usually performed only on small spacecraft. Acoustic noise tests are performed on large spacecraft, in a reverberant chamber, to demonstrate the space vehicle can withstand the acoustically-induced vibration environment during launch. However, random vibration tests during qualification at lower, equipment levels eliminate problems at spacecraft level.

Random vibration testing is performed on a shaker (see Figure 17.5 for a typical set-up). The vibration input is applied simultaneously across a spectrum of 20–2000 Hz for 2 min (Q) or 1 min (A).

The acoustic test is performed with the spacecraft in launch configuration on an adaptor mechanically decoupled from the chamber floor so that chamber vibrations are not transmitted into the spacecraft. Figure 17.6 shows the configuration of the SMOS spacecraft in the Estec acoustic facility. Typically a spectrum from 30 to 8000 Hz is applied for 2 min (Q) or 1 min (A). Where flight spacecraft equipment will operate during launch, its performance should be demonstrated during the random vibration or acoustic test.

Shock test (Q)—the spacecraft is subjected to inputs representative of the shocks induced into structures as a result of (a) shroud jettison and spacecraft separation from the launch vehicle, and (b) release, by pyrotechnic mechanisms, of appendages such as solar panels and antennas, and their subsequent latching into operational orientation.

Pressure test (Q, A). This subjects pressurized subsystems to 150% of the maximum design pressure for at least 5 min, and the test is cycled three times. Any system designed to contain fluids will undergo a *Leakage Test* (Q, A), being pressurized at maximum design pressure for 20 min. During that time the leak rate is determined and all joints and fittings individually checked for leaks.

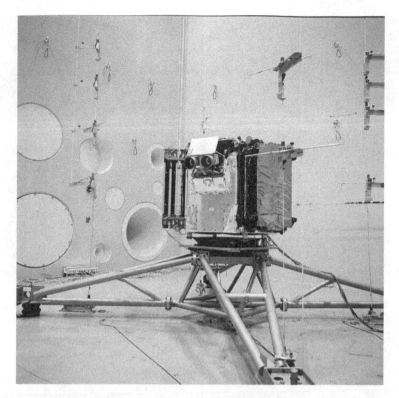

Figure 17.6   The SMOS (Soil Moisture & Ocean Salinity) payload in the Large European Acoustic Facility (LEAF) at Estec. The MIRAS Instrument antenna 'wings' are stowed in launch configuration. The test item is mounted on a stable, isolated stand. The acoustic horns can be seen in the chamber wall. (Photo: ESA, Anneke Le Floc'h)

Physical properties test (Q, A)—the mass, centre of gravity location and moments of inertia of the spacecraft are determined. These results are needed by the launch vehicle/ascent analysis, as inputs to the attitude control design and also for finalizing the test set-up and analysing the results of other mechanical tests.

Thermal vacuum/vacuum temperature cycling tests (Q, A). This is a performance test that characterizes and verifies *electrical* functionality in the vacuum of space under specified thermal conditions. The test is performed at upper and lower temperature extremes and during transients. The temperature profile is cycled between temperatures determined by the qualification or acceptance test limits. For qualification, the temperature is cycled between hot and cold extremes four times with the first two cycles and the last cycle dwelling twelve hours at each extreme. The third cycle 'soaks' the spacecraft at hot and then cold conditions for three days. Acceptance testing is performed with one less hot/cold cycle.

The spacecraft equipment is operated through all operational modes at each hot and cold state and its performance measured. Care will be taken to ensure that internal equipment dissipations do not cause the upper test limits to be exceeded, and, at the other extreme, that imposed cold environments do not subject any sensitive equipments to levels lower than their cold design limits.

Figure 17.7 The SMOS payload in the Large Space Simulator at Estec. The three MIRAS antenna 'wings' are deployed to in-flight configuration. The multi-segment sun-simulator can be seen in the background. Above the test specimen is its suspension rig. (Photo: ESA, Anneke Le Floc'h)

Thermal balance test (Q). This simulates the mission thermal environment (solar radiation, Earth albedo, deep space views, internal dissipations), and is usually performed in a solar simulation chamber (under vacuum) with lamps simulating solar radiation. Shrouds around most of the chamber surfaces are filled with liquid (cold) or heated gaseous nitrogen which simulate hot and/or cold views to space. SMOS can be seen in the large space simulator, ready for thermal balance testing, in Figure 17.7.

The prescribed 'test cases' (see Section 17.4) are run for a number of mission phases (sunlight, eclipse, beginning or end of life, transfer orbit or on-station cases) and for different operating modes of the spacecraft (main or redundant equipments powered, in various combinations, for example). Here, the objective is to verify that all parts of the spacecraft are controlled within specified temperature limits by the thermal control subsystem (blankets, radiators, heaters, paints and tape finishes), and to measure and record the actual temperature distributions.

A note of explanation may be helpful here. The *thermal balance* test verifies thermal control performance by determining temperatures during the test. It can be performed on a thermal model spacecraft built with equipments that are sufficiently thermally representative but not fully functional. The *thermal vacuum* test verifies electrical performance at temperature extremes, and therefore requires fully-functional equipments. Thermal *cycling* induces controlled thermal stresses that might detect component failures.

Electromagnetic compatibility tests (Q,A). These are performed to determine whether the spacecraft performance can be adversely affected by electromagnetic interference from external sources, or whether any of its internal equipment emits stray signals that can affect its own performance, or interfere with external elements such as the launch vehicle and launch site systems (e.g. radars and other RF systems). The system is operated

Figure 17.8 The SMOS payload in the Maxwell EMC Chamber at Estec. Support stands for the antenna subsystem are non-metallic. (Photo: ESA, Anneke Le Floc'h)

in a chamber lined with RF absorbent material (see Figure 17.8) so that signals external to the test facility, and those emitted by the spacecraft, are absorbed and not reflected back into the spacecraft. Measured emissions from the spacecraft are compared against specified maxima (based on launch site criteria). The spacecraft is then operated in its most sensitive modes when irradiated with levels that are representative of those encountered during its life cycle.

Radio frequency compatibility testing is similar to EMC testing, but with particular emphasis on interferences caused by or in RF uplinks and downlinks.

Magnetic field measurements (Q, A)—this is a specific test, which is only required if the spacecraft carries magnetically-sensitive items such as a magnetometer. It measures permanent, stray and induced magnetic fields. The test facility provides the capability to 'zero out' the Earth's magnetic field while low level fields emanating from the spacecraft are measured. Then the facility generates controlled, uniform magnetic fields to measure induced fields in metallic structures, and to calibrate magnetometer responses.

Mission and operations validation (Q, A) Tests are performed to verify the interfaces between the spacecraft and the mission ground control systems. The spacecraft, in its test facility, will be monitored and operated remotely by the Mission Operation Centre via a communications network. A simulated mission will be 'run' to verify the ground-to-spacecraft interfaces, mission operations sequences and spacecraft database parameters. This validation test provides operators with vital experience and familiarity with spacecraft behaviour.

17.8 MODEL PHILOSOPHY

Generally the AIV engineer will quickly find, in following the process described above, that there is simply not enough time to perform all verification activities in a serial manner.

Further, subjecting a spacecraft and its constituent parts to a long series of extensive test will over-work it; it will simply be 'life expired' by the time testing is completed. The solution, to evolve a workable programme, is to develop a *model philosophy*. What hardware do we need to achieve *all* the necessary activities, in a timely manner and at acceptable cost? Inherent in the choice of model philosophy is accepting that some activities must take place in parallel to avoid an unduly long programme.

Working through the model philosophy evolves the optimum number and characteristics of sets of physical hardware to be manufactured. The aim is to achieve a high confidence in the ability to verify the product within a practical timescale, but with a suitable weighting of costs and risks.

We have seen the need to first qualify the *design*, and then to acceptance-test the flight *hardware*. We noted earlier that for qualification we need to demonstrate that the design is adequate—with *margins*. When it comes to testing, that means subjecting the hardware to environments more severe than the predicted in-flight case, i.e. more severe test levels and longer test durations. That, in turn, means that the spacecraft can be over-tested, and fatigue or wear will become a concern.

Thus, at spacecraft level, the first intuitive decision would be to manufacture, assemble and test at least two spacecraft. A *Development Model* (DM) would undergo qualification tests, before a *Flight Model* (FM) is manufactured and assembled for acceptance test and launch.

Dividing the tasks further, separate Development Models might be manufactured and tested for different design disciplines—a *Structural Model* (SM), a *Thermal Model* (TM) and an *Electrical Model* (EM), for example. These Development Models might be followed by a *Qualification Model* (QM). Indeed, that was the philosophy in the 1960/70s. ESA's GEOS (Geostationary Scientific Satellite, for wave and particle field measurement) programme included an SM, a TM, and an EM. There was even a separate Qualification Model (QM), and eventually an FM. To be fair, the analytical methods and computing power/speed/capacity achievable today were not available then, and test was the surest method of verification.

However, the more hardware models employed, the higher the cost of manufacture and test. Generally mechanical and electrical development testing and some qualification activities can be performed on different models in parallel before Flight Model testing begins. The programme needs to incorporate opportunities for changes resulting from Development Model tests to be incorporated into the Flight Model manufacturing schedule. There are ways to optimize the programme and costs, as long as there is sufficient confidence of finding design and manufacturing errors before it is too late. These include:

- Planning to complete as much verification as possible at equipment and subsystem levels. Then the spacecraft-level activities provide final verification that the lower level equipments work together to fulfil the system level requirements.
- On completion of equipment level development and qualification model testing (but not before) giving the go-ahead for both (a) manufacture of flight model equipments and (b) integration and test of the development spacecraft.
- Optimization of the verification test programme on each spacecraft development model or qualification model after identifying the key objectives—for example, a structural test model is intended only to achieve structural qualification, not thermal or electrical performance.

A common solution now is to implement a *Protoflight Model* (PFM) philosophy. In this the objectives—at spacecraft level—of qualification and acceptance, are combined in a single model. This will be the flight spacecraft. It will be subjected to qualification test levels but only for acceptance durations—i.e. in some respects the flight hardware is exposed to overtesting in the severity of test, but the effects are mitigated by keeping the exposure time to a minimum. Importantly, this approach can be adopted only if full qualification, in all respects, is achieved at equipment level.

So what determines if a protoflight philosophy is possible? Consider a geostationary communications satellite. Its mission is to operate in a very well known environment. Much experience has been gained of operating spacecraft there, and the operational scenarios have been well-rehearsed. In the commercial world of communications, there is a high level of reuse of the same design of equipments and of spacecraft configurations. But, the communications spacecraft is likely to be modular. A 'bus' or service module will be a 'build to print' copy of earlier builds, adequately qualified. The payload module will however be mission-specific. Payload data processing and ground-coverage antennas will be customized for different applications (point-to-point telephony, broadcast, military versus commercial, and so on). Straightforward acceptance testing is not enough, but a protoflight approach looks a possibility. But what else needs to be considered?

- Can the payload configuration be adequately qualified at payload module or lower levels?
- Is the configuration qualified for launch on the particular launcher? In all probability, a communications satellite manufacturer or operator will ensure that design and test specifications encompass a range of likely launcher types so that they are not restricted to one launcher and can sign up for the lowest cost launch, or can switch launchers to maintain commercial advantage. So perhaps structural qualification has been achieved elsewhere, but thermal qualification is needed for the new payload configuration.
- Should there be more emphasis on other methods, such as analysis?

Occasionally, a spacecraft may be launched that is fully identical to an earlier one. An existing customer may need the same payload configuration at a second geostationary station. In this situation, the spacecraft would need only acceptance tests, as its design will have been qualified by *similarity* to the earlier mission.

Ultimately, however, a new generation of communication satellite is needed. A step increase in capacity, power and lifetime requirements leads to a new structural design and configuration. Periodically, new technology will be used to keep the mass and volume down and to improve performance. In either case, the contractor will have to demonstrate qualification, using development models and his own resources, before a commercial customer will finalize an order for a flight spacecraft of the new generation.

The same reasoning, leading to a protoflight approach, will apply to, say, a meteorology or Earth-observation spacecraft in polar orbit. The bus might be very similar to a previously qualified design, but the payload will probably be different. Conversely, a scientific spacecraft design will generally be 'new'. Mission scientists want to explore new frontiers, with an ever-increasing complexity of mission objectives and specifications. The hardware, especially in terms of electrical and software complexity will be demanding. In this case, a protoflight solution is *unlikely*, and development models will be needed.

It will be understood, then, that the decisions needed to formulate a model philosophy, and the verification programme, are based on many considerations. It is not possible to give hard and fast rules, but it is vital that the planning process encompasses the complete programme, from component level to spacecraft level, and indeed beyond, taking into account requirements derived from the ground segment, the launcher authorities and the mission objectives themselves.

17.9   BUILD STANDARDS AND APPLICATIONS

We have seen that a number of development models can be built for specific test objectives. They must be built to pre-determined standards of manufacture (*build standards*) according to their verification and test objectives.

17.9.1   Structure Model

The primary purpose of the *Structure Model* (or alternatively 'Structural Test Model') is to validate the modelling of the behaviour of the spacecraft structure, primarily through a series of measurements under imposed loads.

Initial tests such as static load tests and/or modal survey measurements are performed on the 'bare' spacecraft structure, to test that structure and to validate modelling of the structure subsystem design. The structure must be manufactured to full flight standard.

The structure will then be fitted with a complement of units and equipment, to represent the flight spacecraft launch configuration - that is, at predicted launch mass with that mass correctly distributed throughout the structure. The simplest approach is to use mass dummies for each piece of equipment, which works well at 'black box' level. The mass and centre of gravity of each unit is representative of the flight item. The attachment interfaces (via bolts and washers) to the structure will be flight standard design, installed to flight procedures.

Launch environments are imposed in acoustic noise and vibration tests to further validate modelling of the launch configuration and to verify acceptable structural behaviour under launch conditions, including the important demonstration that the spacecraft will survive the launch phase.

Spacecraft level tests are particularly important to the launcher authorities. They require convincing proof that the spacecraft design is qualified for their launch vehicle. The launch phase is of course critical—break-up or high-amplitude vibration of the spacecraft will severely affect the launcher trajectory, possibly leading to a catastrophic disintegration of the launch vehicle and very bad publicity for the launch agency!

For structurally significant equipments such as deployable antennas and their hold-down mechanisms, it is sensible to install structurally *and functionally* representative hardware, since:

- Their vibrational resonances and dynamic behaviours should be present in the spacecraft under test, so that the results can be fed back into the modelling. This includes fuel 'slosh' under vibration. Hence consideration needs to be given to including 'fillable' fuel and pressurant tanks in propulsion systems.

- It is an opportunity to qualify appendage designs for launch in a representative situation—installed on the spacecraft. It is as important to verify that hold-down mechanisms do not release under vibration as it is to verify that they will release correctly after the launch phase—by testing them after vibration and acoustic noise tests.
- It is an opportunity to perform deployment tests and measurements at spacecraft level, and that includes shock testing. When an appendage is deployed (by bolt or cable cutters for instance) and also when it locks into its final position, a shock is transmitted through the spacecraft structure. Such shocks need to be quantified and shown to be non-detrimental.

The Structure Model will be fitted with test instrumentation including accelerometers, strain gauges and displacement sensors to provide test data.

17.9.2   Thermal Model

In a similar way the *Thermal Model* validates the mathematical modelling of the thermal behaviour of the spacecraft by testing the thermal control subsystem in a Thermal Balance test.

The build specification must include a flight-standard structure (for correct thermal conduction), with flight standard thermal finishes (for internal and external radiation exchange). Flight-type thermal control hardware needs to be present but heater mats are sufficient to simulate the heat dissipation of electrical circuits. Units and equipments need to be at least representative of their thermal capacity.

The Thermal Model will be populated with many thermocouples for the test. In themselves they can affect the thermal balance situation—a large bundle of cabling can provide a noticeable conduction path that needs to be taken into account in the test case analysis. The Thermal Model may also provide a convenient test vehicle for specific subsystem equipments that themselves actively control temperatures or are critically temperature dependent. Examples include cryostats, louvres and shutters.

Structure and Thermal Models provide useful tools for the AIT engineers to rehearse and prove ground handling procedures and to verify facility interfaces.

17.9.3   Life Test Models

Life Testing is an important verification method - not at spacecraft level but for mechanisms and moving parts in particular. Every mechanism will have a specified 'life'—the number of operations of a switch or valve, or a number of years of continuous operation for a solar array drive motor, for example.

To verify life, a flight standard model of the mechanism is placed into a representative environment (usually a small thermal vacuum chamber) as early in the programme as possible. It is then operated for a multiple of its specified number of flight operations or of its specified lifetime—including margins for qualification. It is useful to add instrumentation that will measure wear or changes in performance in situ, as long as it does not influence performance. Alternatively the test will be interrupted periodically to measure performance and to inspect for degradation.

17.9.4   Electrical Functional Models

Early verification of the system level functionality and performance requires an *Electrical Model* of some description. This can be an *Engineering Model*, close to flight representative, with a flight-type structure, harness and engineering or qualification model equipments. Redundant units are not generally needed. Components need not be to full space-qualified standard, but military specifications are usual. This Engineering Model approach might be used in the case of a complex and new design.

In many less demanding cases the electrical test model is put together on a 'table top', and referred to as a *Functional Test Bed*, or *Electrical Verification Model*. This is a simple representation of the spacecraft mounting surfaces, with harness lengths being approximately correct. Electrical and electronic equipments are added as soon as they can be delivered from lower level testing. Software simulations can be used to stand in for missing units until the complete functionality of the spacecraft is built up. This model is then used to:

- Verify electrical performance, integrate software and run software validation sequences, rehearsing integration and ground test procedures in the process.
- Validate electrical test equipment and procedures, develop and automate test sequences and train engineering staff.
- Trouble-shoot problems found with the flight spacecraft, and to check out corrective action. This can continue after launch to diagnose in-flight anomalies and validate software upgrades before uplinking them to the flight spacecraft, or to rehearse revised operational command sequences before performing them in-orbit.

17.9.5   Flight or Protoflight Model

The Flight (or Protoflight) Model is of necessity built to full flight standard (high reliability parts, fully redundant units, correct finishes, and more)—it is of course the hardware that will be launched.

17.9.6   Other models

There is often a significant advantage in introducing other 'models' into the AIV programme to narrow down design or analysis options. These include:

- A simple *Configuration Model* representing the size and shape of a spacecraft, that can help to verify facility interfaces and transport movements, and to develop handling procedures.
- A more representative Configuration Model, that can be used to visualize harness routing, to plan equipment locations, and to demonstrate accessibility for attachment bolts or test connections.
- An *Antenna Test Model*, which is built on a simple but metallic representation of the spacecraft, and fitted with transmit and receive antennas, sensitive sensors and detectors. Powering the model in an open-air test range or anechoic facility will quickly identify the most significant problems areas of cross modulation and interference.

These models are relatively cheap to build and can provide much needed confidence at an early stage of the programme. Where interactions are complex, early test data highlights the most important areas for analysis and eliminates the non-viable options from further analysis.

MODEL PHILOSOPHY: A CASE STUDY

Envisat (launched March 2002) was a totally new design of spacecraft. It was too large for transportation as a single entity. In the early stages of the project new test facilities were under construction for the Ariane 5 generation of large spacecraft, but existing facilities, although not large enough could not be ruled out.

Hence, Envisat was 'modular'—a Service Module (SM) and a Payload Module (PLM). The PLM was a new and complex design, with a high probability of interface problems. Functionality and electromagnetic compatibility needed to be verified early.

A thermal balance development test was initially planned, needing a representative PLM structure. Thus a fully-representative Engineering Model (EM) of the PLM was needed. Then, to fully explore electrical performance, the EM PLM needed to be coupled with a fully-functional Service Module.

However, the Service Module, as a direct development of the SPOT 4 Earth Observation spacecraft, did not need an Engineering Model. A hybrid approach was the solution for this stage of the AIT programme. A Protoflight Service Module (PFM SM) would support the EM test programme. i.e. **the Envisat EM spacecraft was (PFM SM)+(EM PLM)**

It was then shown that analytical techniques were good enough to eliminate the early thermal balance test. The EM Payload Module therefore used lower-cost materials of slightly below flight standard. For example, EM equipment panels were aluminium honeycomb, in place of the carbon-skinned honeycomb flight design.

This Engineering Model configuration was not however sufficient for necessary structural qualification. A Structure Model Spacecraft (StM, to distinguish it from SM) was needed. Hence, an StM of both the Service Module and Payload Module would be built. i.e. **The Envisat StM spacecraft was (StM SM)+(StM PLM)**

Of this development model hardware, the Protoflight Service Module would be made ready for flight. It would join with new Flight Model PLM. There was however one final change. After a cost review, it was decided to refurbish the StM PLM structure and to use for the FM PLM. i.e. **The Flight spacecraft was (PFM SM from EM spacecraft) + (FM PLM, with refurbished StM PLM structure)**

17.10   GROUND SUPPORT EQUIPMENT

There is one more significant cost and planning element to consider. No spacecraft AIT programme can be performed without *Ground Support Equipment* (GSE)—that is, the equipment needed to handle and to test the spacecraft up to launch.

Designing, procuring, testing and verifying the GSE itself is a significant part of the AIV process. It can be schedule critical—it will be required in the AIT programme when the first spacecraft components arrive, but the GSE design cannot be finalized until the spacecraft design is approved.

17.10.1 Mechanical Ground Support Equipment (MGSE)

A wide range of MGSE is needed to hold, lift, move, store and transport flight hardware, protecting it from damage and degradation at all times, and ensuring safety for personnel working with it or in the vicinity. MGSE is required at all levels of the programme from equipment level to spacecraft level. Here we discuss primarily the spacecraft level MGSE.

MGSE must be compatible with the attachment interfaces on the hardware, and with all the test facilities to be encountered in the AIT programme. Equipment includes:

- Adapters to securely attach the spacecraft to test facility interfaces or to other MGSE. This will normally be at the spacecraft launch vehicle attachment surface, so that interface on the MGSE must be of the same high standard as on the launcher.
- Integration trolleys, stands and turnover devices (see for example Figure 17.9) that are used during assembly and integration, functional tests throughout the AIV phase, and during launch preparation. Between them, they need to provide access to all parts of the spacecraft by rotating the spacecraft in any attitude from horizontal to vertical. Wheels, or air skates, will allow local movement of these devices within facilities.

Figure 17.9 GOCE (Gravity field and steady-state Ocean Circulation Explorer) spacecraft attached to its Integration and Turn-over Stand for detailed inspection of solar cells. (Photo: ESA, Anneke Le Floc'h)

- Lifting equipment for use with facility cranes—beams, hooks and slings and project-specific interfaces to the spacecraft. It is often necessary to lift a partially-assembled spacecraft in which the centre of gravity is off-axis. The lifting beam therefore needs a mechanism to offset the centre of lift to compensate and to maintain the spacecraft axis vertical.
- Transport containers—to move the spacecraft (or sometimes Service and Payload Modules separately) between facilities, ensuring that cleanliness is maintained. They may be purged and pressurized at a slight over-pressure with nitrogen, in which case external temperature changes must be taken into account for safety. They will include facilities for monitoring and recording the internal environment, and to record any adverse shocks encountered during transport. Containers can also be used to store the spacecraft if necessary—for example, if the launch is delayed.

  Clearly, containers must be designed in conjunction with the proposed transportation scenario, and be compatible with any cargo aircraft (see Figure 17.12 later), road vehicle, barge or railway it is to use. Special transport scenarios might apply. International standards demand a double-walled container if the spacecraft includes radioactive sources. Large spacecraft may have to be moved at night (see Figure 17.10).
- Deployment Rigs—to support deployable solar arrays, booms and antennas in a way that compensates for the Earth's gravity field. They will be used both statically during alignment measurement and adjustment, and dynamically during deployment tests.
- Special-to-test equipment, examples being a thermal test adapter that simulates the Payload Module mechanical and thermal interfaces during a Service module test, and vice versa, and vibration or acoustic noise test adapters.
- Covers and protective devices and sometimes, special tools.

Figure 17.10   The Herschel spacecraft container is loaded to its transporter — at night and in the rain! (Photo: ESA, Anneke Le Floc'h)

17.10.2 Fluids Ground Support Equipment (FGSE)

FGSE is required to service the propulsion subsystem, to load and drain simulated propellant and pressurant during the test programme, and to load (and in a contingency situation to drain) the flight propellant during launch preparations. FGSE also serves to pressurize the propulsion system, for leak testing and for launch. Other FGSE may be required in particular programmes for loading and draining cryogenic fluids.

17.10.3 Electrical Ground Support Equipment (EGSE)

EGSE provides all the power supplies and uplink data to the spacecraft for ground testing, and receives and analyses all downlink data.

Typical EGSE for spacecraft testing provides services and functions to:

- Power the spacecraft, simulating solar arrays and batteries.
- Deliver (uplink) commands and ranging signals, and receive (downlink) telemetry.
- Receive, demodulate, decode and distribute housekeeping data via hardline or radio-frequency links. Automatically check all data against high and low specification limits, and look for unexpected states (e.g. those not in accordance with the expected result of an uplinked mode-change command).
- Display status parameters for an immediate overview of test and spacecraft status. Record, archive and playback all test activities and results.
- Distribute data, routinely and on-demand, to remote workstations for monitoring by payload and subsystem specialists. Check validity of and respond to commands from remote workstations for stimuli and mode changes by uplink to the spacecraft.
- Provide stimuli signals to attitude sensors; receive downlink data and measure responses at the interfaces to attitude and orbit control actuators (thrusters, reaction wheels); provide closed-loop simulation and processing of Attitude and Orbit Control subsystem performance.

All of these functions run under the control of central processors and are performed through validated, repeatable, consistent automatic test sequences.

17.10.4 GSE logistics

How much Ground Support Equipment is needed? The number of each item is determined by the AIT programme schedule and model philosophy. Where a development model is tested in parallel with the assembly of a flight model, two sets of much of the MGSE will be needed. However, a significant cost saving can be made if one transport container can be shared between models. Similarly, two sets of EGSE will be required if an electrical model is to be tested in parallel with a flight model.

Generally for a commercial, communications satellite programme the contractor will be expected to provide all GSE required to fulfil the programme's needs. A scientific spacecraft programme will require project-specific GSE designs, in accordance with the needs of the development programme and GSE will be a project cost. There is one general exception to the latter statement. FGSE for propulsion systems is now generally provided

Figure 17.11   Marshalling equipment for shipment. (Photo: ESA, Anneke Le Floc'h)

by specialist contractors who undertake propulsion-related tasks and provide their own equipment as part of their contract.

For every movement of the spacecraft many boxes, large and small, are needed to protect the test equipment, spare parts, tools and documents travelling with it between test facilities or to the launch site (see Figure 17.11).

17.10.5   GSE verification

It remains to add that every item of GSE, large or small, must be properly tested and verified before it is used with spacecraft hardware. Lifting and handling equipment will be proof-load tested. Interfaces will be carefully measured and tested to prevent damage to flight hardware interfaces. Indeed, verification of GSE follows essentially the same processes as for flight hardware.

17.11   CHECKPOINTS IN THE AIV PROGRAMME

The spacecraft project will include a number of formal reviews in which the customer will assess progress in design—and in the AIV planning. The AIV Programme will therefore, typically, interact with the formal reviews as follows:

- *Preliminary Design Review* (PDR). Is the system level design ready for the lower level design process to begin? The PDR also looks at the initial Verification Planning.
- *Critical Design Review* (CDR). Is the design ready for manufacture or assembly? The review will give the go-ahead for manufacture of engineering and qualification model hardware. Are the Verification Plan, the model philosophy and the verification matrix complete? Are the test plans and procedures ready? Are GSE designs ready?

- *Qualification Review* (QR). Are the qualification tasks complete, the test results satisfactory and the analyses validated? If a protoflight philosophy has been adopted, what qualification tasks are, by agreement, deferred to the protoflight model spacecraft, and are the plans and procedures ready? In this case, a 'Delta-Qualification' review might be held or open items closed-out in the later Flight Acceptance Review.
- *Flight Acceptance Review* (FAR). Has the (proto) flight spacecraft passed all its tests, and are flight prediction analyses completed? Is the spacecraft ready for shipment to launch site?
- *Flight Readiness Review* (FRR). Have the final activities and tests been completed? The FRR includes not only the spacecraft, but the launch vehicle and ground segment.

17.12   VERIFICATION CLOSEOUT

We now return to the *verification* process. Throughout the AIT programme many test reports, analysis reports, inspection records and review-of-design reports will have been generated and reviewed. Anomalies and discrepancies will have been subject to investigation and perhaps retest, always controlled by *Product Assurance* (PA) procedures. It is the task of the verification engineer, in conjunction with PA, to perform the 'verification closeout' process.

If a specification item is verified by a single method (test *or* analysis, say), the associated report is sufficient to close out that requirement—it has been verified. If two or more methods are involved (test *and* analysis perhaps), then an additional verification report is needed to confirm closeout. If all is well, the verification matrix line item is closed. If not, the process repeats, perhaps after hardware or software modifications, with more testing, or more analysis.

Some items in the matrix will be held open until launch site preparations are complete, but by the FRR every item in the verification matrix will be closed out.

17.13   LAUNCH PREPARATION

Figure 17.12 shows the Herschel spacecraft in the process of shipment to the launch site. In an ideal case, the spacecraft will be delivered to the launch site 'ready to go'. However, in practice, there are a number of tasks still to be performed. Flight pyrotechnics and other hazardous devices will not normally be fitted during transportation. The propulsion system will not be fuelled and pressurized until the last practical opportunity (and the filling and pressurizing process includes final checks for leakage). Similarly, batteries will be conditioned and charged to flight state only shortly before launch.

An Integrated System Check and a full inspection survey will be performed to verify no damage or performance degradation during transportation. All alignments will be checked. Mass properties will be remeasured, in the final case for the fuelled spacecraft, which is therefore performed in the launch site's Hazardous Processing Facility.

Particular care is needed to maintain or restore the 'Flight Acceptance' status of the spacecraft where parts are disassembled to facilitate launch site operations—for example, removal of batteries for charging, removal of thermal blankets or close-out panels for access to test connectors or pyrotechnic installation. The spacecraft must be returned to flight readiness by very strict adherence to flight quality processes and procedures.

Figure 17.12 Shipment to the launch site: Herschel, in its container, is loaded into an Antonov transport aircraft. Note container air conditioners, and the bank of gas bottles for purging the interior. (Photo: ESA, Anneke Le Floc'h)

During these final days on the ground, the schedule is coordinated with the preparation of the launch vehicle, and the spacecraft will eventually make the journey to be attached to its launcher. The sequence of events depends to some degree on which launcher is chosen. At this time the FRR will assess the readiness of the *complete* system for launch. It is a mission readiness review for the spacecraft, launcher and ground segment. Only a few tasks will remain to be performed, one of which (for the spacecraft) is the 'last minute' removal of red tag items and the fitting of green tag items.

Red Tag items. Throughout AIT a number of protective devices will have been attached to flight hardware. They include protective covers (for optical surfaces, protruding items, etc.) and safety devices (pins or plugs that prevent inadvertent operation of pyrotechnics, release mechanisms, etc.). These must be *removed* before flight—at the last moment they are accessible. That may be before or during fairing installation, or through a hatch in the fairing at a later time. Rather obviously, they are coloured red and should have the words 'Remove Before Flight' written on them!

Green Tag items are the opposite, and must be *installed* for flight—they are devices, usually 'arm plugs', that make final connections between hazardous items, such as pyrotechnics, motors and thrusters and their power sources.

17.14 CONCLUSION

The AIV/AIT process is an important part of any spacecraft programme. Just as a spacecraft design evolves to optimize mass, power needs, costs, reliability and more, so AIV/AIT planning optimizes the schedule and cost of the task of verifying that the end product, the flight spacecraft, will achieve its mission objectives.

Equally, just as there is no single design that will meet the needs of several missions, there is no standard AIV/AIT programme. This chapter has discussed many considerations and concepts that will be proposed, discussed and eventually included in the verification programme. The aim is always to deliver a fully compliant spacecraft to orbit, where it will perform a successful mission.

REFERENCES

[1] ECSS-E-ST-10-02C *Spacecraft Engineering, Verification Engineering*. www.ecss.nl (accessed 18 October 2010).
[2] ECSS-E-10-03A *Space Engineering, Testing*. www.ecss.nl (accessed 18 October 2010).

SMALL SATELLITE ENGINEERING AND APPLICATIONS

18

Martin N. Sweeting and Craig I. Underwood

Surrey Space Centre, University of Surrey, Guildford, Surrey

18.1 INTRODUCTION

Early western satellites were necessarily small. However, the need for ever-larger, more capable and more complex satellites led to a natural growth in satellite mass. This trend was first limited by the capabilities of available launchers and then later by finance and technological infrastructure, as space-faring nations required a highly developed techno-logical base with huge investment. This resulted in space becoming limited to a few nations—an exclusive club of space 'haves' with enormous military, economic and cultural advantages over the space 'have-nots'.

Changing world politics and military emphasis in the last twenty years has brought about a quiet revolution in space. Pressure on both civil and military space agency budgets has meant increasingly that fewer (and bigger) satellites have been commissioned and new ideas, technologies and scientific experiments have found it difficult to gain timely access to space.

In the 1970s, advances in *Very-Large Scale Integration (VLSI)* led to the possibility of sophisticated functions being built into very small volumes, with low mass and requiring minimal electrical power. This led to the emergence of the modern small satellite—or *microsatellite*—which, in turn, led to the potential for a dramatic cost-reduction in satellite programmes. This was initially demonstrated by the pioneering Orbiting Satellite Carrying Amateur Radio (OSCAR) and Radio-Sputnik (RS) series of satellites produced by the international amateur radio satellite (AMSAT) community and their Russian counter-parts.

Starting in 1961 with OSCAR-1—a simple radio transmitter in a box—these 'amateur' microsatellites, built by enthusiasts in the USA, USSR, Germany, Australia and Japan, grew in sophistication, until by the end of the 1970s, with OSCAR-8, they had reached a significant level of sophistication in terms of supporting amateur radio communications [1] but still without an onboard computing capability.

Spacecraft Systems Engineering, Fourth Edition.
Edited by Peter W. Fortescue, Graham G. Swinerd and John P. W. Stark.
© 2011 John Wiley & Sons, Ltd. Published 2011 by John Wiley & Sons, Ltd.

It was at this time that staff and students (all amateur radio enthusiasts) at the University of Surrey embarked on a programme to develop a microsatellite for the UK—which resulted in the design, and construction of UoSAT-1 (UoSAT-OSCAR-9), launched by NASA in October 1981 [2].

The 1980s thus ushered in a new era of small sophisticated satellites, characterized by extensive use of microprocessors and other VLSI technologies. Indeed the success of UoSAT-1 lead directly to a second spacecraft (UoSAT-2—UoSAT-OSCAR-11), being designed and built in the record time of just 6 months for launch in March 1984 [3].

UoSAT-1 continued to operate successfully for eight years until its re-entry in 1989. UoSAT-2 remained operational for twenty years—despite its three year mission design life! These two spacecraft established the modern concept of a microsatellite, and Surrey has since become a world leader in providing small-satellite platforms and the associated technological 'know-how' to both existing and emerging space nations through its spin-off company—Surrey Satellite Technology Ltd (SSTL).

In many ways the relationship between a 'microsatellite' and a 'conventional' satellite is similar to that between a microcomputer (i.e. a personal computer or PC) and a conventional computer. That is, a microsatellite is sufficiently low in cost and yet has sufficient utility to enable it to serve a much wider user community than would otherwise be the case.

The mass market for PCs is sustained and exists (in-part), due to the astonishing rate of progress of microelectronic technologies over the last 30 years. The staggering developments in microelectronics, stimulated increasingly by the consumer market rather than military requirements, and the dramatic pace of consumer product development, have caused conventional space technology often to lag considerably behind that taken for granted on the ground. However, these advances have had a direct impact on the power and sophistication of microsatellites through their adoption of *commercial-off-the-shelf (COTS)* microelectronic technologies developed for terrestrial use and adapted to the space environment. The combination of reducing budgets for space and increasing capability of low-power microelectronics have enabled a new breed of highly capable *smaller, faster, cheaper* satellites to realize many space missions—complementing the conventional large satellite systems still necessary for large-scale space science and communications services to small terminals.

Like their PC counter-parts, microsatellites are generally small and relatively inexpensive—typically 1/10th or 1/100th the mass and cost of conventional satellites. This has led to the following widely accepted definitions, given in Table 18.1, and shown schematically in Figure 18.1.

Table 18.1 Classification of spacecraft by mass and cost

Class	Mass (kg)	Cost (£M)
Conventional large satellite	>1000	>100
Conventional small satellite	500–1000	25–100
Minisatellite	100–500	7–25
Microsatellite	10–100	1–7
Nanosatellite	1–10	0.1–1
Picosatellite	<1	<0.1

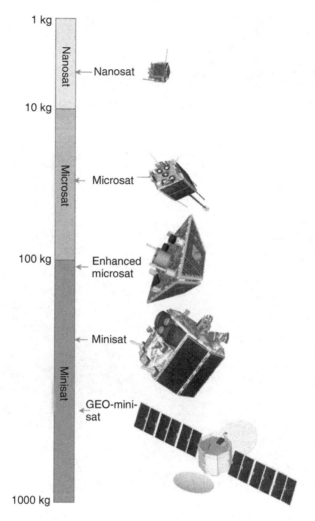

Figure 18.1   Schematic of small satellite classification (Reproduced by permission of SSTL)

Although microsatellites are physically small, they are nevertheless complex and exhibit virtually all the characteristics of a large satellite—but in a microcosm.

In addition to fulfilling communications, remote sensing and space science applications, these miniature satellites are particularly suitable as a focus for the education and training of young scientists and engineers by providing a means for direct, hands-on experience of all stages and aspects (both technical and managerial) of a real satellite mission—from design, construction, test and launch through to orbital operation.

This educational aspect has been given a considerable boost by the *CubeSat* concept, as shown in Figure 18.2—a $10 \times 10 \times 10\,cm$, $1\,kg$ satellite (referred to as a 1U unit)—first proposed in 1998 by Professor Robert Twiggs at Stanford University's Space Systems Development Laboratory [4]. These satellites, and their slightly larger derivatives (so-called 2U and 3U units), have revolutionized practical 'hands-on' space

Figure 18.2 Professor Bob Twiggs with CubeSat (Reproduced by permission of Robert Twiggs)

education and world wide over 70 such projects have been completed or are under development—principally by Universities [4].

Of course, such small satellites cannot cater for all mission objectives—especially where large transmitter powers, antenna arrays, or optical apertures are required for applications such as direct broadcast TV and mobile voice communications from geostationary orbit, or for very high resolution Earth or astronomical imaging. Nevertheless, as on-board processing capabilities have advanced, many missions, hitherto only feasible with large satellites, have become possible with microsatellites at a fraction of the conventional mission cost. In particular, microsatellites make affordable constellations possible (see Section 5.5) and can thus, for example, provide rapid revisit and high temporal resolution for Earth observation missions at a realistic cost. The advent of CubeSat and similarly sized spacecraft opens up the possibility of designing missions using 'swarms' of tiny satellites to form robust networks that can synthesize apertures.

Small, low cost satellites alone do not result in low cost missions, and each element of the qualitative equation below needs to be appropriately apportioned, with typical values given below:

Total Mission cost = satellite cost + launch cost + orbital operations costs over lifetime

100%   70%   20%   10%

Similarly, a small satellite is not necessarily a low cost satellite. Low cost is achieved through the application of a certain *design philosophy* to the entire mission. Indeed, for many, it is not the *size* of the spacecraft as such, but the application of this design philosophy, which distinguishes 'small' satellites from conventional ones.

18.2   SMALL SATELLITE DESIGN PHILOSOPHY

Modern small satellites are built in rapid time-scales (typically 12–18 months), at relatively low-cost, and make maximum use of state-of-the-art COTS technologies to achieve complex functionality, whilst at the same time minimizing dependence on complex mechanisms, deployable structures, and so on.

The satellites are *engineered to cost* specifically to meet their mission objectives during their design lifetime—and no more. The mission objectives are carefully traded against cost to achieve the minimum necessary to achieve the required outcome—typically aiming to achieve 80% of the (conventional) mission capability at 20% of the (conventional) mission cost. Careful management of risk allows more spacecraft missions to be afforded more regularly. The consequences of this design philosophy are contrasted with those of a conventional mission in Figure 18.3.

The rapid development of modern electronic devices means that small satellites often use state-of-the-art technologies with little or no flight heritage. This poses an immediate problem—how do you know if these technologies will work in space if they have never flown before? This is especially true of COTS technologies, which, by their nature, are unlikely to have been designed to operate in the relatively harsh environment of space.

In conventional space missions, there is usually sufficient time and resources to undertake a thorough programme of ground-based testing prior to flight (see Chapter 17), but this is problematical if cost-reduction and short design-to-orbit times are major mission drivers—as they usually are in small-satellite programmes. Thus, a different approach is needed.

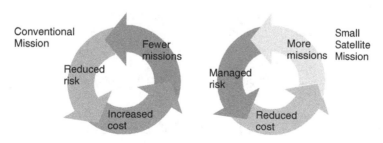

Figure 18.3   Consequences of conventional mission *versus* small satellite mission design philosophy (Reproduced by permission of SSTL)

The general approach to minimizing the risks associated with the use of COTS technologies is to rely more on *sound design practices*, experience and *system-level testing*, rather than on (expensive) programmes of testing individual devices. However, this requires the design-engineers to have a good knowledge of the (space) environment to which their systems will be exposed—and to use appropriate design techniques and margins as necessary. Some of the measures employed may be summarized as follows:

- Minimize the variety of devices/materials.
- Avoid toxic, volatile or potentially explosive substances.
- Keep the interfaces simple.
- Minimize moving parts—use of body cells, use of passive thermal control.
- Use previously-flown designs and components in essential systems.
- Be realistic with safety margins.
- Ensure systems are capable of independent operation—avoid chains.
- Use carefully selected high volume components where possible.
- Avoid software in the loop.
- Knowledgeable use of volume-production, and modern components.
- Use of a layered, failure-resilient system architecture.
- Ensure a thorough burn-in prior to flight.

Solving the technical challenges associated with the design, construction, test and operation of a microsatellite is only half of the story. In parallel with the technical considerations of the mission, *effective project management* is crucial to the realization of a successful low-cost, sophisticated small satellite project.

Affordable small satellites require a very different approach to management as well as to technology if cost, performance and delivery targets are to be met. Several attempts by traditional aerospace organizations to produce such satellites have failed because of the rigidity of management structure and 'mind-set'. Small teams (typically 25 persons), working in close proximity with good communications, and well-informed and responsive management, are essential.

These characteristics are best found in small companies or research teams rather than large aerospace organizations, who may find it difficult to adopt or modify procedures necessary to produce affordable small satellites using staff and structures that are designed for conventional aerospace projects. The main ingredients for a successful small satellite project can be summarized as:

- Highly innovative technical staff.
- Small, motivated teams.
- Personal responsibility for work rigour and quality.
- Good team communications, and close physical proximity.
- Well-defined mission objectives and constraints.
- Technically-competent project management.
- Short timescale (which prevents escalation of objectives).

18.3   SMALL SATELLITE SYSTEM DESIGN

The spacecraft should be designed such that, where possible, essential platform subsystems are fully redundant. However, this does not mean simply duplicating systems, but rather to build a *layered architecture* in which each successive layer of redundancy relies on different systems comprising increasingly well-proven technologies. The upper-layer systems use state-of-the-art high-performance device types—often without *flight heritage*—which give a very high degree of functionality. The lower-layer systems, on the other hand, use device-types which have been flown and tested in previous spacecraft, and which are able to carry out most of the same functions—albeit with a possible loss of performance. In this way, problems caused by an inherent system design fault, or by the systematic failure of a particular device-type, will not be duplicated in the redundant layers. If the newer devices prove successful, they become effectively *flight-qualified* for use in future spacecraft, allowing the capabilities of the satellites to be enhanced generation-by-generation in an evolutionary approach. Table 18.2 shows an example of this type of evolution with respect to the main on-board processors used on some of Surrey's 'UoSAT' microsatellites.

It is the *on-board data handling (OBDH)* system (see also Chapter 13) that is the key to the sophisticated capability of the microsatellite. For example, at the heart of the OBDH system of a current generation UoSAT microsatellite is a 80C386 *on-board computer (OBC)*, which runs a real-time multi-tasking operating system performing various tasks ranging from payload operations, basic *housekeeping* functions, to sophisticated attitude control and orbit maintenance. There is also (usually) a secondary on-board computer to share computing-intensive tasks and to act as a complete back-up if necessary. In addition, all major sub-systems and payloads have their own in-built microcontrollers which provide local intelligence, low-level data processing, and which also support *telemetry and telecommand (TTC)* functions. In this way, the architecture of the microsatellite OBDH system provides similar functionality to that of a modern conventional spacecraft using,

Table 18.2 Small satellite computer processors

Microsatellite	Main On-Board Computer Processors (subsystem microcontrollers and payload processors not included)			
	1802	Z80	80C186	80386EX
UoSAT-2 (1984)	Primary			
UoSAT-3 (1990)	Primary (back-up)	Secondary (experimental)	Primary (experimental)	
UoSAT-5 (1991)		Secondary	Primary	
KITSAT-1 & S80/T (1992)		Secondary	Primary	
PoSAT-1 & HealthSat-2 (1993)			Primary	
Cerise (1995)			Primary	
FASat-Bravo & Thai Phutt (1998)			Secondary	Primary (experimental)
Subsequent Missions (1999–2009)				Primary

for example, the ESA standard OBDH bus—albeit using COTS data bus standards [in this case the controller area network (CAN) bus] with bespoke 'cut-down' software interfaces which avoid the large overheads associated with the *Consultative Committee for Space Data Standards (CCSDS)* TTC formats. However, unlike typical practice with conventional spacecraft, a primary feature of the microsatellite OBDH philosophy is that *all* the primary software on-board the microsatellite is loaded *after* launch and can be upgraded and reloaded at will by the control ground station thereafter.

Normally, the satellite is operated *via* the primary computer and the real-time multitasking operating system. All telecommand instructions are formulated into a 'diary' at the ground station and then transferred to the satellite OBC for execution either immediately or, more usually, at some future time. The OBC can also issue its own telecommands autonomously in response to sensor inputs. Telemetry from on-board platform systems and payloads is gathered and monitored by the OBC and is transmitted to the ground immediately and/or is stored in on-board memory whilst the satellite is out of range of the control station. The OBCs also operate the attitude control systems according to control algorithms that take input from the various attitude sensors and then act accordingly. Thus it is this OBDH environment that allows such a tiny microsatellite to operate in a highly complex, flexible and sophisticated manner, enabling fully automatic and autonomous control of the satellites systems and payloads.

Surrounding the OBDH system are attitude determination and control systems, power generation and conditioning systems, communications systems, as illustrated in Figure 18.4—all of which support the mission payloads housed in a mechanical structure.

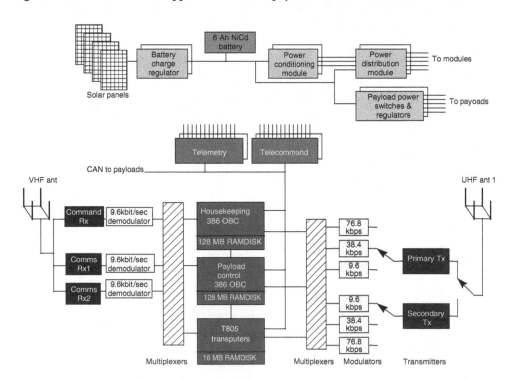

Figure 18.4 Block diagram of typical UoSAT microsatellite system architecture (Reproduced by permission of SSTL)

18.4   COTS COMPONENTS IN THE SPACE ENVIRONMENT

The space environment (see also Chapter 2) can be particularly harmful to COTS devices. Once in orbit, the devices will experience high-vacuum conditions, potentially extreme hot or cold temperature conditions, and relatively high levels of ionizing radiation.

18.4.1   Vacuum effects

Many COTS parts contain plastic materials, which may *out-gas* under vacuum (see Section 2.4.1). These either cannot be flown, or require further encapsulation before they can do so. The plastics used to encapsulate integrated circuits are generally stable under vacuum, and to date, no problems due to use of plastic encapsulated parts have been encountered—despite common usage. However, prior to flight, care must be taken to store these parts under controlled temperature and humidity conditions in order, in particular, to prevent the ingress of water.

Plastic encapsulation is thought to increase the risk of *electrostatic discharge (ESD)* damage (see also Section 16.4.1)—particularly in high orbits such as the geostationary Earth orbit (GEO). However, this effect is insignificant in small spacecraft in low-Earth orbit (LEO).

18.4.2   Thermal control issues

Many COTS parts are only rated to operate at a temperature of between 0 and 70°C. Thus, particular care must be taken with the spacecraft's thermal design. Virtually all microsatellites make use of passive thermal control techniques (see also Chapter 11) to maintain interior temperatures at moderate levels (10–30°C), with interior temperature variations kept to a few degrees Celsius per orbit in order to reduce thermal stress. Where available, COTS parts with extended temperature ranges, e.g. *industrial spec* (−20–+85°C) or *Mil-spec* (−55–+125°C) are preferred. Exterior systems necessarily experience greater thermal cycling during an orbit, with variations of the order of 50–100°C not being unusual.

Prior to flight, extensive thermal-cycle *burn-in* testing is carried out at module level, and mandatory thermal-vacuum testing is performed on the spacecraft as a whole in order to screen the COTS parts for reliability under simulated space conditions.

18.4.3   Effects of ionizing radiation

COTS devices may be particularly susceptible to the deleterious effects of the ionizing radiation environment encountered in space [5]. Thus, particular attention must be paid to the design of COTS-based systems in order to cope with the resultant *total dose effects (TDEs)* and *single event effects (SEEs)* (see also Sections 2.3.2 and 13.7). The trapped radiation belts (*Van Allen belts*) are a very serious threat to satellites, both in terms of radiation dose, and SEEs. There is a single proton belt, comprising high-energy protons, which affects LEO satellites in a region known as the *South Atlantic Anomaly (SAA)* (see Figure 18.5). The electron belts are divided into two, the inner electron belt being

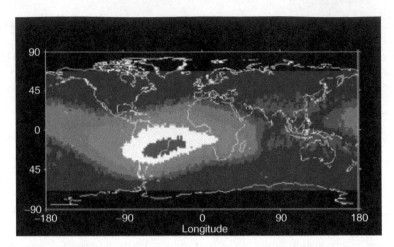

Figure 18.5 South Atlantic Anomaly proton flux at 1330 km altitude measured by KITSAT-1 cosmic-ray experiment (Reproduced by permission of University of Surrey)

more-or-less coincident with the proton belt, and the outer electron belt being at very high altitude. LEO satellites pass through the inner belt electrons in the SAA, and high-inclination LEO satellites will also pass through the outer belt electrons near the poles. GEO satellites orbit within the outer-belt, and elliptical orbits such as *Geostationary Transfer Orbit (GTO)*, and to some extent, *Molniya* orbits, cross both inner and outer belts. Whilst trapped electrons are not a problem as far as SEEs are concerned, they are (potentially) a major source of radiation dose.

In addition to the trapped particles, there are also *galactic cosmic-rays (GCRs)* originating in deep space. These comprise protons and heavy-ions with very high energies (typically a few GeV). Satellites in high-Earth orbits are essentially totally exposed to GCRs, but satellites in LEO receive some shielding effect from the Earth's magnetic field—except at high latitudes. There is also a shielding effect from the solid Earth and its atmosphere. In any case, the flux of GCRs is relatively low, and so they do not constitute a serious source of radiation dose. However, the small heavy-ion content is very effective at causing SEEs.

Solar-flare particles are similar to GCRs, comprising mainly protons with a few percent heavy ions. Their energies tend to be slightly lower than GCRs. The occurrence of flares is correlated with the 11 year solar (Sun-spot) cycle. Major flares occur around the time of *solar maximum*, and can produce very intense particle fluxes at Earth for a day or so, posing a very serious threat to satellites. GEO and other high-altitude satellites are particularly at risk, and one major flare may give the equivalent of a year's radiation dose. Polar, and other high-inclination LEO satellites, will also encounter these particles at high latitudes. As with GCRs, solar flare particles are an effective source of SEEs.

Total dose effects are produced by the accumulated charge and/or displacement damage caused by particles' energy deposition within device structures. For example, changes in threshold voltage and increases in leakage current occur due to hole-trapping within the field and gate oxides of metal oxide semiconductor (MOS) devices exposed to ionizing radiation.

The total dose tolerance of COTS devices varies widely. Some parts fail at less than 5 krad (Si) total dose, whilst others may survive as much as 100 krad (Si). A figure of 5 krad

(Si) can be taken to be a *reasonable* design limit for untested COTS parts. However, where possible, total dose test data on parts should be obtained, either from published sources, or from the results of ground testing using, for example, a Cobalt−60 γ-ray source.

Where a component is likely to receive more than its failure dose within the planned mission lifetime, the use of *spot shielding* by high-density metals (e.g. copper, tungsten or tantalum) should be considered, or if this is still insufficient, the part should be replaced altogether with a *rad-hard* version. Even so, total dose damage will accumulate, and thus design margins must be built into the spacecraft's systems to cope with the expected changes—particularly in terms of voltage level shifts and increased current consumption.

Single-event effects occur due to the charge deposited along the track of an ionizing particle passing through a device structure. These effects may be temporary in nature, or may lead to permanent damage.

SEEs can be reproduced in proton or ion-beam tests, but generally these tests are expensive to perform and require specialist high-energy particle accelerator facilities, which are not widely available. Useful test data may be found in published literature (for example that resulting from the annual Nuclear and Space Radiation Effects Conference [NSREC] published in the *IEEE Transactions on Nuclear Science*), or in on-line radiation effects databases. However, such data is no guarantee that an individual component is safe to fly as there are often considerable batch-to-batch variations in radiation tolerance for COTS devices.

In any case, it should be assumed that SEEs *will* occur, and that the spacecraft systems must be designed with this in mind. SEEs include *single-event upset (SEU)* and *single-event latch-up (SEL)* events. SEUs are unexpected, but impermanent changes in a device's state. They can be corrected by *error-detection and correction (EDAC)*, or majority voting circuits. To prevent the accumulation of SEUs in semiconductor memories, the memory should be *washed* (i.e. the contents read, corrected and re-written) on a regular basis. Figure 18.6 illustrates the occurrence of SEUs recorded in *UoSAT-3*'s program memory,

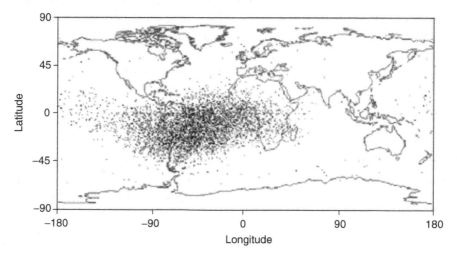

Figure 18.6 Single-event upsets in the S80/T on-board computer program memory at 1330 km altitude (Reproduced by permission of University of Surrey)

as detected and corrected during such a wash cycle. The correlation with the trapped protons of the SAA (see Figure 18.5) is obvious.

Special attention should be paid to whether or not the part is SEL sensitive as this is a destructive effect, which is hard to counter. SELs are usually permanent failures unless the power can be switched off rapidly, whereupon the device may recover. SEL-susceptible parts should be avoided if at all possible. In particular, parts that latch-up readily under proton bombardment should be rejected as unsuitable for use in space.

New threats are emerging for modern fast and complex digital devices, including the *single-event transient (SET)* error, where errors are propagated due to the current spike from a charged particle hit, and the *single-event functional interrupt (SEFI)*, where the device goes into an unexpected non-functional state from which it cannot recover without the power being cycled.

18.4.4 Mechanical environment

Once in orbit, spacecraft experience a *micro-gravity* environment, which imposes relatively little mechanical stress on a spacecraft structure—thermal effects excepted. However, the act of launching a satellite does impart quite severe loads on the satellite's structure—not only from the acceleration of the launch vehicle, but also from the associated vibration and acoustic loads. During the ascent phase, the launch vehicle will undergo stage-separation, achieved through the firing of pyrotechnic devices, which may impart quite severe shock loads on the spacecraft. Thus, as with any spacecraft, microsatellites have to be designed to be mechanically robust. This does not just apply to the structure of the spacecraft. The electrical components also have to be robustly mounted—and particular attention needs to be paid to the use of COTS electronic devices, where there can easily be insufficient strength in the soldered connections to mechanically hold the device under the imparted loads. Devices should not be mounted too high off the printed circuit boards (PCBs), and 'heavy' devices (e.g. large capacitors, crystals, etc.) should have additional support in terms of straps to hold them in place. Plastic (vacuum-rated) *conformal coatings* and foams can also play a useful role in providing extra mechanical support. Vacuum-rated RTV silicone rubber is also useful in supporting wired connections to PCBs.

One factor that has to be considered very carefully is that of *mechanical resonance*. Small satellites often fall in a mass-stiffness range that leads to them having resonant frequencies of the order of a few tens of Hertz. This is just the frequency range where launch vehicles tend to produce large vibrational energies, and so satellites may experience significant amplification (or *Q-factor*) of the imparted loads. The mechanical structure should therefore not be too stiff, and should include mechanisms to damp down vibrations and to dissipate energy.

Any new microsatellite structure must undergo *qualification vibration testing* and *shock testing* which is representative of the intended launch vehicle or vehicles. Once built, the flight-model satellite must also undergo a further set of vibration and shock tests to *acceptance level* to satisfy the launch agency that the satellite is sufficiently robust.

Figure 18.7   UoSAT-1 (1981) showing interior, and exterior with three out of four body-mounted solar panels attached (Reproduced by permission of University of Surrey)

18.5   MICROSATELLITE PLATFORMS

The first modern microsatellites, such as *UoSAT-1* shown in Figure 18.7, used a conventional mechanical structure with electronic module boxes clustered around a central thrust column. Solar panels were body mounted to stringers supported from the end (Z-facet) honeycomb panels. However the need to accommodate a variety of payload customers within a standard launcher envelope, coupled with increased demands on packing density, electro-magnetic compatibility, economy of manufacture and ease of integration, led to the development at Surrey during 1986 of a novel, modular design of a multi-mission microsatellite platform, where the electronics modules actually formed the primary structure of the spacecraft. The same scheme was adopted by US AMSAT and University groups for their 1990-launched microsatellites, and it has subsequently been used by many countries developing their own small spacecraft. Indeed, the modular microsatellite platform has been the mainstay of the commercial/technology-transfer microsatellite programmes undertaken by Surrey's commercial arm SSTL. As of 2009, this platform had been used successfully on over thirty different SSTL missions, each with different payload requirements, allowing the spacecraft to proceed from order-to-orbit in typically 12–18 months.

The SSTL modular microsatellite, shown in Figure 18.8, has no 'skeleton' but rather a series of identical outline machined module boxes, stacked one on top of the other, to form a body onto which solar panels and instruments may be mounted. The modules are held together by tie-rods that pass through the whole stack, and allow some dissipation

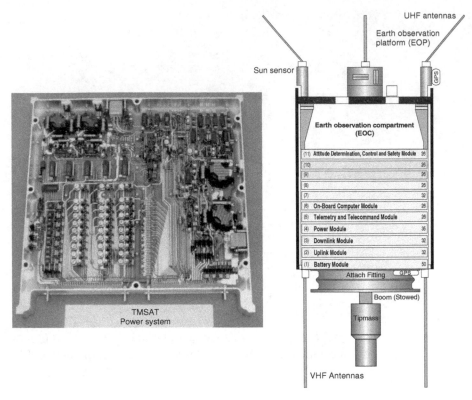

Figure 18.8 Example module box, with schematic showing the construction of a typical SSTL modular microsatellite (Reproduced by permission of SSTL)

of vibrational energy. Whilst this does limit the maximum potential height of the stack, in practice this is not a problem in the context of microsatellite mass constraints. Larger spacecraft can use multiple stacks (see Section 18.6).

At the lower end of the microsatellite mass scale, the $\sim$10 kg AMSAT microsatellites typically have five stacked module trays, forming a cubic-structure with body-mounted solar panels on all six facets. These panels typically provide between 6–16 W of electrical power. Communications, are provided by standard amateur radio uplinks/down-links (e.g. 2 m, 70 cm, 23 cm and/or 13 cm wavelengths) using the AX.25 packet protocol. This protocol was developed in the mid-1980s by AMSAT groups (including Surrey) as a wireless-compatible extension of the standard X.25 wide-area network computer communication protocol. The 'packetized' nature of the transmissions—both to and from the ground—gives the ability to support several 'virtual' links to the spacecraft operating over a single RF carrier. Thus telemetry, telecommand, software up-loads and payload data can all be supported over a single duplex RF link.

Such small, low-cost microsatellites have limited attitude control—usually confined to passive alignment to the local magnetic field achieved via permanent magnets. Sometimes the antennas are black-and-white painted so as to impart a slow spin to aid thermal control. The use of more-or-less omni-directional antennae means that the satellites can carry out their primary role as amateur-radio communication transponders without the need for accurate pointing control or knowledge. However, high-gain communications transponders and Earth-observation payloads do require an Earth-pointing facet or platform. Thus, a typical 50 kg SSTL micro-satellite is maintained to within 1° of nadir by employing a combination of gravity-gradient stabilization using a pyro-released 6 m boom (see also Section 9.4.3), and closed-loop active damping using electromagnets operated by the on-board computer. More recent spacecraft have used momentum wheels instead of gravity-gradient booms to provide even more accurate attitude control (typically ~0.2°)—however, this does introduce moving parts, which are inevitably less reliable. Thus, a gravity-gradient boom is usually retained, ready to be deployed should the wheels fail. Attitude determination is provided by Sun sensors, geomagnetic field sensors (flux-gate magnetometers), and star field cameras, whilst orbital position is determined autonomously to with ±15 m by on-board *Global Positioning System (GPS)* receivers. Electrical power is typically generated by four body-mounted GaAs solar array panels, each generating ~35 W, and is stored in a 7 A-h NiCd rechargeable battery. Communications are supported by very high frequency (VHF), ultra high frequency (UHF), L-band and/or S-band uplinks/downlinks, operating on amateur, commercial or military bands depending on the mission context. These satellites also use fully error-protected AX.25 packet link protocols operating at data rates between 9.6 and 156 kbps.

At the higher end of the microsatellite mass scale, the latest generation of 100–150 kg SSTL 'enhanced microsatellites', is based on the proven flight heritage of the 50 kg microsatellites, but are designed to support missions requiring larger or more demanding payloads—still at relatively low cost. The 600 × 600 × 500 mm bus structure accommodates larger solar panels, and provides flexible internal and external payload accommodation. Nevertheless, this bus fits comfortably on low-cost secondary payload carriers such as the Ariane 5 *Ariane Structure for Auxiliary Payloads (ASAP)*.

Such enhanced microsatellites (see Figure 18.9) give in excess of 100 W orbit average power. A full three-axis attitude control system allows payloads to be pointed at any terrestrial or celestial target with an accuracy of better than 0.1°. A network of on-board computers and embedded controllers automates all telemetry and telecommand,

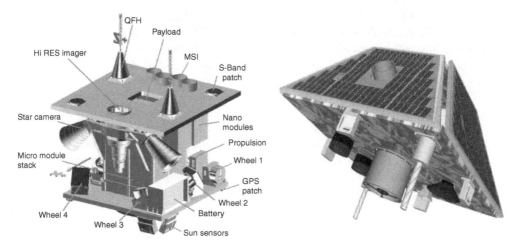

Figure 18.9 High capability 'enhanced' microsatellites (Reproduced by permission of SSTL)

communications, payload control and data management functions. S-band and X-band downlinks provide high-speed (up to 200 Mbps) communications for remote sensing payloads, etc.

18.6   MINISATELLITE PLATFORMS

There has been a steady stream of primarily scientific and military satellites in the few 100 kg 'minisatellite' class. Since the early 1990s a number of such satellites have been built for just a few tens of millions of dollars—low by conventional space standards. However, in response to growing payload demands for power, volume and mass—but within very small-scale financial budgets—there has been increasing interest in flying minisatellites constructed on the same principles as the microsatellites previously described.

SSTL's first mini-satellite platform, shown in Figure 18.10, was designed according to such cost-effective principles—resulting in a basic platform cost of £5–6 million

In an age of decommissioning of nuclear weapons, it is interesting to note that UoSAT-12 was launched successfully by a converted SS-18 (Dnepr) Inter-continental Ballistic Missile (ICBM). Three-axis control was provided by a combination of magnetorquers, momentum wheels and cold gas N_2 thrusters—whilst an experimental electric N_2O *resisto-jet* thruster provided orbit trimming and maintenance demonstrations for future network constellations. Much of the technology proved on this mission has been incorporated into the current-generation of enhanced microsatellites—demonstrating the important inter-play between technology demonstration missions and the development of operational spacecraft.

18.7   NANOSATELLITE PLATFORMS

At the opposite end of the mass scale, the continuing trend in the miniaturization of technology has raised interest in the development of really tiny spacecraft of just a few kg or less. Much of this interest has resided in the USA, particularly at the Aerospace Corporation and amongst the military—the Air Force Office of Scientific Research (AFOSR) and the Defense Advanced Research Projects Agency (DARPA), who stimulated a programme of nanosatellite and picosatellite research in US Universities—most notably at Stanford resulting in the CubeSat concept. Similar interest resides in European Universities including Surrey with its PalmSat initiative. Early examples of such satellites include the 3 kg SPUTNIK-40—a 40th anniversary commemorative one-third scale functional replica of the original Sputnik, which was deployed from the MIR space-station in November 1997. SPUTNIK-40 was built by French students from l'Aeroclub de France (responsible for the radio transmitter), and staff from the Russian Aeronautical Federation (responsible for the structure). This was followed in 1998 by the Technical University of Berlin's 8 kg TUBSAT-N and 3 kg TUBSAT-N1, launched on a Shtil-1 converted missile from a Russian nuclear-powered submarine. The follow-up SPUTNIK-41 was also launched in 1998. However, it is the new millennium that has ushered in the new era of sophisticated nanosatellites and picosatellites, starting with a group of six ∼250 – 500 g picosatellites launched from Stanford's OPAL microsatellite in January 2000. This launch also saw the flight of Arizona State University's 5 kg ASUSAT-1, and the US military JAWSAT nanosatellites, which enjoyed mixed success.

Figure 18.10   SSTL's 312 kg UoSAT-12 minisatellite, launched in 1999 (Reproduced by permission of SSTL)

In June 2000, the UK launched its first nanosatellite—the 6.5 kg SNAP-1, designed and built at Surrey in just nine months. SNAP-1 is a highly-integrated and sophisticated spacecraft carrying advanced microminiature GPS navigation, CMOS video camera technology, on-board computing, propulsion and attitude control technologies. Figure 18.11(a) shows the spacecraft undergoing mechanical acceptance testing. To get a sense of scale, the circular plate it stands on is approximately the size of a dinner plate.

SNAP-1's primary payload is a *machine vision system (MVS)* that enables it to inspect other spacecraft in orbit. The MVS consist of three ultra-miniature wide-angle CMOS video cameras and one narrow-angle CMOS video camera—each weighing just a few

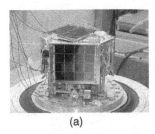
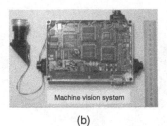
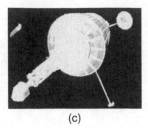

(a)   (b)   (c)

Figure 18.11   (a) The SNAP-1 nanosatellite, (b) the SNAP-1 MVS payload, (c) the Russian Nadezhda search-and-rescue satellite imaged by SNAP-1 in orbit (Reproduced by permission of University of Surrey/SSTL)

grams, together with sophisticated image processing electronics supported by a 220 MHz StrongARM RISC processor (see Figure 18.11b). The MVS has been used to image the Russian Nadezhda satellite (see Figure 18.11c), which carried SNAP into orbit, and in particular was used to monitor the deployment of another SSTL microsatellite into space. Subsequently, the MVS has been used to provide medium (500 m) resolution near-IR images of the Earth from SNAP-1's 650 km altitude, near-polar orbit.

SNAP-1's miniature cold-gas propulsion system, which uses butane as a propellant, has been used to manoeuvre SNAP-1 with respect to the Chinese Tsinghua-1 micro-satellite. Figure 18.12(a) shows the miniature thrusters and fuel reservoir, developed by SSTL and Polyflex Ltd, and Figure 18.12(b) shows a graph of the time history of the altitude of SNAP-1 and Tsinghua-1, as recorded by their respective on-board GPS navigation systems. Tsinghua-1, which has no propulsion, falls naturally under the action of atmospheric drag. SNAP-1 falls rapidly at first (a feature of nanosatellites due to their poor *ballistic ratio*), but climbs rapidly once autonomous orbital manoeuvres are initiated. Intense solar activity meant that SNAP-1 fell more rapidly than planned, and so a second sequence of firings was initiated before SNAP was allowed to fall onto Tsinghua-1's orbit.

Since SNAP-1, many CubeSat missions have been launched, starting with the first batch in June 2003. These satellites have principally been used as practical space engineering education test beds—most often as 'flying cameras'. However, some have also provided significant functionality such as demonstrating new technologies, including micro-electrical-mechanical systems (MEMS), the Automatic Identification System (AIS) signal tracking, tether experiments and space environment monitoring.

Future applications for the nanosatellite include the remote inspection of satellites (e.g. the International Space Station), monitoring of deployments systems in orbit, and demonstrating swam control and formation flying techniques.

18.8   AFFORDABLE LAUNCHES FOR SMALL SATELLITES

A sustained, commercial, low-cost small satellite programme must also be matched by correspondingly inexpensive and regular access to orbit through formal launch service contracts—as it makes little sense to construct sophisticated yet inexpensive microsatellites if the launch costs remain prohibitively high. The majority of microsatellites have been launched as 'piggy-back' or *secondary payloads* accompanying a larger primary payload into orbit—where the primary payload pays for the majority of the launch cost.

(a)

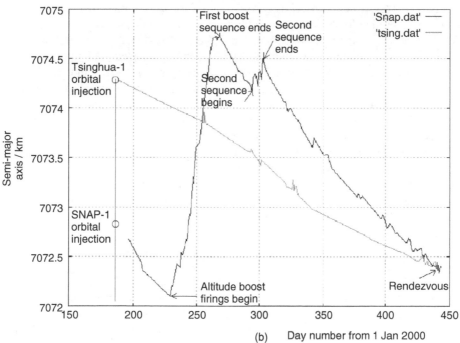

(b)   Day number from 1 Jan 2000

Figure 18.12 (a) SNAP-1 butane cold-gas thrusters and (b) SNAP-1 orbital manoeuvres, showing the SNAP-1 semi-major axis compared with the target spacecraft (Reproduced by permission of University of Surrey/SSTL)

| Delta | Ariane | Tsyklon | Zenit | SS18/Dnepr | Cosmos | Athena |

Figure 18.13 Example launch vehicles used for small satellite missions (Reproduced by permission of SSTL)

Such launch services are available on a diversity of expendable boosters, as shown in Figure 18.13 (see also Chapter 7).

Early microsatellites (particularly the amateur radio satellites) were launched virtually for free on a 'favour' basis by the USA and former-USSR (now Confederation of Independent States—CIS), but these launch opportunities were infrequent and unpredictable. The breakthrough came in 1988 when Arianespace developed the Ariane Structure for Auxiliary Payloads (ASAP) ring specifically to provide, for the first time, regular and affordable launch opportunities for 50 kg microsatellites into both LEO and GTO on a commercial basis. Figure 18.14 shows two UoSAT and four US AMSAT microsatellites mounted on the first ASAP, which flew with the French SPOT-2 spacecraft as primary payload. More than 20 microsatellites were launched via the Ariane-4 ASAP, and Ariane-5 has an enhanced ASAP capability. However, whilst it has been key in providing microsatellite launches world-wide, Ariane alone cannot now provide the number of launch opportunities into LEO needed to meet the burgeoning growth of small satellites. Alternatively, inexpensive launch options from the CIS (on Tsyklon, Zenit, and Cosmos) are now being used increasingly for microsatellites/minisatellites, with the microsatellites often directly attached to the primary payload (see Figure 18.15).

Within the last few years, the large stockpiles of ICBMs in the CIS have become available for use as small launchers through the de-militarilization programme (e.g. SS-18/Dnepr; SS-19/Rockot; SS-25/START). Indeed, SSTL co-operated with ISC Kosmotras (Moscow) to convert the SS-18 ICBM into the first Dnepr small satellite launcher for the successful launch of the UoSAT-12 minisatellite from a silo at Baikonur in April 1999, as illustrated in Figure 18.16.

For CubeSats, a number of commercial suppliers offer launch opportunities, including CalPoly, with their P-POD deployer (co-developed with Stanford), Innovative Solution in Space (ISIS) and the University of Toronto Institute for Aerospace Studies (UTIAS), with their X-POD deployer that can be tailored to nanosatellites as well as CubeSats.

18.9 IN-ORBIT OPERATIONS

The last element of the low cost small satellite equation is *in-orbit operations* (see also Chapter 14). Compact and low-cost mission control ground-stations have been developed to operate microsatellites once in orbit. An example is shown if Figure 18.17. These

Figure 18.14   Micro-satellites mounted as auxiliary payloads on the ASAP ring (Reproduced by permission of SSTL)

ground-stations are based on PCs and are highly automated—interacting autonomously with the microsatellite in orbit—to reduce manpower requirements and to increase reliability. For example, the SSTL Mission Control Centre at Surrey currently operates multiple satellites in LEO and one in *medium Earth orbit (MEO)* with just a single operator. Such facilities rely on the spacecraft themselves having a high level of *autonomy*, as, in practice, the satellites only spend a small fraction of their time in range of any single ground-station (typically 1 h per day for a LEO polar orbiter over a mid-latitude ground-station). Thus, the investment in ground-station and satellite software should not be underestimated. Indeed, the total amount of software on board a typical microsatellite is of the order of several hundreds of kbytes—perhaps an order of magnitude greater than that in a typical conventional spacecraft. Similarly, the ground-station tracking and data handling software comprises many thousands of lines of code.

Figure 18.15   Micro-satellites mounted directly on the primary payload spacecraft (Reproduced by permission of SSTL)

Figure 18.16   Silo launch of a minisatellite using a converted SS-18 ICBM (Reproduced by permission of SSTL)

Figure 18.17 Mission control ground station at SSTL's Tycho House site, Guildford, UK (Reproduced by permission of SSTL)

18.10 SMALL SATELLITE APPLICATIONS

18.10.1 Small satellites for communications

Satellite communications have become synonymous with large geostationary satellites for transparent real-time wideband services. Satellites in low Earth orbits are closer to the user and the consequent reduction in transmission loss and delay-time are attractive, holding out the promise of less expensive ground terminals and regional frequency reuse. Nevertheless, the communications characteristics associated with a LEO constellation pose quite different and demanding problems, such as varying communications path and links, high Doppler shifts, and hand-over from satellite to satellite. The use of early internet-like communication techniques was pioneered on microsatellites by the US military: GLOMR, (1985), MACSAT (1990) and at Surrey: UoSAT-2 (1984), UoSAT-3 (1990), to provide world-wide non-real-time digital data store-and-forward e-mail connectivity—especially to remote regions where existing the telecommunications infra-structure is inadequate or non-existent [6]. Two SSTL microsatellites (HealthSat-1 and 2) were procured by SatelLife (USA) to provide routine e-mail communications for medical teams and aid workers in the Third World. As another example, PoSAT-1 was used to provide military e-mail communications for Portugal during the Bosnia crisis. Figure 18.18 shows a polar scientist using portable hand-held equipment to communicate home via an SSTL microsatellite.

The minisatellite platform provides a means of carrying real-time communications transponders into geostationary orbit.

Figure 18.18   Using hand-held equipment to send e-mails from remote areas via microsatellites (Reproduced by permission of SSTL)

18.10.2   Small satellites for space science

Microsatellites and minisatellites can offer a very quick turnaround and relatively inexpensive means of exploring well-focused, small-scale science objectives (e.g. detecting X-rays, monitoring the space radiation environment, updating the international geomagnetic reference field, etc.)—examples being the UK's ARIEL series and the 74 kg AMPTE-UKS (1984), the 286 kg Swedish VIKING (1986) and ASTRID series, and the 62 kg Danish OERSTED (1999). They can also provide a means of *proof-of-concept* prior to the development of large-scale instrumentation in a fully complementary manner to expensive, long-gestation, large-scale space science missions. This not only yields scientific data quickly, but also provides opportunities for young scientists and engineers to gain 'real-life' experience of satellite and payload engineering—an invaluable experience for later large-scale missions. For example, a doctoral student can initiate a programme of research, propose and build an instrument, and retrieve orbital data for analysis and presentation for a thesis within a normal period of post-graduate study. Eight Surrey microsatellites have carried payloads to monitor the near-Earth radiation environment (see Figure 18.5). Ground-based numerical models have been validated with flight data, and simultaneous measurements have been made of the radiation environment and its induced effects upon on-board systems. A collaborative micro-satellite mission with Chile, FASat-Bravo (1998), carried UV-imaging cameras and UV-radiometers as part of the *Ozone Layer Monitoring Experiment (OLME)*, which provided unique data on the ozone concentrations and structure in the Earth's polar regions, as shown in Figure 18.19.

18.10.3   Small satellites for technology verification

Microsatellites also provide an attractive, low-cost and rapid means of demonstrating, verifying and evaluating new technologies or services in a realistic orbital environment

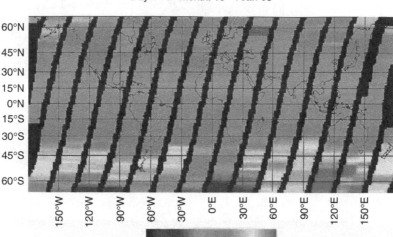

Figure 18.19   Total ozone concentration (in Dobson Units - DU) measured by the
FASat-Bravo Ozone Layer Monitoring Experiment (Reproduced by permission of
University of Surrey)

and within acceptable risks, prior to a commitment to a full-scale, expensive mission.
The UK's Defence Evaluation and Research Agency's (now known as QinetiQ) STRV-1
(1994) series of micro-satellites are prime examples—designed to test new technologies
in the harsh environment of GTO. For example, one of their experiments tested the
performance of new solar cells. Satellites depend upon the performance of solar cell
arrays for the production of primary power to support on-board housekeeping systems and
payloads throughout their 7–15 years operational lifetime in orbit. Knowledge of the long-
term behaviour of different types of cells in the radiation environment experienced in orbit
is, therefore, essential. Unfortunately, ground-based, short-term radiation susceptibility
testing does not necessarily yield accurate data on the eventual in-orbit performance of the
different cells and hence there is a real need for evaluation in an extended realistic orbital
environment. UoSAT-5 (1991) carried a pre-cursor *Solar Cell Technology Experiment
(SCTE)* designed to evaluate the performance in LEO of a range of 27 samples of GaAs,
Si and InP solar cells from a variety of manufacturers, as shown in Figure 18.20.

18.10.4   Small satellites for Earth observation

Microsatellites have really brought about a revolution in Earth observation. Conventional
Earth observation and remote sensing satellite missions are extremely costly—$500
million is not unusual. Thus, there are relatively few such missions and the resulting data,
whilst providing impressive spatial and spectral resolution, yield poor temporal resolution
(revisit) of ground targets due to the small numbers of these spacecraft actually in orbit.
A new opportunity for remote sensing using inexpensive small satellites has come with
the availability of (1) high-density two-dimensional-array semiconductor charge-coupled

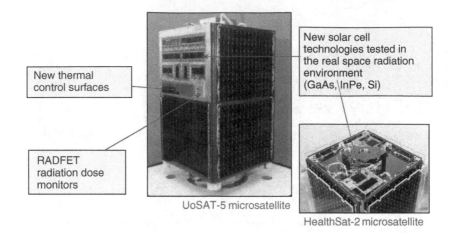

New thermal
control surfaces

New solar cell
technologies tested in
the real space radiation
environment
(GaAs,\InPe, Si)

RADFET
radiation dose
monitors

UoSAT-5 microsatellite

HealthSat-2 microsatellite

Figure 18.20   New technology demonstration experiments on UoSAT-5 (Reproduced by permission of SSTL)

device (CCD) optical detectors (as used in consumer video and digital still cameras), and (2) low-power consumption yet computationally powerful microprocessors. In fact, UoSAT-1 (1981) and -2 (1984) both carried experimental first-generation 2D-CCD Earth-imaging cameras. These paved the way for the first operational cameras on board UoSAT-5 (1991), which was able to image the oil well fires in Kuwait resulting from the 1st Gulf war—the only privately owned Earth-imaging satellite able to do so. The Tsinghua-1 microsatellite launched in June 2000 provides remarkable 35 m resolution images in four spectral bands (compatible with LANDSAT) with the capability of $\pm 15°$ (± 200 km) off-nadir imaging coverage upon demand—all at a total mission cost of £3.5 million, launched into orbit. Minisatellites and enhanced microsatellites have been able to achieve even better results. For example, the 103 kg Korean KITSAT-3 satellite, launched in 1999, produced 17 m ground-resolution images in red, green and near-IR bands, whilst the Disaster Monitoring Constellation (DMC) and RapidEye enhanced microsatellites have shown that fully commercial Earth Observation constellations are viable, with up to ~7 m ground resolution multi-spectral and 4 m ground resolution panchromatic imaging capability. Examples of small-satellite Earth-imaging capability are shown in Figure 18.21.

18.10.5   Small satellites for military applications

Much of the use of small-satellite technology over the years has been driven by the requirements of the military—particularly in the USA and former USSR. Whilst these spacecraft have tended to be conventional—albeit small—there has been increasing interest in taking a more cost-effective approach.

The demands of military-style satellite procurement and the cost-effective approach to microsatellite engineering might, at first sight, appear incompatible. However, whilst retaining the essential characteristics of low cost and rapid response, a military version of the SSTL microsatellite platform with deployable solar panels has been developed to support various military payloads. The main difference between the 'commercial' and

(a)

(b)

Figure 18.21 DMC constellation imagery: (a) 32 m resolution three-band multi-spectral image of Helmand Province, Afghanistan, Spring 2006 — used by the UN Office on Drugs and Crime (UNODC) for opium poppy crop estimates (draped over a digital elevation model). (b) 4 m resolution panochromatic image of Cairo Airport, Egypt. (Reproduced by permission of SSTL/DMCii)

'military' versions of the platform is in the specification and procurement of components and, particularly, in the amount of paperwork that traces hardware and procedures. An optimum trade-off between the constraints of a military programme and economy has been sought, which results in an 'increase factor' for cost and timescale of approximately 1.5 when compared to the 'commercial' microsatellite procurement process.

Practical examples of where small satellites and groups of small satellites have been employed in military scenarios include:

- US TRANSIT programme, where small satellites were employed within a military navigation system.
- US MACSAT and Portuguese PoSAT missions, where satellites were used by small divisions for digital store-and-forward communications.
- US MICROSAT constellation of seven satellites for battlefield communications.
- French military CLEMENTINE and CERISE spacecraft, which have been employed in electronic intelligence gathering.
- TechSAT-21 mission, which will employ small satellites for radar imaging and geo-location.

The first use of the SSTL military microsatellite platform was on the CERISE mission, designed and built for the French MoD and launched into a 700 km low Earth orbit by Ariane in July 1995 (see Figure 18.22). After a year of perfect operations, CERISE made history as the first operational satellite to be (knowingly) struck by a piece of space debris

Figure 18.22   The CERISE spacecraft, built for the French Defence Ministry (Reproduced by permission of SSTL)

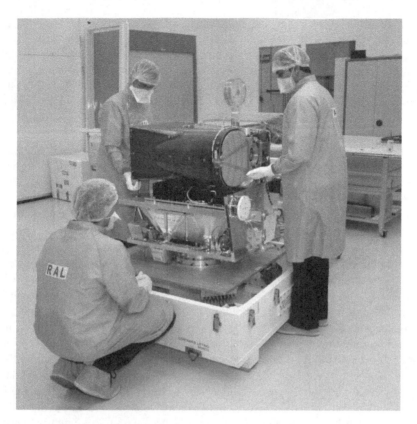

Figure 18.23 TOPSat — a high-resoultion imaging satellite. (Reproduced by permission of the Science and Technology Facilities Council (STFC) UK)

(a rocket fragment) which severed its stabilization boom. However, due to the flexibility of the microsatellite systems, SSTL engineers were able to re-stabilize CERISE by uploading new attitude control algorithms, returning it to operations. A second microsatellite for the French MOD (CLEMENTINE) was launched into LEO in 1999 and a third microsatellite (called PICOsat) was launch successfully on 30th September 2001 for the United States Air Force (USAF), carrying advanced technology payloads for the US Department of Defense.

The use of microsatellites to provide agile, 2.8 m resolution imaging for military applications was demonstrated in 2005 on the TOPSat microsatellite, built jointly by SSTL, RAL (Rutherford Appleton Laboratory) and QinetiQ. Figure 18.23 shows the TOPsat spacecraft, giving an idea of the relatively large optical bench needed to obtain such high resolution images.

18.11 PICOSATELLITES AND RECENT ADVANCES IN MINIATURIZATION

The drive towards small hand-held computers and mobile communications equipment has stimulated even further miniaturization—specifically in low-power microprocessors, efficient transmitters, low mass and high capacity battery technology, and low power GPS receivers.

This has been followed by specific developments of mobile multimedia systems such as digital cameras, spurring low cost, low power, high density data storage as well as high resolution imaging sensors. Recent trends have also tended towards full integration of electronics onto a single chip (*System-on-a-Chip*), which is particularly applicable to picosatellite technology (*Spacecraft-on-A-Chip*). Other technological advances in these areas can be summarized as follows.

Micro-Electro-Mechanical Systems (MEMS) use integrated circuit manufacturing techniques to develop miniature mechanical assemblies together with electronics, and have found early mass market utilization in air-bag deployment sensors and ink-jet printer heads.

Custom devices can be manufactured to order, and in conjunction with application-specific integrated circuit (ASIC) technologies will soon enable entire spacecraft avionics systems, including sensors and actuators, to be assembled within 'sugarcube' volumes. Already, inertial guidance platforms of the order just a few mm across are available, as are some micro-propulsion systems. Much of this work is taking place in US research facilities such as Sandia National Laboratories. Wafer-scale integration of electronic and MEMS technology expands the opportunities, and is the basis of a highly advanced and integrated nanosatellite/picosatellite proposed by the Aerospace Corporation in the US.

Some technologies are driven by space commercialization, including advances in ground segment systems and highly efficient, radiation-tolerant multi-junction solar cells.

Over the next decade, the outcome of these technologies should be the availability of tiny, sophisticated and intelligent satellites which can be mass-produced cheaply. Such satellites could (potentially) be launched by the thousand, and could have sufficient on-board and distributed intelligence to be self-organizing—essentially synthesizing functions through mass action. Quite what the applications of such satellites will be remains to be seen.

18.12 CONCLUSION

The modern concept of the micro-satellite appeared in 1981 with the application of advanced microprocessors and other COTS technologies to the kind of low-cost satellite platforms developed by amateur radio enthusiasts during the 1970s. The 1990s saw the rapid commercialization of these technologies, and now low-cost microsatellites are playing an important role in enabling emerging space nations to carry out increasingly sophisticated space missions including communications, Earth observation, space science and technology demonstration. It would be wrong to directly relate low-cost to the small size of the satellites. Cost-effectiveness comes through the application of a particular design philosophy—not only to the satellites, but to all aspects of the mission. It is this design philosophy which really defines the concept of a 'small' satellite mission. The contrast between the small satellite approach, and that of conventional satellites, has meant that the *smaller, faster, cheaper* ideal has seen most success in small companies and research groups, rather than in large aerospace organizations.

As we enter the new Millenium, rapid progress in the miniaturization of COTS electronics and the advent of MEMS technology has ushered in a new era of even smaller nanosatellites and picosatellites which are opening up new mission scenarios—often involving the use of clusters of satellites to synthesize a particular function. The next decade is likely to see viable satellites mass-produced at the 'chip' level.

REFERENCES

[1] Davidoff, M. (1990) *The Satellite Experimenter's Handbook*, American Radio Relay League, Newington, CT, USA. Chapters 2, 3 and 4.
[2] Sweeting, M.N. (1982) UoSAT—an investigation into cost-effective spacecraft engineering, *J. Inst. Electr. & Radio Eng.*, **52**, (8,9), 363–378.
[3] Sweeting, M.N. (1987) The University of Surrey UoSAT-2 Spacecraft Mission, *J. Inst. Electr. & Radio Eng.*, **57**, (5), (Supplement), S99–S115.
[4] Helvajian, H. and Janson, S.W. (2009) *Small Satellites: Past, Present, and Future*, The Aerospace Press, AIAA, Resto, VA, USA, 2009, Chapter 5.
[5] Underwood, C.I. (1996) Single Event Effects in Commercial Memory Devices in the Space Radiation Environment, PhD Thesis, University of Surrey, Guildford, Surrey, UK.
[6] Ward, J.W. (1991) Microsatellite for global electronic mail networks, *Electron. & Commun. Eng. J.*, **3**, (6), 267–272, December 1991.

BIBLIOGRAPHY

Wertz, J.R and Larson, W.J. (1996) *Reducing Space Mission Cost*, Microcosm Press, Torrance, Calif. and Kluwer Academic Publishers, Dordrecht, The Netherlands.
Boden, D.G. and Larson, W.J. (1996) *Cost-Effective Space Mission Operations*, McGraw-Hill, New York.
Small Satellites Home Page, http://centaur.sstl.co.uk/SSHP/ (accessed 3rd June 2010).

19 PRODUCT ASSURANCE

Geoffrey Hall

Moreton Hall Associates, Maidenhead

19.1 INTRODUCTION

Formalized *Product Assurance* (PA), and its associated terms—Reliability, Quality, etc., came into being as a result of expensive failures, which upon investigation, could have been foreseen and prevented. These usually occurred on large projects typically funded by governments, or sometimes by wealthy individuals. Taxpayers did not like to see their money wasted. The idea emerged that some early checking would be worthwhile even if it added to cost, because the later cost of failure was so large.

19.1.1 The origins of Product Assurance

Reliability, as a formal discipline, emerged during the era of railways when the lifetimes of rolling stock had to be determined to plan replacement of bearings and other wear-out items before a breakage or worse, a loss of life, occurred. Regular inspections were scheduled to monitor wear and tear and this resulted in fewer derailments. Later, the use of electrical machines increased during the industrial era, and they too manifested failures—under-sizing of parts, overwork in application, environmental stresses, etc. Still later, Werner von Braun developed the V2 missile and probability-of-success (reliability) was introduced instead of the percent-of-many statistics used previously.

Quality Assurance (QA), as a formal discipline, emerged from the experience of operating warships at sea in the 1930s and 1940s. The increasing use of electronics for communications and radar meant that ships used about 4000–5000 valves (vacuum tubes). These failed so often that engineers were replacing them on a more-or-less continuous basis. The ship would be unavailable for service for much of its time. Via its Navy Department, the US government visited valve manufacturers and established quality standards

Spacecraft Systems Engineering, Fourth Edition.
Edited by Peter W. Fortescue, Graham G. Swinerd and John P. W. Stark.
© 2011 John Wiley & Sons, Ltd. Published 2011 by John Wiley & Sons, Ltd.

for unit production, and enforced them through recurring inspections and audits. In time, rising prices forced a clear two-part quality objective:

1) reduce failures reaching the user,
2) cut internal waste.

19.1.2   Evolution of Product Assurance

In the decades following, pressures to deliver higher, more sustainable quality have increased as a response to several factors:

- Rising maintenance costs
- Relentless pursuit of 'more-for-the-$'
- Reliability / Performance incentives
- Just-in-time delivery requirements

- Fast changes in market demand
- Longer guarantees and warranties
- Laws on 'fitness-for-purpose'
- Laws on product liability

- Buyers demand Certification to ISO 9000/9001/9002 standards[1]

---

[1]Note that Certification of an organization to ISO 9000 series standard(s) does not guarantee that the customer will get what they ordered—for example the supplier may go bankrupt. If the customer does get what he ordered, it will be of good quality and match the customer's needs. This implies that the buyer must look at the robustness of the supplier's accounts as well as any ISO accreditation before placing an order.

Initiatives by quality experts have led to Quality Circles aimed at introducing continuous improvement, for example, Zero-Defect methods where an operator who does something wrong, and alerts his supervisor to it, is not blamed or victimized. Both 'Total Quality Management' and 'Re-engineering the Corporation' are aimed at management efficiency with quality being a major part of the excellence objective.

19.1.3   Product Assurance for space systems

Space systems differ from other forms of engineering in several respects. For example, in a typical launch vehicle, fuel-feed turbines operate at very high rotation rates, with a temperature of 2000°C+ on one side, and fuel/oxidizer at cryogenic temperatures on the other side. The vehicle also sustains continuous high acceleration and vibration, and upper stages operate in vacuum (see Chapter 2).

Spacecraft are not maintainable (except the Hubble Space Telescope, STS Shuttle and ISS), so spare equipment has to be carried on-board despite the mass penalty (requirement for a more expensive launcher). The spacecraft must not fail irrevocably from an anomaly, so recovery must be pre-planned in design to include a *Failure Detection, Isolation and Recovery* (FDIR) function. Deep space probes are a special case—the total time taken for transmitted telemetry (showing a fault) to reach the operator's console, to be assessed and for a corrective command to be sent back to the spacecraft can be huge. Even then, its effect has to be confirmed through telemetry returned from the spacecraft. Three ground-spacecraft distances must be traversed by the radio signals. Deep space probes must therefore survive failures for many hours, without any intervention by an operator.

High quality parts and materials and ultra-clean manufacturing processes are routinely applied to prevent failures resulting from ageing effects or contamination. The space environment is one of vacuum, radiation and large temperature changes in which components and materials must survive and function properly over long periods.

19.2 PRODUCT ASSURANCE IN A PROJECT

19.2.1 General

Engineers like to use the latest technology—higher efficiency, less consumption of power, lower mass, and so on. Project managers on the other hand like to be always slightly ahead of schedule, and with a margin of unused resources to solve unforeseen engineering problems.

In a spacecraft project, PA acts as the 'conscience' of the project team. The PA role is not to counter innovation, nor to use up resources unnecessarily, but it does focus on achieving required performance in orbit throughout the planned mission lifetime, not just at the start. All elements on a flight spacecraft must be qualified for the application and environment. Where a new part is proposed by engineering, it has to go through a qualification process at a cost to the project. If a particular process is difficult to execute, or critical to mission success, the engineers who perform it must be properly trained and proficiency-tested to the necessary standard. The process outcome is also inspected.

The *product assurance engineer* makes reliability predictions using approved failure rates, promulgates PA standards, witnesses key tests and inspects workmanship.

The '*good product assurance engineer*' also investigates the validity of failure rates, checks that the PA standards are applicable and are applied, checks the calibration records of the test equipment and employs the workmanship standard for inspection.

The general principle here is:

G.P.1—For every PA task, and to the greatest extent possible, verify the assumptions relied upon by the PA task being performed.

It is sometimes said that: 'if anything can go wrong, it will go wrong' (Murphy's Law) The PA function exists both to prevent and to correct problems. The cost of prevention is usually less than correcting a problem after it has occurred.

PA engineers constantly have finely balanced judgments to make. These include:

- insisting on a re-test, when it adds to costs and results in delay,
- inspector's rejection of a device, which similarly adds to costs,
- disposition as 'scrap' can be costly when no spares are available, because a new device has to be purchased or manufactured, and this results in a project schedule delay—itself adding to cost,
- an error in a flight control procedure or contingency procedure can be argued to be a failing of PA, and yet it is the spacecraft operations team who creates the procedures and checks them out with spacecraft simulators,
- an in-orbit failure that ends a spacecraft mission can lead to a large insurance claim. Insurers always receive a presentation of a detailed post-failure investigation report. They also question experts—including the PA staff—who are accountable!

Whenever and wherever something does go wrong, PA staff usually get the blame.

G.P.2—Seek out and understand all unknowns and be fully satisfied that (a) they are characterized by a PA activity and (b) adverse effects are mitigated, before signing off any document.

This principle applies as much in design, manufacture, assembly, integration and test as it does to anomalies that occur in orbit.

19.2.2   Product Assurance definitions and principles

The following form the foundation of all PA activities:

Product Assurance (definition)—A discipline devoted to the study, planning and implementation of activities intended to assure that the design, controls, methods and techniques in a project result in a satisfactory level of quality in a product.

Quality (definition)—The totality of features and characteristics of a product or service that bear on its ability to satisfy stated or implied needs.

G.P.3—Basis of Quality (principle)—The ultimate Quality of a product depends on control over design as well as manufacture and test.

G.P.4—Proof of Quality (principle)—Quality must be evident (contained within verifiable evidence), and demonstrable (capable of demonstration).

These general principles were promulgated by NATO since 1968 in the *AQAP-1 Quality Standard* [1]. These AQAP standards have long since been superseded, first by the British Standard: BS 5750 [2] in 1979, this being adopted later by the International Standards Organisation ISO as a basis for the ISO 9000 series standards [3]. Even so, the AQAP principles are just as applicable today.

19.2.3   PA objectives and project phases

Figure 19.1 shows PA objectives in each project phase, as derived from the top objective: 'Performance shall be as required throughout planned life'. See also Chapter 17.

The design and manufacture phases show the term '*Qualify Design*'. Its purpose is to show that there is a defined baseline from which the end-product (e.g. the spacecraft) can be built repeatedly to the same (qualified) standard. In the subsequent phase—launch and commissioning—this term mutates to '*Qualify Product*'. When the spacecraft is in orbit, it must be checked out and its behaviour characterized sufficiently well for the operators to 'know their spacecraft'. This In-Orbit-Test activity (see also Chapter 14) can result in updates to the *Flight Control Procedures* (FCP) and *Contingency Recovery Procedures* (CRP). At the end-of-life (EOL), whether planned or premature, the task is '*Qualify Mission*'—that is, the mission is assessed in retrospect to determine the extent to which it met the users' requirements and to derive lessons from the project experience. The objective is to apply experience to improve the processes of design, manufacture, test, in-orbit operations.

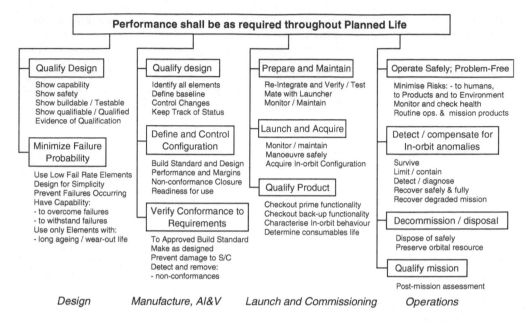

Figure 19.1 PA objectives vs. project phases

In the design phase, '*Minimize Failure Probability*' includes both prevention (use of high quality parts, materials and processes, withstanding failures) and correction (overcoming failures using spare/redundant equipment).

The Manufacturing and AI&V phases include '*Define and Control Configuration*' to ensure that all persons work to the same up-to-date baseline (to avoid wasted effort). This can be summarized as

The Baseline + All Approved Changes = Current Configuration.

'*Verify Conformance to Requirements*'—sometimes called Reconciliation—shows that the 'as-built' product matches the 'as-designed' product (as updated with approved changes). This contributes part of the *proof-of-quality* evidence.

At the launch site, '*Prepare and Maintain*' enables the project manager to be satisfied that the spacecraft is in good condition up to the moment of launch. If any critical parameter is out-of-limits, the project manager can order a launch hold. On-orbit operations are performed to '*Operate Safely; Problem Free*', with anomalies compensated by a CRP.

19.2.4 PA role in risk management

Risk management is a management responsibility (see Table 19.1). The activity as formalized in standards is often delegated to PA staff because they are perceived to have the right skills in 'what-if' studies, probability theory, probabilistic risk assessments and success criteria.

The main features dealt with by project risk management are:

- Risks are forces, or 'jolts', that have an adverse effect on the project plan.
- Risk attributes include: likelihood or probability and type/magnitude of effect.

Table 19.1   Typical Standards Tree (ESA), based upon the European Cooperation for Space Standardization (ECSS) standards

Management ECSS-M-ST-	Engineering ECSS-E-ST-	Product Assurance ECSS-Q-ST-
10. Project Breakdown Structure	10. System Engineering	10. *Not in Use*[1]
20. Project Organization	20. Electrical and Electronics	20. Quality Assurance
30. Project Phasing and Planning	30. Mechanical	30. Dependability
40. Configuration Management	40. Software	40. Safety
50. Info/Documentation Management	50. Communications	50. *Not in Use*[1]
60. Cost and Schedule Management	60. Control Systems	60. EEE Components Control
70. Integrated Logistics Support	70. Ground Systems and Operations.	70. Materials, Materials Procurement and Processes
		80. Software PA

[1]Planned topics merged into other headings

- Risks can manifest themselves in many ways: a threat of legal action; a delay in delivery of a critical item; a potential failure of a test; unachievable qualification, etc.
- Risks are the forewarning of 'if it can go wrong, then it will go wrong' problems.
- Risks must be sought out, recorded, categorized and tracked.
- Continuing effort must be applied to reduce or eliminate identified risks.

In most instances where risk management is applied in a project, the responsible staff create a risk register, listing and defining the discovered risks. Periodically, the register is reviewed to see what action has been applied to reduce each risk, and what reduction in likelihood and/or reduction in effect has been achieved. Newly added risks are also reviewed, and some identified risks are combined if they are very closely related. To help the project manager digest the information in the risk register, identified risks are often colour coded to guide attention, typically as:

Red:	unresolved, likely, imminent and/or severe in effect
Orange:	potentially severe effects and/or growing likelihood
Green:	small likelihood and/or effect
Grey:	retired risk, no longer a threat

Project risk management aims to reduce risk as the project progresses so that the end product of the required standard can be delivered within cost and schedule constraints.

19.2.5   PA — A project-integrated activity

The standards of space agencies, procurement agencies and manufacturers are typically organized under three main headings—*Management, Engineering and Product Assurance* as illustrated in Table 19.1. *Configuration and Data Management*, (C & DM) shown in 'management', is sometimes included within 'product assurance'.

The separation into three branches can sometimes lead to distorted viewpoints e.g. project managers can think of PA as just a layer of expense on top of the project, and treat it as an overhead like, say, a legal department, using this viewpoint to squeeze the PA budget. It can be difficult to 'sell' the concept of 'removal of potential problems' as a valuable contributor to the project. In fact, PA is very much a project-integrated activity as depicted in Figure 19.2.

19.3   RELIABILITY/DEPENDABILITY

19.3.1   Basics

Established over many decades, the term *Reliability* was used to describe and include all the performance attributes that customers wanted in a product. Reliability engineering became the engineering discipline that ensured those attributes were designed-in, built-in

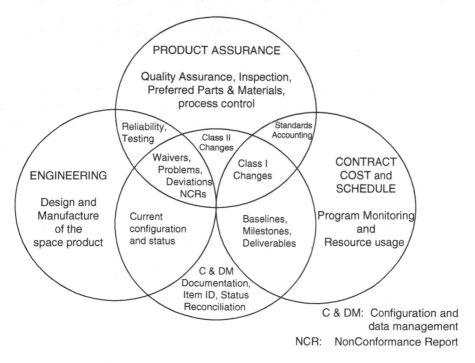

Figure 19.2   Product Assurance in a project context

and verified in the product delivered to the customer. Reliability engineers have always worked to achieve a high probability of delivering the mission products, have ensured that failures could be tolerated or recovered from, and have provided safe modes to prevent the complete loss of the spacecraft.

The term *Dependability* came into common usage because additional attributes (beyond those covered by reliability) became desirable and required by customers. These additional requirements included high *Availability, Maintainability* and *Security* (immunity from interference). Dependability has now become the accepted term that covers both Reliability, and these additional customer requirements.

Dependability (definition [4]) The collective term used to describe the availability performance and its influencing factors: reliability performance, maintainability performance and maintainability support performance. Note that dependability is used only for general descriptions in non-quantitative terms.

Common terms in continued use include:

> *Qualitative Reliability (definition)*— The ability of an item to perform a required function under stated conditions for a specified period of time.
> *Quantitative Reliability (definition)*— The probability that an item will perform a required function under given conditions for a given time interval (equivalent to numerical reliability).
> *Design Reliability or Predicted Reliability (definition)*— The predicted quantitative reliability arrived at during design, using approved input data on failure rates and design implementation (this process assumes that the ground control is 100% perfect and that the spacecraft is exactly as designed)
> *Operational Reliability (definition)*— The quantitative reliability achievable or achieved in-orbit, inclusive of any unpredicted or imperfect behaviour and response to it.

Having synthesized a design, designers analyse it to demonstrate that it will satisfy input requirements on performance, mass, stability, launcher fairing envelope etc. Reliability engineers analyse the design to demonstrate that it will satisfy input requirements on mission lifetime, recoverability from anomalies and removal of *Single Point Failures* (SPF). Sometimes, reliability engineers also trace possible failure cases back to cause(s)—when needed as an input by the operations team.

19.3.2   Related topics

Closely associated with reliability are other terms as mentioned above, namely availability, maintainability and security. Note that various alternative terms are used for 'Security'. Serviceability, Survivability, Safety can be used in this context by different authors in different countries; civilian/military sources can use differing terminology even within a particular country.

> *Availability (definition)*— The *ability* of item to be in a state to perform a required function under given conditions at a specified instant of time.

This definition describes an instantaneous form of availability. In most specifications, availability is more often required as an overall availability throughout the mission lifetime. To respond to this, the analyst integrates the instantaneous form of availability defined above over the mission duration. To perform such an integration, assumptions must be made about outage likelihood and duration. These can, in themselves, become special topics of investigation in probabilistic models.

Maintainability (definition)— The ability of an item to be maintained in, or restored to, a state in which it can perform a required function, when maintenance is performed under given conditions and using stated procedures and resources.

Maintenance applies to spacecraft that are retrievable, and later refurbished and re-launched. It also applies to missions such as the *Hubble Space Telescope* and the *International Space Station* (ISS). The pre-planned, on-board redundant units included in unmanned spacecraft are usually covered by reliability activities.

Security (definition)— The ability of an item to resist interference from intended or unintended external influences and remain functioning as required under the control of the designated control authority. Note that interference can refer to many kinds of external influence, such as meteor or debris impact, physical attack, unauthorized telecommanding, signal-jamming, misappropriation of mission products, and so on.

19.3.3 Reliability theory

The predicted or design reliability is obtained from a reliability model of the spacecraft, developed using series chains of essential hardware/functional elements and parallel elements where sparing or redundancy exists. These models are known as reliability block diagrams, an example of which is shown in Figure 19.3.

In this illustration, all four elements are needed for the device to function. Two-for-one (2/1) redundancy exists at Element 3. Either 1A & 2A or 1B & 2B (or both) are needed, but (for example) 1A and 2B alone do not connect the chain. Element 4 is always needed to ensure function. In this example, individual reliabilities are known inputs to the model. More usually, each 'box' is made up of parts whose failure rates are known, and the reliability is computed from them.

Failure Rate (definition)—The expected number of failures of like devices in a given environment that would occur in a particular time [1] interval. The units of failure rate are fits: failures in 10^9 hours.[2]

[1]Other life exposure measures can also be used such as full cycles, duty cycles or actuations.

[2]This very large number of hours is chosen to encourage failure rates in integer values only.

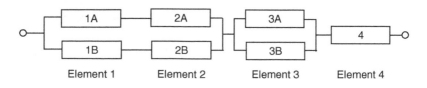

Figure 19.3 Example of reliability block diagram

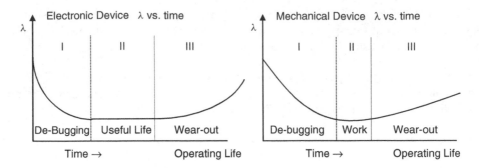

Figure 19.4   Part type failure rate variation with time

Although simply defined, the failure rate of any particular device is not fixed, but varies with time, with stress and with environment. Figure 19.4 illustrates a typical variation of failure rate with time. These shapes are sometimes referred to as bathtub curves. Here λ denotes the failure rate.

For electronic parts, the constant failure rate regime is targeted to get the desired low, sustained values. The wear-out regime occurs when continuing physical and chemical processes have advanced enough to fail the part. For mechanical parts, the duration of the 'work' section of the curve is shorter, relatively speaking, but still useful. Wear-out is just that—surfaces suffer from wear and eventually lead to device failure. Variation of failure rate with stress and environment is also an aspect that the designers as well as parts procurement professionals must take into account, as:

— failure rates begin to rise well inside the part manufacturer's rated maxima,
— high temperatures increase failure rates (Arrhenius's Law quantifies this),
— low temperatures increase failure rates,
— incident radiation increases failure rates.

Derating, good thermal control, use of radiation-hardened (*rad-hard*) parts and physical screening all help to reduce these adverse effects.

After any weak parts have been screened out, the failure rate, as it is defined and illustrated above, is more-or-less constant during the operational working life. This means that during each equal time-interval, a fixed fraction of such parts fail. The reliability, R, or probability of successfully functioning through the required life span T, is then related to the failure rate λ as follows:

$$R = \exp(-\lambda T)$$

Consider an example in which a 'box' is made up of n_1 type 1 parts with failure rate λ_1, n_2 type 2 parts with failure rate λ_2, and so on. For the 'box' to perform successfully during the required life T, all parts within the 'box' must perform successfully. The composite reliability of the box is the product of all the reliabilities of the individual parts, so that

$$R_{\text{type1}} = \exp(-\lambda_1 T)\exp(-\lambda_1 T)\ldots\ldots\exp(-\lambda_1 T), \qquad n_1 \text{ times},$$

$$R_{\text{type2}} = \exp(-\lambda_2 T)\exp(-\lambda_2 T)\ldots\ldots\exp(-\lambda_2 T), \qquad n_2 \text{ times},$$

...and so on.

The total reliability is then given by

$$R_{total} = R_{type1} \, R_{type2} \, R_{type3} \ldots$$

Note that R_{total} is a product of simple exponential functions, so that the arguments or indices add up to form a single 'exp' function with a composite argument as follows:

$$R_{total} = \exp(-(n_1\lambda_1 + n_2\lambda_2 + \ldots.)T) = \exp(-\lambda_{total}T).$$

Adding the indices in this way to obtain λ_{total} is known as the *parts count method*.

Where a constant failure rate is not applicable, several other distributions are available to match the known reliability characteristics of the part, in particular *normal, lognormal* or *gamma-function* distributions. Where a very unusual reliability characteristic is needed to match a part, the use of the *Weibull distribution* may be considered. This distribution relies on the exponential function as a base with three shape parameters γ, σ, β to enable a good fit to characteristics. The *Weibull* distribution is expressed as

$$R = \exp[-\{(T - \gamma)\sigma\}^{\beta}].$$

Referring again to Figure 19.3, and using $\lambda_{\#\#}$ as the applicable box failure rates, we find the reliability of the device for a life span of T, to be given by

$$
\begin{aligned}
R_{device} = \{ & 1 - (1 - \exp[-\lambda_{1A}T]\exp[-\lambda_{2A}T]) \\
& \times (1 - \exp[-\lambda_{1B}T]\exp[-\lambda_{2B}T]\} \quad \text{1 \& 2 both \textbf{not} failed} \\
& \times (1 - \exp[-\lambda_{3A}T]\exp[-\lambda_{3B}T]) \quad \text{3A \& 3B both \textbf{not} failed} \\
& \times \exp[-\lambda_4 T] \qquad\qquad\qquad\qquad\quad \text{4 must work(= \textbf{not} fail)}
\end{aligned}
$$

The combining of terms follows standard probability theory, that is, multiplication '×' is used for *both/and*; '+' for *either/or*; '1−() ' for *not ()*.

19.3.4 Reliability enhancement

As part of their objective to 'minimize failure probability', reliability engineers employ several methods to enhance reliability:

Use of redundancy greatly increases numerical reliability. Say, a piece of equipment has a reliability of 0.9. If an identical equipment is added as a spare, the reliability of the pair rises to $0.99 = 1 - (1-0.9)(1 - 0.9)$ i.e. the probability of both *not* failing. The gain in reliability is even more, if the partner unit is switched off until it is needed. See also Section 19.3.5 for exceptions to this rule.

Design diversity is the deliberate use of dissimilar units that can each perform the same function. This is a more expensive option than adding one identical unit. This diversity approach is used where the unit may be novel or where the application is safety-critical or even to obtain in-flight qualification of two different units.

Effects limitation is aimed at stopping the propagation of a failure to any related functional units or to proximate units (in case of overheating). Resettable fuses, auto-switching to back-up units, invoking of safe modes and thermal shielding are all options available to designers.

Derating of parts can reduce their failure rates and so enhance reliability. There are also beneficial side-effects of derating. For example, the reduced failure rates are closer to those used in reliability prediction, and so validate the predictions. Also, unwanted effects such as overheating are reduced. However, some parts (e.g. microwave field-effect transistors (FET)) cannot be derated as they operate successfully only at their rated power levels.

Radiation screening is used where certain kinds of electronics are employed. The kinds that are susceptible to radiation damage are complementary metal oxide semiconductors (CMOS), on-board processors and random access memories (RAM). Sometimes, it is enough to locate the electronics away from the surface of the spacecraft and use its exterior as a screen.

Handling/assembly controls are employed throughout manufacturing facilities to avoid introducing incipient damage to parts.

Inspection and/or testing of procured parts is a routine activity often referred to as *Incoming Inspection*. Typically, a sample of a batch of procured parts is tested to verify function and note whether variation in part-to-part characteristics are within allowed tolerances. Most or all may be also visually inspected.

Testing to demonstrate reliability is a very rare activity. The mission lifetimes of spacecraft in orbit are often 15–20 years and even accelerated life-testing can take an extremely long time to prove the necessary reliability. Some suppliers do submit their products to continuous life-testing as part of their sales promotion. Teldix in Heidelberg, Germany, for example, has a momentum wheel rotating in an evacuated glass chamber in its reception area. This has been rotating for many years.

19.3.5   Design analysis techniques

The reliability engineer analyses the design from the perspective of reliability and failure potential and relates this to input requirements (usually from a customer). Table 19.2 identifies common analysis techniques, and their advantages and drawbacks.

Note that redundant units are usually switched OFF to reduce power consumption and increase reliability. *Contingency analysis* can check the correctness of these kinds of decision throughout the spacecraft. For example, consider a receiver which is ON to receive telecommands. If it fails, the FMECA remedy is 'switch to redundant receiver'. Contingency analysis flags up that no command can be received by the spacecraft if the redundant receiver is OFF. The redundant receiver is therefore always ON in hot redundancy to the primary receiver. The logic here relies on thinking beyond the point of first failure (which is often ignored).

19.4   PARTS

19.4.1   Basics

A part is one of the basic elements of a spacecraft. The spacecraft is built up from its constituent parts. At one time, these were thought to be the only items that could fail—a view supported by the 'parts count method' for reliability prediction. In the early years

Table 19.2 Design analysis techniques

Technique	Used for	Advantages	Disadvantages
Failure states probabilities	Making reliability studies. Probability of success = 1—prob. of failure.	Can influence the design process.	Usefulness of any states less than 'success' is not considered.
Worst case analysis (WCA)	Showing performance at tolerance extremes.	Adds confidence to products made in production lines.	Expensive to do. Statistical validity is questionable.
Failure modes effects and crit-icality analysis (FMECA)	Tracing all single origin faults to effects, observables and remedy. Flags single point failures.	Forces thinking into the 'what-if' failures domain. Input to safety.	Labour intensive. Timing is sometimes too late to have a benefit.
Fault tree analysis (FTA)	Tracing identified system failure modes back to single/multiple causes.	Useful input to the compilers of flight ops. procedures.	Labour intensive.
Contingency analysis (CA)	Validating the practical aspects of the remedy output of FMECA and the FTA diagnostics.	Useful input to the safe mode design and operations team.	None.
Sneak circuit analysis (SCA)	Finding unwanted circuits in designs created to be requirements-compliant.	Can be useful in one company that data-stores its experience.	Not useful across an industry; needs a custom database.

of the space era, part failures accounted for between 6–16% of all anomalies occurring in-orbit. Part-related anomalies are now much rarer, but do still occur—for example, in high-power GHz field-effect transistors (FET), travelling wave tube amplifiers (TWTA), and solar cells. Each of these part-types has passed through several technology upgrades over the decades. Each new iteration brings with it new types of failure.

19.4.2 Part types

Part (definition)—The lowest level of assembly, beyond which further disassembly irrevocably destroys the item.

Parts are sometimes referred to as components although this can cause confusion. In the USA, the term component is used to mean elements assembled from parts. In Europe both terms are in use to mean parts, as defined above.

There are essentially two main categories of part, summarized as:

Electrical, electronic, and electromechanical (EEE)	Mechanical
Wires, connectors, resistors, capacitors, transistors, integrated circuits, opto-couplers, switches, relays, etc.	Structural elements, honeycomb floors, thrust tubes, solar array substrates, antenna fixings, etc.

19.4.3   Failures in parts

Just as in reliability studies, parts experts look at failed parts and try to understand how and why they failed. From this information, they establish controls aimed at preventing each kind of failure. Some typical part failure problems are categorized in Tables 19.3 and 19.4.

Some of the drivers that provoke failures of parts in orbit are summarized in Table 19.5.

19.4.4   Parts procurement and controls

The earlier approach of using mostly *Hi-Rel* or *MIL-STD*, or alternative specialist production line parts has become impossible to sustain for a number of reasons:

- rising costs vs. commercial competition pressures,
- lack of Hi-Rel production lines,
- availability of *Commercial-Off-The-Shelf* (COTS) parts—acceptable for space use.
- availability of *Civil Avionic/Military* (CAM) standard parts.

It is now common for spacecraft to be built using some COTS and CAM parts where the sources are known to produce a stable quality of parts, and these parts have been assessed

Table 19.3   Typical part failure problems

Part types	Problems encountered
Passive parts	
Fuses	Metal crystallizes; in-vacuo behaviour differs; rating error.
Resistors	Residual solder flux; surface cracks $\Rightarrow$ leak paths.
Capacitors	Electrolyte leak; non-polar $\Rightarrow$ polar; contamination $\Rightarrow$ resistive reactance Ω.
Active parts	
RF power transistor	Local thin metallization $\Rightarrow$ metal transport with power on.
IC passivation layer	Local thinning $\Rightarrow$ electrical short through passivation.
Transistor lead bond	Current spike to make bond $\Rightarrow$ eutectic alloy spread-out.
Processors and RAM	Cosmic rays $\Rightarrow$ Single Event Upsets (SEU); soft/hard errors.

Table 19.4   Some electromechanical and mechanical part problems

Problem types	Mitigation
Relays experience arcs at all make/breaks	Avoid contact degradation by using a high temperature non-burn material such as platinum (Pt) for contacts. Limit total number of actuations. Limit total switched current × time per actuation.
Armature movement sometimes obstructed	Establish no-change geometry (especially clearance between armature and case) that guarantees movement at all foreseen temperatures.
Gyros can experience leakage or migration of bearing lubricant.	Critical control of dimensions during manufacture.
Oils can leak even from sealed bearings. Vacuum assists leakage.	Noise spectrum is a very good quality indicator. This signature should not change gyro to gyro.

Dry lubricants do not behave the same way in vacuum (graphite becomes abrasive!)

Table 19.5   Triggering causes that can lead to part failures

Environment effects	Problems caused
Heat and temperature	Arrhenius' law: higher temperature ⇒ higher failure rate.
Differential expansion	Causes internal strains within parts at extremes of temperature ⇒ fracture. Exception: thermostats rely on differential expansion to make/break the circuit controlled.
Radiation effects	Electronic switching degrades.
Other	Presence of contaminants ⇒ premature failure. Power overstress ⇒ premature failure. Vibration dislodges loose (part) materials.

as suitable for use in the space environment. Even so, great care is taken to ensure that all the parts that are procured are of the required quality. Various *procurement systems* for ordering parts of the appropriate quality exist and these detail the kinds of tests and tolerances that the parts must satisfy. Buyers maintain their own systems for these, often relying on the standards of NASA, ESA, CNES, JAXA and RKA for guidance. However, caution is needed, as summarized in the following principle:

G.P.5—*Parts procurement* standards are not all interchangeable country-to-country.

Examples of different *quality levels* at which parts can be procured are shown in Table 19.6. Note that reliability engineers also use *quality factors* to adjust the part failure rates used in reliability predictions.

Table 19.6 Parts/components procurement levels

Procurement system	Quality level	Quality factor
ESCC[1]	Level B	1
ESCC[1]	Level C	2
ER-MIL[2]	FR Level S	1
ER-MIL[2]	FR Level R	2
MIL-S-19500	Jan S	1
MIL-S-19500	Jan TXV	2
MIL-M-38510	Class S	1
MIL-M-38510	Class B	2

[1]European Space Components Coordination
[2]Established Reliability MIL

Given the size and application of a spacecraft, an internal parts procurement authority (or external buying agency) can list the types and quantities of about 60% of parts needed and source them.

The spacecraft project approach is guided by the Parts Procurement Manager. Early on, this manager establishes a *Preferred Parts List* (PPL) for the project. Generally, all parts on the list are qualified for space use or are qualifiable as COTS or CAM. The engineers responsible for design and manufacture match their needs to the PPL as far as possible. As a result, they identify further parts that they need which are not yet included on the PPL.

Following this 'mismatch' there follows a period of negotiation, mainly focused on which of these (many) further parts can be qualified within an acceptable cost and time-frame. Sometimes, the qualification is simply an extension of an existing qualification to a new application or different temperature envelope. In such cases, a delta-qualification is carried out to establish the acceptability of the part in its new application/environment. Occasionally, a complete program of qualification for the required part is undertaken. This can be very expensive, and often the cost has to be shared between two or more projects. This cost-sharing does take place when an upgrade to an equipment/subsystem is being introduced, and the upgrade costs can be amortized over succeeding spacecraft in an existing production line.

One very important duty of the parts procurement manager is to identify the parts that must be ordered well in advance—so-called *long-lead* parts. This is also co-ordinated with other long-lead items (not necessarily parts).

Ultimately, there is a de-facto parts list for the project, to which the following principle applies:

*G.P.*6—All parts/items to be used in a flight spacecraft must be qualified.

19.5 MATERIALS AND PROCESSES

19.5.1 Basics

Materials, and the processes that modify them, result in basic elements of the spacecraft that are not EEE parts. They can be finished mechanical parts such as structural elements

Table 19.7 Material functional properties and constraints

Functional properties	Constraints
	Common constraints
Strength	Outgassing
Adhesion	Radiation resistance
Elastic moduli	Inertness
Thermal stability	Reproducibility
Ductile properties	Tolerance to processing
Malleability	Matching coefficients of expansion
Cuttable without shatter	Man-rated constraints
Reflectivity	Low toxicity
Absorptivity/transmittance	Low odour
Energy content:	Non-flammable
-propellant, electrolyte, springs	Low outgassing/offgassing

or a means of holding other elements together, such as adhesive tape, fixings, solder or potting compound.

Selection, processing and application of materials requires just as much care as does the procurement and application control of parts.

19.5.2 Properties of materials

Table 19.7 lists the main properties of materials that are of special importance to their use in spacecraft.

Outgassing of materials is a problem that is particular to space applications (see also Chapter 2). When materials are removed from air and placed in a vacuum chamber, the material loses mass. This can result in significant changes to its properties, and the release of volatile condensable materials (which can disable optical devices).

Materials for space use are subject to initial screening that requires their mass-loss in vacuum to be no greater than 1%, and the *Collected Volatile Condensable Materials* (CVCM) to be no greater than 0.1%. This CVCM constraint refers to the material emitted from a sample at $+125°C$ and collected on a surface at $+25°C$ over a 24 hour period.

19.5.3 Applications and processing

There is a large volume of data stored by manufacturers of equipments and whole spacecraft. National agencies also hold considerable data on materials that are mostly available to spacecraft builders. As with parts, there is usually a project specialist Materials and Processes engineer who negotiates with the design and production engineers concerning which materials can be used in which applications.

Experience obtained over many years has led to the prohibition of certain materials from use in spacecraft manufacture. This accrued wisdom does not completely exclude the use of such materials, but a very strong case and special assessments have to be made where their use is unavoidable. Examples of such prohibited materials are given in Table 19.8.

Table 19.8 Prohibited materials and materials to be avoided

Material	Type of problem
Flexible adhesives	Loses elasticity; fractures when solidified
Polyvinyl chloride backing tapes	Unstable properties
Cellulose, paper, fabrics, etc	Unstable properties
Varnishes and coatings which rely on solvent evaporation for hardening	Properties change in vacuum and with temperature cycling
Canada balsam; organic glasses in high precision equipment	Unstable
Most oils and greases	Material migration
Graphite	Abrasive in vacuum
Cadmium, zinc and tin	Dendrite growth; risk of shorting
Most paints	Outgassing unstable properties
Polyvinyl chloride, cellulose and acetates, Plastic film	Unstable properties
Potting	Fracture at temperature extremes
Polyester laminates	Unstable properties
Rubbers using poly-sulfides; plasticizers; chlorinated	Unstable properties
Polyvinyl chloride (PVC) thermoplastic; polyvinyl acetate; butyrate; several polyamides	Unstable properties

Note also that some materials that are accepted for use in space require special attention when being processed into their final state. Typical examples are soldering and specialist machining of metals. Beryllium produces poisonous oxide when machined.

19.5.4 Materials problems

Table 19.9 describes some typical problems with materials that have been encountered in past missions. The experience gained through these occurrences have led to further principles and the reinforcement of existing principles.

The main principles in materials and processes engineering are:

*G.P.*7—'Cleanliness is next to Godliness' in space engineering.
*G.P.*8—The processes and application are just as important as the material.
*G.P.*9—The only guarantee of quality from a supplier is *exact* repeatability.

19.5.5 Special case: materials problems in parts

The materials used in the fabric of parts have the potential to produce yet further problems. Some examples are outlined in Table 19.10.

Table 19.9 Examples of material problems that occurred in orbit

Space Shuttle thermal tiles	The Space Shuttle has over 24 000 tiles, each with a different shape—an extreme configuration management problem. Early launches lost tiles during ascent and re-entry. An extra layer of bonding was required. This added to weight and thus reduced payload capability.
Atomic gas	Cleaning fluid can leave traces of contaminant on surfaces. The space environment encourages the release of atomic hydrogen, which disrupts the molecular lattice. The effect shows as embrittlement in metals. This can lead to fracture, and can result in a catastrophic failure in springs.
Growth of dendrites	Cadmium, zinc and tin have all exhibited dendrite (whisker) growth at corner sites when in the presence of an impurity and/or an electric field. This can lead to short circuits.
GOES lamps	Specialist lamps of an exact light frequency were used on a series of GOES satellites. The first lamps were successful. Later lamps failed in orbit before end of duty life. Part way through the production cycle, the lamp manufacturer had changed the supplier of the filament material for the lamps. The newly supplied raw material was of a different composition and its life characteristics had not been checked for application in the space environment.

Table 19.10 Examples of materials problems within parts

Dendrite growth in digital integrated circuits	Temperature + electrical bias + moisture = dendrites. These needle-like growths (of molybdenum) occur at corner sites in metal/substrate interfaces. They can lead to cross-track shorts and capacity changes—important in some part types.
Galvanic corrosion	Dissimilar metals + moisture + warmth = voltage couple. The moisture passes through plating and forms an electrolyte. The sustained emf causes corrosion.
Stress corrosion	Mechanical stress opens tiny fissures in material. Fissures form sites where impurities can gather and leave material without its protective coating at these sites.
Constituents have unmatched coefficients of expansion	Large temperature excursions—in/out of eclipses—generate severe strains within the part that can lead either to fracture or unstable electrical behaviour.

19.5.6 Materials program

Materials are purchased in a highly controlled way and share many practices with parts procurement. The sources are pre-approved to produce materials of the space-acceptable quality required. Each delivery of materials is inspected and checked for the slightest change from previous deliveries, is logged, dated and source-identified in inventory. For a spacecraft production line, the required materials are purchased in bulk—this implies that the storage environment must be highly controlled to maintain the material in good condition and uncontaminated. See also Traceability in Section 19.6.

The spacecraft project approach is conducted by the Materials and Processes Manager. Early on, this manager sets up a *Preferred Materials List* (PML) for the project. Generally, all materials on the PML are *qualified* for applications in space when processed using approved methods. The engineers responsible for design and manufacture attempt to ensure that their needs correspond with the PML. However, occasionally they will identify further materials that are required which are not yet included on the PML.

Following this disparity there follows a period of discussion, considering which of these further materials can be qualified within an acceptable cost and time. Occasionally, the designers can be persuaded to adopt an existing preferred material already on the PML. Sometimes, the qualification is just an extension of an existing qualification to a slightly different application or thermal environment. In such cases, and similar to the procedure adopted in dealing with parts, a delta-qualification is performed to establish the acceptability of the material in its new application/environment. Occasionally, a complete program of qualification for the required material is necessary. This can be required for, say, a scientific spacecraft where a payload instrument relies on a particular component to function, and this component includes materials that are not yet qualified. Such qualification can be very expensive, requiring specialist laboratory equipment, vacuum chambers and time and effort. It can be difficult to share the cost between two or more projects, where only one is ever likely to need the component. However cost-sharing does take place when an upgrade to an equipment/subsystem is being introduced, and the upgrade costs can be amortized over succeeding spacecraft in an existing production line.

The materials and processes manager has to identify the materials that must be ordered well in advance—so-called *long-lead* items. This early-ordering is also co-ordinated with other long-lead procurements.

Ultimately, there is effectively a materials, processes and application list for the project.

19.6   PRODUCT ASSURANCE IN MANUFACTURING, AI&V

19.6.1   Basics

This phase comprises Manufacturing, Assembly, Integration and Verification (AI&V). The *product assurance* effort in this phase is often called *Quality Assurance* (QA). When quality was a term referring only or mainly to the build standard of the delivered product, this was very appropriate. If the product failed, it was called unreliable, or broken, but rarely was it referred to as of poor quality. In the ISO 9000 series standards and widely in the assurance sector, quality now means the 'totality of features or characteristics that bear on its ability...etc.' (see definition in Subsection 19.2.2). This implies that quality includes most, if not all of the assurance topics. Even so, Quality Assurance is still a widely used term and it is employed here. The sense in which the term QA is used here is summarized by the following:

- QA is mainly focused on manufacturing/test,
- QA reach extends back into design,

- QA reach extends forward into deployment and use,
- QA role is monitoring, audit and verification,
- QA activity contributes to the evidence of quality required by the customer.

The main activities involved in the QA process are given in Table 19.11, and these are briefly described in the subsections below.

In Section 19.2.1 the so-called 'Murphy's Law' was stated—'If anything can go wrong, it will go wrong.' A good way to think of QA is to imagine all the possible ways that a particular task might be performed wrongly, and ask what intervention or extra check could be made to prevent this. QA is largely a considered response to *G.P.2.*

19.6.2 Process controls and records

All *routine processes* and all *critical processes* must be properly understood and carried out to an acceptable and approved standard (otherwise it is possible for someone to omit them or perform them wrongly).

This implies that they must all be fully documented, which includes—preparation; resources required; method of execution and witnesses if necessary. In particular cases, it will be necessary to rehearse and exercise the processes to show correct and consistent results. The controls used must be concise and clear enough that they can be signed off by the responsible PA officer and engineering staff.

19.6.3 Training and certification

All persons must have adequate qualifications and experience. Note that it is not only qualifications that are required. Proficiency is the objective. Whenever a new or modified activity has to be accomplished, (re)training is mandatory for the operative who is

---

Table 19.11 Main Quality Assurance disciplines

1.	Software—a special case (see Section 19.9)
2.	Process controls and records
3.	Training and certification
4.	Traceability
5.	Metrology and calibration
6.	Non-conformances control
7.	Alerts
8.	PPHSS (see Subsection 19.6.8)
9.	Trends and statistics
10.	Design margins verification
11.	Procurement controls
12.	Control of changes
13.	Monitoring, inspection and audit
14.	Qualification and acceptance

to perform the task. The training must itself meet standards, and be delivered only by experienced and approved educators. Those executing critical processes must:

- have demonstrated proficiency,
- hold a certificate signed by an authority to show proficiency,
- receive regular checks and training updates.

19.6.4   Traceability

There are three main kinds of *Traceability* in a space project—the ability to:

- trace use of tools/measuring devices/test equipment (see Subsection 19.6.5),
- trace any part or material back to its original procurement and supplier,
- trace any part or material to where it is used. This is helpful when whole batch problems occur—as when an *alert* is received (see Subsection 19.6.7).

19.6.5   Metrology and calibration

It is quite easy for an engineer to use a test instrument to take a reading, even obtaining an automated print-out. Such readings (particularly if they meet the specification) acquire great significance. However, what is not so obvious is whether the test instrument gives a genuine reading. Regular calibration of all measurement devices and instrumentation is essential good practice.

The magnitude of errors (in tools and instruments) must be known. The total error in the worst case includes:

- the error of the calibration process,
- the error in the calibration standard,
- the error in the user reading of the tool or instrument.

Recall that all outputs should be produced to a repeatable baseline. Therefore, all the checks in aggregate should be sufficient—if they indicate a pass, the device is acceptable. Very rarely, the production of devices is so difficult that the produced items are not consistent. In these cases, just the best (on test, or on measurement) must be selected for use in the flight spacecraft. Whilst the metrology may be good, this is poor engineering practice. In these cases, the baseline is not repeatable.

19.6.6   Non-conformances control

It is a customer requirement that a manufacturer/supplier has a system in place for detecting, recording, assessing and processing, to a satisfactory conclusion, all non-conforming material. *Non-conformances* can arise from an inspection, test outcome, failure to calibrate, contract error, or a design review (action item). Every non-conformance is recorded in a *Non-Conformance Report (NCR)*. It is the responsibility of the *Material Review Board (MRB)* to see through the processing of the non-conformance to its conclusion. Table 19.12 illustrates this processing scheme. The NCR is first categorized according to its severity

Table 19.12 Non-conformance NCR processing scheme

Area	Classification (acronym)	Description	Re-test and check	Notify buyer	Buyer action needed
Severity	(NCR)	Non-conformance found and Report issued with unique ID.			
	(MRB)	Material Review Board con venes; assesses significance of NCR			Yes[1]
	Minor	Disposition locally			
	Major	Serious. Involve customer at once		Yes	Yes
Concession	(RFW)	Request for Waiver of requirement that cannot be satisfied.		Yes	Yes
	(RFD)	Request for Deviation from specification of an in-work end-item		Yes	Yes
	None	Concession not needed			
Disposition	Use-as-is	Correction of no benefit and no value. No RFW and no RFD needed.			
	Rework[2]	Rework to full compliance with specification of the item.	Yes	Yes	
	Rework[3]	Rework beyond specification for use subject to approved deviation	Yes	Yes	Yes
	Scrap	Must not be used. segregate and send to materials reclamation.	Yes	Yes	

[1]Only when 'major' classification is declared.
[2]Will meet specification after rework/retest/inspection.
[3]Will not meet specification after rework/retest/inspection.

of effect on the spacecraft programme. A decision is made as to whether the customer should be asked for a Concession (to waive a requirement, or allow a manufactured item to deviate from a specification). Finally, the appropriate action (one of a limited number of possibilities) is taken. This is referred to as a Disposition.

19.6.7 Alerts

Alerts deal with 'found' problems, mainly with a part or material from a known supplier. Most space communities have a knowledge-pooling facility to notify others of problems that have been encountered with a supplier. ESA and CNES have alert systems in use, as do their counterparts in North America. The purpose of an alert is to disseminate to the wider user community any adverse experience of a particular batch of parts or batch of material.

The *Traceability* functions (see Subsection 19.6.4) are particularly helpful, as they speed up the search for the suspect items. The purchasing records can quickly determine whether or not the suspect parts or materials have been purchased. Similarly, the 'where used' records will identify if any remedial work is needed on flight hardware. Retrofitting new elements is something to be avoided wherever possible as it may require repetition of earlier testing—at a cost.

19.6.8 Preservation, Packaging, Handling, Storage and Shipping (PPHSS)

PPHSS is collectively about environmental controls to prevent damage to the spacecraft, or indeed to any bought-in finished units. The general requirements are as follows:

- packing materials must be inert with respect to the packed item,
- containers must be fitted with environmental sensors to monitor temperature, pressure, humidity, shock, and vibration,
- handling must be non-injurious to humans, and non-damaging to the spacecraft,
- shipping (transportation) must meet the collective requirements[1] of the owner, sender/conveyor and recipient,
- the storage environment must 'preserve'. In practice this usually means the use of an atmosphere of dry nitrogen (or similar) and temperature control. Batteries are not usually fitted to a stored spacecraft.

[1]Some spacecraft conveyed by aircraft are now sent fully fuelled

19.6.9 Trends and statistics

Monitoring of trends and statistics over time can be of significant help to the QA team. Typical data worth monitoring are:

- inspection accept/reject statistics, with special note taken of rising numbers of rejects per inspection,
- NCR occurrence profile with time. A rise in such occurrences should be expected, followed by a sustained fall,
- NCR disposition statistics, being alert to excessive reworks or excessive scrap.

Trends and statistics can also be used to:

- give early warning of a developing process problem,
- initiate a problem diagnosis,
- monitor PA effectiveness,
- improve PA cost-effectiveness.

19.6.10 Design margins verification

All designers synthesize designs which will meet the requirements of:

- Form: e.g. cuboid with light metal casing,
- Fit: mechanical interfaces; electrical power, electrical signal interfaces,
- Function: all inputs generate required outputs,

The launch phase and space environment also present challenges to designers, so their designs must also meet these requirements:

- Environment: operation in thermal environment, vacuum, acceleration, vibration, shock, etc.

Analyses of the designs are made to demonstrate that these requirements are met. During the Manufacture and AI&V phases, the as-built designs undergo tests to show that they do indeed meet these requirements. The designer generally wants to avoid any failure in testing of the as-built design. To ensure this success in testing, the designer will slightly over-design the device. This is common in most branches of engineering where failures in service can ruin the reputation of a device manufacturer. On the other hand, extreme over-design can result in unmanageable costs. The space engineering community has established various levels of design robustness that enable devices to reach orbit in good condition and function in-service for long periods.

The project Parts, Materials/Processes and Applications will all be qualified for the flight spacecraft (recall *G.P.*6, that all parts/items to be used in a flight spacecraft must be qualified). In this phase, it is the assemblies, equipments, subsystems and the whole spacecraft that are qualified.

Whatever level of assembly is being performed, there are predicted stress exposures for the device. The designer's planning for the various levels of exposure are based on the following environmental load levels:

- the Expected levels: stress ranges predicted by analysis,
- the Acceptance levels: just beyond worst case expected,
- the Qualification levels: proof of a Margin beyond Acceptance,
- the Design Point: just beyond Qualification level (to avoid failure in test).

Typically, thermal Acceptance levels are $+/-5°C$ beyond worst case expected and are applied for a known duration that will not over-expose the device to energy input. Qualification levels can be up to $+/-10°C$ beyond worst case expected and last rather longer than Acceptance level tests (see also Chapter 11) There are also defined threshold levels for vibration and shock stresses at Acceptance, Qualification etc.

'Qualification Models' are those devices subjected to these Qualification-level and duration tests. They are usually not used in flight spacecraft. Spacecraft Qualification Models are hardly ever made currently because of cost, but occasionally, a planned series of military satellites will be preceded by a Qualification Model spacecraft

'ProtoFlight Models' represent a compromise between meeting the proof-of-margin qualification requirement and limitation of energy input. Such Protoflight Models are devices subjected to Qualification Level Tests for Acceptance Duration (see also Chapter 17).

The general principle is to avoid damaging the test article by limiting the total energy input of all testing. It is therefore very important to log all test exposures of devices. These test exposure logs also form part of the quality evidence the customer requires.

*G.P.*7—Log all test exposures (levels, durations, environment) and limit total energy input.

Whatever the level of assembly, the ultimate aim is to have the flight items qualified. The status of 'Qualified for flight' is achieved through the controls over parts, materials/processes and applications during build, and testing to demonstrate performance and margins in simulated space environments. Once a device has flown successfully, it is referred to as 'Flight Qualified' but only to the extent of its 'Qualification Envelope'—temperature range, maximum vibration, shock, acceleration, and so on.

Qualification is achieved by one of two main methods:

- similarity: comparison with like, qualified, items,
- testing: environmental exposure (thermal vacuum, vibration table).

'Qualification by Similarity' is becoming progressively more common, especially where COTS and CAM devices are adopted. However, failures in-orbit are often attributable to qualification by similarity that is poorly done, which is referred to as 'Flawed Qualification by Similarity'.

19.6.11 Procurement controls

This is a specialist domain (see Parts, Materials). Procurement Documents must define:

- item, quality, quantity and receive inspection/checks,
- accept/reject criteria and supplier approval status.

Procurement planning must take account of:

- contractual documentation of space vehicle,
- lead times and contingency plans,
- suitable storage of bought items.

19.6.12 Control of changes

Control of changes is a critically important activity for several reasons:

- it ensures that all changes are properly examined by someone other than the proposer of the change,
- all changes are authorized by a senior project officer,
- approved changes are disseminated to all project team members,
- all project team members work to the same baseline.

Changes begin in a variety of ways, for example, with a rejection by an inspector, a Failure Review Board (FRB), a Test Review Board (TRB), a Material Review Board (MRB), a Non-Conformance Report (NCR), or a contract error. If the appropriate action is a change, then a change proposal/request, with justification, is compiled and issued for discussion. In some contracts, a change notification is originated by the customer and a change request is originated by the manufacturer. In both cases, the change is classified in one of two ways:

- Class I, which has no effect on performance, cost, schedule, interchangeability, or
- Class II, which does affect performance, cost, schedule, interchangeability. This classification always involves the customer in the decision-making process.

19.6.13 Audits and inspections

Audits are made to check the PA systems of subcontractors and suppliers. Audits can also be used to assess a manufacturer's own PA (self evaluation). The main function of the audit is to verify that the PA system does perform properly as presented and accepted. Most spacecraft manufacturers have in place an ISO 9000 series accreditation that requires a regular audit of its quality function. Even so, project audits within a contract still take place.

Formal *inspections* take place to verify the successful completion of a step or stage of manufacture. A successful inspection contributes to the quality evidence that has to be made available to the customer. Inspections must be carried out at every point where:

- a critical or very difficult process has been completed,
- discovery of a failure at a later stage would be costly in time or money,
- the customer requires it.

Inspections that are not successful must generate a formal NCR.

19.6.14 Qualification and acceptance

All flight-items must be *Qualified* for flight. The nearer the launch becomes, the more pressure grows to sign off the last few unqualified items as 'qualified'. The PA manager needs a cool head at this stage. The status 'qualified' must be used extremely carefully. Items are designated as *'Qualified'* only:

- for the defined environment (mechanical, thermal, etc),
- for the defined application,
- when intended use is within the qualified use envelope,
- when calculations, tests, part/material sources and processes are all satisfactory,
- when testing has demonstrated needed margins.

Items are *'Accepted'* for flight use only:

- when the item is made to the qualified design/build standard,
- when all NCRs are properly closed,
- when it passes Acceptance Level tests satisfactorily.

ProtoFlight is a special case (see Subsection 19.6.10) and is often used at spacecraft level for the first flight-spacecraft in a series. Later models are simply acceptance tested.

19.7 SAFETY

19.7.1 Basics

The overall objective of the *Safety* programme is to prevent accidents and to identify and control all hazards or threats to people, the system and the mission. Safety requirements apply to all programme phases; these requirements span ground and flight hardware, software and documentation. They also aim to protect people from 'human-induced' hazards.

Hazard analysis has a general form as exemplified by the following:

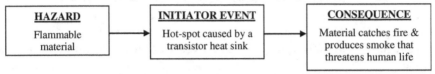

For unmanned missions, safety is largely a matter of industrial protection and includes handling of toxic fluids, protection against high-voltages and RF supplies, transport and storage of flight equipment. Potential for personal injury should not be dismissed completely—on some occasions, spacecraft have fallen or tipped over in the manufacturer's factory. Inappropriate usage of high-pressure air hoses has also caused injury severe enough to hospitalize the victim.

For crewed missions, protection of the crew and accompanying scientists is the main objective. This applies in countdown, during launch, docking, while on station, during EVA, undocking, re-entry and landing, and return to ground base station.

19.7.2 Categories of threat to safety

Hazards are classified as catastrophic, critical or minor, depending on their consequences, as shown in Table 19.13.

19.7.3 Hazard reduction process

One aspect of safety is to identify hazards—another altogether is how to deal with them. The objective is always hazard reduction, ideally to the point where it can be eliminated altogether.

The following *Hazard Reduction* precedence applies:

1) Eliminate hazard (e.g. remove flammable material).
2) Design for minimum hazard.
3) Control hazard (minimizes residual risk):
   - Minimize probability of a hazardous event
   - Implement devices that warn of such events
   - Special procedures that must be followed by users to avoid the development of a hazard into an adverse consequence. This means that the hazard and initiating event cannot be eliminated by design.

Table 19.13 Hazard severity categories

Effect on:	Hazard severity categories		
	I. Catastrophic	II. Critical	III. Minor
Human	Loss of life; life threatening or permanently disabling injury or occupational illness	Temporary disabling injury; temporary occupational illness	
Interfacing manned space flight system	Loss of element	Loss of or major damage	Minor damage
Ground facilities		Loss of or major damage	Minor damage
Launch site	Loss of facilities		Minor damage
Public or private property		Loss of, or major damage	Minor damage
Environment	Long-term detrimental effects	Short-term detrimental effects	

19.7.4 Emergency, warning and caution

These terms relate to threats to human life in crewed missions. Analysis needs to be done which aims to define the criteria for each such situation, where and when they can occur and what must be done to respond to them. This uses inputs from hazard analysis and warning-time analysis, where the latter includes the time interval needed to detect and respond fully to the initiating event.

It is also now a well established requirement that humans must have a means of escape from uncontrolled life-threatening situations. In the case of the space shuttle, this was not available in the 1986 Challenger launch disaster, nor was it available in the 2003 Columbia re-entry accident. It is, however, available on the ISS, which comprises several interconnected modules that can be isolated from each other.

The outputs of emergency, warning and caution studies include various safing functions and safety features that must be included in the design. Some of these are necessary only during certain periods of the mission, but are typically available throughout.

The *emergency, warning, caution* and *safing* function is required to flag hazardous situations and render them safe. Categories of such situations may be summarized as follows:

Emergencies:	potentially catastrophic and/or critical hazardous events have occurred.
Warning situations:	potentially catastrophic and/or critical hazardous events are imminent.
Caution situations:	potentially catastrophic and/or critical hazardous events may occur as a result of a known condition.

19.7.5   Fault tolerance requirements

There are special criteria and requirements for the mandatory implementation of fault-tolerant designs. Single failure criteria can be summarized as:

(a) No single failure shall have a catastrophic or critical hazardous consequence.
(b) No single operator error shall have a catastrophic or critical hazardous consequence.
(c) No combination of a single failure and a human operational error shall have a catastrophic hazardous consequence.

Some examples of requirements placed upon safety-critical functions and items include:

Compliance with fault tolerance requirements for catastrophes.
For software, multiple ($> = 2$) versions are created by different development teams—this eliminates systematic errors that could exist in identical software instances. Hardware functional back-up is acceptable if only one software instance is available.
Parts must be of Quality Level B (best). Generally, COTS and CAM parts are not acceptable.
Verification of requirements implementation. This includes verification by test, by inspection and by documentation review. All safety requirements must have a response in the safety analysis, the design documentation and the design implementation in hardware/software/procedures.
Formal acceptance of residual risk.

19.7.6   Comparison of safety and reliability

Safety and *Reliability* both deal with the causes and consequences of something unexpected occurring. *Safety* deals with protection of personnel and the investment. *Reliability* focuses on the success of the mission (as measured by the delivery of intended mission products).

19.7.7   Safety is an essential burden

It should be noted that implementing the safety requirements places a huge additional burden on the project. More than any other aspect, protection of human life is regarded as the most important issue. The presence of humans in space requires a living environment, storage of food and water, periodic dependable re-supply and removal of waste products.

In addition numerous safety features must be included and the spacecraft's exterior surfaces must be proof against the most likely meteor and debris impact events. All of this adds considerable mass to the orbiting facility, and reduces the volume and mass available for any high-priority payload experiments.

19.8 PRODUCT ASSURANCE IN OPERATIONS

In spacecraft operations, the main objective is to reduce as much as possible any risk to the flow of mission products and/or risk to the spacecraft itself (see also Chapter 14).

Operations Control Centres (OCC) have extensive computing facilities to process telemetry and send telecommands, leased lines to connect to ground stations and direct communication with those requiring the mission products. OCC personnel also use the orbit evolution data of the spacecraft under their control in order to plan necessary orbit-control manoeuvres. The task of controlling the spacecraft is assisted by the following common features:

- *Telemetry channels* are displayed to facilitate a (selectable) inspection of each area of interest such as power, AOCS, TT&C, payload status, and so on. The telemetry channels show the values of parameters in engineering units when the device monitored is both on and on-line. If it is on, but off-line, the values are often suppressed (but can still be viewed if needed). These values are continuously monitored and automatically checked against predetermined defined limits. If a limit is breached, an alarm is triggered and the affected channel display changes colour (e.g. to red). Both soft and hard limits are often included to help the operations controller's decision-making process.
- So called '*mode equations*' (using Boolean logic) handle the suppression or the display of telemetry channels. These equations also have a protective purpose. Some commands are unsafe to send when the spacecraft is in particular conditions. The mode equations prevent the command being sent in these situations. A further alarm is generated.
- There are usually at least two *safe-modes* on the spacecraft: (a) to permit continuity of service in the face of a failure, and (b) to seek the sun if an encountered failure is too difficult to correct automatically. From each safe mode (a known state), operators can begin recovery of the normal operational status using Contingency Recovery Procedures (CRP), bypassing the fault that led to the safe mode.

The operators are given extensive training in the design of the spacecraft and its foreseen behaviour. There are software simulators for the particular spacecraft, and operators can practice Flight Control Procedures (FCP) and Contingency Recovery Procedures (CRP). Flight Dynamics experts provide the necessary orbit evolution and manoeuvre plans for execution by the operators.

The FCPs and CRPs are generated with the objective of avoiding 'creative thinking under time-pressure'. Everything is pre-planned, as far as it is possible to be so. Because of all the checks and controls that exist in an OCC, it is rare to encounter an operator error that damages the spacecraft. However, sometimes the flow of mission products may be temporarily interrupted by command errors.

19.9   SOFTWARE PRODUCT ASSURANCE

19.9.1   Basics

Software differs from hardware in several respects:

- it weighs nothing,
- it can be copied multiple times at negligible cost,
- it requires a host (a processor and maybe an operating system) to function,
- it is very easy to modify,
- it is virtually impossible to test thoroughly,
- host hardware failure (e.g. through a SEU) can cause software failure, and
- if software fails (for any reason), its behaviour and that of the hardware it controls can rapidly become chaotic.

To develop dependable software originating from a new requirement from a user (either human or other part of a system), controls are needed at all stages of the software cycle. Software engineering standards (see for example, Reference [5]) typically require a software development process which includes the following phases:

User Requirements Capture,
Software Requirements Definition,
Architectural Design,
Software Detailed Design and Production,
Transfer to Buyer, and
Operations and Maintenance.

This method is helpful when completely new software is to be developed. However, it is sometimes necessary or desirable to use previously developed/used software. This can be 'as is' in a similar application or with modification/enhancement. When used as a module in a larger software development, the re-used software module maybe used 'as is' or modified. Such re-use is only practical when the software to be re-used has itself been designed for re-use, and is fully documented.

In any but the very smallest software development, most organizations that develop software have a well-established set of standards to which they work. These include extensive Software Engineering standards that deal with modular design, language selection, coding, structured programming, object oriented designs verification, validation, and so on. These are complemented by Software Product Assurance standards [6] that can conveniently be organized by function.

19.9.2   Software PA programme

Organization and Responsibility, and *Software PA programme* management activities are mainly aimed at ensuring that competent persons are available to undertake the work and that the PA function is properly resourced. For efficiency reasons, the PA experts may be embedded within the development team, but they must be able to both accept, and to criticize/reject. These organization and management responsibilities also deal with regular reporting, inspection points, audits, software problem detection, recording and resolution.

Risk management in software development is similar to that function in hardware development, and includes identifying and characterizing risks (to performance, qualification, certification and cost/schedule) and applying effort to eliminate them or to reduce their significance.

Supplier selection and control can apply to the buyer/customer or to a software contractor who is choosing a subcontractor. In both cases, the organization requiring the software development must be satisfied that the contractor has competent staff, that they are available for the project and that the necessary PA controls will be in place throughout the development. Where a contractor may be engaging a subcontractor, it is the contractor's duty to ensure that the subcontractor's standards are at least as good as those required by the end customer.

Procurement controls are aimed mainly at defining exactly what is required, options for extensions/enhancements, acceptance methods and criteria, operations and maintenance support.

Tools and supporting environment(s) are essential to any development of software. Selection and approval of support tools such as requirements analysers, coding standards checkers, and testing environments are part of the management function for the contract itself and within the project.

Assessment and improvement is a process is to be found in any good software contractor's management scheme. Where this is effective and routinely applied throughout the contractor's organization, this can be used to confirm that the software development project itself has such a function. Where it does not form part of the contractor's management, it must be established and PA-monitored for effectiveness.

19.9.3   Software Process Assurance

The *Software Development Life Cycle* forms part of the overall life cycle from concept to end of application life. Within the development life cycle, the following must be defined: phases, phase input(s) and output(s), completion status, milestones, dependencies, responsibilities and customer action at milestones.

General Software Engineering Requirements are those used for planning of development, specifications, customer documentation, configuration management, verification testing, validation activities and maintenance.

Individual Software Engineering Processes must be governed by standards which define development phases such as those identified in Subsection 19.9.1 above, from user requirements capture through the whole process to operations and maintenance.

19.9.4   Software product quality assurance

Product quality objectives and the metrics used are chosen to be applicable directly to the project and as such are unique to the project. There is no standard set of objectives and metrics. Both the objectives and metrics must be quantitative, and the means to obtain their values defined explicitly. The PA engineer is responsible for ensuring the measurement process is correctly executed throughout the development cycle.

Product quality requirements form a set of requirements in their own right. These may well be derived from some of the user requirements, and will generally have been approved by the customer. The main characteristics of these software quality requirements are that they shall be: correct, unambiguous, complete, consistent, verifiable and traceable. The software must be designed to facilitate testing, and the required testing must be capable of being executed as planned. Note that modifying the testing baseline can lead to the passing of a test that proves nothing, which will not satisfy an alert customer.

Software intended for re-use is sometimes commissioned by a customer, but more often, a contractor develops it for multiple applications in his software inventories. Software intended for re-use must go through some extra steps: documentation must specify the platforms on which it is expected to operate, testing on all such platforms must be performed and the developer must certify that the tests were carried out to a named standard and were successful.

Standard ground segments for operational systems are also heavily software driven. For a new facility, the hardware and software would certainly be procured as a compatible pair. For an existing ground facility, the acquisition of new software may demand changes to the hardware/operating system environment. Whatever approach is taken, there are some essential requirements that must be satisfied in the acquisition process:

- performance,
- maintenance,
- durability and technical consistency with the operational equipment,
- assessment of the product against requirements, including criticality,
- available support documentation,
- acceptance and warranty conditions,
- conditions of installation, preparation, training and use,
- maintenance conditions, including the possibilities of evolutionary development
- copyright constraints,
- availability—readiness, installation timeframe, training,
- compatibility with existing facility and spacecraft control needs, and
- site operational constraints

Firmware can also sometimes be required, dedicated to a particular application. PA controls are included in the plan, that is keyed to the development life cycle. Since the firmware is embedded in particular hardware, that hardware has to be marked permanently in such a way as to identify uniquely the hardware and its firmware.

19.10   PA IN TECHNOLOGY DEVELOPMENTS

NASA, US DoD and ESA have adopted a standardized approach to defining the readiness of technologies to be used in spacecraft missions—see Table 19.14. The common stages in development of new technology overlap some levels as shown in Table 19.15.

Throughout, PA staff must aim to qualify the technology. This aim can affect decisions in the early TRL stages of development. 'Qualifiability' is a challenge in all these activities, as emphasized by the following examples:

- *Atomic Clocks* are needed in navigation satellites for their extremely accurate time measurements. To qualify such a clock on the ground requires test apparatus that

Table 19.14 Technology readiness levels

Technology readiness level	Description
TRL 1	Basic principles observed and reported
TRL 2	Technology concept and/or application formulated
TRL 3	Analytical and experimental critical function and/or characteristic proof-of-concept
TRL 4	Component and/or breadboard validation in laboratory environment
TRL 5	Component and/or breadboard validation in relevant environment
TRL 6	System/subsystem model or prototype demonstration in a relevant environment (ground or space)
TRL 7	System prototype demonstration in a space environment
TRL 8	Actual system completed and 'flight qualified' through test and demonstration (ground or space)
TRL 9	Actual system 'flight proven' through successful mission operations

Table 19.15 Technology development stages

Development stage	TRL1	TRL2	TRL3	TRL4	TRL5	TRL6	TRL7	TRL8	TRL9
Basic technology research.	■	■							
Research to prove feasibility		■	■	■					
Technology development.			■	■	■				
Technology demonstration.					■	■	■		
System/subsystem dev't.							■	■	
System test, launch and ops.								■	■

is much more accurate than the clock being tested. This is a very challenging task. Manufacturers refer to such test configurations as a 'test solution', acknowledging the difficult problem they have solved.

- *Large Deployable Antennas* with multiple struts and ribs are typically folded-up and packed during the launch phase and only deployed when in orbit. These are especially difficult to qualify on the ground (see also Chapter 15). Gravitational force is always present, and however the on-ground deployment testing is performed it is subject to forces that are not present in orbit. The intended communications service is often critically dependent on the geometrical accuracy of the deployed antennae. The customer rightly demands that it should be proven. Testing in one orientation, and in the inverted orientation, can give some confidence by assuming that the zero-*g* configuration would be expected to be somewhere between the two measured on-ground deployed positions. Even so, it is difficult to prove that the dynamics of deployment, as observed in ground testing, will be the same as that which will occur in-orbit. In reality, a zero-g qualification is not possible on ground. Ultimately, operating an instrumented test device in orbit may be the only way to obtain genuine qualification.

- *Pointing Accuracy Requirements* on spacecraft continue to become more and more demanding, especially in astronomy missions. The James Webb Telescope is expected to achieve a pointing accuracy of 10^{-6} deg. or 0.0036 arc-sec. Demonstrating this on ground is an extremely challenging task. It may only be possible to demonstrate stability of pointing, rather than absolute pointing accuracy.

19.11 THE ASSURANCE MESSAGE

The text of this chapter and the general principles mentioned are all intended to help the product assurance engineer carry out the job effectively.

This author has been consulting in the space insurance sector for over two decades, and one problem that crops up repeatedly is the occurrence of failures through inadequate or flawed Qualification by Similarity. Such a qualification method is not wrong, but the 'similarity' has to be close for the qualification to be valid. All of the parameters need to be close, not just those that are headlined. 'This unit is in-flight qualified' is, by itself, just not good enough. Also, to be considered are such parameters as the application, the expected temperature envelope, the tolerance to power fluctuations, and so on. Is its use the first on the selected launcher? If so, will it withstand the vehicle's vibration and g-force environment?

Whatever task product assurance engineers perform, they should always ask themselves—what does this task actually contribute? Has the risk been reduced? Do we need any extra tasks or tests that have not yet been planned? And if a test is invalidated, they should insist that it is repeated and be prepared for the complaints from the project manager.

Finally, and in a slightly paranoiac frame of mind, it should always be remembered that if it goes wrong, the product assurance function on a project will most likely be attributed with the blame, and the PA engineer(s) will be required to present some of the findings of the Enquiry Board!

REFERENCES

[1] NATO AQAP-1 Quality Standard first published 1968, adopted from the US DoD MIL-Q-9858 Rev A (1963).
[2] British Standards Institute document BS 5750 Quality Systems, first created 1979.
[3] International Standards Organisation (ISO) documents ISO 9000/9001/9002, Quality Management Systems. First produced as ISO 9000 in 1987, this document is derived directly from BS 5750.
[4] Dependability Definition 191-02-03 given in *'Dependability and Quality of Service'* document IEC 60050-191 Vocabulary. Published by Technical Panel T.P.59 of the International Electrotechnical Commission, Switzerland.
[5] Guide to the Software Engineering Standards, ESA PSS-05-01, Issue 1, ESA Board for Software Standardization and Control, March 1995.
[6] Software Product Assurance Requirements. ESA standards document: ECSS-Q-ST-80C, March 2009.

20 SPACECRAFT SYSTEM ENGINEERING

Adrian R. L. Tatnall[1], John B. Farrow[2], Massimo Bandecchi[3] and C. Richard Francis[3]

[1]*Aeronautics and Astronautics, Faculty of Engineering and the Environment, University of Southampton*
[2]*International Space University, Strasbourg*
[3]*European Space Research and Technology Centre (ESTEC), European Space Agency*

20.1 INTRODUCTION

The first chapter of this book gave an introduction to the system view of the spacecraft, and the relationship between the subsystems and the overall system was explained. In the main body of the text the subsystems have been discussed and the interfaces between them have been considered. All of these aspects are drawn together in this final chapter in which the *system engineering* process required to ensure the overall spacecraft meets the mission objectives is explained and illustrated with a discussion of a specific spacecraft—the European Space Agency's *CryoSat*. There are many other examples that could have been chosen but this particular example provides a good demonstration of the breadth and complexity of mission drivers which arise in the search for optimum solutions to such challenging mission objectives.

System engineering is often characterized as both an art and a science. This apt characterization is appropriate because good system engineering requires the creativity and knowledge of systems engineers, but it also requires systems management or the application of a systematic disciplined approach. In this chapter the systematic disciplined approach is considered in more detail, and there is emphasis on the methodology of systems engineering, particularly in the section on the use of *concurrent engineering*.

The *mission requirements* are the top level requirements on the spacecraft. They are quantitative in nature and must be specified by the customer or user, and they are an assessment of the performance required to meet the mission objectives. For the spacecraft system design these requirements are translated into engineering parameters. This translation can be complex, depending on the particular application.

Spacecraft Systems Engineering, Fourth Edition.
Edited by Peter W. Fortescue, Graham G. Swinerd and John P. W. Stark.
© 2011 John Wiley & Sons, Ltd. Published 2011 by John Wiley & Sons, Ltd.

For a communication spacecraft, the translation is relatively straightforward since the user coverage and data requirements can readily be used to define the satellite parameters. The link budget described in Chapter 12 is at the heart of this process. On a satellite such as *CryoSat*, however, described in Section 20.4, the translation of the user requirements in terms of geophysical parameters, such as ice thickness changes of a few centimetres, into instrument specifications is complex and involves many system algorithms. In this case the process involves assumptions about other related parameters, such as the level of processing required and the orbital data available from other sources. During the design process these assumptions may prove to be incorrect and the engineering requirements will have to be modified. It may also be necessary to modify the requirements as a result of a clearer understanding of the impact they have on the spacecraft design.

This iterative process is essential to ensure that the most relevant and realistic requirements are used for the spacecraft design. There are plenty of examples where the engineering requirements have become 'tablets of stone' at the start of the design and the overall system has suffered because of unwillingness to question them as the design has evolved. It is always necessary to define how much quality is needed, or how much 'science' is enough in order to hold down mission costs and avoid unnecessarily restrictive requirements. In other examples, technological constraints such as the inability to space qualify critical parts or processes, may dictate a revision of requirements.

There is another issue that can determine the design approach taken. The procurement of commercial communication satellites is targeted towards meeting market needs perceived to be arising through increased capacity requirements. Since a commercial organization is generally driven by the need to satisfy customers' needs, in competition with other suppliers, timeliness is of the essence. This commercialism is demonstrated by Intelsat which became a private company in 2001 and has since been acquired by Serafina Holdings; an entity formed from equity investors. Private equity firms are keen to be involved with satellite operators because their business guarantees steady cash flows.

As a result of these commercial considerations, the principal mission requirements for communication satellites are specified at the outset. Engineering solutions must be proposed to solve these specific requirements, although clearly some level of discussion takes place between contractors and customer to define them precisely. These programmes therefore do not go through the iterative process associated with missions such as *Envisat*, which are principally experimental in nature and used to establish the user market. Instead, their design iterations are more directed towards having a major impact on the timescale of the programme, and its cost. Even in this market, however, there needs to be an appreciation by the user community of the implications of their requirements on the overall design. It is this appreciation that has led to the pressure in the USA and Europe to reduce the scale of the spacecraft proposed and designed for science and Earth observation applications over the past 10 years. Further discussions of these issues are given in Section 20.4, where the rationale for the relatively inexpensive satellite, *CryoSat*, is described.

20.2 SYSTEM ENGINEERING

20.2.1 Programme phases

The overall process of system engineering may be conveniently understood by considering the way in which space programmes are broken down into individual phases.

The total time from the initial conception of a complex mission to its launch and operation can extend over a period of 10 years, and sometimes even longer when delays are introduced due to funding limitations or the mission itself has to be of long duration. The *Rosetta* mission was approved in 1993, and was launched in March 2004. The spacecraft will rendezvous with the comet 67P/Churyumov-Gerasimenko in May 2014, with the end of its planned mission in December 2015. The *New Horizons* mission to Pluto and the Kuiper Belt finally gained financial approval in 2003, and was launched in 2006. The spacecraft will fly-by Pluto in 2015 before going on to encounter objects within the Kuiper Belt from 2016 to 2020. Many missions fall by the wayside through failure to gain funding approval. As a measure of the competition that exists, in response to ESA's *Cosmic Vision 2015-2025* call for proposals, 50 proposals were received and in 2010 three were selected for further definition.

Several factors influence the duration of the planning cycle for the new mission selection. Firstly, political and financial constraints can impede the smooth flow of a programme. Secondly, launch opportunities can be limited, particularly for interplanetary missions. The *Giotto* spacecraft, for example, presented a once in a lifetime opportunity to encounter Comet Halley and this imposed critical deadlines for development and launch. In some cases there will be the need to demonstrate or develop new technology before the feasibility of the mission can be adequately assured. This generally has most impact upon the 'engineering' timescale of a programme.

The spacecraft programme is traditionally divided into several distinct phases, as outlined below. It is worth noting that this format is not adopted for small satellites and it may be significantly shortened in the case of commercial programmes, particularly if existing buses are used. It should also be said that the definitions used in different organizations (e.g. ESA and NASA) can vary.

Phase A (feasibility)

This phase, lasting typically 8–12 months, is termed the feasibility or outline design phase. Its objectives include:

- selection of an optimum (and cost effective) system concept from the range of options under consideration;
- demonstration of the feasibility of the project by design and analysis;
- definition of a technical solution to the extent necessary to generate and substantiate realistic performance, schedule, planning and cost data for all subsequent phases.

This Phase is sometimes split further into a *pre-Phase A*, or *Phase 0*, wherein specific aspects of the system are considered in outline. The use of concurrent engineering facilities, described in Section 20.3, is particularly useful in this stage of the design. It enables the user to be involved in establishing the system requirements. The cost of performing this phase is relatively low, but the cost implications arising from it are large. The ultimate cost of the mission will depend on decisions made during the early phases of the design.

In a spacecraft Phase A study, the customer normally provides preliminary information on mission, launcher and payload requirements, as well as a target performance specification. System engineering plays a central role in the selection of a preferred concept in these early phases as indicated by the flow diagram in Figure 20.1 (by their very

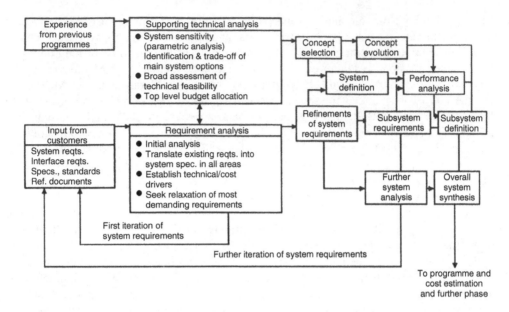

Figure 20.1   Phase A system engineering flow diagram

nature some of the design tasks arc iterative rather than sequential). The system engineer provides an overview of the entire system in order to advise the programme manager and the customer on the overall technical feasibility. The *preliminary requirements review* (PRR) is held at the end of Phase A. The outcome of this review is used to judge the readiness of the project to move into Phase B.

Once the programme has gained approval or a contract is awarded to a *prime contractor*, the subsequent phases are summarized below.

Phase B (detailed definition)

This phase will typically last 12–18 months. A significant extension of the phase may occur in large agency (ESA, NASA, etc.) programmes if financial and political consider-ations slow down the process of full mission commitment. This phase is associated with detailed design and definition. Specific activities include:

- definition of the system and subsystems designs in sufficient detail to allow the main design and development phase C/D to proceed with the minimum of problems;
- production of subsystem requirements and design specifications, subsystem and equipment design and development plans, programme schedules and a full proposal for the subsequent C/D phases;
- initiation of advanced C/D activities such as ordering of long-lead items or detailed design of critical parts.

The reviews in this definition phase are the *system requirements review* (SRR), the *system design review* (SDR) and the non-advocate or independent review.

Although the tasks are similar in any space programme, there are some differences between agencies in the way that the phases are defined. In ESA, for example, Phase B is referred to as the preliminary definition phase and the *preliminary design review* (PDR) is held at the end of this Phase.

There is also a tendency, in ESA, towards splitting the Phase B into Phase B1 and B2. The former is regarded as part of the programme preparation while the latter is part of the implementation phase and is thus combined contractually with the Phase C, D and, often, E1 as well. The distinction between the two parts of the phase B is that during the phase B1 the prime contractor prepares tender invitations for all of the subsystems and equipments which could be open to competition, and, together with ESA, evaluates the proposals. The selected subcontractors are then brought into the team during the phase B2.

Phase C/D (design, development, manufacture, integration and verification)

This is the longest phase, and may extend for three to five years. It encompasses development, manufacture, integration and verification. The specific activities (see also Chapter 17) include:

- completion of all design and analyses;
- preparation of manufacturing drawings and special procedures;
- completion of all development and qualification testing;
- manufacture of the flight hardware and acceptance testing.

The *preliminary design review* (PDR), *critical design review* (CDR), *test readiness review* (TRR) and *flight readiness review* (FRR) all take place in this phase.

Phase E (the mission operations and data analysis)

The Phase E starts after the spacecraft has passed the flight readiness review, and is normally divided into two parts: Phase E1 is concerned with getting the satellite (and the rest of the system) into a condition when it can start routine operations, which constitute the Phase E2 (see also Chapter 14). Phase E1 normally includes the shipment to the launch site and preparations for launch (the 'launch campaign'). This is followed by the launch and early orbit phase (LEOP), which is the critical first few days of operations after launch. This includes mission critical activities such as deployment of appendages. At the end of LEOP the satellite should be in a stable, self-sufficient condition, although payload equipment may not yet be switched on.

LEOP is followed by a commissioning phase, where the spacecraft, the payload and any required ground infrastructure are brought into an operational condition, are checked out and are optimized, tuned and, if necessary, debugged. Calibration and validation activities may also take place during this phase.

At the end of commissioning, routine operations start, and this marks the beginning of the Phase E2. Routine operations rely on ground support staff (and, potentially consumables) and thus incur an annual cost. This can be expensive, although early system design to maximize autonomy, for example by means of intelligent failure detection, isolation and recovery (FDIR), can significantly reduce this.

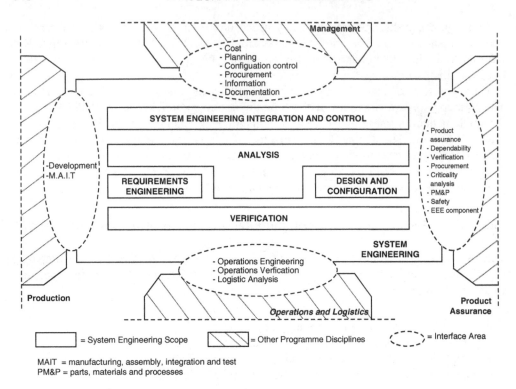

Figure 20.2 System engineering boundaries. (Figure reproduced from [1] by permission of ESA on behalf of the ECSS members)

20.2.2 System engineering techniques

The boundaries of the system engineering discipline, its relationship with the production, operations, product assurance and management disciplines are given in Figure 20.2 taken from the European Cooperation for Space Standardization [1]. System engineering encompasses the following functions (see also Chapters 17 and 19):

- requirement engineering including requirements analysis and validation, requirement allocation, and requirement maintenance;
- analysis in order to resolve requirements conflicts, decompose and allocate requirements during functional analysis, and assess system effectiveness (including analysing risk factors). It is also used to complement testing evaluation and to provide trade studies for assessing effectiveness, risk, cost and planning;
- design and configuration in order to define a physical architecture, and its complete system functional, physical and software characteristics;
- verification, to demonstrate that the deliverables conform to the specified requirements, including qualification and acceptance;
- system engineering integration and control to ensure the integration of the various engineering disciplines and participants throughout all the project phases.

These functions require the techniques identified in Table 20.1 to be used.

Table 20.1   System engineering techniques

• Requirements identification/analysis	• Concept selection
• System specification	• Budget allocation
• Options identification	• Performance analysis
• Mission assessments	• System optimization
• Trade-offs	• Interface specification
• Feasibility assessment	• System definition
• Cost comparison	• Cost estimation

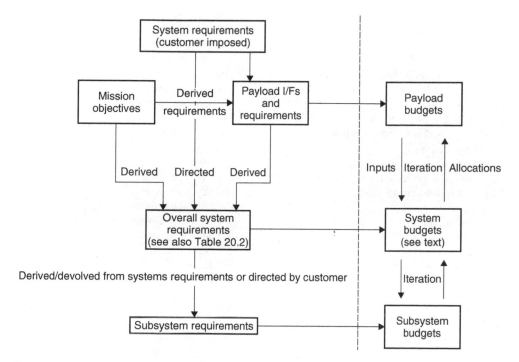

Figure 20.3   Expansion/translation of top-level requirements to system and subsystem level

At the start of the programme, the requirements imposed by the customer or end user are likely to be defined at a fairly high level, possibly covering only the overall mission objectives and the main payload interfaces. Figure 20.3 shows how it is then necessary to expand these top-level requirements into specifications covering the entire range of system and subsystem engineering parameters. It also shows the importance of establishing, in parallel, technical budget data.

Table 20.2 is intended to act as a checklist of the full range of parameters that are likely to be specified in later, more detailed phases of a programme.

In the early conceptual design phases of a new mission, choices may have to be made between system-level options for the class of orbit, the launcher type, the propulsion

Table 20.2 Check-list of system requirements

Mission requirements (Chapters 4, 5, 14)	*Physical requirements* (Chapters 3, 9, 10)
Launch windows	Axes definition
Orbit (transfer; operating	Configuration constraints
-nominal and back-up)	Dimensions
Operations	Mass
-Launch & early orbit phase	Mass properties
-Operational phase	Internal torques
-End-of-life	Disturbances
Lifetime	Power/energy
(Retrieval/repair/re-supply)*	
Autonomy	*Performance requirements* (Chapters 5, 8, 9,
Reliability/availability	10, 11, 12, 13, 14)
Ground segment	Orbit maintenance
	Ranging accuracy
Environmental requirements (Chapters 2,	Timing accuracy
7, 8, 11, 16, 17)	Pointing accuracy
Ground activities	Measurement accuracy
Launch & ascent conditions	Stability
Transfer & operating orbit environment	Pointing range
(Re-entry, descent)*	Slew rate
Structural/thermal inputs, loads, ranges	Data rate
Environmental protection	Data storage capacity
Cleanliness/contamination	Onboard processing
EMC	Link budget margins
DC magnetic fields	Telemetry/telecommands
Radiation	Strength/stiffness
Spacecraft charge	Thermal control
Atomic oxygen*	Reliability
Autonomy	
PA requirements (Chapters 2, 17, 19)	
Reliability	*AIV programme requirements* (Chapter 17)
Availability	Schedule
Maintainability	Model philosophy
Safety	Test philosophy
Parts, materials, processes	GSE requirements
Cleanliness	Facilities usage
Storage, handling, transport	
Configuration management	*Cost contraints*
Software	

*For some missions only
Note: AIV: assembly, integration and verification; EMC: electromagnetic compatibility; PA: product assurance.

system for orbit adjustment if required, the attitude control concept, the data retrieval mode, and the spacecraft configuration.

The case study described later in this chapter illustrates many of these factors. Here we briefly consider a single example—the choice of orbit for an astronomy mission. This highlights some of the key points that must be taken into account in concept selection and optimization.

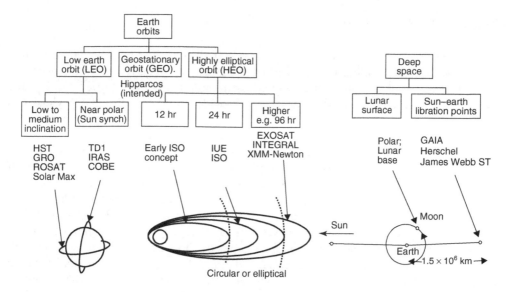

Figure 20.4 Orbit options for astronomy missions

Figure 20.4 is a tree diagram showing the possible orbits about the Earth, and beyond, which could be adopted for an astronomy mission. With the exception of the lunar surface, all of the listed orbits have actually been selected as indicated by the mission names in the diagram.

The main point to emerge from it is that the choice of orbit for this class of mission is by no means clear-cut. Much depends on the existing infrastructure in the form of launcher availability, data retrieval capabilities or existing spacecraft bus designs. For example NASA's major observatories—the *Hubble Space Telescope* (HST) and the *Gamma Ray Observatory* (GRO)—had to be in a circular low earth orbit (LEO) in order that they could be launched/serviced by the space shuttle, and *TDRSS* could be used for data retrieval. However, ESA's astronomy missions, such as *Integral* and the X-ray Multi-Mirror Mission (*XMM-Newton*) selected highly elliptical orbits to guarantee long periods of uninterrupted observation away from trapped radiation in the Earth's proton and electron belts. Other missions such as *GAIA, Herschel* and the *James Webb Space Telescope* have selected orbits around a point about 1.5 million km from Earth in the direction away from the Sun known as the L_2 Lagrangian point (see Chapter 4). In this orbit, advantage can be taken of the fact that the benign thermal and radiation environments are ideal for long-duration observations. In addition there is continuous solar power and a continuous communications link. Other spacecraft, such as NASA's *Kepler* spacecraft, are in orbits around the Sun trailing the Earth so that a star field can be observed continuously for several years.

Reuse of existing designs of spacecraft equipment offer very significant savings compared to new developments—for example, the satellite bus used for *Venus Express* was almost a copy of that used for *Mars Express*. The cost of the satellite bus, the launcher and ground segment operations over the mission life feature very significantly in the total programme cost and can have a dominant effect on early trade-offs. Clearly the ensuing spacecraft system and subsystem design for LEO, HEO or other options can differ greatly as a result of the very different environmental constraints and orbit conditions.

20.2.3 Design drivers

The payload is, in most cases, the single most significant driver of the satellite design. The purpose of the satellite bus is to provide the support required for the payload to ensure that it can be operate in the required orbit and environment. Power, heating and cooling, structure and communication are all provided to ensure that the payload can operate satisfactorily and relay its data back to ground. The propulsion subsystem, the Attitude and Orbit Control subsystem (AOCS) and the mission analysis provide the means of getting the payload into the right position to make its measurements. In the case of *Giotto*, the spacecraft that intercepted Halley's Comet, for example, the retrograde orbit of Halley around the sun meant that any interception had to occur at extremely high velocity (∼70 km/s) and the spacecraft had to have a means of protection against particles moving at this speed. The design of the bumper shield therefore became a key design driver. It also became necessary to spin the spacecraft to provide gyroscopic rigidity to ensure that it remained in the right orientation for the payload to view the nucleus and transmit data to the Earth. On the other hand *Rosetta*, the next ESA cometary mission, is due to rendezvous with the comet 67P/Churyumov-Gerasimenko and move along with it orbiting around the nucleus at a few centimetres per second, as it draws closer to the Sun. This comet has a prograde orbit. It is therefore possible for the spacecraft to have a three-axis stabilized configuration. In both cases the design of the spacecraft stems from the requirements and constraints imposed by the payload.

20.2.4 Trade-offs

It is common to make use of trade-off tables to 'score' the alternative options in early concept studies. Major evaluation criteria for such trade-offs include:

- cost, which is generally a dominant factor;
- satisfaction of performance requirements (e.g. image quality in an Earth observation mission);
- physical characteristics, notably mass, size and power which, in turn, impact on cost and feasibility;
- availability of suitable hardware technology and timescales for any predevelopment;
- compatibility with launcher, ground segment and other system elements, and the complexity of interfaces;
- flexibility to encompass alternative mission options;
- reliability and availability.

If some of these criteria are considered more important than others then a weighted trade-off can be performed. Regardless of whether a trade-off is weighted or not, it should only be used as a guide. It is impossible to guarantee that a trade-off is entirely objective and that the evaluation criteria are exhaustive and independent. Cost, for example, is influenced by all the criteria above and its use as an independent parameter is highly questionable.

Only some of these factors lend themselves to quantitative evaluation and hence direct comparison to select a preferred solution. Other factors in the list are fairly subjective and rely more on broad engineering judgment than on the results of parametric analyses and optimization methods. Still others may be quantitative in nature but, in the early stages of a programme the estimation of the respective values may be too coarse to allow a single solution to be selected with confidence from the range of options. However, despite its shortcomings, a trade-off analysis is often helpful in identifying design drivers, and in focusing resources on issues that may threaten feasibility. Finally, it should be noted that, as in the example of orbit selection for an astronomy mission, there may not necessarily be a single preferred choice and two or more alternative approaches may be equally viable.

20.2.5 Budgets

An important system engineering tool is that concerned with system budgeting. At spacecraft level the main system budgets to be compiled include:

- mass and mass properties (e.g. centre of mass, inertias);
- ΔV and associated propellant quantity;
- power and energy;
- pointing and stabilization;
- aligment/boresight error (e.g. of a payload sightline with respect to nominal spacecraft axes);
- telemetry and telecommand budgets;
- data rate and storage capacity (if appropriate);
- communication link budgets;
- reliability.

Almost invariably mass and power are at a premium in space programmes, but just how critically they are constrained depends on the type of mission. The design drivers—such as payload power demand and data rate for *CryoSat*, or pointing and stabilization for an astronomical telescope—must be established from the outset. Devolving the key system requirements and budgets to subsystem level, and iterating these as necessary, is intimately linked to other steps in the system engineering processes in the early programme stages. These steps traditionally include the identification of alternative concepts, trade-offs of these options against the main evaluation criteria and the shortlisting of preferred solutions. Parametric analyses at system and subsystem level are necessary to establish system sensitivities and technical drivers.

Note that the system requirements listed in Table 20.2, and the associated budgets, are applicable to almost all types of spacecraft. If one were to consider a specific mission, it would be necessary to extend the list of mission and payload requirements still further. For example, a remote sensing payload is generally characterized by parameters including spectral and spatial resolution, dynamic range, signal-to-noise ratio, calibration accuracy, and the area and repeat frequency of coverage. Only a few of these parameters are themselves subject to compilation of budgets but they all play a part, in a very interactive manner, in establishing the optimum system design.

The system engineering tasks must be applied throughout the overall programme and not just in the early phases. In particular the following aspects must be addressed:

- compliance of all physical and performance parameters with the system requirements;

- devolution of the system-level specification and budgets to subsystem level;
- resolution of conflicts at detailed design level to establish the most cost-effective compromise solution satisfying overall mass, risk, cost and other targets;
- definition, monitoring and control of interfaces between all system elements and between subsystems;
- detailing of operational procedures for the mission;
- management and maintenance of budgets for mass, power, fuel and other key resources;
- analysis of 'system-level' factors such as reliability, EMC and chemical cleanliness.

20.3 CONCURRENT ENGINEERING

20.3.1 Introduction

Concurrent Engineering (CE) is a set of techniques in which the design, development, procurement and manufacturing of a product is carried out by near-real-time teamwork. In addition, there is a high level of integration of tools between all the disciplines involved in the process, and the same design facility or environment is shared by all the parties involved in the product life-cycle. The aim of this approach is to drastically reduce the time to market and the cost of development, and to increase the quality of the product.

Concurrent Engineering has been active in academia and industry for more than 30 years. The concept was introduced in the aeronautics industry, and it has been continuously developed and extended into several other industrial sectors and into the broader range of the whole product life-cycle.

A related sub-discipline, *collaborative and distributed engineering*, though maintaining the characteristics of near-real-time activities and tool integration, puts the focus on the involvement of different remote sites. This is mostly concerned with multi-company projects, management of suppliers, etc. in which team integration is somehow looser.

To explain the principles of CE, a specific illustrative implementation will be used in this section—that of the ESA *Concurrent Design Facility* (CDF), which is sited at ESA ESTEC in the Netherlands [2].

Depending mainly on the context and the application environment, various definitions of concurrent engineering have, up to now, been introduced and used in literature. The definition of CE adopted in the ESA CDF is:

> '*Concurrent Engineering is a systematic approach to integrated product development that emphasises the response to customer expectations. It embodies team values of co-operation, trust and sharing in such a manner that decision making is by consensus, involving all perspectives in parallel, from the beginning of the product life-cycle.*'

20.3.2 Concurrent Design

Experience shows that space system engineering is particularly suited to the early formulation of the space mission and to conceptual design. In this case *Concurrent Design*

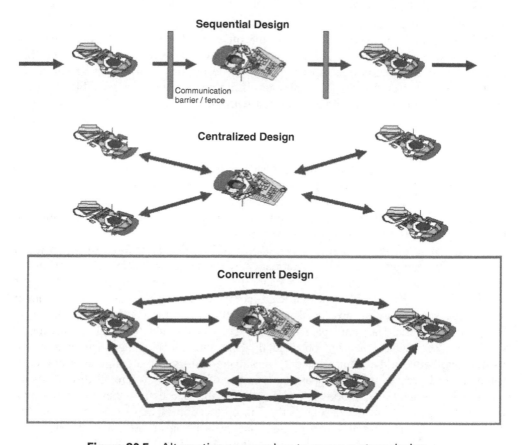

Figure 20.5 Alternative approaches to space system design

(CD) represents an alternative to other approaches. The three main types of design methodology are:

- Sequential
- Centralized
- Concurrent

which are schematically represented in Figure 20.5.

The most 'classical' method is the *sequential* approach. The overall design passes, during the various design steps, from a technical domain specialist (working in isolation from the rest of the design team) to another, in successive time intervals—in other words a sequence of specialists working 'in series'. The removal of design inconsistencies, to guarantee the design convergence, is obtained by the process of iteration. Due to lack of communication among the specialists (that is why it is also known as the 'over-the-fence' approach), incorrect assumptions may be adopted, the main system parameters are not monitored in real-time, and consequently several iterations may be necessary to reach the ultimate goal. This can lead to a very lengthy process. This method has some advantages, such as the flexibility in the use of manpower resources and the fact that it is a well-tried and routine process. On the other hand it reduces the opportunity to find interdisciplinary solutions and to create system awareness in the specialists. Furthermore, in such a 'decentralized' approach, all the design data and models are dispersed among the specialists. It is then very difficult, if not impossible, to re-assemble all of this knowledge, for instance to resume the design with modified requirements at a later time. Last but not least, the time required for performing studies using the classical approach (6–9 months) was incompatible with the drive towards a shorter time-scale from concept to flight.

An improvement to the sequential approach is represented by the *centralized design method*, where the various technical domain specialists provide subsystem design information and data to a core team of one or more system engineers. Their task it is to analyse and check the design at system level, promoting and encouraging communication between specialists when appropriate or required.

An alternative to the classical approach is offered by *Concurrent Engineering* in which the design method provides better performance by taking full advantage of modern information technology (IT). The enabling factor for the CE approach has been the rapid evolution of IT. The complete design team, composed of the various technical domain specialists, starts working on the different aspects of the project at the beginning of the design process. The rapid gain of design consistency and the final acquisition of overall project convergence are obtained by constant, direct communication and data interchange between the team members, working in real-time in a common design environment. The process still encompasses a number of iterations where all disciplines are requested to present their proposals and results in rapid succession, thus minimizing the risk of divergence from the desired path. The co-location of experts from various disciplines to elucidate a preliminary design concept is not a novelty in itself. However, in the past the process and personal interactions were limited to basic brainstorming sessions, because any numerical analysis required the use of tools which could not provide results in real time, nor easily be interconnected or co-located with the team. Nowadays, with perhaps few exceptions, most of the analysis needed for the preliminary design phases can be performed in real time on a personal computer (PC) or on a lap-top.

The concurrent approach is particularly effective in improving the design process of complex systems in which many disciplines are involved, such as in the case of space mission design.

As all the aspects of the spacecraft mission design are studied at the same time with the use of the concurrent design approach, the degree of project design consistency improves rapidly. The time period required for the study is also dramatically shorter than in the case of a classical design approach. Each discipline is, in fact, always in communication with all the others, and all design decisions, when not directly taken at team level, are quickly communicated to the whole team for immediate verification. In the concurrent approach every design issue, starting with the revision of mission requirements, is discussed collectively. Hence, there is a general awareness of the decisions taken. Therefore, all the design team members can constantly follow the same design path, avoiding the occurrence of incompatible approaches to the various subsystems' design, and therefore reducing the time and workload required by the mission study. The cornerstone of success in CE is communication.

20.3.3 The ESA CDF

Each year, the European Space Agency (ESA) performs a number of pre-Phase A (or Level 0) assessment studies. The purpose of these studies is to assess the feasibility of a new space mission from the technical, programmatic and economic points of view. This is normally achieved by producing a preliminary conceptual design of the mission and space system. The study results are used to support the mission selection process. If the mission is accepted, the study report is used as an input to the industrial Phase-A design studies. Such pre-Phase A studies are normally performed in-house at the European Space Research and Technology Centre (ESTEC), by technical-support specialists.

The implementation of the ESA Concurrent Design Facility (CDF) started in 1998 at ESTEC on an experimental basis with the initiative and support of the ESA General Studies Programme (GSP). In the space domain, there were a few other examples at that time such as the NASA/JPL Project Design Center (PDC), used by Team X for conceptual mission design—see Figure 20.6.

For ESA the objectives of the initiative were to:

- create a mission design environment in which the conceptual design of space missions could be performed in a more effective way;
- apply the practice of concurrent engineering to a number of test cases to evaluate the potential of such an approach in the various phases of space mission development;
- gather the information needed to evaluate the resources required to create a permanent facility available to all ESA programmes.

The CDF is in effect an *Integrated Design Environment* (IDE) based on the concurrent engineering methodology. At ESA, CD is primarily used to assess the technical, programmatic and financial feasibility of future space missions and new spacecraft concepts (e.g. internal pre-phase A or Level-0 assessment studies).

Specific multidisciplinary CD applications are:

- new mission concept assessment;
- space system trade-offs and options evaluation;
- new technology validation at system/mission level.

Additionally the ESA CDF is now used for many other multi-disciplinary applications, such as:

- payload instrument preliminary design;
- System of System (SoS) architectures - an SoS is a service and end-user-oriented architecture, especially conceived for security applications, whereby disparate and independent space assets are exploited in an integrated fashion to provide higher number and quality end-user services;
- space exploration scenarios;
- technology infusion and road mapping;
- reviews of industrial Phase A studies;
- scientific requirements definition and consolidation;
- anomaly investigation (e.g. design inconsistencies found during testing or operations);
- education and training.

(a)

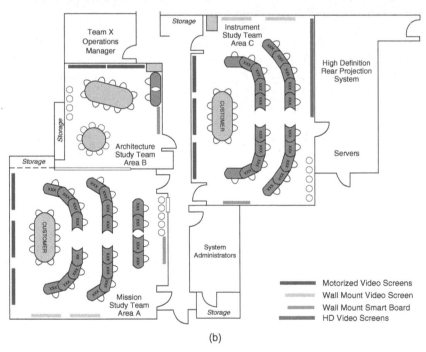

(b)

Figure 20.6   The JPL Project Design Center (PDC) operated by Team X in Pasadena CA, USA. (a) The Mission Study Team Area A. (b) PDC layout. (Reproduced by permission of NASA JPL-Caltech)

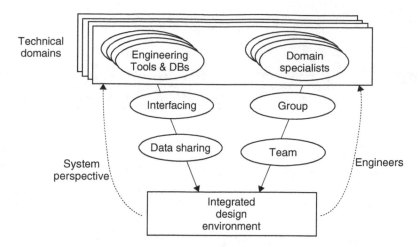

Figure 20.7   The ESA approach to the creation of an Integrated Design Environment

The approach adopted for the ESA CDF

The first step towards a more permanent application of CE in ESA required the re-organization of existing tools and human resources in a more effective manner, by means of the creation of an *Integrated Design Environment* (IDE) - see Figure 20.7. This consists of one component that is concerned with the interfacing of all the engineering tools and databases (DB) brought in by the various technical domains, in addition to another involving the implementation of real-time mechanisms for data sharing. This combination comprises the so-called *Integrated Design Model* (IDM).

It is necessary to *regroup* the necessary expertise represented by the technical domain specialists and proceed with their training to work together as a *team*. These specialists are not permanently and exclusively assigned to CDF activities, as this is just one of the several tasks they perform in a matrix-organization. One of the advantages of working in the concurrent environment is that the specialists get used to keeping the 'system perspective' in mind.

The concurrent engineering approach being adopted for the ESA CDF is based on five key elements:

- the process;
- a multidisciplinary team;
- an integrated design model;
- the facility;
- the hardware and software infrastructure.

The process

The conceptual model of the design process is shown in Figure 20.8, which highlights the fact that a space system has many interdependencies between components. This implies that the definition and evolution of each component has a design impact on other components, and that any change will propagate through the system. Early assessment

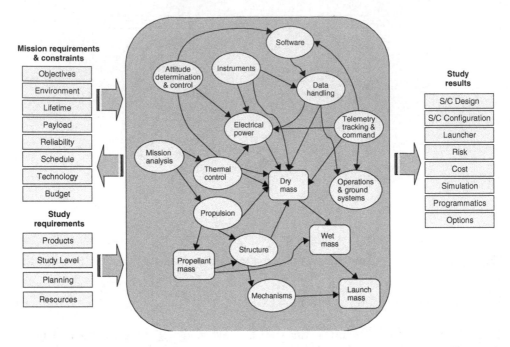

Figure 20.8 Conceptual model of the mission and spacecraft design process. The ovals represent the disciplines, the boxes represent aggregated key parameters, the arrows are interactions and data exchange. Each discipline contributes, directly or indirectly to the definition of the main mission parameters (or *key parameters*)

of the impact of changes is essential to ensure that the design process converges to an optimized solution. The CE approach is intended to provide the means to maximize the interaction among disciplines and to improve the means of achieving this early review and verification, step-by-step. The process starts with a preparation phase in which some representatives of the engineering team (team leader, system engineer, selected specialists) and of the Customer meet to refine and formalize the mission requirements, to define the constraints, to identify design drivers, and to estimate the resources needed to achieve the study objectives.

The design process begins with the study 'kick-off'. CDF activities are conducted in a number of plenary meetings, referred to as 'sessions', in which representatives of all space engineering domains participate, from the early phases (requirements analysis) to the end of the design (costing).

The customer is invited to participate in all sessions along with other specialists of their choice (e.g. study scientist, project controller), so that they can contribute to the formulation of the study assumptions, answer questions from the team and follow the evolution of the design. This includes the possibility to discuss and correct in real-time any aspect of the design not in line with their expectations.

The first design session starts with the customer presenting the mission requirements and constraints to the team. In subsequent sessions, each specialist presents the proposed option or solutions for his or her domain, highlighting and discussing the implications for the other domains. Out of the debate, a baseline is retained and the related parameter values recorded in a shared database. With the consensus of the customer, the mission requirements and constraints might need to be modified or refined as a consequence of the engineering process outcome.

One key factor is the ability to conduct a process that is not dependent on the path followed. At any stage it must be possible to take advantage of alternative paths or use 'professional estimates' to ensure that the process is not blocked by lack of data or lack of decisions. It is an iterative process that addresses all aspects of the system design in a rapid and complete fashion. The concurrent design process conforms to the so-called *Spiral Model* illustrated in Figure 20.9. A spiral pattern is drawn in the plane of all engineering solutions which is subdivided into sectors, each one dedicated to a specific technical discipline. Each discipline is therefore responsible for the calculation of all the design parameters included in that portion of the plane.

Each design iteration represents a rapid turn-around, in which, one-by-one, all the disciplines present the design (sizing, configuration etc.) of the subsystem they represent. The design process consists of a number of successive passes across progressively refined solutions, until all key parameters have reached the required target, i.e. all mission requirements and constraints have been satisfied. Each axis represents the value of a selected design key parameter, corresponding to mission or system requirements. It could, for instance, be the total mass of the system, the launch date, the mission cost, or the mission and system performances. On each axis, the target value of the respective parameter can be marked as shown in Figure 20.9. In this example with successive iterations the solution approaches the target T of the key parameter A.

A certain number of spiral models can be 'drawn and executed', depending on the number of design options that need to be analysed and traded. Normally the spiral pattern (i.e. the design option) that features the 'best convergence' becomes the baseline design that is carried through further iterations.

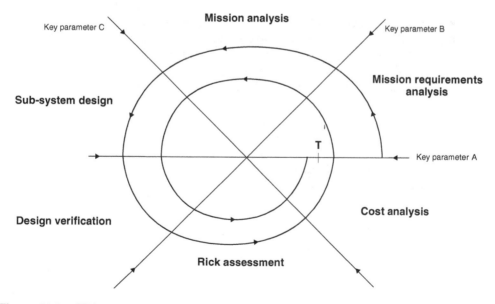

Figure 20.9 CE Iterative process: Spiral Model representation. In the example, point T is the target value of key parameter A

The acceptance of the proposed subsystem design solution and the decisions on the various trades are taken collectively by the team. The debate concerning the solutions proposed by one discipline is completed by the authorization to publish the related data in the central repository (namely the data exchange), so making it available to the other domains.

This process must be repeated until the impact of the change has been propagated to the point where design iterations no longer have a significant effect. In other words, when the proposed system design satisfies all the requirements.

The team

Human resources are by far the most important element of the whole process! A group of engineering specialists working together in one room, using sophisticated tools, are all essential elements but this is not sufficient to create a collaborative environment. On the contrary, it may become a place where conflicts are amplified. Above all else, the group of specialists must work as a team.

A fundamental part of the concurrent engineering approach is to create a highly motivated, multi-disciplinary team that performs the design work in real-time. The challenge, the novelty of the method, the collective approach, the co-operative environment, the intense and focused effort and a clear, short term goal are all essential elements that contribute to personal motivation.

Table 20.3 lists the technical disciplines typically involved in a CDF mission assessment study.

For each discipline, a 'position' is created within the facility and assigned to an expert within that particular technical domain. Each position is equipped with the necessary tools for design modelling, calculations and data exchange.

Table 20.3 Technical disciplines involved in a CDF mission assessment

Discipline/position
Systems
Payload instruments
Mission analysis
Propulsion
Attitude and orbit control
Structures/configuration
Mechanisms/pyrotechnics
Thermal
Electric power
Command and data handling
Communications
Ground systems and operations
Simulation
Cost Analysis
Risk Assessment
Programmatics

The choice of disciplines involved depends on the level of detail required and on the specialization of the available expertise. On the other hand, the number of disciplines has to be limited, especially in the first experimental study, to avoid extended debate and to allow fast turn-around of design iterations. Normally the size of the team should allow a full iteration to be run (i.e. all disciplines presenting their analyses) within a session, which in the ESA CDF lasts four hours. The ideal team is a balanced mix of young and more experienced engineers.

The process is conducted by a *Team Leader* or Facilitator. This role is also crucial for the success of the concurrent approach. A good team leader is normally a talented system engineer with skills in human resources management, who is able to conduct the design process dynamically and in real-time, while motivating people and maintaining a fast turn-around.

The model

The spacecraft design is based on mathematical models, which make use of custom software and the CDF Integrated Design Model (IDM). By this means, a consistent set of design parameters can be defined and exchanged throughout the study, and any changes which may have an impact on other disciplines can immediately be identified and collectively assessed. In this way, a number of design iterations can be performed, and different design options can easily be analysed and compared.

A parametric approach supports fast modification and analysis of new scenarios, which is essential for the real-time process. It acts as a means to establish and fix the ground rules of the design and to formalize the boundaries of responsibility of each domain. Once a specific model is established it is then used to refine the design and to introduce further levels of detail.

The first activity in the modelling process is to acquire or establish the model suited to the mission scenario, before it is parameterized to enable the iterative design process. The CDF IDM is a tool developed in-house, which allows the integration of all of the subsystem discipline tools and parameters in a consistent and effective design environment. The model also represents the means to capture the technical knowledge and to document the 'engineering views' that are required to run the design processes. Figure 20.10 shows the architecture of the IDM adopted in the ESA CDF.

The model consists of three parts:

— *Domain Specific Tools* and *Databases* (e.g. tools for CAD, mission analysis, simulation, cost modelling and estimation; databases of launchers, materials, equipment, etc.),
— a multifunctional *Intermediate Layer*, simply based on several MS Excel workbooks and models,
— a central data repository, called the *Data Exchange*.

The domain specific tools brought by each expert have to be integrated into the infrastructure of the facility.

The intermediate layer is the integration element of the IDM because it provides the data links among the disciplines and the interface between the domain specific tools. The

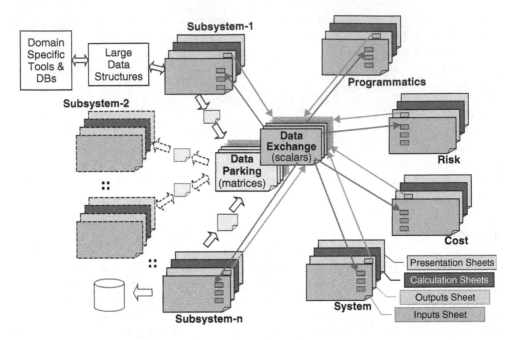

Figure 20.10   The CDF Integrated Design Model (IDM)

presentation workbook sheets contain a summary of the numeric results and a graphical representation of the specific design to be used for presentation during the design process and as part of the report at the end of the study. In practice, multiple sheets are used for the modelling process to enhance the clarity of the major parts of the design and to facilitate display during design sessions.

The data exchange is the core of the IDM, containing the most up-to-date version of the shareable data and linked, in turn, to all technical domain workbooks.

The CDF IDM template has been distributed to ESA partners, industry and academia to facilitate the creation of their concurrent design centres. The principle of a standardized exchange of data, as applied in the IDM, has been proposed as a potential standard in the context of the European Cooperation for Space Standardization (ECSS).

The facility

The team of specialists meets in the *Concurrent Design Facility* (CDF), located within ESA ESTEC, to perform design sessions.

The accommodation comprises a main design room as illustrated in Figure 20.11 plus a project design room and a support design room. The equipment and layout of the CDF is designed to facilitate the design process, and the interaction, co-operation and the involvement of the specialists. In particular, the disciplines with the most frequent interaction, or other affinities (e.g. data/model sharing), are located close to each other. In any case the facility can be reconfigured in different layouts and the disciplines can be relocated or changed depending on the specific mission study.

Figure 20.11 The new CDF, inaugurated at ESTEC on April 8, 2008. (Photo by Anneke Le Floc'h, reproduced by permission of ESA)

At the front of the facility, large projection screens are used to project the display of each workstation so that specialists can present design options or proposals, and highlight any implications imposed on, or by, other domains. Video conferencing equipment is also installed to allow team members to participate in sessions from remote sites.

The hardware and software infrastructure

The infrastructure required to implement the concurrent design facility described above can be summarized as:

- software tools for the generation of the model;
- integration of the domain models, with a means to propagate data between models in real time;
- a means to incorporate domain specific tools for modelling and/or complex calculations;
- a documentation support system;
- a storage and archive capability.

The infrastructure must allow its users to:

- work remotely from the facility both within ESTEC and in other centres, for instance the European Space Operations Centre (ESOC) in Darmstadt, Germany, or ESA HQ in Paris;
- exchange information easily between the normal office working environment and the facility environment.

Although driven by the constraints identified above, the choice of tools has, in fact, proven to be satisfactory when looking to the future. Using tools that are already part of ESA's infrastructure brings many benefits (e.g. low investment cost, reduced learning curve, compatibility between office and facility automation tools).

For the system model, the choice of Excel spreadsheets was driven not only by its availability and the existing skills of the team, but also by the fact that earlier work had indicated it to be an ideal platform. A fundamental decision was taken to split the system model into components that mirror the domain of expertise of each of the team members, so allowing work to be performed on the modelling independently and in parallel, and without the reliance on a single modelling expert. This raised the need for a mechanism to exchange relevant data between domains in a controlled manner. This problem was solved by preparing a shared workbook to integrate the data to be exchanged, with macros to handle the propagation of new data in a controlled way.

Benefits of using Concurrent Design

The use of CD in the ESA CDF has demonstrated many advantages, and there are many further benefits that CE could bring when applied to later phases of the project life-cycle. For a typical pre-Phase A study:

- the study duration has reduced from 6–9 months to 3–6 weeks;
- the corresponding cost has reduced by a factor of two;
- the number of studies that are performed per year has increased;
- the use of CD has resulted in an improvement in the quality of these technical assessments by providing consistent and complete mission designs, including technical feasibility, programmatic, cost and risk aspects.

Through the use of the centralized approach of the CDF, a data base of knowledge has been accumulated within the organization. The CDF results provide detailed information to ESA decision-making bodies to aid their assessment of which missions are selected for further industrial implementation. Furthermore this more detailed assessment of the new potential missions allows critical issues to be discovered and highlighted well in advance in the project life-cycle, and consequently this reduces the risk of engineering changes being required later or of mistakes occurring.

20.3.4 Summary

The use of concurrent engineering has become embedded in the application of space systems engineering, and there are now many concurrent design facilities around the world. In this section one of these facilities, the ESA CDF, located at ESA ESTEC in the Netherlands, is described in detail to show how it is implemented within an organization. The ESA CDF has now been in existence for over 10 years and its involvement in all phases of a space programme development continues to grow. It has enabled significant improvements in overall performance by reducing the study duration and cost, and by allowing a considerable increase in the number of studies that can be performed. At the same time, the quality has improved and the output of the CDF has become part of any follow-on industrial activity. The ESA CDF is now an essential tool for ESA decision making, and in the associated risk management processes.

20.4   A CASE STUDY: CRYOSAT

20.4.1   Background

CryoSat was the first of ESA's Earth Explorer satellites—the first to be selected and the first to be launched. Unfortunately, the launch failed after 300 s and the remains of the satellite, destroyed during a fiery re-entry, landed in the very same polar ice it had been intended to measure from orbit. Its successor, *CryoSat*-2, was successfully launched exactly 4.5 years later. How did all this come about?

In 1991 ESA launched *ERS*-1, with its pair of sophisticated radars, into a Sun-synchronous orbit which extended into the polar regions. But by this time plans were already well advanced for a much more ambitious successor, called *POEM*-1, which would carry a collection of 19 Earth-observation instruments into a Sun-synchronous polar orbit. The name *POEM*-1 was an acronym for the first Polar Orbiting Earth-observation Mission. Since both *ERS*-1, about to be launched, and *ERS*-2, which was approved in 1988 as a gap-filler, were also polar orbiting Earth-observation missions this name was evidently poorly chosen. However, there is no evidence that this contributed to its demise in 1992!

Instead it was the huge size and associated cost which eventually brought the decision to separate the environmental research part of its payload from the operational meteorological instruments. This led to two satellites, *EnviSat* carrying the first set, and *MetOp* the second. Both would be based on the *Polar Platform* (PPF) design, initially foreseen as a serviceable, multi-role spacecraft. In the end, *EnviSat*, the first to get going, grew significantly larger than when it was first split off from the full *POEM*-1 concept. *MetOp*, started later, also grew, and deviated significantly from the PPF design until only the general arrangement and a reduced-size *EnviSat*-style solar array remained. Even the Service Module was derived from *SPOT*-5 rather than the PPF, which had been based on the earlier *SPOT*-4.

EnviSat was eventually launched in 2002, the single passenger on a mighty Ariane 5 launcher, its 8.5 tonnes precluding any other reasonably accessible launcher. *MetOp* is quite a bit smaller, at 4.5 tonnes and the first of the three was launched by a Soyuz launch vehicle in 2006.

Long before these launch events however, it was becoming clear that the tide was turning against such huge satellites. *EnviSat* could boast that its large suite of instruments could make simultaneous measurements with widely different instruments which could offer more than the necessarily reduced payload of a smaller satellite. However, the cost of this was high. In terms of development time it was well over a decade between the basic instrument designs and the first results appearing. In financial terms, the cost was sobering, and this was compounded by the high cost of the Ariane 5, with no other co-passenger to share it with. There was, in addition, the risk of all this riding on a single launcher into orbit.

To exacerbate these difficulties, *EnviSat*'s synergistic measurement capabilities were matched when the NASA *A-Train Constellation* was devised. This is a series of more-or-less independent missions following each other closely in the same orbit, with only a few minutes separating them. The first, *Aqua*, was launched in 2002, and as of 2010 there were four satellites with a total of 15 instruments. This concept offers *EnviSat*'s multi-instrument approach without many of the drawbacks.

In the late 1990s, with the mood now rather firmly in favour of smaller, less ambitious missions, ESA and its scientific advisory groups made a fundamental shift in the way missions would be selected and implemented. A programme was defined which provides a financial envelope sufficient for several small missions, as well as for some instrument development and the continuation of the ageing fleet of satellites (particularly *ERS*-2 and *EnviSat*) past their initial funding allocations. It was called the Earth Observation Envelope Programme (EOEP), and one of its more interesting features is that it comes in several slices, each being individually agreed (several years apart) but with missions having the possibility to split their funding between slices.

With this mechanism it became possible to define a really effective way to select missions which could be much more decoupled from the need to find compromises, and which could win broad political support. In fact two classes of mission were initially specified in the programme definition for the first slice of the EOEP. The first was foreseen as larger missions that could continue to be regarded as flagship, observatory missions, and which might also pave the way for future operational missions. These were the Core missions. Then there were the Opportunity missions, which were seen as smaller missions which could be put together quickly and cheaply, intended to respond to specific scientific investigations. These would be selected later by an open, competitive process following a call for proposals. Interestingly, this differentiation was dropped with the second and third slices of the EOEP, and all missions are now selected by the mechanism used for these Opportunity missions.

The first of these calls for proposals for Earth Explorer Opportunity Missions was made in July 1998. Proposals were required to define a mission which would address an important scientific issue and which could be completed and launched within three years and cost ESA no more than €100 million, including post-launch operations. This was a very tall order, considering that *EnviSat* (which had not yet reached the launch pad) was costing about €2 billion. Both its launch and ground segment cost, individually, would break the cost limit placed on these missions. In fact the situation for the proposers was even more stringent since the costs incurred by ESA itself would have to come from this sum too. So the proposals were required to be costed at less than €80 million.

Despite the severe cost cap, 27 proposals were made, and these entered an intensive evaluation process, considering both scientific and technical merits. In May 1999 ESA presented a paper to the Earth Observation Programme Board (PB-EO) which endorsed the findings of the Earth Science Advisory Committee (ESAC). It proposed that *CryoSat*, the mission ranked highest by ESAC, should be implemented as the first Earth Explorer Opportunity Mission to be launched. It also proposed that the *Soil Moisture and Ocean Salinity* mission (*SMOS*) should be prepared as the second, with flexibility maintained for it to replace *CryoSat* should it develop problems. A third mission, the *Atmosphere and Climate Explorer* (*ACE*) was also to be maintained in 'hot standby' to replace either *CryoSat* or *SMOS*.

Each of the missions, as proposed, was obliged to include some sources of funding outside ESA. In the case of *CryoSat* this was relatively modest and covered the scientific support for the mission. *SMOS*, on the other hand, required substantial national contributions from France and Spain which included major elements of the satellite and the ground segment. While this had financial advantages it added considerable programmatic complexity.

Both of these top-ranked proposals shared an important, winning argument. It became clear that the scientific desirability of each mission was not the deciding factor: each proposal made such a compelling case that ideally they would all have been funded. Instead the deciding question became the practical feasibility of meeting the mission objectives within the strict limits of what was available. This recognition of the real constraints is an important aspect of systems engineering. *SMOS* achieved this through building a consortium of funding bodies, but the *CryoSat* case was different.

CryoSat was proposed by Prof. Duncan Wingham, of University College London (UCL), supported by an international team, which crucially included the industrial companies who had built the preceding European radar altimeter satellites. The team were able to demonstrate that virtually all of the satellite could be assembled by reusing designs from previous missions, with little extra development required. At a stroke this gave confidence that the strict cost and time limits were achievable.

20.4.2 Mission characteristics

CryoSat is a mission designed to measure small changes in the thickness of the Earth's ice fields, which are mainly the floating sea-ice and the ice sheets such as Greenland and Antarctica. It also makes measurements of smaller ice caps and larger mountain glaciers. Unusually, the measurement requirements are specified as a rate of change, since we are less interested in the annual signal than in year-to-year changes. And the magnitude of the specified rate of change is small, set at about 10% of the expected variation due to natural fluctuations.

This imposes requirements on the type of measurements to be made, the physical stability of the system, control of the measurement configuration and consistency in the data processing system. The programmatic constraints imposed on the Earth Explorer Opportunity Missions, on the other hand, could be simply characterized as the need for a relatively short development cycle with a stringent cost ceiling. This has become a common need in the development of space missions and the approach taken in this case is pertinent to many similar missions.

The original idea was that the 'rules' could be simplified for the Earth Explorer Opportunity Missions, to enable them to be implemented more quickly and cheaply. This was thought to mean less rigorous quality, less reliable hardware and perhaps less redundant systems. In reality it was very difficult to implement two sets of standards in a procurement organization where much expertise is drawn from outside the project team. Furthermore, much of the equipment selected had heritage from other projects and so had an intrinsic quality level. So inevitably *CryoSat* developed into a project with similar quality levels to other ESA missions. This meant that cost and schedule savings had to be achieved in other ways. One way was to reduce the size of the teams, in ESA and in industry. While attractive from the accountant's point of view, this did mean that individual work-loads were rather high; the establishment of a strong team-spirit helped maintain motivation and effectiveness. The other available tool was the decision to develop a proto-flight system, described later. The combination of these methods enabled a 'low-cost' mission to be developed within the normal way of doing business.

20.4.3   Precision measurements from space

The *CryoSat* satellite is the part of the system which makes measurements. The fundamental measure is the distance from the satellite to the surface below and for this a radar altimeter is used. Given that the mission was proposed in 1998 when the enormous potential of the *ERS* radar altimeters over icy surfaces was being realized, this was a natural choice. The other main alternative would have been a laser altimeter and this selection was made in the case of the US *ICEsat* mission, which unfortunately failed just a few months before *CryoSat*-2 made it into orbit in 2010, despite the efforts of the *ICESat* team to conserve their lasers by only operating a few months each year. In fact laser and radar altimeters are really complementary in this application and it is a great pity that simultaneous measurements could not be made.

CryoSat's radar altimeter is called SIRAL, a contraction of *SAR and Interferometric Radar Altimeter*, and this indicates where improvements have been made in the instrument concept. As well as a conventional mode, which offers continuity with earlier missions, the radar can also operate in synthetic aperture mode. This increases the along-track resolution, enabling it to more readily distinguish the narrow 'leads' of open water between sea-ice floes. Over the rough terrain at the edges of the major ice sheets this increased along-track resolution is further augmented by across-track interferometry using the second antenna and receive channel. The derived angle of arrival of the radar echoes allows a more precise identification of the point from which the echo came.

SIRAL makes measurements of the range to the surface which is very precise: each measurement has an uncertainty of a few tens of centimetres (the standard deviation, measured in-orbit, in its SAR-Interferometric mode is 18 cm). The averaging of many such measurements brings the system performance to the level needed to satisfy the mission objectives. However the precise measurement of range alone is insufficient. The position of the satellite at the time of each measurement is needed to convert this simple measure of range to something scientifically meaningful, that is, the height of the surface above some known reference. Loosely we may talk of height above sea-level, but in this demandingly precise application we refer to height above a *reference ellipsoid*, an exactly defined, oblate spheroidal surface which closely approximates the shape of the Earth. In order to determine the satellite's position, and thus its height above this reference ellipsoid, measurements from some further payload equipment are needed.

CryoSat includes a DORIS (Determination of Orbit and Radiopositioning Integrated by Satellite) receiver, a special radio receiver which picks up signals from a network of more than 50 transmitting stations spread evenly over the Earth. By measuring the Doppler shift of these signals the range-rate to each one is determined. The accuracy of the orbit which may be computed from such a data-set obviously depends on several factors, not least the precision and accuracy of the Doppler shift measurements. In this, DORIS excels and since the early 1990s the DORIS system has been the foremost means of routine, high-precision orbit determination (by 2010 systems based on space-borne GPS receivers could match the performance). The DORIS receiver is augmented by a passive laser retro-reflector, which allows precise range measurements to be made by ground-based laser ranging stations.

The final item in this collection of high-precision payload equipment is a set of star trackers. These are needed to complement the SIRAL interferometer measurement. SIRAL measures the angle of arrival of the echo in its own reference frame, that is, with respect to the line joining the centres of the two antennas, the *baseline*. Before that information can be used to identify the exact position on the Earth we must know the orientation of that baseline, and in order to meet the mission objectives this measure must also be precise—within 30 arc-seconds.

The scientific mission objectives not only defined the type and accuracy of the measurements which had to be made (which led to the payload selection) but also where and when. This 'where and when' is encapsulated in the definition of a specific orbit for *CryoSat*. The orbital inclination is 92° (and therefore retrograde) with a mean altitude of about 720 km, the exact altitude being defined by the required track repeat characteristics. This orbit is not Sun-synchronous (see Chapter 5) in which the orbit plane keeps a fixed orientation with respect to the Sun. Such a synchronous orbit would have undesirable tidal aliasing characteristics. Furthermore, this orbit requirement had impacts on the satellite design.

20.4.4 Designing the system

The full *CryoSat* system has several parts, and although those which remain on the Earth will not be described here, we shall outline the overall architecture as this has impacts on the design of all of the elements. Programmatic constraints were dominant in this part of the system definition and led to the minimum configuration required to satisfy the mission objectives. The key feature is that a single ground station is used for *CryoSat*, both for command and control and for downlink, processing and distribution of the science data. The ESA ground station at Kiruna was selected. As well as enabling the sharing of resources with other on-going ESA missions, this choice resulted in manageable requirements to cope with the three to four consecutive orbits per day where contact with the ground station is not possible. For stations at lower latitudes this would be a larger number, while for the higher latitude of Svalbard the number of these 'blind' orbits is reduced to zero, but the opportunity to share costs with other ESA missions would not exist.

The design of the *CryoSat* satellite was determined, as is always the case, by a number of key factors. From the science mission objectives came the payload complement and its requirements, the orbit and the minimum lifetime. Programmatic constraints included the need to operate from a single ground station, launch on a 'small' (and therefore low-cost) launcher, extensive on-board autonomy, a low-cost design and a decision to forego the normal approach of building precursor 'proof-of-concept' models of the satellite (the so-called Structure Model, Engineering Model, etc.—see Chapters 8 and 17).

These driving factors do not lead inevitably to the *CryoSat* design (indeed during the competitive feasibility study phase, another, entirely different, concept was developed) but many of the main features of the *CryoSat* design can be traced to these drivers. In many missions the heritage from other programmes has a role to play in the design of a particular spacecraft. In this case, the *CHAMP* and *GRACE* satellites, which were designed against similar orbit and programmatic constraints, had a significant bearing on the final design of *CryoSat*.

Instead of a rather ponderous deduction of *CryoSat*'s design from the mission's requirements and constraints, in the following discussion *CryoSat* will be reversed engineered to show how it responds to these drivers. The spacecraft is shown in Figure 20.12. We shall start with the most obvious feature: its shape.

Figure 20.12   The *CryoSat* configuration. (Reproduced by permission of ESA/P. Carril)

The required orbit is not Sun-synchronous. Every day the orbit plane shifts almost 3 min earlier with respect to the Sun; in eight months the orbital plane drifts through all local times. This means that the direction from which sunlight falls on the satellite is constantly changing. The operation of the SIRAL demands that its antennas point towards the Earth's surface to within a few tenths of a degree, and furthermore that the two antennas are arranged side by side (across-track) as the satellite flies. This means that rotating the satellite to face the Sun is out of the question. Mechanisms on satellites are very costly—a rule of thumb suggests that each one costs about €1 million—so the *CryoSat* geometry was arranged such that every orbit has enough sunlight on one or both of the solar panels to maintain an overall positive energy balance.

Of course this assumes that the solar panels have the intrinsic capability to generate enough power in the first place, if the Sun were shining directly on them. The requirement to fit *CryoSat* inside the fairing of a 'small' launcher placed absolute constraints on the size of the panels, thus removing one of the key degrees of freedom in this equation. That only left one parameter to ensure sufficient power generation—the efficiency of the solar cells. Thus *CryoSat*, a low-cost mission, ended up pioneering the use of new, high-efficiency solar cells in low-Earth orbit. It should be noted that this choice was still cheaper than introducing fold-out panels and emphasizes the goal of system engineering, and that the best overall solution may require some areas to have additional complexity and cost in order to reduce the overall complexity and cost of the mission.

The two antennas of the SIRAL are accommodated near the front of the satellite, side by side. The antennas are elliptical in outline in order to fit within the launcher fairing—this slightly complicates the scientific data processing but has no impact on performance. We have already implied that both the structural stability and knowledge of this assembly are vital to the mission performance, and the design has a number of special features to ensure this. In the calm environment of space the principal enemy of stability is heat, which causes expansion. Astronomy missions have the same enemy but can select orbits with a very benign thermal environment. This is clearly not possible for an earth observation spacecraft. To guard against this enemy, therefore, the first line of defence is to use materials which are least susceptible to it. Mainly this requires the use of carbon fibre reinforced plastic (CFRP) which has a coefficient of thermal expansion close to zero. This is used for the antennas themselves and for the substantial 'bench' upon which they are mounted, and the mounts for the star trackers fitted to this bench. Invar, a low-expansion metal originally used for the pendulums of clocks, was also used. However the majority of the satellite is aluminium, so another vital part of the defence is isolation. The sensitive antenna bench is attached to the rest of the satellite by a three-point quasi-isostatic suspension, which minimizes the transfer of thermal distortions and heat variations. The theme of isolation is carried further by the wrapping of the antenna bench structure as so obviously illustrated in Figure 20.12. Not only is the bench and all its attachments well wrapped in multi-layer insulation, but even the antenna apertures are protected from the hot Sun by an exotic single-layer insulation, coated with microwave-transparent germanium.

20.4.5 The payload: re-use and innovation

Like all of the equipment on *CryoSat*, the SIRAL radar altimeter is derived from existing equipment; here a conventional pulsewidth-limited altimeter called Poseidon-2, which is flying on the US-French *Jason* mission, is used. SIRAL is a single frequency Ku-band radar which uses the full deramp range compression technique. It also features some new design characteristics which enable it to provide data which can be more elaborately processed on ground. A high pulse repetition frequency and pulse-to-pulse phase coherence are needed for the along-track SAR processing. The across-track interferometry needs a second complete receiver chain, including the second antenna. These features make this instrument unique.

The electronics of the radar are divided into a number of separate units, principally for ease of manufacturing and testing. One large unit houses all the digital electronics, and the remaining boxes contain radio frequency circuitry and the transmitter's power supply. Several innovations were necessary compared to the Poseidon-2 equipment, most notably in the need for significantly increased transmitter power, which led to the development of a complete new transmitter section, and in the provision of the second receive path.

A less obvious but far more pervasive change was the new requirement for phase stability, introduced by both the SAR and interferometric functions. Phase had never been an issue with previous altimeters but for SIRAL it is critical. In consequence, the SIRAL development has seen much analysis, measurement, characterization, optimization and tuning, as well as the introduction of some special means of calibrating phase performance in-flight.

The antenna subsystem has not been immune to this. It was developed as a discrete item and consists of two Cassegrain antennas mounted side-by-side on the rigid antenna bench. Both antennas are identical, but one is used both to transmit and receive whereas the other is only used to receive echoes. The Cassegrain design offers particular advantages for the SIRAL as the resulting waveguide lengths are much shorter than those required for the more common, front-fed, design. The entire assembly went through a measurement campaign which challenged the capabilities of the test facility due to the exacting phase measurement requirements.

The SIRAL on the original *CryoSat* was non-redundant, but for *CryoSat*-2 the decision was quickly made that, for a mission offering a second chance, there should be no possibility of mission loss through a single-point failure, and so the SIRAL became fully redundant. This, of course introduced a number of other changes: two sets of connections were required for power, for the MilBus 1553 used for command and control, for the 1355 connections used for high speed data and for the reference oscillator input from the DORIS. Extra switching and commands were needed, as well as a means of toggling and monitoring the waveguide switch used for selection of which unit was connected to the antennas. This simple change had many ramifications, and these also extended into many areas of the ground segment.

The low-resolution mode of the SIRAL operates in the same way as a conventional pulsewidth-limited altimeter and uses a single receive channel. The rate at which the radar pulses are transmitted is low (relatively speaking—it is 1970 per second) and the echoes are transformed from the time domain to the spectral domain and averaged on-board. The data rate at which science data are generated in this mode is therefore low, at 51 kbps. This mode will be used over ice sheet interiors, where the slopes of the surface are small. It will also be used over the ocean.

In the SAR mode, which also uses a single receive channel, the along-track horizontal resolution of the altimeter is improved by exploiting the Doppler properties of the echoes, during the on-ground processing. The result is equivalent to decomposing the main antenna beam into a set of 64 narrower synthetic beams along-track. The footprints of the different sub-beams over a flat surface are adjacent rectangular areas ~250 m wide along-track and as large as the antenna footprint across-track (up to 15 km). Consequently, a larger number of independent measurements are available over a given area, and this property is used to enhance the accuracy of the measurements over sea ice. In order to ensure the coherence between echoes from successive pulses, the pulse repetition frequency is about 10 times higher than for the low resolution mode. The instrument operates in bursts, with a group of 64 pulses transmitted together, then a pause during which the echoes arrive. The echoes are stored on-board in the time domain, without any averaging. Therefore the data rate is significantly higher, at 11.3 Mbps.

The SAR-Interferometric mode (SARIn) is used mainly over the margins of the ice sheets, where the surface slopes are high. The combination of SAR and interferometry makes it possible to accurately determine the arrival direction of the echoes both along and across the satellite track, by comparing the phase of one receive channel with respect to the other. In this mode, both receive channels are active and the corresponding echoes are stored in the time domain. The data rate is about twice as high as for SAR mode. In order to cope with abrupt height variations, the range-tracking concept for this mode has to be particularly robust. In SIRAL, this is ensured by using narrow-band tracking pulses, transmitted in-between successive wide-band measurement bursts.

CryoSat includes a set of three identical star trackers, which are the only means of determining the orientation of the SIRAL interferometric baseline. They are also the principal three-axis attitude measurement sensor in the nominal operating mode. They are lightweight, low power consuming, fully autonomous devices. They are accommodated such that the Sun and Moon can each blind only one head at any time; this makes the whole sensor system one-failure tolerant.

The star trackers are fully autonomous, with no need for ground update of star maps. In the acquisition mode, which takes 2–3 s, the star tracker calculates a coarse attitude by matching triangle patterns of stars with patterns stored in its catalogue. After two consecutive successful initial acquisitions it autonomously jumps to attitude update mode in which it uses information from the last attitude update to calculate a precise attitude, at a rate of up to 1.7 Hz.

The star tracker algorithm is optimized to use rather faint stars, around magnitude 5. These stars, which are barely visible to the naked eye except at dark sites, are far more numerous than the brighter stars and provide many triangles for use in the pattern matching process in all directions of the sky.

20.4.6   What makes it tick?

All of the data generated by the scientific payload of *CryoSat* have to be recorded on-board as the satellite is only in contact with its single ground station for brief periods. Typically there are 10 passes of duration 5–10 min each day, occurring on consecutive orbits. These contacts are followed by a gap of three to four 'blind orbits'. To handle the large data volume, a data recorder of capacity 256 Gbits is installed. Following the modern trend, this is realized as solid-state memory—literally thousands of RAM chips. The unit is derived from similar equipment on *Mars Express* and, of course, comprehensive memory management and data handling functions are built in. It can continue recording data as it replays it into the data-link to the ground station.

This downlink is a potential bottleneck as the total contact time is relatively short. To avoid this, the downlink data rate is high at around 100 Mbps. Again this is built on heritage, this time from *MetOp*, with the frequency and bandwidth reused from an allocation originally given to *EnviSat*.

CryoSat is an unusual satellite in that it has virtually no moving parts, the only exceptions are a couple of valves in the propulsion system. This has led to savings in cost as well as testing. One area where this lack is particularly noticeable is in the attitude and orbit control subsystem, where gyroscopes and reaction wheels are normally commonplace.

Attitude control (see also Chapter 9) for *CryoSat* is innovative since it principally uses two of the payload equipments; another example of re-use. The star tracker provides real-time measurements of the satellite orientation with respect to the stars, which together with the DORIS time and orbit information allows the on-board software to calculate the orientation with respect to the Earth. Measurement is half of the problem—it is also necessary to produce torques which will turn the satellite as needed to keep the Earth-pointing within the required tolerance. *CryoSat*'s main means of generating such torques is to use electro-magnets interacting with the Earth's magnetic field. These devices, called magnetic torquers, are simply multiple turns of wire wrapped around a ferrite core, powered by a controllable electric current from the main computer. Magnetic torquers cannot produce torque around the direction of the Earth's field itself, and this direction constantly changes with respect to the satellite as it moves along its orbit. So a 'backstop' control guards against excessive pointing errors—a set of small cold-gas thrusters. These are very small indeed, with a thrust of 10 mN—about the same as the weight of 1 cm^3 of water. Pre-launch simulation showed that these would fire for a total of about 3 s per orbit which would lead to a consumption of about 3 g of gas per day. In reality however, the gas consumption is significantly less, at about 6 g per week. Although this is not much, it will, together with the gas used by the two 40 mN thrusters used to maintain the orbit against air-drag, eventually consume the 35 kg of pressurized nitrogen on-board.

The attitude control system has other sensors too, used during the initial stabilization after separation from the launcher, and in emergencies. These are a set of magnetometers and an ingenious sensor, the combined Earth-Sun sensor (CESS), which measures the temperature difference between a black and a mirror surface on each face of the satellite. A clever piece of software then calculates the direction to both Sun and Earth.

A point related to attitude control is the question of what to do if the satellite has a serious failure and has to enter a so-called 'safe-mode'. The idea here is that the satellite needs a robust attitude control mode which it can maintain with minimal resource usage and which guarantees that the solar arrays continue to generate enough power to keep the satellite alive. For satellites with rotating solar arrays (such as *ERS*-1) this is a real design issue since the solar array has to be moved to a specific position and then the satellite pointing adjusted to keep it in the Sun. This is likely to be a different orientation from the normal Earth-pointing modes.

For *CryoSat*, with its fixed, body-mounted solar arrays, the approach is simpler. The satellite changes to a more robust control mode, using only the CESS and magnetometers for attitude measurement, and the magnetic torquers for attitude control. The satellite remains Earth pointing, and the nadir panel remains the primary radiator.

20.4.7 Putting it together

One of the key means by which the *CryoSat* programme was able to compress schedule and costs lay in a bold early decision. The idea was that by embracing existing equipment designs and building-in conservative margins, it would be possible to directly build a proto-flight satellite. No test articles would be built.

The savings inherent in such an approach were very persuasive, both in time and in equipment. However, it was obvious that the benefit of test models, particularly the engineering model (EM), goes beyond merely testing the hardware. The test model allows unglamorous but essential work, such as test procedure debugging, to be done away from the critical path. So for *CryoSat* it was decided to build a virtual satellite in software. This has been so useful for various aspects of testing that it has been cloned several times, a feat rather difficult to perform with a hardware version.

In some ways, of course, the original *CryoSat* acted as the EM for *CryoSat*-2, but this would not be completely correct since *CryoSat*-2, despite being a rebuild, accumulated a large number of changes compared to the original. The redundant SIRAL was an obvious example, but there were almost a hundred significant changes. The effect of all of these was that virtually every test procedure needed to be checked and a large number rewritten.

The proto-flight *CryoSat* was ready for final testing in mid-2004, and the same process was followed for *CryoSat*-2 which moved into the test facilities at IABG in Otto-brunn, near Munich, in September 2008. Environmental testing is an inescapable ordeal and involves a long programme of 'torture' for the satellite. *CryoSat*-2 had to face it again—the satellite design had changed sufficiently to render the previous testing invalid, and in the case of the mechanical testing, a different launch vehicle had been selected for the new satellite, so requiring a new set of qualification tests (see also Chapters 17 and 19).

So *CryoSat*-2 endured mass properties measurement, vibration testing, acoustic testing, EMC testing, thermal vacuum and thermal balance testing in a vacuum chamber, RF auto-compatibility testing, flash testing to check the solar array performance, launcher-related radiated emission and susceptibility testing and finally a shock separation test using the actual flight adapter, brought to Munich by the launch provider.

In-between times, *CryoSat*-2 went through many standard health tests, both as diagnostics and for status checking. There were also some System Validation Tests to verify that the mission control centre at ESOC was able to flawlessly monitor and control it.

20.4.8 Getting it up there

The original *CryoSat* was launched with a Rockot launch vehicle. This is a converted SS-19 intercontinental ballistic missile (ICBM), with a versatile, restartable upper stage called Breeze-KM. The launch, from Plesetsk, some 800 km north of Moscow, occurred on 8th October 2005. At 300 s after launch the control system encountered an incorrect sequence of two commands and as a result the second stage engine suffered a catastrophic failure, which resulted in the mission being terminated. *CryoSat*, the unused Breeze KM stage and the empty second stage then continued a ballistic trajectory, reentering, and finally exploding at about 40 km altitude near the North Pole.

The subsequent inquiry clearly identified the fault and remedial measures to ensure that it could not recur. Rockot has subsequently had an unbroken safety record and has launched two more recent Earth Explorers (*GOCE* and *SMOS*). Nevertheless, *CryoSat*-2 was launched on a different launcher, a Dnepr, which is a converted SS-18 ICBM. The reason for this change was unrelated to the earlier failure. It was simply a question of availability since the Breeze-KM stages are in short supply, while Dnepr uses the entire SS-18 with little change.

The launch time is selected so that the orbital plane is at right angles to the direction to the Sun, and the Dnepr launcher performs a direct injection, without a cruise phase. The launch, from Baikonur in Kazakhstan, is in a southerly direction. The first stage falls on land, in a deserted region of Turkmenistan, while the fairing and second stage fall in the Indian Ocean. Dnepr is an unusual launch vehicle—as a result of its ICBM heritage the third stage performs a 180° turn after separation and its engines extend on arms to point back along its flanks. It then flies backwards, dragging the payload behind it, which remains protected by a Gas Dynamic Shield (GDS). Two seconds before payload separation the GDS is released, followed by severing of the connection to the satellite. The upper stage moves away under the power of its engines, so that separation springs are not needed. The deployment is shown in Figure 20.13. This method yields a very precise injection. As a result of this though, *CryoSat* could not have its telemetry transmitter on while in the GDS so that, even though the vehicle was in view of the Malindi ground station, no contact could be achieved until the transmitter was switched on some 14 s after separation. Following the experience with the original *CryoSat* exactly four years and six months earlier, the acquisition of this signal generated a certain amount of relief and good cheer.

Figure 20.13   Dnepr injection of CryoSat. (Reproduced by permission of ESA/P. Carril)

20.5   CONCLUSION

The discussion in this chapter illustrates that whilst space systems engineering has a relatively long heritage and many of the fundamental technologies discussed in this book remain the same, there have been some important advances. One of these, Concurrent Engineering, utilizes the huge improvements in information technology to speed up and enhance the iterative, interdisciplinary nature of space systems engineering. The other section in this chapter on the *CryoSat* spacecraft demonstrates that whilst there are an increasing number of tools available to optimize a spacecraft design, pragmatic decisions have to be made with a real mission operating within a limited budget, in order to achieve a successful outcome.

REFERENCES

[1] ECSS-E-ST-10C Space Engineering—System Engineering General Requirements.
[2] CDF Info Pack (2011), http://esamultimedia.esa.int/docs/cdf/CDF_INFOPACK_2011.pdf (accessed 14 February 2011).

Index

Spacecraft Systems Engineering, Fourth Edition.
Edited by Peter W. Fortescue, Graham G. Swinerd and John P. W. Stark.
© 2011 John Wiley & Sons, Ltd. Published 2011 by John Wiley & Sons, Ltd.